VOLUME FOUR HUNDRED AND FORTY

METHODS IN ENZYMOLOGY

Nitric Oxide, Part F Oxidative and Nitrosative Stress in Redox Regulation of Cell Signaling

METHODS IN ENZYMOLOGY

Editors-in-Chief

JOHN N. ABELSON AND MELVIN I. SIMON

Division of Biology
California Institute of Technology
Pasadena, California

Founding Editors

SIDNEY P. COLOWICK AND NATHAN O. KAPLAN

VOLUME FOUR HUNDRED AND FORTY

METHODS IN ENZYMOLOGY

Nitric Oxide, Part F Oxidative and Nitrosative Stress in Redox Regulation of Cell Signaling

EDITED BY

ENRIQUE CADENAS
Professor and Chairman
Molecular Pharmacology and Toxicology
School of Pharmacy
University of Southern California
Los Angeles, CA 90089-9121

LESTER PACKER
Department of Molecular Pharmacology and Toxicology
School of Pharmacy
University of Southern California
Los Angeles, CA 90089-9121

AMSTERDAM • BOSTON • HEIDELBERG • LONDON
NEW YORK • OXFORD • PARIS • SAN DIEGO
SAN FRANCISCO • SINGAPORE • SYDNEY • TOKYO
Academic Press is an imprint of Elsevier

ELSEVIER

Academic Press is an imprint of Elsevier
525 B Street, Suite 1900, San Diego, California 92101-4495, USA
84 Theobald's Road, London WC1X 8RR, UK

This book is printed on acid-free paper. ∞

Copyright © 2008, Elsevier Inc. All Rights Reserved.

No part of this publication may be reproduced or transmitted in any form or by any means, electronic or mechanical, including photocopy, recording, or any information storage and retrieval system, without permission in writing from the Publisher.

The appearance of the code at the bottom of the first page of a chapter in this book indicates the Publisher's consent that copies of the chapter may be made for personal or internal use of specific clients. This consent is given on the condition, however, that the copier pay the stated per copy fee through the Copyright Clearance Center, Inc. (www.copyright.com), for copying beyond that permitted by Sections 107 or 108 of the U.S. Copyright Law. This consent does not extend to other kinds of copying, such as copying for general distribution, for advertising or promotional purposes, for creating new collective works, or for resale. Copy fees for pre-2008 chapters are as shown on the title pages. If no fee code appears on the title page, the copy fee is the same as for current chapters. 0076-6879/2008 $35.00

Permissions may be sought directly from Elsevier's Science & Technology Rights Department in Oxford, UK: phone: (+44) 1865 843830, fax: (+44) 1865 853333, E-mail: permissions@elsevier.com. You may also complete your request on-line via the Elsevier homepage (http://elsevier.com), by selecting "Support & Contact" then "Copyright and Permission" and then "Obtaining Permissions."

> For information on all Elsevier Academic Press publications
> visit our Web site at www.books.elsevier.com

ISBN-13: 978-0-12-373967-4

PRINTED IN THE UNITED STATES OF AMERICA
08 09 10 11   9 8 7 6 5 4 3 2 1

**Working together to grow
libraries in developing countries**

www.elsevier.com | www.bookaid.org | www.sabre.org

ELSEVIER   BOOK AID International   Sabre Foundation

Contents

Contributors   xv
Volumes in Series   xxi

Section I. Molecular Methods   1

1. Mass Spectrometric Characterization of Proteins Modified by Nitric Oxide-Derived Species   3
Anna Maria Salzano, Chiara D'Ambrosio, and Andrea Scaloni

1. Introduction   4
2. Reagents   6
3. Preparation of Nitrated BSA   6
4. In-Gel Protein Digestion and Mass Spectrometric Analysis   6
5. Data Analysis   7
6. MALDI-TOF Peptide Mass Fingerprinting   8
7. LC-ESI-IT Fragment Fingerprinting upon Collisional Fragmentation   9
8. Concluding Remarks   9
Acknowledgments   12
References   12

2. Detecting Nitrated Proteins by Proteomic Technologies   17
Yoki Kwok-Chu Butt and Samuel Chun-Lap Lo

1. Introduction   18
2. Methods   20
3. Conclusions   30
Acknowledgments   30
References   30

3. Using Tandem Mass Spectrometry to Quantify Site-Specific Chlorination and Nitration of Proteins: Model System Studies with High-Density Lipoprotein Oxidized by Myeloperoxidase   33
Baohai Shao and Jay W. Heinecke

1. Introduction   34
2. High-Density Lipoprotein Biology   36
3. Advantages of HDL as a Model System   36

v

4.	Advantages of LC-ESI-MS/MS When Analyzing Posttranslational Modifications of Proteins	37
5.	Isolating HDL, ApoA-I, and MPO	38
6.	Oxidative Reactions	38
7.	Proteolytic Digestion of Proteins	39
8.	Liquid Chromatography–Electrospray Ionization Mass Spectrometry (LC-ESI-MS and MS/MS)	39
9.	A Combination of Tryptic and Glu-C Digests Provides Complete Sequence Coverage of ApoA-I	40
10.	HOCl or MPO Preferentially Chlorinates Tyrosine 192 in Lipid-Free ApoA-I	40
11.	Reagent ONOO$^-$ and MPO Nitrate All the Tyrosine Residues in Lipid-Free ApoA-I, but Tyrosine 192 Is the Main Target	44
12.	HOCl Quantitatively Converts All Three Methionine Residues in ApoA-I to Methionine Sulfoxide	48
13.	HOCl Generates Hydroxytryptophan and Dihydroxytryptophan Residues in ApoA-I	53
14.	Reactive Nitrogen Species Generates Nitrotryptophan Residues in Lipid-Free ApoA-I	57
15.	The YXXK Motif Directs ApoA-I Chlorination	57
16.	Quantitative Analysis of Posttranslational Modifications of Proteins	58
17.	ApoA-I Oxidation Impairs Cholesterol Transport by the ABCA1 System	58
18.	Conclusions	59
	Acknowledgments	60
	References	60

4. Influence of Intramolecular Electron Transfer Mechanism in Biological Nitration, Nitrosation, and Oxidation of Redox-Sensitive Amino Acids   65

Hao Zhang, Yingkai Xu, Joy Joseph, and B. Kalyanaraman

1.	Introduction	66
2.	Methods	67
3.	Results	70
	Acknowledgments	90
	References	90

5. Protein Thiol Modification by Peroxynitrite Anion and Nitric Oxide Donors   95

Lisa M. Landino

1.	Introduction	96
2.	Protein Cysteine Oxidation by ONOO$^-$	97

3.	Methodology to Detect Protein Thiol Modification	97
4.	Iodoacetamide Labeling of Proteins after Oxidant Treatment	98
5.	HPLC Separation of Fluorescein-Labeled Peptides	99
6.	Detection of Interchain Disulfides by Western Blot	100
7.	Repair of Protein Disulfides by Thioredoxin Reductase and Glutaredoxin Systems	101
8.	Thioredoxin Reductase Repair of Protein Disulfides	102
9.	Quantitation of Protein Disulfides by Measuring NADPH Oxidation	103
10.	Quantitation of Total Cysteine Oxidation Using DTNB	103
11.	Glutaredoxin/Glutathione (GSH) Reductase Repair of Protein Disulfides	104
12.	Detection of Protein S-Glutathionylation	105
13.	Thiol/disulfide Exchange with Oxidized Glutathione	105
14.	Protein Thiol Modification by Nitric Oxide Donors	106
15.	Conclusions	108
	References	108

6. Indirect Mechanisms of DNA Strand Scission by Peroxynitrite   111

Orazio Cantoni and Andrea Guidarelli

1.	Introduction	112
2.	Materials and Methods	112
3.	Results and Discussion	116
	Acknowledgments	119
	References	119

7. Nitric Oxide: Interaction with the Ammonia Monooxygenase and Regulation of Metabolic Activities in Ammonia Oxidizers   121

Ingo Schmidt

1.	Ammonia-Oxidizing Bacteria	122
2.	Aerobic Ammonia Oxidation	122
3.	Anaerobic Ammonia Oxidation with Nitrogen Dioxide as Oxidant Releases NO as Product	123
4.	Aerobic Ammonia Oxidation with Nitrogen Dioxide as Oxidant	125
5.	Regulation of Metabolic Activities in Ammonia Oxidizers	127
6.	Nitric Oxide Induces Denitrification in *N. europaea*	128
7.	Nitric Oxide Induces the Biofilm Formation of *N. europaea*	129
8.	Nitric Oxide Is Required in *N. europaea* to Restore Ammonia Oxidation after Chemoorganotrophic Denitrification	132
	References	133

8. **Chemiluminescent Detection of *S*-Nitrosated Proteins: Comparison of Tri-iodide, Copper/CO/Cysteine, and Modified Copper/Cysteine Methods** — 137

Swati Basu, Xunde Wang, Mark T. Gladwin, and Daniel B. Kim-Shapiro

1. Introduction — 138
2. Methods for Detection of *S*-Nitrosothiols — 139
3. Chemiluminescent-Based Detection of *S*-Nitrosothiols — 140
4. Advantages and Disadvantages — 146
5. Comparisons and Validations — 147
6. Conclusions — 151
References — 153

9. ***S*-Nitrosothiol Assays That Avoid the Use of Iodine** — 157

Lisa A. Palmer and Benjamin Gaston

1. Introduction — 158
2. *S*-Nitrosothiol Synthesis — 158
3. *S*-Nitrosothiol Assays — 160
4. *S*-Nitrosothiols in Health and Disease: The Importance of Getting the Assay Right — 171
5. Summary — 171
References — 171

10. **Analysis of Citrulline, Arginine, and Methylarginines Using High-Performance Liquid Chromatography** — 177

Guoyao Wu and Cynthia J. Meininger

1. Introduction — 178
2. The HPLC Apparatus — 179
3. Chemicals and Materials — 179
4. General Precautions in Sample Preparation and HPLC Analysis — 180
5. Analysis of Citrulline and Arginine in Physiological Samples — 181
6. Analysis of Methylarginines in Physiological Samples — 185
7. Conclusion — 188
Acknowledgments — 188
References — 188

11. **Quantitative Proteome Mapping of Nitrotyrosines** — 191

Diana J. Bigelow and Wei-Jun Qian

1. Introduction — 192
2. Multidimensional LC-MS/MS Provides Large Data Sets for Identification of Nitrotyrosine-Modified Proteins — 193

3.	Retention of Complexity in Samples Prepared from Global Proteomic Analysis	195
4.	Confident Identification of Nitrotyrosine-Containing Peptides	197
5.	Comparative Quantitation of Nitrotyrosine-Modified Peptide/Proteins	198
6.	Summary	203
	References	203

Section II. Cellular Methods 207

12. Protein S-Nitrosation in Signal Transduction: Assays for Specific Qualitative and Quantitative Analysis 209
Vadim V. Sumbayev and Inna M. Yasinska

1.	Introduction	210
2.	Examples Demonstrating the Contribution of Protein S-Nitrosation to Intracellular Signal Transduction	210
3.	Assays for Analysis of S-Nitrosation of Specific Cellular Proteins	214
4.	Conclusions	217
	Acknowledgments	218
	References	218

13. Determination of Mammalian Arginase Activity 221
Diane Kepka-Lenhart, David E. Ash, and Sidney M. Morris

1.	Introduction	222
2.	Principle of Assay	222
3.	Preparation of Cell and Tissue Extracts	223
4.	Buffers, Reagents and Other Materials for Assay Protocol I	224
5.	Assay Protocol I	225
6.	Buffers, Reagents, and Other Materials for Assay Protocol II	226
7.	Assay Protocol II	226
8.	Limitations of Assay	227
9.	Determination of Arginase Activity in Cultured Cells	228
	Acknowledgment	229
	References	229

14. Measurement of Protein S-Nitrosylation during Cell Signaling 231
Joan B. Mannick and Christopher M. Schonhoff

1.	Introduction	232
2.	Biotin Switch Technique	233

3. Protocol for Analyzing Protein *S*-Nitrosylation in Tissue
   Samples Using the Biotin Switch Assay — 235
4. Chemical Reduction/Chemiluminescence — 236
5. Protocol for Chemical Reduction/Chemiluminescence
   Measurements of *S*-Nitrosylation of Immunoprecipitated Proteins — 238
6. Conclusion — 240
References — 240

15. **Pivotal Role of Arachidonic Acid in the Regulation of Neuronal Nitric Oxide Synthase Activity and Inducible Nitric Oxide Synthase Expression in Activated Astrocytes** — 243

    Orazio Cantoni, Letizia Palomba, Tiziana Persichini, Sofia Mariotto, Hisanori Suzuki, and Marco Colasanti

    1. Introduction — 244
    2. Materials and Methods — 245
    3. Results and Discussion — 246
    Acknowledgments — 250
    References — 250

16. **Red Blood Cells as a Model to Differentiate between Direct and Indirect Oxidation Pathways of Peroxynitrite** — 253

    Maurizio Minetti, Donatella Pietraforte, Elisabetta Straface, Alessio Metere, Paola Matarrese, and Walter Malorni

    1. Introduction — 254
    2. Direct Reactions of Peroxynitrite with Biological Targets — 255
    3. Peroxynitrite Homolysis: *Indirect* Radical Chemistry — 256
    4. Red Blood Cells as an Experimental Model to Test the Fate of Peroxynitrite in a Biological Environment — 257
    5. Red Blood Cell Modifications Induced by Extracellular Peroxynitrite Decay — 260
    6. Red Blood Cell Modifications Induced by Intracellular Peroxynitrite Decay — 260
    7. Peroxynitrite-Dependent Phosphorylation Signaling of RBC — 261
    8. Peroxynitrite-Induced Biomarkers of RBC Senescence — 262
    9. Peroxynitrite-Induced Biomarkers of RBC Apoptosis — 264
    10. Methods — 265
    11. Data and Statistics — 268
    Acknowledgments — 269
    References — 269

17. **Detection and Proteomic Identification of *S*-Nitrosated Proteins in Human Hepatocytes**   273

Laura M. López-Sánchez, Fernando J. Corrales, Manuel De La Mata, Jordi Muntané, and Antonio Rodríguez-Ariza

    1. Introduction   274
    2. Preparation of CSNO   276
    3. Preparation of Primary Human Hepatocytes and Cell Culture   276
    4. Treatment of Hepatocytes and Sample Preparation   277
    5. Biotin Switch Assay   277
    6. Detection and Purification of Biotinylated Proteins   278
    7. Final Considerations   279
    Acknowledgments   280
    References   280

18. **Identification of *S*-Nitrosylated Proteins in Plants**   283

Simone Sell, Christian Lindermayr, and Jörg Durner

    1. Introduction   283
    2. Generation of Protein Nitrosothiols   285
    3. Blocking Reaction of Free Thiols   287
    4. Reduction of Nitrosothiols and *S*-Biotinylation   288
    5. Affinity Purification of Biotinylated Proteins by NeutrAvidin   289
    6. Modified Techniques Related to the Biotin Switch Assay   290
    References   291

19. **Identification of 3-Nitrotyosine-Modified Brain Proteins by Redox Proteomics**   295

D. Allan Butterfield and Rukhsana Sultana

    1. Introduction   296
    2. Materials   297
    3. Method   298
    4. Comments   305
    Acknowledgment   305
    References   305

20. **Slot-Blot Analysis of 3-Nitrotyrosine-Modified Brain Proteins**   309

Rukhsana Sultana and D. Allan Butterfield

    1. Introduction   309
    2. Materials   311

3.	Solutions	312
4.	Sample Preparation for 3-NT Determination	312
5.	Comments	313
	Acknowledgment	314
	References	314

21. Detection Assays for Determination of Mitochondrial Nitric Oxide Synthase Activity; Advantages and Limitations 317

Pedram Ghafourifar, Mordhwaj S. Parihar, Rafal Nazarewicz, Woineshet J. Zenebe, and Arti Parihar

1.	Introduction	318
2.	Colorimetric Nitric Oxide Synthase Assay	319
3.	Determination of Mitochondrial Nitric Oxide Synthase Activity Using Radioassay	320
4.	Spectrophotometric Determination of Mitochondrial Nitric Oxide Synthase Activity	322
5.	Polarographic Nitric Oxide Synthase Assays	322
6.	Chemiluminescence Assay	323
7.	Fluorescent-Based Nitric Oxide Detection Assays	326
8.	Conclusion	331
	References	332

Section III. Organism Methods 335

22. Assay of 3-Nitrotyrosine in Tissues and Body Fluids by Liquid Chromatography with Tandem Mass Spectrometric Detection 337

Naila Rabbani and Paul J. Thornalley

1.	3-Nitrotyrosine (3-NT) in Physiological Systems	338
2.	Measurement of 3-NT	339
3.	Liquid Chromatography with Tandem Mass Spectrometric Detection (LC-MS/MS) Assay of 3-NT Residues and 3-NT-Free Adducts: Thornalley Group Method	340
4.	Estimates of 3-NT Residues and Free 3-NT in Plasma and Red Blood Cells under Basal Conditions and Effect of Disease	342
5.	3-Nitrotyrosine Residues in Lipoproteins	351
6.	3-Nitrotyrosine Residues and Free Adduct in Cerebrospinal Fluid	353
7.	3-Nitrotyrosine Residue Content of Tissues	353
8.	Concluding Remarks	354
	Acknowledgment	355
	References	355

23. Nitrite and Nitrate Measurement by Griess Reagent in Human Plasma: Evaluation of Interferences and Standardization   361

Daniela Giustarini, Ranieri Rossi, Aldo Milzani, and Isabella Dalle-Donne

1. Introduction   362
2. Experimental Procedures   364
3. Results   366
4. Discussion   373
Acknowledgment   378
References   378

24. Detection of Nitric Oxide and Its Derivatives in Human Mixed Saliva and Acidified Saliva   381

Umeo Takahama, Sachiko Hirota, and Oniki Takayuki

1. Introduction   382
2. Formation of Reactive Nitrogen Oxide Species (RNOS) in Mixed Whole Saliva and the Bacterial Fraction   382
3. Detection of RNOS in Mixed Whole Saliva and the Bacterial Fraction   384
4. Formation of RNOS in Acidified Saliva   390
5. Detection of RNOS in Acidified Saliva   392
6. Concluding Remarks   393
References   394

25. Imaging of Reactive Oxygen Species and Nitric Oxide *In Vivo* in Plant Tissues   397

Luisa M. Sandalio, María Rodríguez-Serrano, María C. Romero-Puertas, and Luis A. del Río

1. Introduction   398
2. Imaging Reactive Oxygen Species and Nitric Oxide *In Vivo* by Confocal Laser Microscopy   399
3. Plant Tissue Preparation and Procedure   402
4. Conclusions   406
Acknowledgments   406
References   406

26. Examining Nitroxyl in Biological Systems   411
Jon M. Fukuto, Matthew I. Jackson, Nina Kaludercic, and Nazareno Paolocci

1.	Introduction	412
2.	Nitroxyl Donors	412
3.	Biological HNO Chemistry	419
4.	Use of HNO Donors in Biological Studies	423
5.	Nitroxyl Pharmacological Effects: *In Vivo* and *In Vitro* Studies	424
6.	Summary	426
	References	427

Author Index   *433*
Subject Index   *459*

Contributors

D. Allan Butterfield
Center of Membrane Sciences, Sanders-Brown Center on Aging and Department of Chemistry, University of Kentucky, Lexington, Kentucky

David E. Ash
Department of Chemistry, Central Michigan University, Mt. Pleasant, Michigan

Swati Basu
Department of Physics, Wake Forest University, Winston-Salem, North Carolina

Diana J. Bigelow
Cell Biology and Biochemistry Group, Division of Biological Sciences, Pacific Northwest National Laboratory, Richland, Washington

Yoki Kwok-Chu Butt
The Proteomic Task Force, Department of Applied Biology and Chemical Technology, the Hong Kong Polytechnic University and the State Key Laboratory of Chinese Medicine and Molecular Pharmacology, Shenzhen, China

Orazio Cantoni
Istituto di Farmacologia e Farmacognosia, Università degli Studi di Urbino "Carlo Bo," Urbino, Italy

Samuel Chun-Lap Lo
The Proteomic Task Force, Department of Applied Biology and Chemical Technology, the Hong Kong Polytechnic University and the State Key Laboratory of Chinese Medicine and Molecular Pharmacology, Shenzhen, China

Marco Colasanti
Dipartimento di Biologia, Università di Roma Tre, Rome, Italy

Fernando J. Corrales
Hepatology and Gene Therapy Unit, Universidad de Navarra, Pamplona, Spain

Chiara D'Ambrosio
Proteomics and Mass Spectrometry Laboratory, ISPAAM, National Research Council, Naples, Italy

Isabella Dalle-Donne
Department of Biology, University of Milan, Milan, Italy

Manuel De La Mata
Liver Research Unit, Hospital Universitario Reina Sofía, Córdoba, Spain

Luis A. del Río
Departamento de Bioquímica, Biología Celular y Molecular de Plantas, Estación Experimental del Zaidín, CSIC, Granada, Spain

Jörg Durner
Institute of Biochemical Plant Pathology, Helmholtz Zentrum München, German Research Center for Environmental Health, Munich-Neuherberg, Germany

Jon M. Fukuto
Interdepartmental Program in Molecular Toxicology, UCLA School of Public Health, Los Angeles, California and Department of Pharmacology, UCLA School of Medicine, Center for the Health Sciences, Los Angeles, California

Benjamin Gaston
Department of Pediatrics, University of Virginia School of Medicine, Charlottesville, Virginia

Pedram Ghafourifar
Department of Surgery, The Ohio State University College of Medicine, Columbus, Ohio

Daniela Giustarini
Department of Evolutionary Biology, University of Siena, Siena, Italy

Mark Gladwin
Critical Care Medicine Department, Clinical Center, National Institutes of Health, Bethesda, Maryland and Pulmonary and Vascular Medicine Branch, National Heart Lung and Blood Institute, National Institutes of Health, Bethesda, Maryland

Andrea Guidarelli
Istituto di Farmacologia e Farmacognosia, Università degli Studi di Urbino "Carlo Bo," Urbino, Italy

Jay W. Heinecke
Department of Medicine, University of Washington, Seattle, Washington

Sachiko Hirota
Department of Nutritional Science, Kyushu Women's University, Kitakyushu, Japan

Matthew I. Jackson
Interdepartmental Program in Molecular Toxicology, UCLA School of Public Health, Los Angeles, California

Joy Joseph
Department of Biophysics and Free Radical Research Center, Medical College of Wisconsin, Milwaukee, Wisconsin

Nina Kaludercic
Division of Cardiology, Department of Medicine, Johns Hopkins Medical Institutions, Baltimore, Maryland

B. Kalyanaraman
Department of Biophysics and Free Radical Research Center, Medical College of Wisconsin, Milwaukee, Wisconsin

Diane Kepka-Lenhart
Department of Microbiology and Molecular Genetics, University of Pittsburgh School of Medicine, Pittsburgh, Pennsylvania

Daniel B. Kim-Shapiro
Department of Physics, Wake Forest University, Winston-Salem, North Carolina

Lisa M. Landino
Department of Chemistry, The College of William and Mary, Williamsburg, Virginia

Laura M. López-Sánchez
Liver Research Unit, Hospital Universitario Reina Sofía, Córdoba, Spain

Christian Lindermayr
Institute of Biochemical Plant Pathology, Helmholtz Zentrum München, German Research Center for Environmental Health, Munich-Neuherberg, Germany

Walter Malorni
Drug Research and Evaluation, Istituto Superiore di Sanità, Rome, Italy

Joan B. Mannick
Departments of Medicine and Cell Biology, University of Massachusetts Medical School, Worcester, Massachusetts

Sofia Mariotto
Dipartimento di Scienze Neurologiche e della Visione, Sezione di Chimica Biologica, Università degli Studi di Verona, Verona, Italy

Paola Matarrese
Drug Research and Evaluation, Istituto Superiore di Sanità, Rome, Italy

Cynthia J. Meininger
Cardiovascular Research Institute and Department of Systems Biology and Translational Medicine, Texas A&M Health Science Center, College Station, Texas

Alessio Metere
Departments of Cell Biology and Neurosciences, Istituto Superiore di Sanità, Rome, Italy

Aldo Milzani
Department of Biology, University of Milan, Milan, Italy

Maurizio Minetti
Departments of Cell Biology and Neurosciences, Istituto Superiore di Sanità, Rome, Italy

Sidney M. Morris
Department of Microbiology and Molecular Genetics, University of Pittsburgh School of Medicine, Pittsburgh, Pennsylvania

Jordi Muntané
Liver Research Unit, Hospital Universitario Reina Sofía, Córdoba, Spain

Rafal Nazarewicz
Department of Surgery, The Ohio State University College of Medicine, Columbus, Ohio

Lisa A. Palmer
Department of Pediatrics, University of Virginia School of Medicine, Charlottesville, Virginia

Letizia Palomba
Istituto di Farmacologia e Farmacognosia, Università di Urbino "Carlo Bo," Urbino, Italy

Nazareno Paolocci
Department of Clinical and Experimental Medicine, General Pathology and Immunology Section, University of Perugia, Perugia, Italy and Division of Cardiology, Department of Medicine, Johns Hopkins Medical Institutions, Baltimore, Maryland

Arti Parihar
Department of Surgery, The Ohio State University College of Medicine, Columbus, Ohio

Mordhwaj S. Parihar
Department of Surgery, The Ohio State University College of Medicine, Columbus, Ohio

Tiziana Persichini
Dipartimento di Biologia, Università di Roma Tre, Rome, Italy

Donatella Pietraforte
Departments of Cell Biology and Neurosciences, Istituto Superiore di Sanità, Rome, Italy

Wei-Jun Qian
Division of Biological Sciences, Environmental Molecular Sciences Laboratory, Pacific Northwest National Laboratory, Richland, Washington

Naila Rabbani
Protein Damage and Systems Biology Research Group, Clinical Sciences Research Institute, Warwick Medical School, University of Warwick, University Hospital, Coventry, United Kingdom

Antonio Rodríguez-Ariza
Liver Research Unit, Hospital Universitario Reina Sofía, Córdoba, Spain

María Rodríguez-Serrano
Departamento de Bioquímica, Biología Celular y Molecular de Plantas, Estación Experimental del Zaidín, CSIC, Granada, Spain

María C. Romero-Puertas
Departamento de Bioquímica, Biología Celular y Molecular de Plantas, Estación Experimental del Zaidín, CSIC, Granada, Spain

Ranieri Rossi
Department of Evolutionary Biology, University of Siena, Siena, Italy

Anna Maria Salzano
Proteomics and Mass Spectrometry Laboratory, ISPAAM, National Research Council, Naples, Italy

Luisa M. Sandalio
Departamento de Bioquímica, Biología Celular y Molecular de Plantas, Estación Experimental del Zaidín, CSIC, Granada, Spain

Andrea Scaloni
Proteomics and Mass Spectrometry Laboratory, ISPAAM, National Research Council, Naples, Italy

Ingo Schmidt
Mikrobiologie, Universität Bayreuth, Bayreuth, Germany

Christopher M. Schonhoff
Department of Biomedical Sciences, Tufts University Cummings School of Veterinary Medicine, North Grafton, Massachusetts

Simone Sell
Institute of Biochemical Plant Pathology, Helmholtz Zentrum München, German Research Center for Environmental Health, Munich-Neuherberg, Germany

Baohai Shao
Department of Medicine, University of Washington, Seattle, Washington

Elisabetta Straface
Drug Research and Evaluation, Istituto Superiore di Sanità, Rome, Italy

Rukhsana Sultana
Sanders-Brown Center on Aging and Department of Chemistry, University of Kentucky, Lexington, Kentucky

Vadim V. Sumbayev
Medway School of Pharmacy, University of Kent, United Kingdom

Hisanori Suzuki
Dipartimento di Scienze Neurologiche e della Visione, Sezione di Chimica Biologica, Università degli Studi di Verona, Verona, Italy

Umeo Takahama
Department of Bioscience, Kyushu Dental College, Kitakyushu, Japan

Oniki Takayuki
Department of Bioscience, Kyushu Dental College, Kitakyushu, Japan

Paul J. Thornalley
Protein Damage and Systems Biology Research Group, Clinical Sciences Research Institute, Warwick Medical School, University of Warwick, University Hospital, Coventry, United Kingdom

Xunde Wang
Pulmonary and Vascular Medicine Branch, National Heart Lung and Blood Institute, National Institutes of Health, Bethesda, Maryland

Guoyao Wu
Department of Animal Science, Texas A&M University, College Station, Texas and Cardiovascular Research Institute and Department of Systems Biology and Translational Medicine, Texas A&M Health Science Center, College Station, Texas

Yingkai Xu
Department of Biophysics and Free Radical Research Center, Medical College of Wisconsin, Milwaukee, Wisconsin

Inna M. Yasinska
Medway School of Pharmacy, University of Kent, United Kingdom

Woineshet J. Zenebe
Department of Surgery, The Ohio State University College of Medicine, Columbus, Ohio

Hao Zhang
Department of Biophysics and Free Radical Research Center, Medical College of Wisconsin, Milwaukee, Wisconsin

Methods in Enzymology

Volume I. Preparation and Assay of Enzymes
Edited by Sidney P. Colowick and Nathan O. Kaplan

Volume II. Preparation and Assay of Enzymes
Edited by Sidney P. Colowick and Nathan O. Kaplan

Volume III. Preparation and Assay of Substrates
Edited by Sidney P. Colowick and Nathan O. Kaplan

Volume IV. Special Techniques for the Enzymologist
Edited by Sidney P. Colowick and Nathan O. Kaplan

Volume V. Preparation and Assay of Enzymes
Edited by Sidney P. Colowick and Nathan O. Kaplan

Volume VI. Preparation and Assay of Enzymes *(Continued)*
Preparation and Assay of Substrates
Special Techniques
Edited by Sidney P. Colowick and Nathan O. Kaplan

Volume VII. Cumulative Subject Index
Edited by Sidney P. Colowick and Nathan O. Kaplan

Volume VIII. Complex Carbohydrates
Edited by Elizabeth F. Neufeld and Victor Ginsburg

Volume IX. Carbohydrate Metabolism
Edited by Willis A. Wood

Volume X. Oxidation and Phosphorylation
Edited by Ronald W. Estabrook and Maynard E. Pullman

Volume XI. Enzyme Structure
Edited by C. H. W. Hirs

Volume XII. Nucleic Acids (Parts A and B)
Edited by Lawrence Grossman and Kivie Moldave

Volume XIII. Citric Acid Cycle
Edited by J. M. Lowenstein

Volume XIV. Lipids
Edited by J. M. Lowenstein

Volume XV. Steroids and Terpenoids
Edited by Raymond B. Clayton

VOLUME XVI. Fast Reactions
Edited by KENNETH KUSTIN

VOLUME XVII. Metabolism of Amino Acids and Amines (Parts A and B)
Edited by HERBERT TABOR AND CELIA WHITE TABOR

VOLUME XVIII. Vitamins and Coenzymes (Parts A, B, and C)
Edited by DONALD B. MCCORMICK AND LEMUEL D. WRIGHT

VOLUME XIX. Proteolytic Enzymes
Edited by GERTRUDE E. PERLMANN AND LASZLO LORAND

VOLUME XX. Nucleic Acids and Protein Synthesis (Part C)
Edited by KIVIE MOLDAVE AND LAWRENCE GROSSMAN

VOLUME XXI. Nucleic Acids (Part D)
Edited by LAWRENCE GROSSMAN AND KIVIE MOLDAVE

VOLUME XXII. Enzyme Purification and Related Techniques
Edited by WILLIAM B. JAKOBY

VOLUME XXIII. Photosynthesis (Part A)
Edited by ANTHONY SAN PIETRO

VOLUME XXIV. Photosynthesis and Nitrogen Fixation (Part B)
Edited by ANTHONY SAN PIETRO

VOLUME XXV. Enzyme Structure (Part B)
Edited by C. H. W. HIRS AND SERGE N. TIMASHEFF

VOLUME XXVI. Enzyme Structure (Part C)
Edited by C. H. W. HIRS AND SERGE N. TIMASHEFF

VOLUME XXVII. Enzyme Structure (Part D)
Edited by C. H. W. HIRS AND SERGE N. TIMASHEFF

VOLUME XXVIII. Complex Carbohydrates (Part B)
Edited by VICTOR GINSBURG

VOLUME XXIX. Nucleic Acids and Protein Synthesis (Part E)
Edited by LAWRENCE GROSSMAN AND KIVIE MOLDAVE

VOLUME XXX. Nucleic Acids and Protein Synthesis (Part F)
Edited by KIVIE MOLDAVE AND LAWRENCE GROSSMAN

VOLUME XXXI. Biomembranes (Part A)
Edited by SIDNEY FLEISCHER AND LESTER PACKER

VOLUME XXXII. Biomembranes (Part B)
Edited by SIDNEY FLEISCHER AND LESTER PACKER

VOLUME XXXIII. Cumulative Subject Index Volumes I–XXX
Edited by MARTHA G. DENNIS AND EDWARD A. DENNIS

VOLUME XXXIV. Affinity Techniques (Enzyme Purification: Part B)
Edited by WILLIAM B. JAKOBY AND MEIR WILCHEK

VOLUME XXXV. Lipids (Part B)
Edited by JOHN M. LOWENSTEIN

VOLUME XXXVI. Hormone Action (Part A: Steroid Hormones)
Edited by BERT W. O'MALLEY AND JOEL G. HARDMAN

VOLUME XXXVII. Hormone Action (Part B: Peptide Hormones)
Edited by BERT W. O'MALLEY AND JOEL G. HARDMAN

VOLUME XXXVIII. Hormone Action (Part C: Cyclic Nucleotides)
Edited by JOEL G. HARDMAN AND BERT W. O'MALLEY

VOLUME XXXIX. Hormone Action (Part D: Isolated Cells, Tissues, and Organ Systems)
Edited by JOEL G. HARDMAN AND BERT W. O'MALLEY

VOLUME XL. Hormone Action (Part E: Nuclear Structure and Function)
Edited by BERT W. O'MALLEY AND JOEL G. HARDMAN

VOLUME XLI. Carbohydrate Metabolism (Part B)
Edited by W. A. WOOD

VOLUME XLII. Carbohydrate Metabolism (Part C)
Edited by W. A. WOOD

VOLUME XLIII. Antibiotics
Edited by JOHN H. HASH

VOLUME XLIV. Immobilized Enzymes
Edited by KLAUS MOSBACH

VOLUME XLV. Proteolytic Enzymes (Part B)
Edited by LASZLO LORAND

VOLUME XLVI. Affinity Labeling
Edited by WILLIAM B. JAKOBY AND MEIR WILCHEK

VOLUME XLVII. Enzyme Structure (Part E)
Edited by C. H. W. HIRS AND SERGE N. TIMASHEFF

VOLUME XLVIII. Enzyme Structure (Part F)
Edited by C. H. W. HIRS AND SERGE N. TIMASHEFF

VOLUME XLIX. Enzyme Structure (Part G)
Edited by C. H. W. HIRS AND SERGE N. TIMASHEFF

VOLUME L. Complex Carbohydrates (Part C)
Edited by VICTOR GINSBURG

VOLUME LI. Purine and Pyrimidine Nucleotide Metabolism
Edited by PATRICIA A. HOFFEE AND MARY ELLEN JONES

VOLUME LII. Biomembranes (Part C: Biological Oxidations)
Edited by SIDNEY FLEISCHER AND LESTER PACKER

VOLUME LIII. Biomembranes (Part D: Biological Oxidations)
Edited by SIDNEY FLEISCHER AND LESTER PACKER

VOLUME LIV. Biomembranes (Part E: Biological Oxidations)
Edited by SIDNEY FLEISCHER AND LESTER PACKER

VOLUME LV. Biomembranes (Part F: Bioenergetics)
Edited by SIDNEY FLEISCHER AND LESTER PACKER

VOLUME LVI. Biomembranes (Part G: Bioenergetics)
Edited by SIDNEY FLEISCHER AND LESTER PACKER

VOLUME LVII. Bioluminescence and Chemiluminescence
Edited by MARLENE A. DELUCA

VOLUME LVIII. Cell Culture
Edited by WILLIAM B. JAKOBY AND IRA PASTAN

VOLUME LIX. Nucleic Acids and Protein Synthesis (Part G)
Edited by KIVIE MOLDAVE AND LAWRENCE GROSSMAN

VOLUME LX. Nucleic Acids and Protein Synthesis (Part H)
Edited by KIVIE MOLDAVE AND LAWRENCE GROSSMAN

VOLUME 61. Enzyme Structure (Part H)
Edited by C. H. W. HIRS AND SERGE N. TIMASHEFF

VOLUME 62. Vitamins and Coenzymes (Part D)
Edited by DONALD B. MCCORMICK AND LEMUEL D. WRIGHT

VOLUME 63. Enzyme Kinetics and Mechanism (Part A: Initial Rate and Inhibitor Methods)
Edited by DANIEL L. PURICH

VOLUME 64. Enzyme Kinetics and Mechanism
(Part B: Isotopic Probes and Complex Enzyme Systems)
Edited by DANIEL L. PURICH

VOLUME 65. Nucleic Acids (Part I)
Edited by LAWRENCE GROSSMAN AND KIVIE MOLDAVE

VOLUME 66. Vitamins and Coenzymes (Part E)
Edited by DONALD B. MCCORMICK AND LEMUEL D. WRIGHT

VOLUME 67. Vitamins and Coenzymes (Part F)
Edited by DONALD B. MCCORMICK AND LEMUEL D. WRIGHT

VOLUME 68. Recombinant DNA
Edited by RAY WU

VOLUME 69. Photosynthesis and Nitrogen Fixation (Part C)
Edited by ANTHONY SAN PIETRO

VOLUME 70. Immunochemical Techniques (Part A)
Edited by HELEN VAN VUNAKIS AND JOHN J. LANGONE

Volume 71. Lipids (Part C)
Edited by John M. Lowenstein

Volume 72. Lipids (Part D)
Edited by John M. Lowenstein

Volume 73. Immunochemical Techniques (Part B)
Edited by John J. Langone and Helen Van Vunakis

Volume 74. Immunochemical Techniques (Part C)
Edited by John J. Langone and Helen Van Vunakis

Volume 75. Cumulative Subject Index Volumes XXXI, XXXII, XXXIV–LX
Edited by Edward A. Dennis and Martha G. Dennis

Volume 76. Hemoglobins
Edited by Eraldo Antonini, Luigi Rossi-Bernardi, and Emilia Chiancone

Volume 77. Detoxication and Drug Metabolism
Edited by William B. Jakoby

Volume 78. Interferons (Part A)
Edited by Sidney Pestka

Volume 79. Interferons (Part B)
Edited by Sidney Pestka

Volume 80. Proteolytic Enzymes (Part C)
Edited by Laszlo Lorand

Volume 81. Biomembranes (Part H: Visual Pigments and Purple Membranes, I)
Edited by Lester Packer

Volume 82. Structural and Contractile Proteins (Part A: Extracellular Matrix)
Edited by Leon W. Cunningham and Dixie W. Frederiksen

Volume 83. Complex Carbohydrates (Part D)
Edited by Victor Ginsburg

Volume 84. Immunochemical Techniques (Part D: Selected Immunoassays)
Edited by John J. Langone and Helen Van Vunakis

Volume 85. Structural and Contractile Proteins (Part B: The Contractile Apparatus and the Cytoskeleton)
Edited by Dixie W. Frederiksen and Leon W. Cunningham

Volume 86. Prostaglandins and Arachidonate Metabolites
Edited by William E. M. Lands and William L. Smith

Volume 87. Enzyme Kinetics and Mechanism (Part C: Intermediates, Stereo-chemistry, and Rate Studies)
Edited by Daniel L. Purich

Volume 88. Biomembranes (Part I: Visual Pigments and Purple Membranes, II)
Edited by Lester Packer

VOLUME 89. Carbohydrate Metabolism (Part D)
Edited by WILLIS A. WOOD

VOLUME 90. Carbohydrate Metabolism (Part E)
Edited by WILLIS A. WOOD

VOLUME 91. Enzyme Structure (Part I)
Edited by C. H. W. HIRS AND SERGE N. TIMASHEFF

VOLUME 92. Immunochemical Techniques (Part E: Monoclonal Antibodies and General Immunoassay Methods)
Edited by JOHN J. LANGONE AND HELEN VAN VUNAKIS

VOLUME 93. Immunochemical Techniques (Part F: Conventional Antibodies, Fc Receptors, and Cytotoxicity)
Edited by JOHN J. LANGONE AND HELEN VAN VUNAKIS

VOLUME 94. Polyamines
Edited by HERBERT TABOR AND CELIA WHITE TABOR

VOLUME 95. Cumulative Subject Index Volumes 61–74, 76–80
Edited by EDWARD A. DENNIS AND MARTHA G. DENNIS

VOLUME 96. Biomembranes [Part J: Membrane Biogenesis: Assembly and Targeting (General Methods; Eukaryotes)]
Edited by SIDNEY FLEISCHER AND BECCA FLEISCHER

VOLUME 97. Biomembranes [Part K: Membrane Biogenesis: Assembly and Targeting (Prokaryotes, Mitochondria, and Chloroplasts)]
Edited by SIDNEY FLEISCHER AND BECCA FLEISCHER

VOLUME 98. Biomembranes (Part L: Membrane Biogenesis: Processing and Recycling)
Edited by SIDNEY FLEISCHER AND BECCA FLEISCHER

VOLUME 99. Hormone Action (Part F: Protein Kinases)
Edited by JACKIE D. CORBIN AND JOEL G. HARDMAN

VOLUME 100. Recombinant DNA (Part B)
Edited by RAY WU, LAWRENCE GROSSMAN, AND KIVIE MOLDAVE

VOLUME 101. Recombinant DNA (Part C)
Edited by RAY WU, LAWRENCE GROSSMAN, AND KIVIE MOLDAVE

VOLUME 102. Hormone Action (Part G: Calmodulin and Calcium-Binding Proteins)
Edited by ANTHONY R. MEANS AND BERT W. O'MALLEY

VOLUME 103. Hormone Action (Part H: Neuroendocrine Peptides)
Edited by P. MICHAEL CONN

VOLUME 104. Enzyme Purification and Related Techniques (Part C)
Edited by WILLIAM B. JAKOBY

VOLUME 105. Oxygen Radicals in Biological Systems
Edited by LESTER PACKER

VOLUME 106. Posttranslational Modifications (Part A)
Edited by FINN WOLD AND KIVIE MOLDAVE

VOLUME 107. Posttranslational Modifications (Part B)
Edited by FINN WOLD AND KIVIE MOLDAVE

VOLUME 108. Immunochemical Techniques (Part G: Separation and Characterization of Lymphoid Cells)
Edited by GIOVANNI DI SABATO, JOHN J. LANGONE, AND HELEN VAN VUNAKIS

VOLUME 109. Hormone Action (Part I: Peptide Hormones)
Edited by LUTZ BIRNBAUMER AND BERT W. O'MALLEY

VOLUME 110. Steroids and Isoprenoids (Part A)
Edited by JOHN H. LAW AND HANS C. RILLING

VOLUME 111. Steroids and Isoprenoids (Part B)
Edited by JOHN H. LAW AND HANS C. RILLING

VOLUME 112. Drug and Enzyme Targeting (Part A)
Edited by KENNETH J. WIDDER AND RALPH GREEN

VOLUME 113. Glutamate, Glutamine, Glutathione, and Related Compounds
Edited by ALTON MEISTER

VOLUME 114. Diffraction Methods for Biological Macromolecules (Part A)
Edited by HAROLD W. WYCKOFF, C. H. W. HIRS, AND SERGE N. TIMASHEFF

VOLUME 115. Diffraction Methods for Biological Macromolecules (Part B)
Edited by HAROLD W. WYCKOFF, C. H. W. HIRS, AND SERGE N. TIMASHEFF

VOLUME 116. Immunochemical Techniques
(Part H: Effectors and Mediators of Lymphoid Cell Functions)
Edited by GIOVANNI DI SABATO, JOHN J. LANGONE, AND HELEN VAN VUNAKIS

VOLUME 117. Enzyme Structure (Part J)
Edited by C. H. W. HIRS AND SERGE N. TIMASHEFF

VOLUME 118. Plant Molecular Biology
Edited by ARTHUR WEISSBACH AND HERBERT WEISSBACH

VOLUME 119. Interferons (Part C)
Edited by SIDNEY PESTKA

VOLUME 120. Cumulative Subject Index Volumes 81–94, 96–101

VOLUME 121. Immunochemical Techniques (Part I: Hybridoma Technology and Monoclonal Antibodies)
Edited by JOHN J. LANGONE AND HELEN VAN VUNAKIS

VOLUME 122. Vitamins and Coenzymes (Part G)
Edited by FRANK CHYTIL AND DONALD B. MCCORMICK

VOLUME 123. Vitamins and Coenzymes (Part H)
Edited by FRANK CHYTIL AND DONALD B. MCCORMICK

VOLUME 124. Hormone Action (Part J: Neuroendocrine Peptides)
Edited by P. MICHAEL CONN

VOLUME 125. Biomembranes (Part M: Transport in Bacteria, Mitochondria, and Chloroplasts: General Approaches and Transport Systems)
Edited by SIDNEY FLEISCHER AND BECCA FLEISCHER

VOLUME 126. Biomembranes (Part N: Transport in Bacteria, Mitochondria, and Chloroplasts: Protonmotive Force)
Edited by SIDNEY FLEISCHER AND BECCA FLEISCHER

VOLUME 127. Biomembranes (Part O: Protons and Water: Structure and Translocation)
Edited by LESTER PACKER

VOLUME 128. Plasma Lipoproteins (Part A: Preparation, Structure, and Molecular Biology)
Edited by JERE P. SEGREST AND JOHN J. ALBERS

VOLUME 129. Plasma Lipoproteins (Part B: Characterization, Cell Biology, and Metabolism)
Edited by JOHN J. ALBERS AND JERE P. SEGREST

VOLUME 130. Enzyme Structure (Part K)
Edited by C. H. W. HIRS AND SERGE N. TIMASHEFF

VOLUME 131. Enzyme Structure (Part L)
Edited by C. H. W. HIRS AND SERGE N. TIMASHEFF

VOLUME 132. Immunochemical Techniques (Part J: Phagocytosis and Cell-Mediated Cytotoxicity)
Edited by GIOVANNI DI SABATO AND JOHANNES EVERSE

VOLUME 133. Bioluminescence and Chemiluminescence (Part B)
Edited by MARLENE DELUCA AND WILLIAM D. MCELROY

VOLUME 134. Structural and Contractile Proteins (Part C: The Contractile Apparatus and the Cytoskeleton)
Edited by RICHARD B. VALLEE

VOLUME 135. Immobilized Enzymes and Cells (Part B)
Edited by KLAUS MOSBACH

VOLUME 136. Immobilized Enzymes and Cells (Part C)
Edited by KLAUS MOSBACH

VOLUME 137. Immobilized Enzymes and Cells (Part D)
Edited by KLAUS MOSBACH

VOLUME 138. Complex Carbohydrates (Part E)
Edited by VICTOR GINSBURG

VOLUME 139. Cellular Regulators (Part A: Calcium- and Calmodulin-Binding Proteins)
Edited by ANTHONY R. MEANS AND P. MICHAEL CONN

VOLUME 140. Cumulative Subject Index Volumes 102–119, 121–134

VOLUME 141. Cellular Regulators (Part B: Calcium and Lipids)
Edited by P. MICHAEL CONN AND ANTHONY R. MEANS

VOLUME 142. Metabolism of Aromatic Amino Acids and Amines
Edited by SEYMOUR KAUFMAN

VOLUME 143. Sulfur and Sulfur Amino Acids
Edited by WILLIAM B. JAKOBY AND OWEN GRIFFITH

VOLUME 144. Structural and Contractile Proteins (Part D: Extracellular Matrix)
Edited by LEON W. CUNNINGHAM

VOLUME 145. Structural and Contractile Proteins (Part E: Extracellular Matrix)
Edited by LEON W. CUNNINGHAM

VOLUME 146. Peptide Growth Factors (Part A)
Edited by DAVID BARNES AND DAVID A. SIRBASKU

VOLUME 147. Peptide Growth Factors (Part B)
Edited by DAVID BARNES AND DAVID A. SIRBASKU

VOLUME 148. Plant Cell Membranes
Edited by LESTER PACKER AND ROLAND DOUCE

VOLUME 149. Drug and Enzyme Targeting (Part B)
Edited by RALPH GREEN AND KENNETH J. WIDDER

VOLUME 150. Immunochemical Techniques (Part K: *In Vitro* Models of B and T Cell Functions and Lymphoid Cell Receptors)
Edited by GIOVANNI DI SABATO

VOLUME 151. Molecular Genetics of Mammalian Cells
Edited by MICHAEL M. GOTTESMAN

VOLUME 152. Guide to Molecular Cloning Techniques
Edited by SHELBY L. BERGER AND ALAN R. KIMMEL

VOLUME 153. Recombinant DNA (Part D)
Edited by RAY WU AND LAWRENCE GROSSMAN

VOLUME 154. Recombinant DNA (Part E)
Edited by RAY WU AND LAWRENCE GROSSMAN

VOLUME 155. Recombinant DNA (Part F)
Edited by RAY WU

VOLUME 156. Biomembranes (Part P: ATP-Driven Pumps and Related Transport: The Na, K-Pump)
Edited by SIDNEY FLEISCHER AND BECCA FLEISCHER

VOLUME 157. Biomembranes (Part Q: ATP-Driven Pumps and Related Transport: Calcium, Proton, and Potassium Pumps)
Edited by SIDNEY FLEISCHER AND BECCA FLEISCHER

VOLUME 158. Metalloproteins (Part A)
Edited by JAMES F. RIORDAN AND BERT L. VALLEE

VOLUME 159. Initiation and Termination of Cyclic Nucleotide Action
Edited by JACKIE D. CORBIN AND ROGER A. JOHNSON

VOLUME 160. Biomass (Part A: Cellulose and Hemicellulose)
Edited by WILLIS A. WOOD AND SCOTT T. KELLOGG

VOLUME 161. Biomass (Part B: Lignin, Pectin, and Chitin)
Edited by WILLIS A. WOOD AND SCOTT T. KELLOGG

VOLUME 162. Immunochemical Techniques (Part L: Chemotaxis and Inflammation)
Edited by GIOVANNI DI SABATO

VOLUME 163. Immunochemical Techniques (Part M: Chemotaxis and Inflammation)
Edited by GIOVANNI DI SABATO

VOLUME 164. Ribosomes
Edited by HARRY F. NOLLER, JR., AND KIVIE MOLDAVE

VOLUME 165. Microbial Toxins: Tools for Enzymology
Edited by SIDNEY HARSHMAN

VOLUME 166. Branched-Chain Amino Acids
Edited by ROBERT HARRIS AND JOHN R. SOKATCH

VOLUME 167. Cyanobacteria
Edited by LESTER PACKER AND ALEXANDER N. GLAZER

VOLUME 168. Hormone Action (Part K: Neuroendocrine Peptides)
Edited by P. MICHAEL CONN

VOLUME 169. Platelets: Receptors, Adhesion, Secretion (Part A)
Edited by JACEK HAWIGER

VOLUME 170. Nucleosomes
Edited by PAUL M. WASSARMAN AND ROGER D. KORNBERG

VOLUME 171. Biomembranes (Part R: Transport Theory: Cells and Model Membranes)
Edited by SIDNEY FLEISCHER AND BECCA FLEISCHER

VOLUME 172. Biomembranes (Part S: Transport: Membrane Isolation and Characterization)
Edited by SIDNEY FLEISCHER AND BECCA FLEISCHER

VOLUME 173. Biomembranes [Part T: Cellular and Subcellular Transport: Eukaryotic (Nonepithelial) Cells]
Edited by SIDNEY FLEISCHER AND BECCA FLEISCHER

VOLUME 174. Biomembranes [Part U: Cellular and Subcellular Transport: Eukaryotic (Nonepithelial) Cells]
Edited by SIDNEY FLEISCHER AND BECCA FLEISCHER

VOLUME 175. Cumulative Subject Index Volumes 135–139, 141–167

VOLUME 176. Nuclear Magnetic Resonance (Part A: Spectral Techniques and Dynamics)
Edited by NORMAN J. OPPENHEIMER AND THOMAS L. JAMES

VOLUME 177. Nuclear Magnetic Resonance (Part B: Structure and Mechanism)
Edited by NORMAN J. OPPENHEIMER AND THOMAS L. JAMES

VOLUME 178. Antibodies, Antigens, and Molecular Mimicry
Edited by JOHN J. LANGONE

VOLUME 179. Complex Carbohydrates (Part F)
Edited by VICTOR GINSBURG

VOLUME 180. RNA Processing (Part A: General Methods)
Edited by JAMES E. DAHLBERG AND JOHN N. ABELSON

VOLUME 181. RNA Processing (Part B: Specific Methods)
Edited by JAMES E. DAHLBERG AND JOHN N. ABELSON

VOLUME 182. Guide to Protein Purification
Edited by MURRAY P. DEUTSCHER

VOLUME 183. Molecular Evolution: Computer Analysis of Protein and Nucleic Acid Sequences
Edited by RUSSELL F. DOOLITTLE

VOLUME 184. Avidin-Biotin Technology
Edited by MEIR WILCHEK AND EDWARD A. BAYER

VOLUME 185. Gene Expression Technology
Edited by DAVID V. GOEDDEL

VOLUME 186. Oxygen Radicals in Biological Systems (Part B: Oxygen Radicals and Antioxidants)
Edited by LESTER PACKER AND ALEXANDER N. GLAZER

VOLUME 187. Arachidonate Related Lipid Mediators
Edited by ROBERT C. MURPHY AND FRANK A. FITZPATRICK

VOLUME 188. Hydrocarbons and Methylotrophy
Edited by MARY E. LIDSTROM

VOLUME 189. Retinoids (Part A: Molecular and Metabolic Aspects)
Edited by LESTER PACKER

VOLUME 190. Retinoids (Part B: Cell Differentiation and Clinical Applications)
Edited by LESTER PACKER

VOLUME 191. Biomembranes (Part V: Cellular and Subcellular Transport: Epithelial Cells)
Edited by SIDNEY FLEISCHER AND BECCA FLEISCHER

VOLUME 192. Biomembranes (Part W: Cellular and Subcellular Transport: Epithelial Cells)
Edited by SIDNEY FLEISCHER AND BECCA FLEISCHER

VOLUME 193. Mass Spectrometry
Edited by JAMES A. MCCLOSKEY

VOLUME 194. Guide to Yeast Genetics and Molecular Biology
Edited by CHRISTINE GUTHRIE AND GERALD R. FINK

VOLUME 195. Adenylyl Cyclase, G Proteins, and Guanylyl Cyclase
Edited by ROGER A. JOHNSON AND JACKIE D. CORBIN

VOLUME 196. Molecular Motors and the Cytoskeleton
Edited by RICHARD B. VALLEE

VOLUME 197. Phospholipases
Edited by EDWARD A. DENNIS

VOLUME 198. Peptide Growth Factors (Part C)
Edited by DAVID BARNES, J. P. MATHER, AND GORDON H. SATO

VOLUME 199. Cumulative Subject Index Volumes 168–174, 176–194

VOLUME 200. Protein Phosphorylation (Part A: Protein Kinases: Assays, Purification, Antibodies, Functional Analysis, Cloning, and Expression)
Edited by TONY HUNTER AND BARTHOLOMEW M. SEFTON

VOLUME 201. Protein Phosphorylation (Part B: Analysis of Protein Phosphorylation, Protein Kinase Inhibitors, and Protein Phosphatases)
Edited by TONY HUNTER AND BARTHOLOMEW M. SEFTON

VOLUME 202. Molecular Design and Modeling: Concepts and Applications (Part A: Proteins, Peptides, and Enzymes)
Edited by JOHN J. LANGONE

VOLUME 203. Molecular Design and Modeling: Concepts and Applications (Part B: Antibodies and Antigens, Nucleic Acids, Polysaccharides, and Drugs)
Edited by JOHN J. LANGONE

VOLUME 204. Bacterial Genetic Systems
Edited by JEFFREY H. MILLER

VOLUME 205. Metallobiochemistry (Part B: Metallothionein and Related Molecules)
Edited by JAMES F. RIORDAN AND BERT L. VALLEE

VOLUME 206. Cytochrome P450
Edited by MICHAEL R. WATERMAN AND ERIC F. JOHNSON

VOLUME 207. Ion Channels
Edited by BERNARDO RUDY AND LINDA E. IVERSON

VOLUME 208. Protein–DNA Interactions
Edited by ROBERT T. SAUER

VOLUME 209. Phospholipid Biosynthesis
Edited by EDWARD A. DENNIS AND DENNIS E. VANCE

VOLUME 210. Numerical Computer Methods
Edited by LUDWIG BRAND AND MICHAEL L. JOHNSON

VOLUME 211. DNA Structures (Part A: Synthesis and Physical Analysis of DNA)
Edited by DAVID M. J. LILLEY AND JAMES E. DAHLBERG

VOLUME 212. DNA Structures (Part B: Chemical and Electrophoretic Analysis of DNA)
Edited by DAVID M. J. LILLEY AND JAMES E. DAHLBERG

VOLUME 213. Carotenoids (Part A: Chemistry, Separation, Quantitation, and Antioxidation)
Edited by LESTER PACKER

VOLUME 214. Carotenoids (Part B: Metabolism, Genetics, and Biosynthesis)
Edited by LESTER PACKER

VOLUME 215. Platelets: Receptors, Adhesion, Secretion (Part B)
Edited by JACEK J. HAWIGER

VOLUME 216. Recombinant DNA (Part G)
Edited by RAY WU

VOLUME 217. Recombinant DNA (Part H)
Edited by RAY WU

VOLUME 218. Recombinant DNA (Part I)
Edited by RAY WU

VOLUME 219. Reconstitution of Intracellular Transport
Edited by JAMES E. ROTHMAN

VOLUME 220. Membrane Fusion Techniques (Part A)
Edited by NEJAT DÜZGÜNEŞ

VOLUME 221. Membrane Fusion Techniques (Part B)
Edited by NEJAT DÜZGÜNEŞ

VOLUME 222. Proteolytic Enzymes in Coagulation, Fibrinolysis, and Complement Activation (Part A: Mammalian Blood Coagulation Factors and Inhibitors)
Edited by LASZLO LORAND AND KENNETH G. MANN

VOLUME 223. Proteolytic Enzymes in Coagulation, Fibrinolysis, and Complement Activation (Part B: Complement Activation, Fibrinolysis, and Nonmammalian Blood Coagulation Factors)
Edited by LASZLO LORAND AND KENNETH G. MANN

VOLUME 224. Molecular Evolution: Producing the Biochemical Data
Edited by ELIZABETH ANNE ZIMMER, THOMAS J. WHITE, REBECCA L. CANN, AND ALLAN C. WILSON

VOLUME 225. Guide to Techniques in Mouse Development
Edited by PAUL M. WASSARMAN AND MELVIN L. DEPAMPHILIS

VOLUME 226. Metallobiochemistry (Part C: Spectroscopic and Physical Methods for Probing Metal Ion Environments in Metalloenzymes and Metalloproteins)
Edited by JAMES F. RIORDAN AND BERT L. VALLEE

VOLUME 227. Metallobiochemistry (Part D: Physical and Spectroscopic Methods for Probing Metal Ion Environments in Metalloproteins)
Edited by JAMES F. RIORDAN AND BERT L. VALLEE

VOLUME 228. Aqueous Two-Phase Systems
Edited by HARRY WALTER AND GÖTE JOHANSSON

VOLUME 229. Cumulative Subject Index Volumes 195–198, 200–227

VOLUME 230. Guide to Techniques in Glycobiology
Edited by WILLIAM J. LENNARZ AND GERALD W. HART

VOLUME 231. Hemoglobins (Part B: Biochemical and Analytical Methods)
Edited by JOHANNES EVERSE, KIM D. VANDEGRIFF, AND ROBERT M. WINSLOW

VOLUME 232. Hemoglobins (Part C: Biophysical Methods)
Edited by JOHANNES EVERSE, KIM D. VANDEGRIFF, AND ROBERT M. WINSLOW

VOLUME 233. Oxygen Radicals in Biological Systems (Part C)
Edited by LESTER PACKER

VOLUME 234. Oxygen Radicals in Biological Systems (Part D)
Edited by LESTER PACKER

VOLUME 235. Bacterial Pathogenesis (Part A: Identification and Regulation of Virulence Factors)
Edited by VIRGINIA L. CLARK AND PATRIK M. BAVOIL

VOLUME 236. Bacterial Pathogenesis (Part B: Integration of Pathogenic Bacteria with Host Cells)
Edited by VIRGINIA L. CLARK AND PATRIK M. BAVOIL

VOLUME 237. Heterotrimeric G Proteins
Edited by RAVI IYENGAR

VOLUME 238. Heterotrimeric G-Protein Effectors
Edited by RAVI IYENGAR

VOLUME 239. Nuclear Magnetic Resonance (Part C)
Edited by THOMAS L. JAMES AND NORMAN J. OPPENHEIMER

VOLUME 240. Numerical Computer Methods (Part B)
Edited by MICHAEL L. JOHNSON AND LUDWIG BRAND

VOLUME 241. Retroviral Proteases
Edited by LAWRENCE C. KUO AND JULES A. SHAFER

VOLUME 242. Neoglycoconjugates (Part A)
Edited by Y. C. LEE AND REIKO T. LEE

VOLUME 243. Inorganic Microbial Sulfur Metabolism
Edited by HARRY D. PECK, JR., AND JEAN LEGALL

VOLUME 244. Proteolytic Enzymes: Serine and Cysteine Peptidases
Edited by ALAN J. BARRETT

VOLUME 245. Extracellular Matrix Components
Edited by E. RUOSLAHTI AND E. ENGVALL

VOLUME 246. Biochemical Spectroscopy
Edited by KENNETH SAUER

VOLUME 247. Neoglycoconjugates (Part B: Biomedical Applications)
Edited by Y. C. LEE AND REIKO T. LEE

VOLUME 248. Proteolytic Enzymes: Aspartic and Metallo Peptidases
Edited by ALAN J. BARRETT

VOLUME 249. Enzyme Kinetics and Mechanism (Part D: Developments in Enzyme Dynamics)
Edited by DANIEL L. PURICH

VOLUME 250. Lipid Modifications of Proteins
Edited by PATRICK J. CASEY AND JANICE E. BUSS

VOLUME 251. Biothiols (Part A: Monothiols and Dithiols, Protein Thiols, and Thiyl Radicals)
Edited by LESTER PACKER

VOLUME 252. Biothiols (Part B: Glutathione and Thioredoxin; Thiols in Signal Transduction and Gene Regulation)
Edited by LESTER PACKER

VOLUME 253. Adhesion of Microbial Pathogens
Edited by RON J. DOYLE AND ITZHAK OFEK

VOLUME 254. Oncogene Techniques
Edited by PETER K. VOGT AND INDER M. VERMA

VOLUME 255. Small GTPases and Their Regulators (Part A: Ras Family)
Edited by W. E. BALCH, CHANNING J. DER, AND ALAN HALL

VOLUME 256. Small GTPases and Their Regulators (Part B: Rho Family)
Edited by W. E. BALCH, CHANNING J. DER, AND ALAN HALL

VOLUME 257. Small GTPases and Their Regulators (Part C: Proteins Involved in Transport)
Edited by W. E. BALCH, CHANNING J. DER, AND ALAN HALL

VOLUME 258. Redox-Active Amino Acids in Biology
Edited by JUDITH P. KLINMAN

VOLUME 259. Energetics of Biological Macromolecules
Edited by MICHAEL L. JOHNSON AND GARY K. ACKERS

VOLUME 260. Mitochondrial Biogenesis and Genetics (Part A)
Edited by GIUSEPPE M. ATTARDI AND ANNE CHOMYN

VOLUME 261. Nuclear Magnetic Resonance and Nucleic Acids
Edited by THOMAS L. JAMES

VOLUME 262. DNA Replication
Edited by JUDITH L. CAMPBELL

VOLUME 263. Plasma Lipoproteins (Part C: Quantitation)
Edited by WILLIAM A. BRADLEY, SANDRA H. GIANTURCO, AND JERE P. SEGREST

VOLUME 264. Mitochondrial Biogenesis and Genetics (Part B)
Edited by GIUSEPPE M. ATTARDI AND ANNE CHOMYN

VOLUME 265. Cumulative Subject Index Volumes 228, 230–262

VOLUME 266. Computer Methods for Macromolecular Sequence Analysis
Edited by RUSSELL F. DOOLITTLE

VOLUME 267. Combinatorial Chemistry
Edited by JOHN N. ABELSON

VOLUME 268. Nitric Oxide (Part A: Sources and Detection of NO; NO Synthase)
Edited by LESTER PACKER

VOLUME 269. Nitric Oxide (Part B: Physiological and Pathological Processes)
Edited by LESTER PACKER

VOLUME 270. High Resolution Separation and Analysis of Biological Macromolecules (Part A: Fundamentals)
Edited by BARRY L. KARGER AND WILLIAM S. HANCOCK

VOLUME 271. High Resolution Separation and Analysis of Biological Macromolecules (Part B: Applications)
Edited by BARRY L. KARGER AND WILLIAM S. HANCOCK

VOLUME 272. Cytochrome P450 (Part B)
Edited by ERIC F. JOHNSON AND MICHAEL R. WATERMAN

VOLUME 273. RNA Polymerase and Associated Factors (Part A)
Edited by SANKAR ADHYA

VOLUME 274. RNA Polymerase and Associated Factors (Part B)
Edited by SANKAR ADHYA

VOLUME 275. Viral Polymerases and Related Proteins
Edited by LAWRENCE C. KUO, DAVID B. OLSEN, AND STEVEN S. CARROLL

VOLUME 276. Macromolecular Crystallography (Part A)
Edited by CHARLES W. CARTER, JR., AND ROBERT M. SWEET

VOLUME 277. Macromolecular Crystallography (Part B)
Edited by CHARLES W. CARTER, JR., AND ROBERT M. SWEET

VOLUME 278. Fluorescence Spectroscopy
Edited by LUDWIG BRAND AND MICHAEL L. JOHNSON

VOLUME 279. Vitamins and Coenzymes (Part I)
Edited by DONALD B. MCCORMICK, JOHN W. SUTTIE, AND CONRAD WAGNER

VOLUME 280. Vitamins and Coenzymes (Part J)
Edited by DONALD B. MCCORMICK, JOHN W. SUTTIE, AND CONRAD WAGNER

VOLUME 281. Vitamins and Coenzymes (Part K)
Edited by DONALD B. MCCORMICK, JOHN W. SUTTIE, AND CONRAD WAGNER

VOLUME 282. Vitamins and Coenzymes (Part L)
Edited by DONALD B. MCCORMICK, JOHN W. SUTTIE, AND CONRAD WAGNER

VOLUME 283. Cell Cycle Control
Edited by WILLIAM G. DUNPHY

VOLUME 284. Lipases (Part A: Biotechnology)
Edited by BYRON RUBIN AND EDWARD A. DENNIS

VOLUME 285. Cumulative Subject Index Volumes 263, 264, 266–284, 286–289

VOLUME 286. Lipases (Part B: Enzyme Characterization and Utilization)
Edited by BYRON RUBIN AND EDWARD A. DENNIS

VOLUME 287. Chemokines
Edited by RICHARD HORUK

VOLUME 288. Chemokine Receptors
Edited by RICHARD HORUK

VOLUME 289. Solid Phase Peptide Synthesis
Edited by GREGG B. FIELDS

VOLUME 290. Molecular Chaperones
Edited by GEORGE H. LORIMER AND THOMAS BALDWIN

VOLUME 291. Caged Compounds
Edited by GERARD MARRIOTT

VOLUME 292. ABC Transporters: Biochemical, Cellular, and Molecular Aspects
Edited by SURESH V. AMBUDKAR AND MICHAEL M. GOTTESMAN

VOLUME 293. Ion Channels (Part B)
Edited by P. MICHAEL CONN

VOLUME 294. Ion Channels (Part C)
Edited by P. MICHAEL CONN

VOLUME 295. Energetics of Biological Macromolecules (Part B)
Edited by GARY K. ACKERS AND MICHAEL L. JOHNSON

VOLUME 296. Neurotransmitter Transporters
Edited by SUSAN G. AMARA

VOLUME 297. Photosynthesis: Molecular Biology of Energy Capture
Edited by LEE MCINTOSH

VOLUME 298. Molecular Motors and the Cytoskeleton (Part B)
Edited by RICHARD B. VALLEE

VOLUME 299. Oxidants and Antioxidants (Part A)
Edited by LESTER PACKER

VOLUME 300. Oxidants and Antioxidants (Part B)
Edited by LESTER PACKER

VOLUME 301. Nitric Oxide: Biological and Antioxidant Activities (Part C)
Edited by LESTER PACKER

VOLUME 302. Green Fluorescent Protein
Edited by P. MICHAEL CONN

VOLUME 303. cDNA Preparation and Display
Edited by SHERMAN M. WEISSMAN

VOLUME 304. Chromatin
Edited by PAUL M. WASSARMAN AND ALAN P. WOLFFE

VOLUME 305. Bioluminescence and Chemiluminescence (Part C)
Edited by THOMAS O. BALDWIN AND MIRIAM M. ZIEGLER

VOLUME 306. Expression of Recombinant Genes in Eukaryotic Systems
Edited by JOSEPH C. GLORIOSO AND MARTIN C. SCHMIDT

VOLUME 307. Confocal Microscopy
Edited by P. MICHAEL CONN

VOLUME 308. Enzyme Kinetics and Mechanism (Part E: Energetics of Enzyme Catalysis)
Edited by DANIEL L. PURICH AND VERN L. SCHRAMM

VOLUME 309. Amyloid, Prions, and Other Protein Aggregates
Edited by RONALD WETZEL

VOLUME 310. Biofilms
Edited by RON J. DOYLE

VOLUME 311. Sphingolipid Metabolism and Cell Signaling (Part A)
Edited by ALFRED H. MERRILL, JR., AND YUSUF A. HANNUN

VOLUME 312. Sphingolipid Metabolism and Cell Signaling (Part B)
Edited by ALFRED H. MERRILL, JR., AND YUSUF A. HANNUN

VOLUME 313. Antisense Technology (Part A: General Methods, Methods of Delivery, and RNA Studies)
Edited by M. IAN PHILLIPS

VOLUME 314. Antisense Technology (Part B: Applications)
Edited by M. IAN PHILLIPS

VOLUME 315. Vertebrate Phototransduction and the Visual Cycle (Part A)
Edited by KRZYSZTOF PALCZEWSKI

VOLUME 316. Vertebrate Phototransduction and the Visual Cycle (Part B)
Edited by KRZYSZTOF PALCZEWSKI

VOLUME 317. RNA–Ligand Interactions (Part A: Structural Biology Methods)
Edited by DANIEL W. CELANDER AND JOHN N. ABELSON

VOLUME 318. RNA–Ligand Interactions (Part B: Molecular Biology Methods)
Edited by DANIEL W. CELANDER AND JOHN N. ABELSON

VOLUME 319. Singlet Oxygen, UV-A, and Ozone
Edited by LESTER PACKER AND HELMUT SIES

VOLUME 320. Cumulative Subject Index Volumes 290–319

VOLUME 321. Numerical Computer Methods (Part C)
Edited by MICHAEL L. JOHNSON AND LUDWIG BRAND

VOLUME 322. Apoptosis
Edited by JOHN C. REED

VOLUME 323. Energetics of Biological Macromolecules (Part C)
Edited by MICHAEL L. JOHNSON AND GARY K. ACKERS

VOLUME 324. Branched-Chain Amino Acids (Part B)
Edited by ROBERT A. HARRIS AND JOHN R. SOKATCH

VOLUME 325. Regulators and Effectors of Small GTPases (Part D: Rho Family)
Edited by W. E. BALCH, CHANNING J. DER, AND ALAN HALL

VOLUME 326. Applications of Chimeric Genes and Hybrid Proteins (Part A: Gene Expression and Protein Purification)
Edited by JEREMY THORNER, SCOTT D. EMR, AND JOHN N. ABELSON

VOLUME 327. Applications of Chimeric Genes and Hybrid Proteins (Part B: Cell Biology and Physiology)
Edited by JEREMY THORNER, SCOTT D. EMR, AND JOHN N. ABELSON

VOLUME 328. Applications of Chimeric Genes and Hybrid Proteins (Part C: Protein–Protein Interactions and Genomics)
Edited by JEREMY THORNER, SCOTT D. EMR, AND JOHN N. ABELSON

VOLUME 329. Regulators and Effectors of Small GTPases (Part E: GTPases Involved in Vesicular Traffic)
Edited by W. E. BALCH, CHANNING J. DER, AND ALAN HALL

VOLUME 330. Hyperthermophilic Enzymes (Part A)
Edited by MICHAEL W. W. ADAMS AND ROBERT M. KELLY

VOLUME 331. Hyperthermophilic Enzymes (Part B)
Edited by MICHAEL W. W. ADAMS AND ROBERT M. KELLY

VOLUME 332. Regulators and Effectors of Small GTPases (Part F: Ras Family I)
Edited by W. E. BALCH, CHANNING J. DER, AND ALAN HALL

VOLUME 333. Regulators and Effectors of Small GTPases (Part G: Ras Family II)
Edited by W. E. BALCH, CHANNING J. DER, AND ALAN HALL

VOLUME 334. Hyperthermophilic Enzymes (Part C)
Edited by MICHAEL W. W. ADAMS AND ROBERT M. KELLY

VOLUME 335. Flavonoids and Other Polyphenols
Edited by LESTER PACKER

VOLUME 336. Microbial Growth in Biofilms (Part A: Developmental and Molecular Biological Aspects)
Edited by RON J. DOYLE

VOLUME 337. Microbial Growth in Biofilms (Part B: Special Environments and Physicochemical Aspects)
Edited by RON J. DOYLE

VOLUME 338. Nuclear Magnetic Resonance of Biological Macromolecules (Part A)
Edited by THOMAS L. JAMES, VOLKER DÖTSCH, AND ULI SCHMITZ

VOLUME 339. Nuclear Magnetic Resonance of Biological Macromolecules (Part B)
Edited by THOMAS L. JAMES, VOLKER DÖTSCH, AND ULI SCHMITZ

VOLUME 340. Drug–Nucleic Acid Interactions
Edited by JONATHAN B. CHAIRES AND MICHAEL J. WARING

VOLUME 341. Ribonucleases (Part A)
Edited by ALLEN W. NICHOLSON

VOLUME 342. Ribonucleases (Part B)
Edited by ALLEN W. NICHOLSON

VOLUME 343. G Protein Pathways (Part A: Receptors)
Edited by RAVI IYENGAR AND JOHN D. HILDEBRANDT

VOLUME 344. G Protein Pathways (Part B: G Proteins and Their Regulators)
Edited by RAVI IYENGAR AND JOHN D. HILDEBRANDT

VOLUME 345. G Protein Pathways (Part C: Effector Mechanisms)
Edited by RAVI IYENGAR AND JOHN D. HILDEBRANDT

VOLUME 346. Gene Therapy Methods
Edited by M. IAN PHILLIPS

VOLUME 347. Protein Sensors and Reactive Oxygen Species (Part A: Selenoproteins and Thioredoxin)
Edited by HELMUT SIES AND LESTER PACKER

VOLUME 348. Protein Sensors and Reactive Oxygen Species (Part B: Thiol Enzymes and Proteins)
Edited by HELMUT SIES AND LESTER PACKER

VOLUME 349. Superoxide Dismutase
Edited by LESTER PACKER

VOLUME 350. Guide to Yeast Genetics and Molecular and Cell Biology (Part B)
Edited by CHRISTINE GUTHRIE AND GERALD R. FINK

VOLUME 351. Guide to Yeast Genetics and Molecular and Cell Biology (Part C)
Edited by CHRISTINE GUTHRIE AND GERALD R. FINK

VOLUME 352. Redox Cell Biology and Genetics (Part A)
Edited by CHANDAN K. SEN AND LESTER PACKER

VOLUME 353. Redox Cell Biology and Genetics (Part B)
Edited by CHANDAN K. SEN AND LESTER PACKER

VOLUME 354. Enzyme Kinetics and Mechanisms (Part F: Detection and Characterization of Enzyme Reaction Intermediates)
Edited by DANIEL L. PURICH

VOLUME 355. Cumulative Subject Index Volumes 321–354

VOLUME 356. Laser Capture Microscopy and Microdissection
Edited by P. MICHAEL CONN

VOLUME 357. Cytochrome P450, Part C
Edited by ERIC F. JOHNSON AND MICHAEL R. WATERMAN

VOLUME 358. Bacterial Pathogenesis (Part C: Identification, Regulation, and Function of Virulence Factors)
Edited by VIRGINIA L. CLARK AND PATRIK M. BAVOIL

VOLUME 359. Nitric Oxide (Part D)
Edited by ENRIQUE CADENAS AND LESTER PACKER

VOLUME 360. Biophotonics (Part A)
Edited by GERARD MARRIOTT AND IAN PARKER

VOLUME 361. Biophotonics (Part B)
Edited by GERARD MARRIOTT AND IAN PARKER

VOLUME 362. Recognition of Carbohydrates in Biological Systems (Part A)
Edited by YUAN C. LEE AND REIKO T. LEE

VOLUME 363. Recognition of Carbohydrates in Biological Systems (Part B)
Edited by YUAN C. LEE AND REIKO T. LEE

VOLUME 364. Nuclear Receptors
Edited by DAVID W. RUSSELL AND DAVID J. MANGELSDORF

VOLUME 365. Differentiation of Embryonic Stem Cells
Edited by PAUL M. WASSAUMAN AND GORDON M. KELLER

VOLUME 366. Protein Phosphatases
Edited by SUSANNE KLUMPP AND JOSEF KRIEGLSTEIN

VOLUME 367. Liposomes (Part A)
Edited by NEJAT DÜZGÜNEŞ

VOLUME 368. Macromolecular Crystallography (Part C)
Edited by CHARLES W. CARTER, JR., AND ROBERT M. SWEET

VOLUME 369. Combinational Chemistry (Part B)
Edited by GUILLERMO A. MORALES AND BARRY A. BUNIN

VOLUME 370. RNA Polymerases and Associated Factors (Part C)
Edited by SANKAR L. ADHYA AND SUSAN GARGES

VOLUME 371. RNA Polymerases and Associated Factors (Part D)
Edited by SANKAR L. ADHYA AND SUSAN GARGES

VOLUME 372. Liposomes (Part B)
Edited by NEJAT DÜZGÜNEŞ

VOLUME 373. Liposomes (Part C)
Edited by NEJAT DÜZGÜNEŞ

VOLUME 374. Macromolecular Crystallography (Part D)
Edited by CHARLES W. CARTER, JR., AND ROBERT W. SWEET

VOLUME 375. Chromatin and Chromatin Remodeling Enzymes (Part A)
Edited by C. DAVID ALLIS AND CARL WU

VOLUME 376. Chromatin and Chromatin Remodeling Enzymes (Part B)
Edited by C. DAVID ALLIS AND CARL WU

VOLUME 377. Chromatin and Chromatin Remodeling Enzymes (Part C)
Edited by C. DAVID ALLIS AND CARL WU

VOLUME 378. Quinones and Quinone Enzymes (Part A)
Edited by HELMUT SIES AND LESTER PACKER

VOLUME 379. Energetics of Biological Macromolecules (Part D)
Edited by JO M. HOLT, MICHAEL L. JOHNSON, AND GARY K. ACKERS

VOLUME 380. Energetics of Biological Macromolecules (Part E)
Edited by JO M. HOLT, MICHAEL L. JOHNSON, AND GARY K. ACKERS

VOLUME 381. Oxygen Sensing
Edited by CHANDAN K. SEN AND GREGG L. SEMENZA

VOLUME 382. Quinones and Quinone Enzymes (Part B)
Edited by HELMUT SIES AND LESTER PACKER

VOLUME 383. Numerical Computer Methods (Part D)
Edited by LUDWIG BRAND AND MICHAEL L. JOHNSON

VOLUME 384. Numerical Computer Methods (Part E)
Edited by LUDWIG BRAND AND MICHAEL L. JOHNSON

VOLUME 385. Imaging in Biological Research (Part A)
Edited by P. MICHAEL CONN

VOLUME 386. Imaging in Biological Research (Part B)
Edited by P. MICHAEL CONN

VOLUME 387. Liposomes (Part D)
Edited by NEJAT DÜZGÜNEŞ

VOLUME 388. Protein Engineering
Edited by DAN E. ROBERTSON AND JOSEPH P. NOEL

VOLUME 389. Regulators of G-Protein Signaling (Part A)
Edited by DAVID P. SIDEROVSKI

VOLUME 390. Regulators of G-Protein Signaling (Part B)
Edited by DAVID P. SIDEROVSKI

VOLUME 391. Liposomes (Part E)
Edited by NEJAT DÜZGÜNEŞ

VOLUME 392. RNA Interference
Edited by ENGELKE ROSSI

VOLUME 393. Circadian Rhythms
Edited by MICHAEL W. YOUNG

VOLUME 394. Nuclear Magnetic Resonance of Biological Macromolecules (Part C)
Edited by THOMAS L. JAMES

VOLUME 395. Producing the Biochemical Data (Part B)
Edited by ELIZABETH A. ZIMMER AND ERIC H. ROALSON

VOLUME 396. Nitric Oxide (Part E)
Edited by LESTER PACKER AND ENRIQUE CADENAS

VOLUME 397. Environmental Microbiology
Edited by JARED R. LEADBETTER

VOLUME 398. Ubiquitin and Protein Degradation (Part A)
Edited by RAYMOND J. DESHAIES

VOLUME 399. Ubiquitin and Protein Degradation (Part B)
Edited by RAYMOND J. DESHAIES

VOLUME 400. Phase II Conjugation Enzymes and Transport Systems
Edited by HELMUT SIES AND LESTER PACKER

VOLUME 401. Glutathione Transferases and Gamma Glutamyl Transpeptidases
Edited by HELMUT SIES AND LESTER PACKER

VOLUME 402. Biological Mass Spectrometry
Edited by A. L. BURLINGAME

VOLUME 403. GTPases Regulating Membrane Targeting and Fusion
Edited by WILLIAM E. BALCH, CHANNING J. DER, AND ALAN HALL

VOLUME 404. GTPases Regulating Membrane Dynamics
Edited by WILLIAM E. BALCH, CHANNING J. DER, AND ALAN HALL

VOLUME 405. Mass Spectrometry: Modified Proteins and Glycoconjugates
Edited by A. L. BURLINGAME

VOLUME 406. Regulators and Effectors of Small GTPases: Rho Family
Edited by WILLIAM E. BALCH, CHANNING J. DER, AND ALAN HALL

VOLUME 407. Regulators and Effectors of Small GTPases: Ras Family
Edited by WILLIAM E. BALCH, CHANNING J. DER, AND ALAN HALL

VOLUME 408. DNA Repair (Part A)
Edited by JUDITH L. CAMPBELL AND PAUL MODRICH

VOLUME 409. DNA Repair (Part B)
Edited by JUDITH L. CAMPBELL AND PAUL MODRICH

VOLUME 410. DNA Microarrays (Part A: Array Platforms and Web-Bench Protocols)
Edited by ALAN KIMMEL AND BRIAN OLIVER

VOLUME 411. DNA Microarrays (Part B: Databases and Statistics)
Edited by ALAN KIMMEL AND BRIAN OLIVER

VOLUME 412. Amyloid, Prions, and Other Protein Aggregates (Part B)
Edited by INDU KHETERPAL AND RONALD WETZEL

VOLUME 413. Amyloid, Prions, and Other Protein Aggregates (Part C)
Edited by INDU KHETERPAL AND RONALD WETZEL

VOLUME 414. Measuring Biological Responses with Automated Microscopy
Edited by JAMES INGLESE

VOLUME 415. Glycobiology
Edited by MINORU FUKUDA

VOLUME 416. Glycomics
Edited by MINORU FUKUDA

VOLUME 417. Functional Glycomics
Edited by MINORU FUKUDA

VOLUME 418. Embryonic Stem Cells
Edited by IRINA KLIMANSKAYA AND ROBERT LANZA

VOLUME 419. Adult Stem Cells
Edited by IRINA KLIMANSKAYA AND ROBERT LANZA

VOLUME 420. Stem Cell Tools and Other Experimental Protocols
Edited by IRINA KLIMANSKAYA AND ROBERT LANZA

VOLUME 421. Advanced Bacterial Genetics: Use of Transposons and Phage for Genomic Engineering
Edited by KELLY T. HUGHES

VOLUME 422. Two-Component Signaling Systems, Part A
Edited by MELVIN I. SIMON, BRIAN R. CRANE, AND ALEXANDRINE CRANE

VOLUME 423. Two-Component Signaling Systems, Part B
Edited by MELVIN I. SIMON, BRIAN R. CRANE, AND ALEXANDRINE CRANE

VOLUME 424. RNA Editing
Edited by JONATHA M. GOTT

VOLUME 425. RNA Modification
Edited by JONATHA M. GOTT

VOLUME 426. Integrins
Edited by DAVID CHERESH

VOLUME 427. MicroRNA Methods
Edited by JOHN J. ROSSI

VOLUME 428. Osmosensing and Osmosignaling
Edited by HELMUT SIES AND DIETER HAUSSINGER

VOLUME 429. Translation Initiation: Extract Systems and Molecular Genetics
Edited by JON LORSCH

VOLUME 430. Translation Initiation: Reconstituted Systems and Biophysical Methods
Edited by JON LORSCH

VOLUME 431. Translation Initiation: Cell Biology, High-Throughput and Chemical-Based Approaches
Edited by JON LORSCH

VOLUME 432. Lipidomics and Bioactive Lipids: Mass-Spectrometry–Based Lipid Analysis
Edited by H. ALEX BROWN

VOLUME 433. Lipidomics and Bioactive Lipids: Specialized Analytical Methods and Lipids in Disease
Edited by H. ALEX BROWN

VOLUME 434. Lipidomics and Bioactive Lipids: Lipids and Cell Signaling
Edited by H. ALEX BROWN

VOLUME 435. Oxygen Biology and Hypoxia
Edited by HELMUT SIES AND BERNHARD BRÜNE

VOLUME 436. Globins and Other Nitric Oxide-Reactive Protiens (Part A)
Edited by ROBERT K. POOLE

VOLUME 437. Globins and Other Nitric Oxide-Reactive Protiens (Part B)
Edited by ROBERT K. POOLE

VOLUME 438. Small GTPases in Disease (Part A)
Edited by WILLIAM E. BALCH, CHANNING J. DER, AND ALAN HALL

VOLUME 439. Small GTPases in Disease (Part B)
Edited by WILLIAM E. BALCH, CHANNING J. DER, AND ALAN HALL

VOLUME 440. Nitric Oxide, Part F Oxidative and Nitrosative Stress in Redox Regulation of Cell Signaling
Edited by ENRIQUE CADENAS AND LESTER PACKER

SECTION ONE

MOLECULAR METHODS

CHAPTER ONE

Mass Spectrometric Characterization of Proteins Modified by Nitric Oxide-Derived Species

Anna Maria Salzano, Chiara D'Ambrosio, *and* Andrea Scaloni

Contents

1. Introduction	4
2. Reagents	6
3. Preparation of Nitrated BSA	6
4. In-Gel Protein Digestion and Mass Spectrometric Analysis	6
5. Data Analysis	7
6. MALDI-TOF Peptide Mass Fingerprinting	8
7. LC-ESI-IT Fragment Fingerprinting upon Collisional Fragmentation	9
8. Concluding Remarks	9
Acknowledgments	12
References	12

Abstract

Nitric oxide-derived metabolites have been demonstrated to covalently modify cellular protein repertoire, thus affecting specific enzymatic functions. Among the various redox posttranslational modifications, protein nitration has been broadly recognized by immunological, spectroscopical, and chromatographic methods as a widespread reaction regulating essential phatophysiological processes. With the introduction of matrix-assisted laser desorption and electrospray as soft ionization methods for mass spectrometry of biomolecules, nitration has been investigated directly at the protein level, assigning polypeptide modification sites. Peptide mass fingerprinting and fragment fingerprinting upon collisional fragmentation approaches have been widely used to this purpose. This chapter describes how minimal levels of nitration present on a model protein, namely bovine serum albumin, generated *in vitro* by $ONOO^-$ treatment, were

Proteomics and Mass Spectrometry Laboratory, ISPAAM, National Research Council, Naples, Italy

Methods in Enzymology, Volume 440   © 2008 Elsevier Inc.
ISSN 0076-6879, DOI: 10.1016/S0076-6879(07)00801-4   All rights reserved.

ascertained by integrated mass spectrometry approaches, identifying sites of modification. Critical considerations on the limits of each mass spectrometric ionization methods are also provided.

1. INTRODUCTION

Reactive oxygen species (ROS) and reactive nitrogen species (RNS) are generated at low amounts within organisms living under physiological conditions (Mikkelsen and Wardman, 2003; Moncada et al., 1991). Their moderate levels, controlled by dedicated enzymatic and nonenzymatic biological machineries (Berlett and Stadtman, 1997), play an integral role in the modulation of several cellular functions (Balaban et al., 2005; Schopfer et al., 2003). Interference with electron transport or redox environmental challenges may increase their concentration dramatically, contributing to generate additional reactive species by various pathways, which may cause extensive cellular damage (Thomas et al., 2006). While physiological levels of nitric oxide (NO•) have been demonstrated as essential in regulating signaling cascades, immune responses, and angiogenesis, high levels of NO• and NO•-derived reactive species such as peroxynitrite (ONOO$^-$) and nitrogen dioxide (NO$_2$•) have been associated with nitrosative stress, tissue injury, and progression of inflammatory diseases (Balaban et al., 2005; Thomas et al., 2006). NO•-derived metabolites have been reported to covalently modify cellular protein repertoire, eventually affecting specific enzymatic functions. Major protein modification reactions include nitration of Tyr, nitrosation of Cys, oxidation of Cys, Met, Trp, and Tyr, and formation of cross-linked di-Tyr and carbonyl adducts (Berlett and Stadtman, 1997; Dalle-Donne et al., 2005; Tien et al., 1999). Because direct determination of RNS in biological samples is difficult for several reasons, detection of specific protein modification products has been widely used to highlight its occurrence in an attempt to investigate the role of nitrosative stress in the pathogenesis and progression of diseases (Dalle-Donne et al., 2005; Scaloni, 2006). Thus, a number of analytical procedures have been developed for the direct assessment of S-nitrosation (Greco et al., 2006; Hao et al., 2006), Trp nitrosation/nitration (Sala et al., 2004; Wendt et al., 2003), Cys, Met, and Trp oxidation (Grunert et al., 2003; Taylor et al., 2003; Vascotto et al., 2007; Wagner et al., 2002), and formation of cross-linked di-Tyr (Heinecke et al., 1999) and carbonyl adducts (Thornalley et al., 2003) in protein samples, taking into consideration instability and traceability of each kind of modified amino acid (Dalle-Donne et al., 2005; Scaloni, 2006).

On the basis of its relative stability and specificity of the precursor reactive species involved, 3-nitrotyrosine (Tyr-NO$_2$) has widely been also considered as a valid biological marker of protein RNS insult (Gaut et al., 2002). In vivo nitration of Tyr can be catalyzed by metalloproteins such as myeloperoxidase, eosinophil peroxidase, myoglobin, and several cytochromes, which oxidize nitrite to $NO_2^{\bullet}$; in addition, myeloperoxidase induces nitration by the potent nitrating agent $ONOO^-$, produced by the reaction of $NO^{\bullet}$ with the superoxide anion (Dalle-Donne et al., 2005; Schopfer et al., 2003; Thomas et al., 2006). Nonenzymatic nitration of Tyr involves nitrous acid formed by the acidification of nitrite and a reactive intermediate product of the reaction between carbon dioxide and $ONOO^-$. Currently, biological aging and more than 60 human disorders, including atherosclerosis and tumors, stand in conection with protein nitration (Dalle-Donne et al., 2005; Ischiropoulos, 1998; Schopfer et al., 2003; Thomas et al., 2006).

A large part of the studies on protein nitration have focused on the detection and/or quantification of free $Tyr-NO_2$ either in plasma or in total protein hydrolyzates by gas chromatography/mass spectrometry (MS) (Gaut et al., 2002). Absolute levels reported in these investigations vary considerably and it has been shown that they may be the result of sample preparation artifacts (Scwedhelm et al., 1999). With the introduction of matrix-assisted laser desorption/ionization (MALDI) and electrospray ionization (ESI) as soft ionization methods for MS of biomolecules, nitration has been investigated directly at the protein level, also assigning polypeptide modification sites (Dalle-Donne et al., 2005; Scaloni, 2006). Peptide mass fingerprinting and fragment fingerprinting upon collisional fragmentation (MS/MS) approaches have been widely used to this purpose (Dalle-Donne et al., 2005; Scaloni, 2006). In these studies, high sequence coverage is a prime prerequisite for the comprehensive detection of protein nitration, generally obtained by direct MALDI–time of flight (TOF)–MS and/or liquid chromatography (LC)–ESI-MS/MS analysis of enzymatic digests. Before MS analysis, nitroproteins to be further investigated are generally verified by immunological and spectroscopical methods (Aulak et al., 2004; Wong and van der Vliet, 2002). Specific ultraviolet absorbance at 360 nm and intrinsic fluorescence (excitation 284 nm; emission 410 nm) are also used to selectively reveal $Tyr-NO_2$-containing peptides during chromatographic resolution of protein digests. Thus, MS-based approaches allowed identification of nitration sites in model proteins/peptides modified in vitro by various reagents (Ghesquiere et al., 2006; Jiao et al., 2001; Koeck et al., 2004; Minetti et al., 2000; Petersson et al., 2001; Sarver et al., 2001; Turko and Murad, 2005; Walcher et al., 2003; Wong and van der Vliet, 2002), as well as in proteins from biological tissues/fluids modified in vivo as a result of physiological/pathological conditions (Aslan et al., 2003; Aulak et al., 2001; Casoni et al., 2005; Haqqani et al., 2002; Kanski et al., 2005; Miyagi et al., 2002; Murray et al., 2003; Shao et al., 2005; Zhan and Desiderio, 2006; Zheng et al., 2005); in the latter case, nitroproteins were preventively resolved by bidimensional electrophoresis and identified by immunoblotting with specific antibodies (Aulak et al., 2004). This chapter shows how minimal levels of nitration on a model protein, namely bovine serum albumin (BSA), selectively generated *in vitro* by ONOO⁻ treatment, were ascertained by integrated MS approaches, also identifying sites of modification.

2. Reagents

Bovine serum albumin is from Sigma (St. Louis, MO). ONOO⁻ is from Upstate Biotechnology (Lake Placid, NY). A monoclonal antibody against Tyr-NO$_2$ came from Cayman Chem. (Ann Arbor, MI); goat anti-mouse IgG is from Pierce (Rockford, IL). The enhanced chemiluminescence kit and Hybond C nitrocellulose membranes are from Amersham Biosciences Italia (Milan).

3. Preparation of Nitrated BSA

Bovine serum albumin samples (2 nmol), dissolved in 100 μl of 50 mM NH$_4$HCO$_3$, pH 7, are nitrated by the addition of a 1- to 50-fold molar excess of ONOO⁻; the reaction is carried out overnight at 4 °C. Modified BSA samples are desalted rapidly with a PD10 column (Amersham Biosciences), dried, and analyzed by SDS-PAGE together with unreacted BSA. To prepare samples for further MS analysis, half of the gel is visualized with colloidal Coomassie staining; to ascertain protein nitration, the other half is visualized by Western blotting with a monoclonal anti-Tyr-NO$_2$ antibody. Mass spectrometric characterization is limited to modified samples for whom the minimal ONOO⁻/protein molar excess (10:1) allows visualization of nitrated BSA by Western blotting.

4. In-Gel Protein Digestion and Mass Spectrometric Analysis

Stained bands from SDS-PAGE are excised from the gel, S-alkylated, and digested with bovine trypsin as reported previously (D'Ambrosio et al., 2006; Renzone et al., 2006; Salzano et al., 2007; Vascotto et al., 2007). Gel particles are extracted with 25 mM NH$_4$HCO$_3$/acetonitrile (1:1, v/v) by sonication, and peptide mixtures are concentrated. Samples are desalted using μZipTipC18 pipette tips (Millipore) before MALDI-TOF-MS analysis and/or analyzed directly by μLC-ESI-IT-MS/MS.

Peptide mixtures are loaded on the MALDI target together with CHCA as the matrix, using the dried droplet technique. Samples are analyzed with a Voyager-DE PRO spectrometer (Applera) equipped with a nitrogen laser (337 nm) (D'Ambrosio et al., 2006; Renzone et al., 2006; Salzano et al., 2007; Vascotto et al., 2007). Mass spectra are acquired in reflectron mode; internal mass calibration is performed with peptides from trypsin autoproteolysis. Data are elaborated using the DataExplorer 5.1 software (Applera).

Peptide mixtures are also analyzed by μLC-ESI-IT-MS/MS using a LCQ Deca Xp Plus mass spectrometer (Thermo) equipped with an electrospray source connected to a Phoenix 40 pump (Thermo) (D'Ambrosio et al., 2006; Renzone et al., 2006; Salzano et al., 2007; Vascotto et al., 2007). Peptide mixtures are separated on a capillary ThermoHypersil-Keystone Aquasil C18 Kappa column (100 × 0.32 mm, 5 μm) (Hemel Hempstead, UK) using a linear gradient from 10 to 60% of acetonitrile in 0.1% formic acid, over 60 min, at a flow rate of 5 μl/min. Spectra for peptides are acquired in the range m/z 200–2000. Acquisition is controlled by a data-dependent product ion-scanning procedure over the three most abundant ions, enabling dynamic exclusion (repeat count 2 and exclusion duration 3 min). The mass isolation window and collision energy are set to m/z 3 and 35%, respectively. Data are elaborated using the BioWorks 3.1 software provided by the manufacturer.

5. DATA ANALYSIS

Assignment of MALDI-TOF-MS signals to specific peptides is achieved by GPMAW 4.23 software (Lighthouse Data, Odense, Denmark) (Qian et al., 2005), which generates a mass database output based on the BSA sequence, protease specificity, and static/dynamic mass modifications associated with Cys carbamidomethylation, Met oxidation, and Tyr nitration, respectively. The SEQUEST algorithm is used to identify peptides from MS/MS data (Peri et al., 2001). Peptides are identified automatically by comparison of ion mass spectra against those generated from a database containing BSA (SwissProt code P02769), bovine trypsin, and keratins. Sequest parameters include selection of trypsin with up to two missed cleavage sites and static/dynamic mass modifications reported earlier. Identified peptides are ranked in ascending order according to consensus scores and false positive identifications minimized by filtration against four of the following criteria: Xcorr >2, ΔCn >0.2, Sp >400, rsp <5, ions >30%. Where appropriate, peptide identifications are checked manually to provide for a false positive rate of <1% using Xcorr and ΔCn values described and validated elsewhere (D'Ambrosio et al., 2006; Renzone et al., 2006; Salzano et al., 2007; Vascotto et al., 2007).

6. MALDI-TOF Peptide Mass Fingerprinting

To obtain a qualitative overview on peptides eventually containing Tyr-NO$_2$, protein digests from untreated and nitrated BSA were analyzed in comparison by MALDI-TOF-MS. Common mass signals present in spectra recorded in the 700 to 4800 m/z range allowed covering 75.1% of the whole BSA sequence (Fig. 1.1), including 16/20 of the present Tyr residues. No significant differences between two samples were observed for satellite peaks associated with peptide oxidation products (Δm = +16); accordingly, no results were obtained on ONOO$^-$-mediated oxidation of Met residues. In contrast, a specific MH$^+$ signal at m/z 972.49 occurred only in nitrated BSA (inset in Fig. 1.1), which was tentatively assigned to nitrated peptide 137–143 (Δm = +45 with respect to the nonmodified one). This peptide contained two residues hypothetically subjected to modification, namely Tyr137 and Tyr139. Three additional MH$^+$ ions were also observed in this spectral region at m/z 940.51, 942.52, and 956.49, which were associated with molecular species showing a Δm = +13, +15, and +29 with respect to unmodified peptide, respectively. They were assigned to the nitrene, the amino, and the nitroso derivative of Tyr, respectively (Petersson et al., 2001; Sarver et al., 2001). The whole set of observed peaks was related to the typical pattern of signals already ascribed to the photodecomposition and reduction of nitropeptides during MALDI-TOF-MS analysis (Petersson et al., 2001; Sarver et al., 2001). The use of various matrices, namely dihydroxybenzoic acid and trihydroxyacetophenone, as

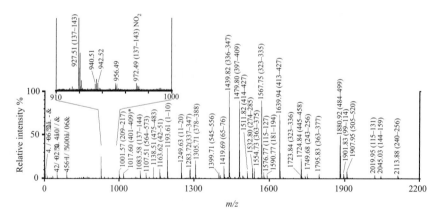

Figure 1.1 Reflectron positive-ion MALDI-TOF spectrum of a whole tryptic digest of nitrated BSA. (Inset) An expanded spectral region where signals associated to nitrated peptide 137–143 were observed, including photodecomposition and reduction products.

well as of different experimental conditions during spectral acquisition (number of laser shots and laser power), did not significantly change the signal profile observed in this region. Furthermore, these experiments did not allow detection of other signals associated with nitropeptides.

7. LC-ESI-IT Fragment Fingerprinting upon Collisional Fragmentation

To reveal additional nitropeptides not detected as a result of eventual photodecomposition phenomena during MALDI-TOF-MS and to assign the modification site in nitrated peptide 137–143, digests from untreated and nitrated BSA were also analyzed in comparison by μLC-ESI-IT-MS/MS. This analysis revealed the unique occurrence in nitrated BSA of concomitant signals at m/z 495.69 and 742.85 (eluting at a retention time of 30.4 min) and concomitant signals at m/z 486.93 and 972.47 (eluting at a retention time of 31.2 min), which were associated with multiply charged ions of two nitropeptides. In particular, the first ones were related to the triply and doubly charged ions of the nitrated peptide 336–347. Tandem mass spectrometry data confirmed the nature of this modified species, assigning the nitration site at Tyr340 (Fig. 1.2, top). The second ones were associated to the doubly and singly charged ions of nitrated peptide 137–143, already detected during MALDI-TOF-MS. Collision-induced dissociation experiments definitively assigned modification at Tyr137 (see Fig. 1.2, bottom). In both cases, major signals corresponding to nonmodified peptides were also observed. These experiments generated information on 17 of 20 Tyr residues present in BSA. Combining these data with those derived by MALDI-TOF-MS analysis, a 90.9% coverage of the whole protein sequence and a complete coverage of all Tyr residues within BSA were obtained.

8. Concluding Remarks

Literature data show that BSA has been chosen as a model protein to develop analytical methods for the assignment of nitration sites in polypeptides (Ghesquiere et al., 2006; Jiao et al., 2001; Petersson et al., 2001; Sarver et al., 2001; Walcher et al., 2003). Liquid- and gas-phase reactants have been used to modify this protein, and a number of BSA derivatives having a different degree of nitration have been produced and characterized. Our results on most reactive tyrosine residues are in good agreement with previous investigations, where poorly nitrated BSA was obtained by treatment with tetranitromethane or ONOO$^-$ (Ghesquiere et al., 2006;

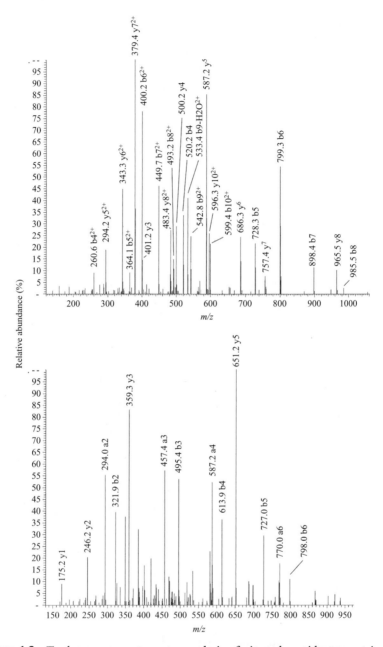

Figure 1.2 Tandem mass spectrometry analysis of nitrated peptides present in the tryptic digest of nitrated BSA as ascertained by μLC-ESI-IT-MS/MS analysis. (Top) The fragmentation mass spectrum of the triply charged ion at m/z 495.69 associated with the nitropeptide 336–347 (RHPEY^{nitro}AVSVLLR). (Bottom) The fragmentation mass spectrum of the doubly charged ion at m/z 486.93 associated with the nitropeptide 137–143 (Y^{nitro}LYEIAR).

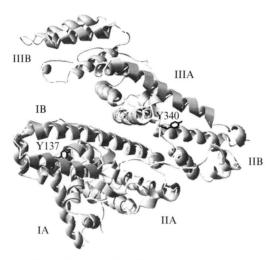

Figure 1.3 A three-dimensional model of the human serum albumin monomer where residues modified by ONOO$^-$ are highlighted with a side chain representation and colored in black. Protein subdomains are also indicated.

Jiao et al., 2001; Petersson et al., 2001; Sarver et al., 2001; Walcher et al., 2003). Figure 1.3 shows the position of Tyr137 and Tyr340 in the X-ray crystal structure of human serum albumin (Sugio et al., 1999), which is 76% identical with BSA and with all tyrosine residues conserved except two. Each of the modified tyrosine residues resides in a known reactive region within the protein, that is, subdomains IB and IIB, respectively. Because several other tyrosine residues are readily accessible on the protein surface, these findings indicate that Tyr nitration seems a highly site-specific reaction, most likely directed by the surrounding amino acid residues by hydrogen bonding, electrostatic interactions, or steric effects.

A number of investigations have been published in which site-specific nitration of various proteins from biological tissues/fluids has been related to pathophysiological conditions (Aslan et al., 2003; Aulak et al., 2001; Casoni et al., 2005; Haqqani et al., 2002; Kanski et al., 2005; Miyagi et al., 2002; Murray et al., 2003; Shao et al., 2005; Zhan and Desiderio, 2006; Zheng et al., 2005). These studies demonstrated the importance of MALDI and ESI mass spectrometry for the assignment of protein nitration. However, different authors underlined how the prompt fragmentation of nitrated peptides in MALDI-TOF-MS reduced their signal intensity, thus resulting in a failure to observe such species in complex peptide mass maps (Petersson et al., 2001; Sarver et al., 2001). We verified this phenomenon also in the work reported in this chapter. On the basis of these considerations, differential peptide mass mapping experiments by ESI approaches have found a broader application, and a large number of nitration sites within electrophoretically resolved proteins have been determined by automatic LC-ESI-MS/MS procedures. In this context, promising ESI methods based on precursor ion scanning or precursor ion-scanning/MS^n analysis of underivatized nitropepides or reduced/derivatized nitropepides have been proposed (Amoresano et al., 2007; Ghesquiere et al., 2006; Petersson et al., 2001). They allowed determination of a modification site in nitrated peptides present in very complex peptide mixtures. Selective trapping of nitrated peptides has been obtained by Tyr-NO$_2$ affinity particles (Nikov et al., 2003; Zhan and Desiderio, 2006). These novel approaches seem particularly useful for the analysis of in vivo protein nitration events, where low degrees of modification are generally observed.

ACKNOWLEDGMENTS

This work was partially supported by grants from the Italian National Research Council (AG-PO04-ISPAAM-C1) and Ministero dell'Istruzione, dell'Università e della Ricerca Scientifica e Tecnologica (FIRB2001 n. RBAU01PRLA).

REFERENCES

Amoresano, A., Chiappetta, G., Pucci, P., D'Ischia, M., and Marino, G. (2007). Bidimensional tandem mass spectrometry for selective identification of nitration sites in proteins. *Anal. Chem.* **79,** 2109–2117.

Aslan, M., Ryan, T. M., Townes, T. M., Coward, L., Kirk, M. C., Barnes, S., Alexander, C. B., Rosenfeld, S. S., and Freeman, B. A. (2003). Nitric oxide-dependent generation of reactive species in sickle cell disease: Actin tyrosine induces defective cytoskeletal polymerization. *J. Biol. Chem.* **278,** 4194–4204.

Aulak, K. S., Koeck, T., Crabb, J. W., and Stuehr, D. J. (2004). Proteomic method for identification of tyrosine-nitrated proteins. *Methods Mol. Biol.* **279,** 151–165.

Aulak, K. S., Miyagi, M., Yan, L., West, K. A., Massillon, D., Crabb, J. W., and Stuehr, D. J. (2001). Proteomic method identifies proteins nitrated in vivo during inflammatory challenge. *Proc. Natl. Acad. Sci. USA* **98,** 12056–12061.

Berlett, B. S., and Stadtman, E. R. (1997). Protein oxidation in aging, disease, and oxidative stress. *J. Biol. Chem.* **272,** 20313–20316.

Balaban, R. S., Nemoto, S., and Finkel, T. (2005). Mitochondria, oxidants, and aging. *Cell* **120,** 483–495.

Casoni, F., Basso, M., Massignan, T., Gianazza, E., Cheroni, C., Salmona, M., Bendotti, C., and Bonetto, V. (2005). Protein nitration in a mouse model of familial amyotrophic lateral sclerosis: Possible multifunctional role in the pathogenesis. *J. Biol. Chem.* **280,** 16295–16304.

Dalle-Donne, I., Scaloni, A., Giustarini, D., Cavarra, E., Tell, G., Lungarella, G., Colombo, R., Rossi, R., and Milzani, A. (2005). Proteins as biomarkers of oxidative/nitrosative stress in diseases: The contribution of redox proteomics. *Mass Spectrom. Rev.* **24,** 55–99.

D'Ambrosio, C., Arena, S., Fulcoli, G., Scheinfeld, M. H., Zhou, D., D'Adamio, L., and Scaloni, A. (2006). Hyperphosphorylation of JNK-interacting protein 1, a protein associated with Alzheimer disease. *Mol. Cell. Proteomics* **5,** 97–113.

Gaut, J. P., Byun, J., Tran, H. D., and Heinecke, J. W. (2002). Artifact-free quantification of free 3-chlorotyrosine, 3-bromotyrosine, and 3-nitrotyrosine in human plasma by electron capture-negative chemical ionization gas chromatography mass spectrometry and liquid chromatography-electrospray ionization tandem mass spectrometry. *Anal. Biochem.* **300,** 252–259.

Ghesquiere, B., Goethals, M., Van Damme, J., Staes, A., Timmerman, E., Vandekerckhove, J., and Gevaert, K. (2006). Improved tandem mass spectrometric characterization of 3-nitrotyrosine sites in peptides. *Rapid Commun. Mass Spectrom.* **20,** 2885–2893.

Greco, T. M., Hodara, R., Parastatidis, I., Heijnen, H. F., Dennehy, M. K., Liebler, D. C., and Ischiropoulos, H. (2006). Identification of S-nitrosylation motifs by site-specific mapping of the S-nitrosocysteine proteome in human vascular smooth muscle cells. *Proc. Natl. Acad. Sci. USA* **103,** 7420–7425.

Grunert, T., Pock, K., Buchacher, A., and Allmaier, G. (2003). Selective solid-phase isolation of methionine-containing peptides and subsequent matrix-assisted laser desorption/ionisation mass spectrometric detection of methionine- and of methionine-sulfoxide-containing peptides. *Rapid Commun. Mass Spectrom.* **17,** 1815–1824.

Hao, G., Derakhshan, B., Shi, L., Campagne, F., and Gross, S. S. (2006). SNOSID, a proteomic method for identification of cysteine S-nitrosylation sites in complex protein mixtures. *Proc. Natl. Acad. Sci. USA* **103,** 1012–1017.

Haqqani, A. S., Kelly, J. F., and Birnboim, H. C. (2002). Selective nitration of histone tyrosine residues *in vivo* in mutatect tumors. *J. Biol. Chem.* **277,** 3614–3621.

Heinecke, J. W., Hsu, F. F., Crowley, J. R., Hazen, S. L., Leeuwenburgh, C., Mueller, D. M., Rasmussen, J. E., and Turk, J. (1999). Detecting oxidative modification of biomolecules with isotope dilution mass spectrometry: Sensitive and quantitative assays for oxidized amino acids in proteins and tissues. *Methods Enzymol.* **300,** 124–144.

Ischiropoulos, H. (1998). Biological tyrosine nitration: A pathophysiological function of nitric oxide and reactive oxygen species. *Arch. Biochem. Biophys.* **356,** 1–11.

Jiao, K., Mandapati, S., Skipper, P. L., Tannenbaum, S. R., and Wishnok, J. S. (2001). Site-selective nitration of tyrosine in human serum albumin by peroxynitrite. *Anal. Biochem.* **293,** 43–52.

Kanski, J., Hong, S. J., and Schoneich, C. (2005). Proteomic analysis of protein nitration in aging skeletal muscle and identification of nitrotyrosine-containing sequences *in vivo* by nanoelectrospray ionization tandem mass spectrometry. *J. Biol. Chem.* **280,** 24261–24266.

Koeck, T., Levison, B., Hazen, S. L., Crabb, J. W., Stuehr, D. J., and Aulak, K. S. (2004). Tyrosine nitration impairs mammalian aldolase A activity. *Mol. Cell. Proteomics* **3,** 548–557.

Mikkelsen, R. B., and Wardman, P. (2003). Biological chemistry reactive oxygen and nitrogen and radiation-induced signal transduction mechanisms. *Oncogene* **22,** 5734–5754.

Minetti, M., Pietraforte, D., Carbone, V., Salzano, A. M., Scorza, G., and Marino, G. (2000). Scavenging of peroxynitrite by oxyhemoglobin and identification of modified globin residues. *Biochemistry* **39,** 6689–6697.

Miyagi, M., Sakaguchi, H., Darrow, R. M., Yan, L., West, K. A., Aulak, K. S., Stuehr, D. J., Hollyfield, J. G., Organisciak, D. T., and Crabb, J. W. (2002). Evidence that light modulates protein nitration in rat retina. *Mol. Cell. Proteomics* **1,** 293–303.

Moncada, S., Palmer, R. M., and Higgs, E. A. (1991). Nitric oxide: Physiology, pathophysiology, and pharmacology. *Pharmacol. Rev.* **43,** 109–142.

Murray, J., Taylor, S. W., Zhang, B., Ghosh, S. S., and Capaldi, R. A. (2003). Oxidative damage to mitochondrial complex I due to peroxynitrite: Identification of reactive tyrosines by mass spectrometry. *J. Biol. Chem.* **278,** 37223–37230.

Nikov, G., Bhat, V., Wishnok, J. S., and Tannenbaum, S. R. (2003). Analysis of nitrated proteins by nitrotyrosine-specific affinity probes and mass spectrometry. *Anal. Biochem.* **320,** 214–222.

Peri, S., Steen, H., and Pandey, A. (2001). GPMAW, a software tool for analyzing proteins and peptides. *Trends Biochem. Sci.* **26,** 687–689.

Petersson, A. S., Steen, H., Kalume, D. E., Caidahl, K., and Roepstorff, P. (2001). Investigation of tyrosine nitration in proteins by mass spectrometry. *J. Mass Spectrom.* **36,** 616–625.

Qian, W. J., Liu, T., Monroe, M. E., Strittmatter, E. F., Jacobs, J. M., Kangas, L. J., Petritis, K., Camp, D. G., 2nd, and Smith, R. D. (2005). Probability-based evaluation of peptide and protein identifications from tandem mass spectrometry and SEQUEST analysis: The human proteome. *J. Proteome Res.* **4,** 53–62.

Renzone, G., Salzano, A. M., Arena, S., D'Ambrosio, C., and Scaloni, A. (2006). Selective ion tracing and MS^n analysis of peptide digests from FSBA-treated kinases for the analysis of protein ATP-binding sites. *J. Proteome Res.* **5,** 2019–2024.

Sala, A., Nicolis, S., Roncone, R., Casella, L., and Monzani, E. (2004). Peroxidase catalyzed nitration of tryptophan derivatives: Mechanism, products and comparison with chemical nitrating agents. *Eur. J. Biochem.* **271,** 2841–2852.

Salzano, A. M., Arena, S., Renzone, G., D'Ambrosio, C., Rullo, R., Bruschi, M., Ledda, L., Maglione, G., Candiano, G., Ferrara, L., and Scaloni, A. (2007). A widespread picture of the *Streptococcus thermophilus* proteome by cell lysate fractionation and gel-based/gel-free approaches. *Proteomics* **7,** 1420–1433.

Sarver, A., Scheffler, N. K., Shetlar, M. D., and Gibson, B. W. (2001). Analysis of peptides and proteins containing nitrotyrosine by matrix-assisted laser desorption/ionization mass spectrometry. *J. Am. Soc. Mass Spectrom.* **12,** 439–448.

Scaloni, A. (2006). Mass spectrometry approaches for the molecular characterization of oxidatively/nitrosatively modified proteins. *In* "Redox Proteomics: From Protein Modifications to Cellular Dysfunction and Diseases" (I. Dalle-Donne, A. Scaloni, and D. A. Butterfield, eds.), p. 59. Wiley, Hoboken.

Schopfer, F. J., Baker, P. R., and Freeman, B. A. (2003). NO-dependent protein nitration: A cell signaling event or an oxidative inflammatory response? *Trends Biochem. Sci.* **28,** 646–654.

Schwedhelm, E., Tsikas, D., Gutzki, F. M., and Frohlich, J. (1999). Gas chromatographic-tandem mass spectrometric quantification of free 3-nitrotyrosine in human plasma at the basal state. *Anal. Biochem.* **276,** 195–203.

Shao, B., Bergt, C., Fu, X., Green, P., Voss, J. C., Oda, M. N., Oram, J. F., and Heinecke, J. W. (2005). Tyrosine 192 in apolipoprotein A-I is the major site of nitration and chlorination by myeloperoxidase, but only chlorination markedly impairs ABCA1-dependent cholesterol transport. *J. Biol. Chem.* **280,** 5983–5993.

Sugio, S., Kashima, A., Mochizuki, S., Noda, M., and Kobayashi, K. (1999). Crystal structure of human serum albumin at 2.5 A resolution. *Protein Eng.* **12,** 439–446.

Taylor, S. W., Fahy, E., Murray, J., Capaldi, R. A., and Ghosh, S. S. (2003). Oxidative post-translational modification of tryptophan residues in cardiac mitochondrial proteins. *J. Biol. Chem.* **278,** 19587–19590.

Thomas, D. D., Ridnour, L., Donzelli, S., Espey, M. G., Mancardi, D., Isenberg, J. S., Feelisch, M., Roberts, D. D., and Wink, D. D. (2006). The chemistry of protein modifications elicited by nitric oxide and related nitrogen oxides. *In* "Redox Proteomics: From Protein Modifications to Cellular Dysfunction and Diseases" (I. Dalle-Donne, A. Scaloni, and D. A. Butterfield, eds.), p. 25. Wiley, Hoboken.

Thornalley, P. J., Battah, S., Ahmed, N., Karachalias, N., Agalou, S., Babaei-Jadidi, R., and Dawnay, A. (2003). Quantitative screening of advanced glycation endproducts in cellular and extracellular proteins by tandem mass spectrometry. *Biochem. J.* **375,** 581–592.

Tien, M., Berlett, B. S., Levine, R. L., Chock, P. B., and Stadtman, E. R. (1999). Peroxynitrite-mediated modification of proteins at physiological carbon dioxide concentration: pH dependence of carbonyl formation, tyrosine nitration, and methionine oxidation. *Proc. Natl. Acad. Sci. USA* **96,** 7809–7814.

Turko, I. V., and Murad, F. (2005). Mapping sites of tyrosine nitration by matrix-assisted laser desorption/ionization mass spectrometry. *Methods Enzymol.* **396,** 266–275.

Vascotto, C., Salzano, A. M., D'Ambrosio, C., Fruscalzo, A., Marchesoni, D., di Loreto, C., Scaloni, A., Tell, G., and Quadrifoglio, F. (2007). Oxidized transthyretin in amniotic fluid as an early marker of preeclampsia. *J. Proteome Res.* **6,** 160–170.

Wagner, E., Luche, S., Penna, L., Chevallet, M., Van Dorsselaer, A., Leize-Wagner, E., and Rabilloud, T. (2002). A method for detection of overoxidation of cysteines: Peroxiredoxins are oxidized *in vivo* at the active-site cysteine during oxidative stress. *Biochem. J.* **366,** 777–785.

Walcher, W., Franze, T., Weller, M. G., Poschl, U., and Huber, C. G. (2003). Liquid- and gas-phase nitration of bovine serum albumin studied by LC-MS and LC-MS/MS using monolithic columns. *J. Proteome Res.* **2,** 534–542.

Wendt, S., Schlattner, U., and Wallimann, T. (2003). Differential effects of peroxynitrite on human mitochondrial creatine kinase isoenzymes: Inactivation, octamer destabilization, and identification of involved residues. *J. Biol. Chem.* **278,** 1125–1130.

Wong, P. S., and van der Vliet, A. (2002). Quantitation and localization of tyrosine nitration in proteins. *Methods Enzymol.* **359,** 399–410.

Zhan, X., and Desiderio, D. M. (2006). Nitroproteins from a human pituitary adenoma tissue discovered with a nitrotyrosine affinity column and tandem mass spectrometry. *Anal. Biochem.* **354,** 279–289.

Zheng, L., Settle, M., Brubaker, G., Schmitt, D., Hazen, S. L., Smith, J. D., and Kinter, M. (2005). Localization of nitration and chlorination sites on apolipoprotein A-I catalyzed by myeloperoxidase in human atheroma and associated oxidative impairment in ABCA1-dependent cholesterol efflux from macrophages. *J. Biol. Chem.* **280,** 38–47.

CHAPTER TWO

DETECTING NITRATED PROTEINS BY PROTEOMIC TECHNOLOGIES

Yoki Kwok-Chu Butt *and* Samuel Chun-Lap Lo

Contents

1. Introduction	18
2. Methods	20
2.1. Sample preparation	20
2.2. Dot blot screening	20
2.3. Two-dimensional gel electrophoresis	21
2.4. Two-in-one 2DE Western blot	23
2.5. Protein identification by MALDI-TOF mass spectrometry	26
2.6. Identification of nitrotyrosine sites by MALDI-TOF MS and LC-MS/MS	28
3. Conclusions	30
Acknowledgments	30
References	30

Abstract

Nitration is a posttranslational modification of tyrosine residues of proteins mediated by peroxynitrite ($ONOO^-$). It commonly occurs in neurological and pathological disorders, which involve nitric oxide (NO)-mediated oxidative stress. Nitration of tyrosine or tyrosyl groups of a protein modulates protein function and initiates signal transduction pathways, which lead to alternation of cellular metabolism and functions. Because of its apparent significance, there is an increasing urge to identify nitrated proteins as a bridge to expand our understanding of their involvement in different biological processes. This chapter describes strategies that could be used for rapid screening and detection of nitrated proteins, subsequent resolution, and identification of nitrated proteins and peptides using proteomic technologies. These include two-dimensional gel electrophoresis coupled with Western blotting and matrix-assisted laser desorption/ionization time of flight mass spectrometry, as well as liquid chromatography-linked tandem mass spectrometry.

The Proteomic Task Force, Department of Applied Biology and Chemical Technology, the Hong Kong Polytechnic University and the State Key Laboratory of Chinese Medicine and Molecular Pharmacology, Shenzhen, China

Methods in Enzymology, Volume 440   © 2008 Elsevier Inc.
ISSN 0076-6879, DOI: 10.1016/S0076-6879(07)00802-6   All rights reserved.

1. INTRODUCTION

Nitric oxide (NO) is an important cellular messenger that participates in reduction–oxidation mechanisms and in signal transduction pathways. One of the better described posttranslational modifications (PTMs) induced by NO is called nitration. As shown in Fig. 2.1, NO reacts rapidly with free superoxide (O_2^-) to yield peroxynitrite ($ONOO^-$), a strong oxidant that converts tyrosine residues in amino acid backbones of proteins into nitrotyrosines in nonenzymatic reactions. The nitro ($-NO_2$) group is added to position 3 of the phenolic ring of a tyrosine residue.

Nitration has been found to associate with many neurodegenerative and inflammatory diseases, including Parkinson's disease, Alzheimer's disease, transplant injection, lung infection, central nervous system and ocular inflammation, shock, and cancer (Beckman and Koppenol, 1996; Ischiropoulos, 1998; Peluffo and Radi, 2007; Reynolds et al., 2007). Hence, there is an increasing need to understand which proteins are nitrated during pathogenesis of these diseases. Detection of protein tyrosine nitration is possible by several different methods. Immunoreactivity with the antinitrotyrosine antibody is one of the most commonly employed tools used to evaluate the presence of protein-bound nitrotyrosines. Protein-bound nitrotyrosines, as well as increased levels of nitrotyrosine, had been documented by immunohistochemical staining as well as immunoprecipitation in tissues from patients with various kinds of disease (Hopkins et al., 2003; Jeng et al., 2005; Pittman et al., 2002). However, these methods are inadequate to elucidate the identities

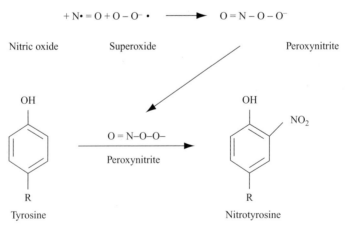

Figure 2.1 Schematic representation of the formation of a peroxynitrite ($OONO^-$) ion from the reaction of nitric oxide (NO) with superoxide (O_2^-) followed by nitration of L-tyrosine by peroxynitrite to form nitrotyrosine.

of proteins that had been nitrated during various cellular activities. However, using proteomic technologies, including various mass spectrometric approaches, nitrated proteins can be detected and identified easily. Two-dimensional gel electrophoresis (2DE), one of the most commonly used proteomic techniques, can be used to resolve proteins in a complex biological mixture. Proteins that are nitrated can be detected with the antinitrotyrosine antibody in a classical Western blotting setup. The immunoreactive proteins can then be pinpointed and identified by mass spectrometric techniques such as matrix-assisted laser/desorption ionization time of flight mass spectrometry (MALDI-TOF-MS). Protein identification is based on matching the list of experimentally measured masses of tryptic digested peptides of the nitrated protein (peptide mass fingerprint, PMF) with virtually digested peptide masses of all amino acid sequences available in computerized databases such as that of NCBI, SWISS-Prot, and TrEmBL. With the use of liquid chromatography-linked tandem mass spectrometry (LC-MS/MS), the tyrosine nitration sites in the nitrated protein can be determined subsequently. This chapter discusses methodologies that can be used to screen the total nitrotyrosine content in the samples rapidly, followed by the identification of nitrated proteins and then their nitrated tyrosine sites. An overview of the methodologies is shown in Fig. 2.2.

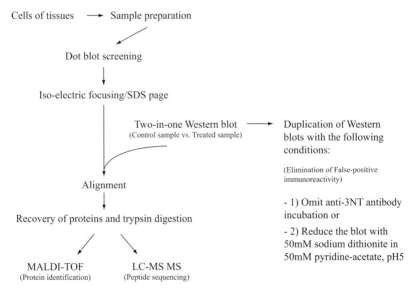

Figure 2.2 Overview of a proteomic approach to pinpoint and identify nitrated proteins in cell and tissue homogenates.

2. Methods

2.1. Sample preparation

For successful 2DE analysis, a well-focused, first-dimensional isoelectric focusing (IEF) protein separation step is essential. A concomitant sample preparative protocol that ensures all proteins are fully solubilized while dissociated from each other is essential. Regardless of whether the sample is a crude lysate or a mixture containing partially purified proteins, the sample solution must contain reducing agents, urea or thiourea, and certain detergents, such as CHAPS, to ensure complete solubilization and denaturation prior to IEF (Rabilloud, 1999). A lysis buffer [6 M urea, 2 M thiourea, and 4% (w/v) CHAPS] is effective in solubilizing a wide range of different samples. In addition, reductant(s) and immobilized pH gradient (IPG) buffer are also frequently added to samples to enhance sample solubility. A sample solution containing high salt or >0.25% (w/v) SDS has been found to interfere with IEF using immobilized pH gradients. Therefore, in order for subsequent successful IEF resolution, these samples should be buffer exchanged with other compatible solvents by dialysis, ultrafiltration, or protein precipitation. In some samples, such as bacteria cells, their high content of nucleic acids increases sample viscosity and eventually yields smears in the 2DE gel background. Importantly, nucleic acids bind to proteins through electrostatic interactions that interfere with the protein focusing process. Hence, samples rich in nucleic acids content should be incubated with a protease-free DNase/RNase mixture (e.g., $0.1\times$ volume of solution containing 1 mg/ml DNase and 0.25 mg/ml RNase) at 4 °C overnight, which reduces the nucleic acids to mono- and oligonucleotides. Further, additional sonication could facilitate breaking up of the DNA and any insoluble material can be spun down by high-speed centrifugation (14,000g for 30 min at 4 °C). Protein concentrations in samples solubilized in lysis buffer are usually determined by a modified Bradford protein assay. The classical Bradford assay is not suitable, as it is known that the urea, thiourea, and CHAPS in the lysis buffer can lead to inaccuracies (Ramagli, 1999).

2.2. Dot blot screening

Dot blot analysis is based on solid-phase immobilization of an antigen(s) onto a nitrocellulose/polyvinylidene fluoride (PVDF) membrane before being probed with mono-specific antibodies. Results of dot blot analysis give much of the same information as that of Western blotting analysis. In addition, it is a fast and high-throughput screening procedure for a batch of samples. Multiple samples, especially those taking in different time slots after treatment, can be analyzed in parallel using a dot blot apparatus. Hence, dot blot screening analysis is a good starting point for the evaluation of total nitrotyrosine content (including free nitrotyrosine and protein-bound nitrotyrosine) in these samples prior to 2DE-coupled Western blotting analysis of nitrated proteins. Using a 96-well vacuum manifold, each sample (with equal volume) containing 10 μg protein per dot (sample well) is filtered onto a nitrocellulose/PVDF membrane by gravity. Unbound proteins are removed by washing three times (5 min each) with 1× Tris-buffered saline with Tween [TBST, 20 mM Tris-HCl (pH 7.6), 137 mM NaCl, 0.02% Tween 20]. After washing, the nitrocellulose/PVDF membrane is removed from the dot-blotting apparatus before blocking with 3% bovine serum albumin (BSA) in 1× TBS for 1 h at room temperature. Nitrated proteins present in the samples are detected using antinitrotyrosine antibodies (1:1000, Upstate Biotechnology, Lake Placid, NY) and horseradish peroxidase-conjugated goat antimouse IgG antibody (1:2000, Santa Cruz Biotechnology Company, CA). Immunoreactivity, visualized using the ECL-Plus Western blot detection kit (Amersham Pharmacia Biotech, Piscataway, NJ), is normalized to the actual amount of proteins loaded on each dot in the membrane as calibrated using BSA with Coomassie staining.

Figure 2.3 shows a dot blot analysis of proteins extracted from a suspension culture of *Arabidopsis thaliana* cells at different time slots post-*Pseudomonas syringae* inoculation. A strong immunoreactivity was seen in protein extract from samples collected at 8, 10, and 12 h post-*P. syringae* inoculation. The degree of immunoreactivity was seen to be time dependent, up to 12 h. Based on this dot blot screening, samples collected at 12 h were chosen to continue on identification of the nitrated proteins using the proteomic approach.

2.3. Two-dimensional gel electrophoresis

Two-dimensional gel electrophoresis separates proteins in two dimensions to allow high resolution of complex mixtures. In the first dimension IEF, proteins are separated by virtue of their isoelectric points on commercially available IPG strips. In the second dimension, these proteins are resolved further by their molecular weights on a vertical SDS-PAGE gel. With a

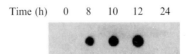

Figure 2.3 Immuno-dot blot analysis of samples extracted from *Pseudomonas syringae* M6 post-infection into *Arabidopsis thaliana* culture cells. Equal volumes of 10-μg protein samples collected at different time intervals (0, 8, 10, 12, and 24 h) were probed using antinitrotyrosine antibodies.

good sample preparation protocol and optimal IEF conditions, 2DE can resolve more than 2000 proteins spots on a 20 × 25-cm gel. Higher resolution with higher protein loading is allowed when the IEF step is performed within narrow pH ranges, for example, across a single pH unit. Optimal focusing conditions will vary with the nature of the sample composition, its complexity, its amount, how the sample is applied, and which pH range is analyzed. The suggested IEF protocol shown in Table 2.1 is a good starting point for 18-cm IPG strips, as it has been modified to optimize performance for most samples. Samples with a high nucleic acid content should be applied using the cup-loading method at the catholic end of an IPG strip. In our hands, the face-up mode of the strip during IEF frequently yields good resolution.

Procedure

1. Incubate IPG strips of linear pH gradient 4–7 (e.g., 18 cm long, Bio-Rad, Hercules, CA) in 350 μl rehydration solution containing the sample proteins for 16 h in a rehydration tray.
2. Place the sample-containing IPG strips in the focusing tray in the face-up mode before applying the movable electrodes onto the IPG strips.
3. Perform the IEF with a PROTEAN IEF Cell (Bio-Rad) using the protocol shown in Table 2.1.
4. After IEF, equilibrate the IPG strips with 1% (w/v) dithiothreitol in the equilibration buffer [50 mM Tris-HCl (pH 8.8), 6 M urea, 39% (v/v) glycerol, 2% (w/v) SDS, and 0.006% (w/v) bromphenol blue] for 15 min before incubating in a solution of 1% (w/v) iodoacetamide in the same equilibration buffer for another 15 min.

Table 2.1 Representative rehydration step and IEF protocol for 18-cm IPG strip

	Ramp	Voltage (V)	Duration (h:min)	kVH
Rehydration (passive)			16:00	
Rehydration (active)		30 (50 mA)	00:30	0.003
Step 1	Rapid	500	01:00	0.06
Step 2	Linear	1000	01:00	0.6
Step 3	Linear	4000	02:00	4.8
Step 4	Linear	8000	04:00	21.6
Step 5	Step-n-hold	8000		80

5. Place the IPG strip on top of a vertical 10% polyacrylamide slab gel containing a 4% stacking gel.
6. Place a filter paper square soaked with 1 μl of molecular weight standards (e.g., prestained molecular weight markers from Bio-Rad) adjacent to the IPG strip.
7. If needed, place another filter paper square soaked with a nitrated BSA as a positive control (see Section 2.4).
8. Seal the strip and filter paper squares with a 0.5% agarose sealing solution.
9. Run the second dimension SDS-PAGE according to Laemmli at a constant current of 30 mA/gel until the bromphenol blue dye reaches the bottom of the gel.
10. Stain the gel(s) according to the "MALDI-TOF mass spectrometry compatible silver staining protocol" as shown in Table 2.2.

2.4. Two-in-one 2DE Western blot

When probing nitrated proteins with 2DE, we suggest performing a "two-in-one" gel procedure followed by Western blotting. A diagrammatic illustration of this "two-in-one" gel–Western blot scheme is illustrated in Fig. 2.4. As elaborated in Section 2.3, equal amounts of control and treated samples are focused on separated IPG strips initially. The strips are then cut in the middle, and strips with the same pH range coverage but having different

Table 2.2 A MALDI-TOF MS compatible silver staining protocol

Step	Solution (250 ml per gel)	Time (min)
1. Fixation (two times)	40% acetic acid, 10% methanol in 250 ml Milli-Q water	15
2. Sensitization	0.2% (w/v) sodium thiosulfate, 6.8% (w/v) sodium acetate, 30% methanol in 250 ml Milli-Q water	30
3. Wash (three times)	250 ml Milli-Q water	5
4. Silver	0.25% (w/v) silver nitrate in 250 ml Milli-Q water	20
5. Wash (two times)	250 ml Milli-Q water	1
6. Development	2.5% (w/v) sodium carbonate, 0.015% formaldehyde in 250 ml Milli-Q water	10
7. Stopping	1.46% (w/v) EDTA in 250 ml Milli-Q water	10

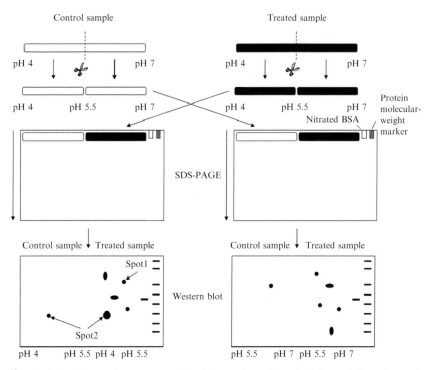

Figure 2.4 Schematic representation illustrating the principles of "two-in-one" Western blot. Using antinitrotyrosine antibodies, spot 1 is only detected in the treated sample. However, spot 2 is detected in both control and treated samples but with different degrees of immunoreactivities.

samples are loaded side by side on the top of a vertical SDS-PAGE slab gel for molecular weight separation. Because the tyrosine nitration within a certain protein may be in an all-or-nothing fashion (as spot 1 in Fig. 2.4) or only altered with different degrees of expression before and after treatment (as spot 2 in Fig. 2.4), "two-in-one" Western blot detection allows comparison of the degree of immunoreactivity of nitrated proteins of the control and treated samples in one Western blot. In this strategy, the two halves of IPG strips represent differential protein expression in control and treated samples run on the same SDS-PAGE and further electrotransfer on the same membrane. Therefore, the same microenvironment of electrophoresis, condition of antibody incubation, and exposure time in film development can be controlled as far as possible in this pair of comparable samples.

2.4.1. Procedure

1. As elaborated in procedures in Section 2.3, soak a nitrated BSA and protein molecular weight marker into separated filter paper squares and

run them adjacent to the IPG strips before the second dimension SDS-PAGE (see Fig. 2.4). Nitrated BSA serves as a positive control for immunodetection.
2. After running the SDS-PAGE, soak the gels in Tris-glycine electrotransfer buffer before electroblotting onto a nitrocellulose/PVDF membrane at 25 V for 10 h at 4 °C.
3. Block the nitrocellulose/PVDF membrane with TTBS containing 3% BSA for at least 4 h at room temperature.
4. Probe the membrane with antinitrotyrosine antibodies (1:1000, Upstate Biotechnology) in 1× TBST containing 1% BSA for 4 h at room temperature.
5. After removing the unbound antibodies, probe the membrane with horseradish peroxidase-conjugated goat antimouse IgG antibody (1:2000, Santa Cruz Biotechnology Company).
6. Detect immunoreactive spots using the ECL-Plus Western blot detection kit (Amersham Pharmacia Biotech).

An example of a "two-in-one" Western blot is shown in Fig. 2.5.

2.4.2. Elimination of false-positive signals

Detection of tyrosine nitration to date is usually achieved by immunoreactivity with antibodies directed against nitrotyrosine residues in Western blot analysis. However, because of the qualities of these antibodies, some artifacts may be generated due to nonspecific cross-reactivities. To minimize this

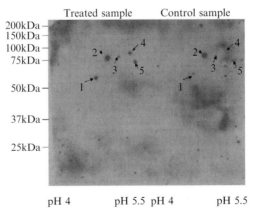

Figure 2.5 "Two-in-one" Western blot of control cells and *Pseudomonas syringae* M6 post-infection into *Arabidopsis thaliana* culture cells. Arrows indicate the immunoreactive protein spots on the Western blot. The expression level of spot 2 seems to be constant in both treated and untreated samples. However, spots 1, 3, 4, and 5 seem to have different expression levels in the two samples.

inadequacy, controls should be done to distinguish specific from nonspecific immunoreactivity. The most employed method is to chemically reduce nitrotyrosine to aminotyrosine with dithionite before incubation with the antinitrotyrosine antibodies. In this case, duplicated Western blots should be performed. Reduction of nitrotyrosine to aminotyrosine is achieved by treating one nitrocellulose/PVDF membrane with 10 mM sodium dithionate in 50 mM pyridine-acetate buffer (pH 5) for 1 h at room temperature (Miyagi et al., 2002). After reduction, the membrane is rinsed with distilled water and equilibrated with TBST. The membrane reduced previously with dithionate and the nonreduced membrane are processed subsequently in identical fashions using methodology described in Section 2.4. Truly nitrated proteins should only be seen in the nonreduced membrane. Any immunoreactivity on the reduced membrane could be the result of nonspecific binding of the antinitrotyrosine antibodies. In addition, control experiments in which the antinitrotyrosine antibodies incubation step was omitted should be performed to see if there is any nonspecific binding of these antibodies to proteins in the samples or to the nitrocellulose/PVDF membrane.

2.5. Protein identification by MALDI-TOF mass spectrometry

Once the nitrated proteins have been detected and pinpointed, they are identified by MALDI-TOF mass spectrometry. Protein identification is based on matching the list of experimentally measured masses of tryptic digested peptides of the nitrated protein, PMF, with virtually digested peptide masses of all amino acid sequences available in computerized databases. Modern day MALDI-TOF MS has sensitivity at the 50 fM level or less. Hence, extra care should be taken in every sample preparation step to avoid contamination, especially keratins from fingers and exposed parts of skin. All buffers should be prepared carefully, using chemicals of the highest purity. The nitrated proteins are excised individually from the corresponding gels and destained before performing in-gel digestion with sequencing grade-modified trypsin (Promega) to generate unique PMFs for protein identification. Trypsin cleaves peptide bonds after a lysine or arginine in the C-terminal and generates peptides with a size within 1000 to 3000 Da that are well suited to MS analysis. For MALDI-TOF MS, we use an Autoflex (Bruker Daltonics, Billerica, MA), which is equipped with a nitrogen laser (377 nm) and operated in the reflectron mode over a mass range of 1000 to 3000 Da. The tryptic digested mixture of peptide fragments obtained from each nitrated protein is dotted onto a 384 position, 600-μm AnchorChip target plate (Bruker Daltonics) in the presence of a matrix. The PMF spectrum obtained is calibrated using a mass calibration kit (Bruker Daltonics, Germany) with an external calibration standard. The reported spectrum is usually the summation of 200 laser shots. Identification of the protein is through searching with the measured PMFs through various databases such as the NCBI, SWISS-Prot, and TrEMBL using the Mascot search program (www.matrixscience.com). All searches are performed with a mass tolerance of 50 ppm.

Procedure

1. Excise the immunoreactive spots from the corresponding silver-stained gel and place into Eppendorf tubes.
2. Destain the silver-stained excised gel spot with destaining solution (30 mM potassium ferricyanide and 64 mM sodium thiosulfate) for 15 min, followed by washing with 25 mM ammonium bicarbonate twice.
3. Dehydrate the gel with 100% acetonitrile and dry under a vacuum.
4. Incubate the gel with reducing solution (25 mM ammonium bicarbonate containing 10 mM dithiothreitol) for 45 min at 56 °C and subsequently with alkylating solution (25 mM ammonium bicarbonate and 55 mM iodoacetamide) for 30 min in the dark.
5. Wash the gel with 50% (v/v) acetonitrile containing 50 mM ammonium bicarbonate for 15 min and shrink again with 100% acetonitrile.
6. Rehydrate the gel with 4 µl of 10 µg/ml modified trypsin (Promega) in 25 mM ammonium bicarbonate before incubation at 56 °C for 2 h.
7. Add a small volume (e.g., 5 µl, just enough to cover the gel) of 50% (v/v) acetonitrile and 1% (v/v) trifluoroacetic acid into the Eppendorf vial. Extract the tryptic-digested peptides from the gel plug by allowing diffusion from the gel plug into the acetonitrile/trifluoroacetic acid mixture. Ultrasonication is usually applied to facilitate the diffusion of peptides.
8. Repeat step 7 two more times and dry the combined extracted peptides using a Speed-Vac.
9. Spot 1 µl of matrix solution [α-cyano-4-hydroxycinnamic acid (HCCA), saturated in acetonitrile: 0.1% TFA (1:1)] on the AnchorChip target.
10. Dissolve the digested peptides with acetonitrite:0.1% TFA (1:2), apply the same volume of digested peptide solution onto the dried HCCA matrix, and dry it at room temperature again.
11. Wash the sample spot with 1% TFA solution by allowing the TFA droplet to sit on top of the peptides–matrix spot for approximately 30 s before sucking the droplet back into the pipette tip again.
12. After evaporation of the TFA, recrystallize the peptides by adding 1 µl of a solution of ethanol:acetone:0.1% TFA (6:3:1 ratio) onto the resulting diffused sample spot. In the course of evaporation, hydrophobicity of the AnchorChip will pull the peptides into the center of the spot.

2.6. Identification of nitrotyrosine sites by MALDI-TOF MS and LC-MS/MS

2.6.1. MALDI-TOF MS analysis of nitrated BSA

In addition to protein identification, according to expected peptide mass ions with increased mass, MALDI-TOF MS analysis of nitrated protein can also be used to investigate the presence of nitrated peptides. As an example, this chapter uses the PMF spectrum of BSA and its nitrated form as an illustration (Fig. 2.6). Compared to an unmodified peptide of m/z 1740.8 digested from BSA (shown in Fig. 2.6B), one can assign the characteristic mass shifts to higher masses of nitration, that is, +13 Da for N^+ ion (m/z 1753.9), +29 Da for NO^+ (m/z 1769.8), and +45 Da for NO_2^+ (m/z 1785.8) to the m/z of the unmodified peptide (shown in Fig. 2.6A). This spectral pattern is known to be characteristic of nitrated peptides. It is attributed that a NO_2 group is introduced to the *ortho* carbon of the

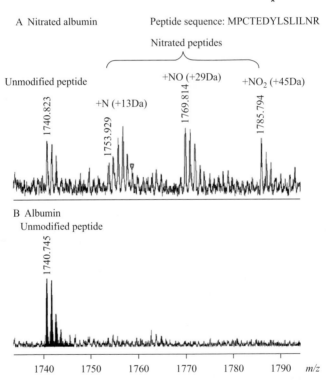

Figure 2.6 MALDI-TOF MS spectrum of trypsin-digested peptide fragments of (A) nitrated BSA and (B) native BSA. Mass ion m/z 1740.8 is the unmodified peptide, whereas mass ions m/z 1785.8, 1769.8, and 1753.9 represent the addition of the NO_2 group to the tyrosine residue and two as well as one oxygen losses from the NO_2 group, respectively. The asterisk on tyrosine (Y) represents the site of nitration.

phenolic ring of tyrosine to generate nitrotyrosine, resulting in a +45-Da mass increase. The two additional modified peaks with mass shifts of +13 and +29 Da result from the loss of one or two oxygen atoms from the nitro group during the ionization process with the laser irradiation of MALDI-TOF (Sarver *et al.*, 2001).

2.6.2. LC-MS/MS analysis of nitrated BSA

Based on the PMF spectrum generated from MALDI-TOF MS analysis, peptides with +13-, +29-, or +45-Da mass shifts, reflecting nitrated peptides, can be determined easily. However, knowing that a peptide is nitrated is one thing while knowing the exact nitration site on the peptide is another. Therefore, the LC MS/MS-based peptide *de novo* sequencing approach has been used for identifying the exact tyrosine nitration site on the nitrated peptides (Greis *et al.*, 1996; MacMillan-Crow *et al.*, 1998; Yi *et al.*, 1997).

For LC MS/MS investigations, electrospray ionization (ESI)/collision-induced dissociation (CID) mass spectra are acquired on an ESI ion trap mass spectrometer (Bruker Daltonic) equipped with a nanoelectrospray ion source. The trypsin-digested peptides are dissolved in distilled water before being loaded onto a 75 μm × 15 cm C18 PepMap 100 column at a rate of 30 μl/min. The peptides are eluted at 300 nl/min with 2 to 45% acetonitrile over 40 min. Peptides eluted from the C18 column are ionized immediately using an electrosprayer. All spectra are obtained in the positive ion mode. An ionization potential of 1200 V is applied in the ion source. The MS/MS analysis is performed with intense ions with an intensity >10,000 units. For CID experiments, the precursor ion is selected before being fragmented in the hexapole collision cell with a collision energy varying from 20 to 40 eV.

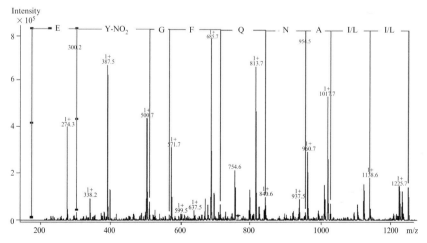

Figure 2.7 ESI/CID spectrum of nitrated BSA peptide of m/z 763.3. The *de novo* sequence is I/L-I/L-A-N-Q-F-G-Y+NO$_2$-E.

Figure 2.7 shows an ESI/CID spectrum of nitrated BSA peptide at m/z 763.3. Based on *de novo* sequencing methodology, the specific tyrosine nitration site Y-NO$_2$ is pinpointed.

Although MS/MS analysis provides the most accurate information on identification of the tyrosine nitration site, this tandem MS-based method is much more complex, is technically challenging, is time-consuming, and has a relatively low throughput. Because the level of protein tyrosine nitration is relatively low *in vivo*, it usually requires greater sample amounts, possibly supplemented with sample enrichment procedures, and a higher sensitivity is needed to identify the tyrosine nitration sites. Other commonly used sample enrichment procedures include immunoprecipitation, solid-phase extraction, liquid chromatographic-based fractionation methods, and affinity chromatographic techniques.

3. Conclusions

Tyrosine nitration on amino acid backbones of proteins is a major indicator of nitrosative stress. It is also an important PTM in the pathogenesis of many human diseases. Proteomic technologies provide a means toward obtaining a comprehensive view of the nitroproteome in different cells and tissues and it promises to broaden our understanding on how NO-mediated modification regulates cellular processes.

ACKNOWLEDGMENTS

The authors thank Dr. Thomas C. Lam for his help with the LC-MS/MS. This work is partly supported by a research grant from the Research Grant Council of Hong Kong (RGC Account No. PolyU 5399/03M; HKPU Account No: B-Q644).

REFERENCES

Beckman, J. S., and Koppenol, W. H. (1996). Nitric oxide, superoxide, and peroxynitrite: The good, the bad, and ugly. *Am. J. Physiol.* **271,** C1424–C1437.

Greis, K. D., Zhu, S., and Matalon, S. (1996). Identification of nitration sites on surfactant protein A by tandem electrospray mass spectrometry. *Arch. Biochem. Biophys.* **335**(2), 396–402.

Hopkins, N., Cadogan, E., Giles, S., Bannigan, J., and McLoughlin, P. (2003). Type 2 nitric oxide synthase and protein nitration in chronic lung infection. *J. Pathol.* **199**(1), 122–129.

Ischiropoulos, H. (1998). Biological tyrosine nitration: A pathophysiological function of nitric oxide and reactive oxygen species. *Arch. Biochem. Biophys.* **356,** 1–11.

Jeng, B. H., Shadrach, K. G., Meisler, D. M., Hollyfield, J. G., Connor, J. T., Koeck, T., Aulak, K. S., and Stuehr, D. J. (2005). Immunohistochemical detection and Western blot analysis of nitrated protein in stored human corneal epithelium. *Exp. Eye Res.* **80**(4), 509–514.

MacMillan-Crow, L. A., Crow, J. P., and Thompson, J. A. (1998). Peroxynitrite-mediated inactivation of manganese superoxide dismutase involves nitration and oxidation of critical tyrosine residues. *Biochemistry* **37**(6), 1613–1622.

Miyagi, M., Sakaguchi, H., Darrow, R. M., Yan, L., West, K. A., Aulak, K. S., Stuehr, D. J., Hollyfield, J. G., Organisciak, D. T., and Crabb, J. W. (2002). Evidence that light modulates protein nitration in rat retina. *Mol. Cell. Proteomics* **1**(4), 293–303.

Peluffo, G., and Radi, R. (2007). Biochemistry of protein tyrosine nitration in cardiovascular pathology. *Cardiovasc. Res.* **75,** 291–302.

Pittman, K. M., MacMillan-Crow, L. A., Peters, B. P., and Allen, J. B. (2002). Nitration of manganese superoxide dismutase during ocular inflammation. *Exp. Eye Res.* **74**(4), 463–471.

Rabilloud, T. (1999). Solubilization of proteins in 2-D electrophoresis: An outline. *Methods Mol. Biol.* **112,** 9–12.

Ramagli, L. S. (1999). Quantifying protein in 2D PAGE solubilization buffers. *Methods Mol. Biol.* **112,** 99–103.

Reynolds, M. R., Berry, R. W., and Binder, L. I. (2007). Nitration in neurodegeneration: Deciphering the "Hows" "nYs". *Biochemistry* **46,** 7325–7336.

Sarver, A., Scheffler, N. K., Shetlar, M. D., and Gibson, B. W. (2001). Analysis of peptides and proteins containing nitrotyrosine by matrix-assisted laser desorption/ionization mass spectrometry. *J. Am. Soc. Mass Spectrom.* **12**(4), 439–448.

Yi, D., Smythe, G. A., Blount, B. C., and Duncan, M. W. (1997). Peroxynitrite-mediated nitration of peptides: Characterization of the products by electrospray and combined gas chromatography-mass spectrometry. *Arch. Biochem. Biophys.* **344**(2), 253–259.

CHAPTER THREE

Using Tandem Mass Spectrometry to Quantify Site-Specific Chlorination and Nitration of Proteins: Model System Studies with High-Density Lipoprotein Oxidized by Myeloperoxidase

Baohai Shao *and* Jay W. Heinecke

Contents

1. Introduction	34
2. High-Density Lipoprotein Biology	36
3. Advantages of HDL as a Model System	36
4. Advantages of LC-ESI-MS/MS When Analyzing Posttranslational Modifications of Proteins	37
5. Isolating HDL, ApoA-I, and MPO	38
6. Oxidative Reactions	38
7. Proteolytic Digestion of Proteins	39
8. Liquid Chromatography–Electrospray Ionization Mass Spectrometry (LC-ESI-MS and MS/MS)	39
9. A Combination of Tryptic and Glu-C Digests Provides Complete Sequence Coverage of ApoA-I	40
10. HOCl or MPO Preferentially Chlorinates Tyrosine 192 in Lipid-Free ApoA-I	40
11. Reagent ONOO$^-$ and MPO Nitrate All the Tyrosine Residues in Lipid-Free ApoA-I, but Tyrosine 192 Is the Main Target	44
12. HOCl Quantitatively Converts All Three Methionine Residues in ApoA-I to Methionine Sulfoxide	48
13. HOCl Generates Hydroxytryptophan and Dihydroxytryptophan Residues in ApoA-I	53
14. Reactive Nitrogen Species Generates Nitrotryptophan Residues in Lipid-Free ApoA-I	57

Department of Medicine, University of Washington, Seattle, Washington

Methods in Enzymology, Volume 440
ISSN 0076-6879, DOI: 10.1016/S0076-6879(07)00803-8

© 2008 Elsevier Inc.
All rights reserved.

15. The YXXK Motif Directs ApoA-I Chlorination   57
16. Quantitative Analysis of Posttranslational Modifications of Proteins   58
17. ApoA-I Oxidation Impairs Cholesterol Transport by the ABCA1 System   58
18. Conclusions   59
Acknowledgments   60
References   60

Abstract

Protein oxidation is implicated in atherogenesis and other inflammatory conditions. Measuring levels of chlorinated and nitrated proteins in biological matrices serves as a quantitative index of oxidative stress *in vivo*. One potential mechanism for oxidative stress involves myeloperoxidase, a heme protein expressed by neutrophils, monocytes, and some populations of macrophages. The enzyme uses hydrogen peroxide to generate an array of cytotoxic oxidants, including hypochlorous acid (HOCl), a potent chlorinating intermediate, and nitrogen dioxide radical, a reactive nitrogen species (RNS). One important target may be high-density lipoprotein (HDL), which is implicated in atherogenesis. This chapter describes liquid chromatography–tandem mass spectrometric methods for quantifying site-specific modifications of proteins that have been oxidized by HOCl or RNS. Our studies center on apolipoprotein A-I, the major HDL protein, which provides an excellent model system for investigating factors that target specific residues for oxidative damage. Our approach is sensitive and rapid, applicable to a wide array of posttranslational modifications, and does not require peptides to be derivatized or labeled with an isotope.

1. Introduction

Many lines of evidence implicate oxidative stress in aging (Finkel and Holbrook, 2000; Stadtman, 1992). Moreover, oxidative modifications of biomolecules have been linked to the pathogenesis of inflammatory diseases, including atherosclerosis, ischemia–reperfusion injury, and cancer (Ames *et al.*, 1993; Berliner and Heinecke, 1996). Sources of oxidants that can damage proteins include macrophages, which secrete the heme enzyme myeloperoxidase (MPO). This enzyme uses hydrogen peroxide (H_2O_2) to execute oxidative reactions in the extracellular milieu (Foote *et al.*, 1983; Harrison and Shultz, 1976). At plasma concentrations of chloride ion (Cl^-), hypochlorous acid (HOCl) is an important end product. HOCl or the MPO-H_2O_2-chloride system converts tyrosine residues of proteins to a stable product, 3-chlorotyrosine (Domigan *et al.*, 1995; Hazen *et al.*, 1996). Because no other mammalian enzyme generates 3-chlorotyrosine at plasma concentrations of halide ions, this abnormal amino acid serves as a molecular fingerprint that implicates the MPO pathway in oxidative tissue damage when it is detected in proteins (Gaut et al., 2002).

Oxidation of low-density lipoprotein (LDL) has been proposed to play a critical role in converting macrophages into cholesteryl ester-laden foam cells (Berliner and Heinecke, 1996; Heinecke, 1998; Witztum and Steinberg, 1991), the cellular hallmark of atherosclerosis. In contrast, high-density lipoprotein (HDL) is thought to protect against foam cell formation, in part by promoting cholesterol efflux (Oram and Heinecke, 2005; Rothblat et al., 1999). Elevated levels of 3-chlorotyrosine have been detected in LDL and HDL isolated from human atherosclerotic lesions, indicating that MPO can damage both lipoproteins *in vivo* (Bergt et al., 2004b; Hazen and Heinecke, 1997; Zheng et al., 2004). Moreover, MPO colocalizes with macrophages in the artery wall (Daugherty et al., 1994).

Myeloperoxidase can also produce nitrogen dioxide radical ($NO_2^•$), a reactive species that converts tyrosine to 3-nitrotyrosine (Klebanoff, 1993; Eiserich et al., 1998; Gaut et al., 2002). Lipoproteins isolated from atherosclerotic lesions are enriched in 3-nitrotyrosine (Leeuwenburgh et al., 1997; Pennathur et al., 2004), indicating that MPO or other pathways that generate reactive nitrogen species (RNS) also operate in the atherosclerotic artery wall. Peroxynitrite ($ONOO^-$) is another nitrating species that might be important in oxidizing lipoproteins during inflammation (Beckman and Koppenol, 1996; Beckman et al., 1990). Like nitrogen dioxide radical, $ONOO^-$ generates 3-nitrotyrosine when it reacts with tyrosine residues.

Staining tissues immunohistochemically with antibodies that recognize protein-bound oxidation products has been widely used to study mechanisms of oxidative damage *in vivo* (Yla-Herttuala et al., 1989). Although this approach has provided important information, its disadvantages include its semiquantitative nature and possible interference from structurally related molecules. Gas chromatography/mass spectrometry (GC/MS) is a more sensitive and specific method for detecting oxidative damage because it can unambiguously quantify trace amounts of analyte in complex biological mixtures (Crowley et al., 1998; Hazen et al., 1997; Heinecke et al., 1999; Leeuwenburgh et al., 1997). However, proteins must first be hydrolyzed to free amino acids, which must then be derivatized, enhancing the complexity of the method. Moreover, neither immunohistochemistry nor GC/MS is able to identify or locate a modified amino acid in a protein.

This chapter describes liquid chromatography (LC)–electrospray ionization (ESI) tandem mass spectrometry (MS)/MS methods for quantifying site-specific modifications in proteins exposed to HOCl or RNS. Using apolipoprotein A-I (apoA-I) and HDL as model systems, this chapter characterizes the modifications of tyrosine, tryptophan, and methionine residues generated in apoA-I by HOCl and RNS.

2. High-Density Lipoprotein Biology

Clinical, genetic, and animal studies have demonstrated a strong inverse relationship between HDL level and risk for coronary artery disease, and various explanations for its protective actions have been offered (Gordon and Rifkind, 1989; Lewis and Rader, 2005). One potential mechanism involves cholesterol removal from the artery wall. More than 30 years ago, Glomset proposed that HDL mediates the transfer of cholesterol from peripheral tissues (such as the artery) to the liver, where the sterol and its oxygenated products are excreted into bile (Glomset, 1968). Subsequent studies demonstrated that HDL accepts cholesterol from macrophage foam cells in this process, which is termed reverse cholesterol transport (Lewis and Rader, 2005).

One important pathway for promoting cholesterol efflux from macrophages in humans and mice involves the interaction of lipid-free or poorly lipidated apoA-I, HDL's major protein with ABCA1, a membrane-associated protein. In this scenario, HDL or apolipoproteins derived from it accept cholesterol from macrophage foam cells in the artery wall and transport it back to the liver for excretion (Glomset, 1968; Oram and Heinecke, 2005; Rothblat *et al.*, 1999). HDL also carries proteins that scavenge lipid peroxides and exert potent anti-inflammatory effects (Navab *et al.*, 2006; Shao *et al.*, 2006b; Vaisar *et al.*, 2007). Many lines of evidence support the proposal that such anti-inflammatory properties contribute to the cardioprotective effects of HDL. There are therefore several reasons why damage to HDL might make it less able to protect the artery wall.

Studies indicate that MPO targets HDL for oxidation in humans with established cardiovascular disease (Bergt *et al.*, 2004b; Pennathur *et al.*, 2004; Zheng *et al.*, 2004), raising the possibility that the enzyme provides a specific mechanism for generating dysfunctional HDL *in vivo*. Moreover, oxidizing apoA-I with the MPO system prevents HDL from removing excess cholesterol from cells by the ABCA1 pathway (Bergt *et al.*, 2004b; Shao *et al.*, 2005a). Analysis of mutant forms of apoA-I and oxidized apoA-I treated with methionine sulfoxide reductase suggest that oxidation of specific tyrosine and methionine residues creates this impairment (Shao *et al.*, 2006a). Biophysical studies and the crystal structure of lipid-free apoA-I suggest that such oxidative damage might prevent apoA-I from assuming conformations that can interact with ABCA1 (Ajees *et al.*, 2006; Shao *et al.*, 2006b).

3. Advantages of HDL as a Model System

High-density lipoprotein is a circulating, noncovalent assembly of protein and lipid that was originally defined by its density on ultracentrifugation (Gotto *et al.*, 1986). ApoA-I, a 28-kDa single polypeptide, accounts for ≈70% of the protein mass of HDL (Davidson and Thompson, 2007). It contains 243 amino acid residues and appears to have eight amphipathic α-helical repeats of 22 amino acids and two amphipathic α-helical repeats of 11 amino acids. In human plasma or serum, HDL exists predominantly as two major density classes: HDL_2 (density 1.063–1.125 g/ml) and HDL_3 (density 1.125–1.21 g/ml).

Most of the apoA-I in plasma is found in HDL_2 and HDL_3, but approximately 5% of the apoA-I in plasma is thought to be lipid free (Davidson and Thompson, 2007). Biophysical studies suggest that lipid-free apoA-I, the ligand for ABCA1, contains at least one bundle of four amphipathic α helices. Analysis of chemically induced intramolecular cross-links suggests that lipid-free apoA-I is a cluster of five α helices and predicts that its N and C termini are close enough to interact. Lipoprotein- or lipid-associated apoA-I exhibits an open elliptical form composed of α-helical helices. In this structure, virtually all of the amino acids of the amphipathic α-helical repeats are exposed to either the aqueous (hydrophilic amino acids) or the lipid (hydrophobic amino acids) phase.

ApoA-I and HDL offer many advantages as model systems for studying the factors that influence the modification of proteins at specific amino acid residues. These include the relatively small size of apoA-I (which facilitates MS/MS sequencing), structures of the lipid-free and lipid-associated forms in crystals and solution, and the ready accessibility of most of the protein's amino acids to reactive intermediates. Moreover, recombinant apoA-I and mutant forms of apoA-I are readily isolated in large quantities (Ryan et al., 2003), greatly facilitating studies of proposed reaction mechanisms (Shao et al., 2006a). Importantly, apoA-I is strongly linked to the cardioprotective effects of HDL, making it possible to associate specific sites of damage with changes in the biological properties of the protein (Davidson and Thompson, 2007; Lewis and Rader, 2005; Shao et al., 2006a,b).

4. Advantages of LC-ESI-MS/MS When Analyzing Posttranslational Modifications of Proteins

Liquid chromatography–mass spectrometry is used increasingly in biological laboratories. ESI interfaced with liquid chromatographs and mass spectrometers is a sensitive, rapid, and accurate method for determining the molecular weights of proteins and peptides (Fenn et al., 1989; Whitehouse et al., 1985). Although direct analysis of intact protein molecules is possible (Kelleher, 2004), MS analysis is typically performed on peptides, which are commonly digested from proteins with an enzyme such as trypsin (Aebersold and Mann, 2003). To unambiguously identify the sequence of a peptide, tandem mass spectrometry or MS/MS is performed. In the first MS step, the mass of the intact precursor peptide is determined.

In the second, the precursor ions are isolated from all other peptide ions and fragmented by collision-induced dissociation, which generates a set of C-terminal and N-terminal peptide ions (termed y and b ions, respectively). The masses of these ions are then determined by MS, yielding MS/MS spectra. These "sequence ladders" make it possible to deduce the complete or partial primary sequence of a peptide. Significantly, this method allows the nature and locations of posttranslational modifications of a protein to be determined unequivocally. It is also simple and direct because it does not require proteins or peptides to be derivatized or labeled with an isotope.

5. Isolating HDL, ApoA-I, and MPO

Blood collected from healthy adults who have fasted overnight is anticoagulated with EDTA and then centrifuged to produce plasma. HDL (density 1.125–1.210 g/ml) is prepared from plasma by sequential ultracentrifugation and is depleted of contaminating lipoproteins (apolipoprotein E and apolipoprotein B100 of LDL) by heparin–agarose chromatography (Mendez et al., 1991). ApoA-I is purified to apparent homogeneity from HDL by ion-exchange chromatography (Mendez et al., 1991). Protein concentrations are determined by the Lowry assay (Bio-Rad, Hercules, CA), with albumin as the standard. MPO (donor:hydrogen peroxide, oxidoreductase, EC 1.11.1.7) is isolated from human neutrophils by lectin affinity chromatography and size-exclusion chromatography (Heinecke et al., 1993; Hope et al., 2000) and is stored at $-20\,^\circ$C. Purified enzyme typically exhibits an A_{430}/A_{280} ratio of 0.8 and is apparently homogeneous on SDS-PAGE analysis. Its concentration is determined spectrophotometrically ($\epsilon_{430} = 0.17\ M^{-1}\ cm^{-1}$) (Morita et al., 1986).

6. Oxidative Reactions

Reactions are carried out at 37 $^\circ$C in phosphate buffer [20 mM sodium phosphate, 100 μM diethylenetriaminepentaacetic acid (DTPA), pH 7.4] containing 5 μM apoA-I (Shao et al., 2005a). To create the MPO-H_2O_2-chloride system, the reaction mixture is supplemented with 50 nM MPO, 0.1 M NaCl, and the indicated concentrations of H_2O_2. For the MPO-H_2O_2-nitrite system, the mixture is supplemented with 50 nM MPO, 100 μM nitrite, and the indicated concentrations of H_2O_2. Reactions are initiated by adding oxidant and are terminated by adding 5 mM methionine (a scavenger of H_2O_2 and HOCl). Peroxynitrite is synthesized from nitrite and H_2O_2 under acidic conditions, and peroxynitrous acid is stabilized by rapidly quenching the reaction with excess sodium hydroxide (Beckman et al., 1994). Concentrations of $ONOO^-$, HOCl, and H_2O_2 are determined spectrophotometrically ($\epsilon_{302} = 1670\ M^{-1}\ cm^{-1}$, $\epsilon_{292} = 350\ M^{-1}\ cm^{-1}$, and $\epsilon_{240} = 39.4\ M^{-1} cm^{-1}$, respectively) (Beckman et al., 1994; Morris, 1966; Nelson and Kiesow, 1972).

7. Proteolytic Digestion of Proteins

Native or oxidized apoA-I is incubated overnight at 37 °C with sequencing grade modified trypsin (Promega, Madison, WI) at a ratio of 25:1 (w/w, protein/enzyme) in 50 mM NH$_4$HCO$_3$, pH 7.8 (Shao et al., 2005a). For digestion with Glu-C, native or oxidized apoA-I is incubated overnight at room temperature with sequencing grade endoproteinase Glu-C (from staphylococcal serine protease Protease V8) (Roche Applied Science, Indianapolis, IN) at a ratio of 10:1 (w/w, protein/enzyme) in 50 mM NH$_4$HCO$_3$, pH 7.8 (Shao et al., 2005b). Digestion is halted by acidifying the reaction mixture (pH 2–3) with trifluoroacetic acid. Digested peptides are transferred to HPLC vials (MicroSolv Technology Co., Eatontown, NJ) and injected into the LC-ESI-MS/MS system.

8. Liquid Chromatography–Electrospray Ionization Mass Spectrometry (LC-ESI-MS and MS/MS)

LC-ESI-MS and MS/MS analyses are performed in the positive ion mode with a Thermo-Finnigan LCQ Deca XP Plus instrument (San Jose, CA) interfaced with an Agilent 1100 Series HPLC system (Santa Clara, CA). The HPLC is equipped with a reverse-phase Vydac C18 MS column (2.1 × 250 mm). The eluting solvents are 0.1% HCOOH in H_2O (solvent A) and 0.1% HCOOH in acetonitrile (solvent B). Trypsin or Glu-C digested peptides are separated at a flow rate of 0.2 ml/min, using a gradient of solvent A and solvent B. Solvent B is increased from 2 to 40% over 60 min. After 5 min at 80% solvent B, the column is finally reequilibrated for 25 min, using 2% solvent B. Nitrogen, the sheath gas, is set at 70 units, and a setting of 10 units of auxiliary gas (also nitrogen) is used. A spray voltage of 5 kV is applied. The heated metal capillary is maintained at 220 °C. The collision gas is helium. Analyses are performed in the positive ion mode with a mass range of 200 to 2000 Da. Most tandem MS spectra are obtained using data-dependent analysis with the following parameters: isolation width, 3 Da; normalized collision energy, 35%; activation time, 30 ms; activation Q, 0.25; default charge state, 3. The instrument defines the mass range in the MS/MS mode.

9. A Combination of Tryptic and Glu-C Digests Provides Complete Sequence Coverage of ApoA-I

Trypsin, a serine endopeptidase that cleaves proteins C-terminally of Lys and Arg residues, is commonly used to convert proteins into peptides. ApoA-I is readily digested with trypsin, and the resulting peptides are analyzed by LC-ESI-MS and MS/MS. ApoA-I contains 243 amino acids without cysteine (Fig. 3.1), including 21 lysine residues and 16 arginine residues. LC-ESI-MS/MS analysis of a tryptic digest detected 25 peptides that covered 216 amino acid residues (≈90%) of apoA-I (Table 3.1). All peptide sequences are confirmed with MS/MS analysis. We are unable to detect single amino acids or very small peptides because they are poorly retained on HPLC columns.

To ensure complete coverage, we use another endoproteinase, Glu-C, to digest apoA-I. Glu-C is a serine endopeptidase, cleaving peptide bonds C-terminally at glutamic acid and, at a 100- to 300-fold lower rate, at aspartic acid (Drapeau, 1977). ApoA-I contains 30 Glu residues. LC-ESI-MS/MS analysis of a Glu-C digest of apoA-I protein detects 20 peptides representing 233 amino acid residues (>95% of apoA-I sequence) (Table 3.2).

Used in concert, tryptic and Glu-C digests of apoA-I cover the entire sequence of the protein.

10. HOCl or MPO Preferentially Chlorinates Tyrosine 192 in Lipid-Free ApoA-I

To determine whether HOCl or the MPO-H_2O_2-chloride system chlorinates tyrosine residues in apoA-I, we expose lipid-free apolipoprotein to HOCl or the MPO-chloride system. Oxidation reactions are performed in the absence of nitrite ion (to prevent nitration) and at neutral pH in

	1	DEPPQSP**W**⁸DRVKDLATV**Y**¹⁸VDVLK
	24	DSGRD**Y**²⁹VSQFEGSALGKQLN
Helix 1	44	LKLLDN**W**⁵⁰DSVTSTFSKLREQLG
Helix 2	66	PVTQEF**W**⁷²DNLEKETEGLRQE**M**⁸⁶S
Helix 3	88	KDLEEVKAKVQ
Helix 4	99	P**Y**¹⁰⁰LDDFQKK**W**¹⁰⁸QEE**M**¹¹²EL**Y**¹¹⁵RQKVE
Helix 5	121	PLRAELQEGARQKLHELQEKLS
Helix 6	143	PLGEE**M**¹⁴⁸RDRARAHVDALRTHLA
Helix 7	165	P**Y**¹⁶⁶SDELRQRLAARLEALKENGG
Helix 8	187	ARLAE**Y**¹⁹²HAKATEHLSTLSEKAK
Helix 9	209	PALEDLRQGLL
Helix 10	220	PVLESFKVSFLSALEE**Y**²³⁶TKKLN
	242	TQ

Figure 3.1 Apolipoprotein A-I. Primary sequence of apolipoprotein A-I and its 10 proposed α-helical amphipathic helices (Brouillette and Anantharamaiah, 1995; Segrest et al., 1994). Amino acid residues shown in bold and underlined are those that can be modified by HOCl or RNS.

Table 3.1 Peptides detected in a tryptic digest of ApoA-I[a]

Position	Sequence	m/z	Detected m/z (charge state)
1–10	DEPPQSPWDR	1226.54	1226.6 (+1), 614.0 (+2)
13–23	DLATVYVDVLK	1235.69	1235.6 (+1), 618.7 (+2)
28–40	DYVSQFEGSALGK	1400.67	1400.6 (+1), 701.3 (+2)
41–45	QLNLK	615.38	615.5 (+1), 308.4 (+2)
46–59	LLDNWDSVTSTFSK	1612.79	1612.7 (+1), 807.4 (+2)
62–77	EQLGPVTQEFWDNLEK	1932.93	1932.8 (+1), 967.8 (+2)
78–83	ETEGLR	704.36	704.4 (+1), 352.9 (+2)
89–94	DLEEVK	732.38	732.3 (+1), 367.0 (+2)
97–106	VQPYLDDFQK	1252.62	1252.6 (+1), 627.1 (+2)
108–116	WQEEMELYR	1283.57	1283.6 (+1), 642.7 (+2)
119–123	VEPLR	613.37	613.5 (+1), 307.7 (+2)
124–131	AELQEGAR	873.44	873.5 (+1), 437.4 (+2)
134–140	LHELQEK	896.48	896.4 (+1), 449.0 (+2), 299.8 (+3)
141–149	LSPLGEEMR	1031.52	1031.5 (+1), 516.5 (+2)
154–160	AHVDALR	781.43	781.5 (+1), 391.5 (+2), 261.6 (+3)
161–171	THLAPYSDELR	1301.65	1301.7 (+1), 651.6 (+2), 434.9 (+3)
174–177	LAAR	430.28	430.4 (+1), 215.8 (+2)
178–182	LEALK	573.36	573.4 (+1), 287.4 (+2)
183–188	ENGGAR	603.28	603.4 (+1), 302.3 (+2)

(continued)

Table 3.1 (continued)

Position	Sequence	m/z	Detected m/z (charge state)
189–195	LAEYHAK	831.44	831.4 (+1), 416.3 (+2), 278.1 (+3)
196–206	ATEHLSTLSEK	1215.62	1215.5 (+1), 608.6 (+2), 406.2 (+3)
207–215	AKPALEDLR	1012.58	1012.6 (+1), 506.9 (+2), 338.5 (+3)
216–226	QGLLPVLESFK	1230.71	1230.6 (+1), 616.3 (+2)
227–238	VSFLSALEEYTK	1386.72	1386.7 (+1), 694.2 (+2)
240–243	LNTQ	475.25	475.2 (+1)

[a] The apoA-I protein was incubated overnight at 37 °C with sequencing grade modified trypsin in 50 mM NH$_4$HCO$_3$ (pH 7.8). Digestion was halted by acidification (pH 2–3) with trifluoroacetic acid. The digested peptides were analyzed by LC-ESI-MS and MS/MS. Peptide sequences were confirmed using MS/MS.

phosphate buffer supplemented with DTPA (to chelate redox-active metal ions) (Heinecke et al., 1986). We use a 10:1 mol/mol ratio of oxidant (HOCl or H$_2$O$_2$) to apoA-I, incubating the mixture for 60 min at 37 °C and terminating the reaction with 5 mM of methionine. There are seven tyrosine residues in apoA-I (Fig. 3.1), and LC-ESI-MS analysis of the tryptic digest of native apoA-I detects all seven peptides predicted to contain tyrosine (Table 3.1). To determine which tyrosine residues have been chlorinated, we use reconstructed ion chromatograms to detect (i) each of the peptides that contain tyrosine and (ii) any tyrosine-containing peptides that have gained 34 amu (addition of 1 chlorine and loss of 1 hydrogen).

LC-ESI-MS and MS/MS analysis of the tryptic digest of oxidized apoA-I detected five peptides whose masses corresponded to the masses of the precursor peptides plus 34 amu, suggesting the formation of chlorinated amino acids (+Cl-H) (Table 3.3). We detected one peptide that gained 50 amu, suggesting the addition of both a chlorine atom and an oxygen atom (34 amu + 16 amu) (Table 3.3). Using LC-ESI-MS/MS analysis, we confirmed each peptide's sequence and showed that its tyrosine had been targeted for chlorination. Figure 3.2B shows the MS/MS analysis of peptide LAEYHAK (which contains Tyr192) in apoA-I that had been exposed to HOCl. Compared with the same peptide in control apoA-I (Fig. 3.2A), every y ion from y_4 to y_6 gained 34 amu, as

Table 3.2 Peptides detected in a Glu-C digest of apoA-I[a]

Position	Sequence	m/z	Detected m/z (charge state)
1–13	DEPPQSPWDRVKD	1568.73	1569.6 (+1), 785.2 (+2), 523.9 (+3)
14–34	LATVYVDVLKDSGRDYVSQFE	2404.20	1203.0 (+2), 802.6 (+3)
35–62	GSALGKQLNLKLLDNWDSVTSTFSKLRE	3120.67	1561.6 (+2), 1041.4 (+3), 781.5 (+4), 625.5 (+5)
63–70	QLGPVTQE	871.45	871.5 (+1)
71–78	FWDNLEKE	1080.50	1080.4 (+1), 541.0 (+2)
81–85	GLRQE	602.33	602.5 (+1), 302.1 (+2)
86–92	MSKDLEE	851.38	851.4 (+1), 426.4 (+2)
93–111	VKAKVQPYLDDFQKKWQEE	2379.23	1190.5 (+2), 794.2 (+3), 596.1 (+4), 477.0 (+5)
114–125	LYRQKVEPLRAE	1501.85	1501.8 (+1), 751.7 (+2), 501.7 (+3), 376.6 (+4)
129–136	GARQKLHE	938.52	938.5 (+1), 470.2 (+2), 313.7 (+3)
140–147	KLSPLGEE	872.47	872.5 (+1), 437.0 (+2)
148–169	MRDRARAHVDALRTHLAPYSDE	2580.28	1290.9 (+2), 861.1 (+3), 646.3 (+4), 517.3 (+5)
170–179	LRQRLAARLE	1225.75	1225.7 (+1), 613.6 (+2), 409.6 (+3), 307.5 (+4)
180–191	ALKENGGARLAE	1228.66	1229.7 (+1), 615.6 (+2), 411.0 (+3)
192–198	YHAKATE	819.40	819.4 (+1), 410.5 (+2)

(continued)

Table 3.2 *(continued)*

Position	Sequence	m/z	Detected m/z (charge state)
199–205	HLSTLSE	786.40	786.4 (+1), 394.0 (+2)
206–212	KAKPALE	756.46	756.5 (+1), 378.9 (+2)
213–223	DLRQGLLPVLE	1252.73	1252.7 (+1), 627.2 (+2)
224–235	SFKVSFLSALEE	1356.70	1356.6 (+1), 679.2 (+2)
236–243	YTKKLNTQ	995.55	995.4 (+1), 498.6 (+2), 332.9 (+3)

[a] The apoA-I protein was incubated overnight at room temperature with sequencing grade endoproteinase Glu-C in 50 mM NH$_4$HCO$_3$ (pH 7.8). Digestion was halted by acidification (pH 2–3) with trifluoroacetic acid. The digested peptides were analyzed by LC-ESI-MS and MS/MS. Peptide sequences were confirmed using MS/MS.

had every b ion from b_4 to b_6, indicating that Tyr192 in the peptide had been converted to 3-chlorotyrosine (Fig. 3.2B).

To determine which tyrosine is most likely to be chlorinated when HOCl or the MPO-H$_2$O$_2$-chloride system oxidizes apoA-I, we incubate various ratios (2:1 to 25:1, mol/mol) of oxidant (HOCl or H$_2$O$_2$) to apoA-I for 60 min at 37 °C and terminate the reaction with 5 mM of methionine. We quantify product yields using the ion current of each precursor and product peptide (within the same sample) and reconstruct ion chromatograms. When apoA-I is exposed to either reagent HOCl or the MPO-H$_2$O$_2$-chloride system, similar patterns of tyrosine chlorination are seen (compare Figs. 3.3A and 3.3B), with Tyr192 being the major site of chlorination (Fig. 3.3). To a lesser extent, Tyr115 and Tyr236 are the other two predominant chlorination sites. We observed a much lower level of chlorination at Tyr18 and Tyr29 and none at Tyr100 and Tyr166 (Fig. 3.3). At a molar ratio of 25:1 of oxidant to apoA-I (125 μM oxidant), ≈30% of the Tyr192 residues are chlorinated by HOCl and ≈25% by the MPO system (Fig. 3.3).

11. Reagent ONOO⁻ and MPO Nitrate All the Tyrosine Residues in Lipid-Free ApoA-I, but Tyrosine 192 Is the Main Target

To determine whether ONOO⁻ or the MPO-H$_2$O$_2$-nitrite system nitrates tyrosine residues in apoA-I, we exposed the lipid-free apolipoprotein to ONOO⁻ or the MPO-nitrite system. Oxidative reactions are carried out in the absence of chloride ion (to prevent chlorination) and at

Table 3.3 Detection by LC-ESI-MS of peptides containing 3-chlorotyrosine in a tryptic digest of lipid-free apoA-I exposed to HOCl or the MPO-H_2O_2-NaCl system[a]

Position	Sequence	Precursor m/z (charge state)	Product m/z (charge state)	Modification
13–23	DLATVYVDVLK	1235.6 (+1), 618.7 (+2)	1269.6 (+1), 635.6 (+2)	Y18+34
28–40	DYVSQFEGSALGK	1400.6 (+1), 701.3 (+2)	1434.6 (+1), 718.2 (+2)	Y29+34
97–106	VQPYLDDFQK	1252.6 (+1), 627.1 (+2)	ud[b]	
108–116	WQEEMELYR[c]	1283.6 (+1), 642.7 (+2)	ud	
108–116	WQEE(M+16)ELYR[d]	1299.5 (+1), 650.4 (+2)	1333.5 (+1), 667.4 (+2)	Y115+34
161–171	THLAPYSDELR	1301.7 (+1), 651.6 (+2), 434.9 (+3)	ud	
189–195	LAEYHAK	831.4 (+1), 416.3 (+2), 278.1 (+3)	865.4 (+1), 433.2 (+2), 289.4 (+3)	Y192+34
227–238	VSFLSALEEYTK	1386.7 (+1), 694.2 (+2)	1420.7 (+1), 711.1 (+2)	Y236+34

[a] ApoA-I (5 μM) was exposed to HOCl or the MPO-H_2O_2-NaCl system (10:1, mol/mol, oxidant:apoA-I) for 60 min at 37 °C in phosphate buffer (20 mM sodium phosphate, 100 μM DTPA, pH 7.4). After the reaction was terminated with L-methionine, a tryptic digest of apoA-I was analyzed by LC-ESI-MS and MS/MS. Peptide sequences and modifications were confirmed using MS/MS.
[b] Undetectable.
[c] No 3-chlorotyrosine is detected in this peptide when the methionine residue is not oxidized.
[d] Methionine (M) is oxidized to methionine sulfoxide.

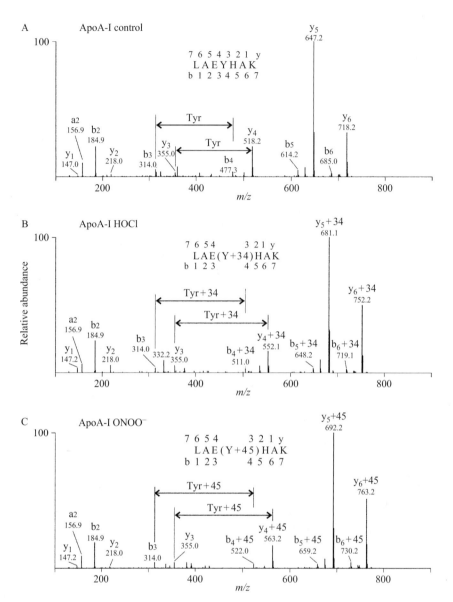

Figure 3.2 MS/MS identification of the major site of tyrosine chlorination and nitration in lipid-free apoA-I exposed previously to HOCl and ONOO$^-$. MS/MS analysis of (A) [LAEYHAK + 2H]$^{+2}$ (m/z 416.3), (B) [LAE(Y-Cl)HAK + 2H]$^{+2}$ (m/z 433.2), and (C) [LAE(Y-NO$_2$)HAK + 2H]$^{+2}$ (m/z 438.9) in apoA-I that had been oxidized with HOCl or ONOO$^-$. Lipid-free apoA-I (5 μM) was exposed to HOCl or ONOO$^-$ (10:1, mol/mol, oxidant:protein) for 60 min at 37 °C in phosphate buffer (20 mM sodium phosphate, 100 μM DTPA, pH 7.4). After the reaction was terminated with L-methionine, apoA-I was digested with trypsin, and the peptides were analyzed by LC-ESI-MS/MS.

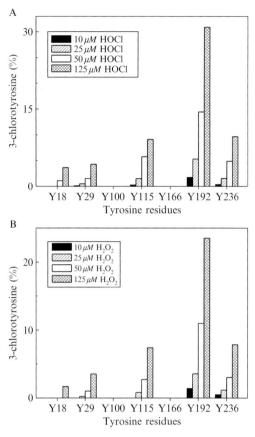

Figure 3.3 Site-specific chlorination of tyrosine residues in lipid-free apoA-I exposed to HOCl or the myeloperoxidase-H_2O_2-chloride system. ApoA-I (5 μM) was exposed to (A) HOCl or (B) the myeloperoxidase-H_2O_2-chloride system at the indicated concentrations of oxidant for 60 min at 37 °C in phosphate buffer (pH 7.4). The myeloperoxidase system was supplemented with 0.1 M NaCl. A tryptic digest of apoA-I was analyzed by LC-ESI-MS and MS/MS. Chlorinated peptides were detected and quantified using reconstructed ion chromatograms of precursor and product peptides. Product yield (%) = peak area of product ion/sum (peak area of precursor ion + peak areas of product ions) × 100. Peptide sequences were confirmed using MS/MS. Results are representative of three independent experiments.

neutral pH in phosphate buffer supplemented with DTPA. To determine which tyrosine residues have been nitrated, we use reconstructed ion chromatograms to detect (i) each of the peptides that contain tyrosine and (ii) any tyrosine-containing peptides that gained 45 amu (addition of 1 NO_2 group and loss of 1 hydrogen). We quantify product yields using the ion current of each precursor and product peptide and reconstruct ion chromatograms.

LC-ESI-MS and MS/MS analysis of the tryptic digest of oxidized apoA-I detected seven peptides whose masses corresponded to those of their precursor peptides plus 45 amu (Table 3.4), suggesting nitration of amino acids ($+NO_2-H$). We also detected one peptide that had gained 61 amu (Table 3.4), suggesting the addition of both a nitro group and an oxygen atom (45 amu + 16 amu). Using LC-ESI-MS/MS analysis, we confirmed each peptide's sequence and showed that its tyrosine had been targeted for nitration. Figure 3.2C shows the MS/MS analysis of peptide LAEYHAK (containing Tyr192) from apoA-I exposed to $ONOO^-$. Compared with the same peptide in control apoA-I (Fig. 3.2A), every y ion from y_4 to y_6 of the exposed peptide had gained 45 amu, as had every b ion from b_4 to b_6, indicating that the peptide's Tyr192 had been converted to 3-nitrotyrosine (Fig. 3.2C).

To determine whether reactive nitrogen species selectively target Tyr192 (as opposed to other tyrosine residues) in apoA-I, we use various ratios of oxidant ($ONOO^-$ or H_2O_2, 2:1 to 10:1, mol/mol, oxidant:apoA-I) for 60 min at 37 °C. When apoA-I is exposed to reagent $ONOO^-$, Tyr192 and, to a lesser extent, Tyr236 are the predominant nitration sites. A much lower level of nitration is observed at the other tyrosine residues (Fig. 3.4A). In contrast, when apoA-I is exposed to the $MPO-H_2O_2$-nitrite system, Tyr192 is still the predominant site of nitration, but all of the other tyrosine residues of the protein are also significantly nitrated (Fig. 3.4B). At a molar ratio of 10:1 of oxidant to apoA-I (50 μM oxidant), ≈40% of Tyr192 residues are nitrated by $ONOO^-$ and ≈50% by the MPO system (Fig. 3.4). At this molar ratio of oxidant (10:1), $ONOO^-$ converts ≈15% of Tyr236 and <6.5% of other tyrosine residues to 3-nitrotyrosine, whereas the $MPO-H_2O_2$-nitrite system converts ≈27% of Tyr236 and >16% of other tyrosine residues to 3-nitrotyrosine (compare Figs. 3.4A and 3.4B). These findings indicate that reagent $ONOO^-$ and the $MPO-H_2O_2$-nitrite system nitrate all seven tyrosine residues in apo A-I, that the major site nitrated by both sources of reactive nitrogen species is Tyr192, and that reagent $ONOO^-$ is a more selective nitrating agent than the $MPO-H_2O_2$-nitrite system. Moreover, chlorination of apoA-I's tyrosine residues by HOCl or the $MPO-H_2O_2$-chloride system is much more selective than nitration.

12. HOCl Quantitatively Converts All Three Methionine Residues in ApoA-I to Methionine Sulfoxide

ApoA-I contains three methionine residues (Met86, Met112, and Met148). Under certain oxidation conditions, two of the three are selectively oxidized. Peroxyl radicals generated by 2,2′-azo-bis(2-amidinopropane)

Table 3.4 Detection by LC-ESI-MS of peptides containing 3-nitrotyrosine in a tryptic digest of lipid-free apoA-I exposed to ONOO⁻ or the MPO-H$_2$O$_2$-NaNO$_2$ system[a]

Position	Sequence	Precursor m/z (charge state)	Product m/z (charge state)	Modification
13–23	DLATVYVDVLK	1235.6 (+1), 618.7 (+2)	1280.6 (+1), 640.9 (+2)	Y18 +45
28–40	DYVSQFEGSALGK	1400.6 (+1), 701.3 (+2)	1445.7 (+1), 723.6 (+2)	Y29 +45
97–106	VQPYLDDFQK	1252.6 (+1), 627.1 (+2)	1297.5 (+1), 649.5 (+2)	Y100+45
108–116	WQEEMELYR	1283.6 (+1), 642.7 (+2)	1328.5 (+1), 665.0 (+2)	Y115+45
108–116	WQEE(M+16)ELYR[b]	1299.5 (+1), 650.4 (+2)	1344.6 (+1), 673.0 (+2)	Y115+45
161–171	THLAPYSDELR	1301.7 (+1), 651.6 (+2), 434.9 (+3)	1346.6 (+1), 674.1 (+2), 449.9 (+3)	Y166+45
189–195	LAEYHAK	831.4 (+1), 416.3 (+2), 278.1 (+3)	876.4 (+1), 438.9 (+2), 293.2 (+3)	Y192+45
227–238	VSFLSALEEYTK	1386.7 (+1), 694.2 (+2)	1431.6 (+1), 716.7 (+2)	Y236+45

[a] ApoA-I (5 μM) was exposed to ONOO⁻ or the MPO-H$_2$O$_2$-NaNO$_2$ system (10:1, mol/mol, oxidant:apoA-I) for 60 min at 37 °C in phosphate buffer (20 mM sodium phosphate, 100 μM DTPA, pH 7.4). After the reaction was terminated with L-methionine, a tryptic digest of apoA-I was analyzed by LC-ESI-MS and MS/MS. Peptide sequences and modifications were confirmed using MS/MS.
[b] Methionine (M112) is oxidized to methionine sulfoxide.

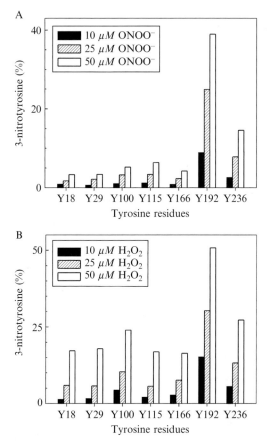

Figure 3.4 Site-specific nitration of tyrosine residues in lipid-free apoA-I exposed to ONOO⁻ or the myeloperoxidase-H$_2$O$_2$-nitrite system. ApoA-I (5 μM) was exposed to (A) ONOO⁻ or (B) the myeloperoxidase-H$_2$O$_2$-nitrite system at the indicated concentrations of oxidant for 60 min at 37 °C in phosphate buffer (pH 7.4). The myeloperoxidase system was supplemented with 100 μM nitrite. A tryptic digest of apoA-I was analyzed by LC-ESI-MS and MS/MS. Nitrated peptides were detected and quantified as described in the legend to Fig. 3.3. Peptide sequences were confirmed using MS/MS. Results are representative of three independent experiments.

hydrochloride (AAPH) selectively oxidize Met 86 and Met 112 (Panzenbock *et al.*, 2000), whereas hydrogen peroxide or chloramine T selectively oxidizes Met 112 and Met 148 (von Eckardstein *et al.*, 1991). To determine which of these methionine residues can be modified by HOCl or RNS, we exposed apoA-I to HOCl, ONOO⁻, or the MPO-system at neutral pH in phosphate buffer containing DTPA for 1 h at 37 °C. After terminating the reaction with methionine, we analyzed proteolytic digests of the protein by LC-ESI-MS/MS. Digesting native apoA-I with trypsin generated two peptides containing two of the protein's three methionine residues (Met112 and Met148) (Table 3.1). The peptide containing Met 86 (QE*M*SK) was not detectable, probably because it is too short to be retained on HPLC columns. However, digestion with Glu-C yielded a readily detectable peptide containing Met 86 (*M*SKDLEE) and also a peptide containing Met148 (Table 3.2). We were unable to detect the predicted peptide containing Met112 (*M*E) in the Glu-C digest due to poor retention on the HPLC column.

After exposing apoA-I to HOCl or the MPO-chloride system and digesting the oxidized protein with trypsin or Glu-C, we observed that all three peptides that contained Met had gained 16 amu (Table 3.5). We quantified product yields using the ion current of each precursor and product peptide and reconstructed ion chromatograms. In the native apoA-I sample, ≈20% of Met112 and ≈5% of Met86 and Met148 had already been oxidized (Fig. 3.5). As the concentration of HOCl or H_2O_2 in the MPO-chloride system increases, all three methionine residues gradually and quantitatively convert to methionine sulfoxide (Figs. 3.5A and 3.5B). At a molar ratio of 5:1 of oxidant to apoA-I (25 μM oxidant), ≈55% of Met86, >90% of Met112, and ≈75% of Met148 are oxidized by HOCl. The corresponding percentages with the MPO-chloride system are ≈90, 100, and ≈95% (Figs. 3.5A and 3.5B). In summary, all three methionine

Table 3.5 Detection by LC-ESI-MS of peptides containing methionine sulfoxide in a tryptic or Glu-C digest of lipid-free apoA-I exposed to HOCl or RNS[a]

Position	Sequence	Precursor m/z (charge state)	Product m/z (charge state)	Modifications
86–92	MSKDLEE[b]	851.4 (+1), 426.4 (+2)	867.4 (+1), 434.3 (+2)	M86+16
108–116	WQEEMELYR[c]	1283.6 (+1), 642.7 (+2)	1299.5 (+1), 650.4 (+2)	M112+16
108–116	WQEEMELYR[c]	1283.6 (+1), 642.7 (+2)	1333.5 (+1), 667.4 (+2)	M112+16, Y115+34[d]
108–116	WQEEMELYR[c]	1283.6 (+1), 642.7 (+2)	1344.6 (+1), 673.0 (+2)	M112+16, Y115+45[e]
141–149	LSPLGEEMR[c]	1031.5 (+1), 516.5 (+2)	1047.5 (+1), 524.4 (+2)	M148+16

[a] ApoA-I (5 μM) was exposed to HOCl, ONOO$^-$, or the MPO system (5:1, mol/mol, oxidant:apoA-I) for 60 min at 37 °C in phosphate buffer (20 mM sodium phosphate, 100 μM DTPA, pH 7.4). After the reaction was terminated with L-methionine, a tryptic digest (for Met112 and Met148) or Glu-C digest (for Met86) of apoA-I was analyzed by LC-ESI-MS and MS/MS. Peptide sequences and modifications were confirmed using MS/MS.
[b] Glu-C digest.
[c] Tryptic digest.
[d] Detected when apoA-I had been exposed to HOCl or the MPO-NaCl system.
[e] Detected when apoA-I had been exposed to ONOO$^-$ or the MPO-NaNO$_2$ system.

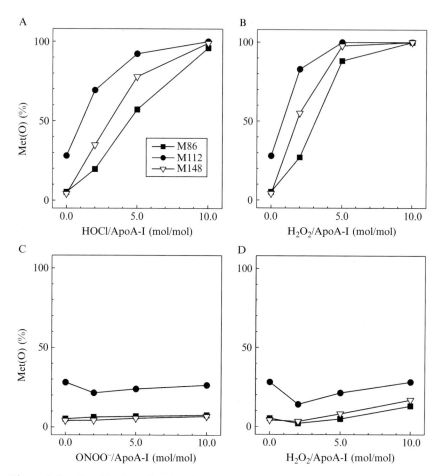

Figure 3.5 Identification of oxidized methionine residues in lipid-free apoA-I exposed to HOCl, ONOO⁻, or the myeloperoxidase-H_2O_2 system. ApoA-I (5 μM) was exposed to (A) HOCl, (B) the myeloperoxidase-H_2O_2-chloride system, (C) ONOO⁻, or (D) the myeloperoxidase-H_2O_2-nitrite system at the indicated molar ratio of oxidant:protein for 60 min at 37 °C in phosphate buffer (pH 7.4). The myeloperoxidase system was supplemented with 0.1 M sodium chloride for chlorination and 100 μM sodium nitrite for nitration. A tryptic digest of apoA-I was analyzed by LC-ESI-MS and MS/MS. Oxidized methionine [methionine sulfoxide, Met(O)] peptides were detected and quantified as described in the legend to Fig. 3.3. Peptide sequences were confirmed using MS/MS. Results are representative of three independent experiments.

residues in apoA-I are oxidized to a high yield of methionine sulfoxide when apoA-I is exposed to HOCl or the MPO-H_2O_2-chloride system, although the latter is more effective. In striking contrast, we detected only a very low level of methionine sulfoxide when we exposed apoA-I to ONOO⁻ or the MPO-H_2O_2-nitrite system (Figs. 3.5C and 3.5D), suggesting that RNS oxidize methionine residues poorly in proteins, although they nitrate tyrosine residues efficiently.

Methionine oxidation may be biologically significant because this amino acid is the most readily oxidized in proteins and because methionine oxidation *in vitro* affects the activities of many proteins (Levine *et al.*, 2000; Vogt, 1995). Indeed, it has been shown that a combination of Tyr192 chlorination and Met oxidation is both necessary and sufficient for depriving apoA-I of its ABCA1-dependent cholesterol transport activity (Shao *et al.*, 2006a).

13. HOCl Generates Hydroxytryptophan and Dihydroxytryptophan Residues in ApoA-I

ApoA-I contains four tryptophan residues (Trp8, Trp50, Trp72, and Trp108), and tryptophan residues are easily modified by HOCl (Hawkins *et al.*, 2003; Jerlich *et al.*, 2000). To determine which tryptophan residues of apoA-I can be modified, we exposed the protein to HOCl or the MPO system at neutral pH in phosphate buffer containing DTPA for 1 h at 37 °C. After terminating the reaction with methionine, we analyzed a proteolytic digest by LC-ESI-MS/MS.

At a molar ratio of 10:1 of oxidant to apoA-I (50 μM oxidant), LC-ESI-MS/MS analysis revealed that all four tryptophan residues had been modified by HOCl or the MPO-H_2O_2-chloride system (Table 3.6). Figure 3.6 shows the MS/MS analysis of the tryptic peptide LLDN*W*DSVTSTFSK, which contains Trp50. Compared with the same peptide in control apoA-I (Fig. 3.6A), every y ion from y_{10} to y_{13} and every b ion from b_5 of the peptide gained 16 amu (Fig. 3.6B) or 32 amu (Fig. 3.6C), indicating that Typ50 had been converted to hydroxytryptophan (+16) or dihydroxytryptophan (+32). To identify the major site of tryptophan oxidation by HOCl or the MPO-chloride system, we exposed lipid-free apoA-I to HOCl or the MPO-chloride system at various ratios of oxidant (HOCl or H_2O_2, 2:1 to 10:1, mol/mol, oxidant:apoA-I) and then quantified product yields using the ion current of each precursor and product peptide and reconstruct ion chromatograms. We observed similar oxidation patterns regardless of whether apoA-I is exposed to reagent HOCl or the MPO-H_2O_2-chloride system (compare Figs. 3.7A and 3.7B). As shown in Fig. 3.7, HOCl or the MPO-chloride system generates high yields of two oxidized tryptophan residues (Trp50 and Trp72) and low yields of the other two (Trp8 and Trp108). At a molar ratio of 10:1 of oxidant to apoA-I (50 μM oxidant), 20 to 40% of Trp50 or Trp72 and 5 to 10% of Trp8 or Trp108 are oxidized by HOCl or the MPO-chloride system (Fig. 3.7).

Table 3.6 Detection by LC-ESI-MS of peptides containing hydroxytryptophan and dihydroxytryptophan in a tryptic digest or Glu-C digest of apoA-I exposed to HOCl or the MPO-H$_2$O$_2$-NaCl system[a]

Position	Sequence	Precursor m/z (charge state)	Product m/z (charge state)	Modification
1–10	DEPPQSPWDR[b]	1226.6 (+1), 614.0 (+2)	1242.7 (+1), 621.9 (+2)	W8+16
46–59	LLDNWDSVTSTFSK[b]	1612.7 (+1), 807.4 (+2)	1628.7 (+1), 815.2 (+2)	W50+16
46–59	LLDNWDSVTSTFSK[b]	1612.7 (+1), 807.4 (+2)	1644.7 (+1), 823.2 (+2)	W50+32
62–77	EQLGPVTQEFWDNLEK[b]	1932.8 (+1), 967.8 (+2)	1948.8 (+1), 975.2 (+2)	W72+16
62–77	EQLGPVTQEFWDNLEK[b]	1932.8 (+1), 967.8 (+2)	1964.6 (+1), 983.3 (+2)	W72+32
35–62	GSALGKQLNLKLLDNW DSVTSTFSKLRE[c]	1561.6 (+2), 1041.4 (+3), 781.5 (+4), 625.5 (+5)	1569.4 (+2), 1046.6 (+3), 785.5 (+4), 628.6 (+5)	W50+16
35–62	GSALGKQLNLKLLDNW DSVTSTFSKLRE[c]	1561.6 (+2), 1041.4 (+3), 781.5 (+4), 625.5 (+5)	1577.3 (+2), 1051.7 (+3), 789.3 (+4), 631.7 (+5)	W50+32
71–78	FWDNLEKE[c]	1080.4 (+1), 541.0 (+2)	1096.5 (+1), 548.9 (+2)	W72+16
71–78	FWDNLEKE[c]	1080.4 (+1), 541.0 (+2)	1112.5 (+1), 557.0 (+2)	W72+32
93–111	VKAKVQPYLDDFQKK WQEE[c]	1190.5 (+2), 794.2 (+3), 596.1 (+4), 477.0 (+5)	1199.0 (+2), 799.2 (+3), 599.9 (+4), 480.2 (+5)	W108+16

[a] ApoA-I (5 μM) was exposed to HOCl or the MPO-H$_2$O$_2$-NaCl system (10:1, mol/mol, oxidant:apoA-I) for 60 min at 37 °C in phosphate buffer (20 mM sodium phosphate, 100 μM DTPA, pH 7.4). After the reaction was terminated with L-methionine, a tryptic digest or Glu-C digest of apoA-I was analyzed by LC-ESI-MS and MS/MS. Peptide sequences and modifications were confirmed using MS/MS.
[b] Tryptic digest.
[c] Glu-C digest.

LC-MS/MS Analysis of Site-Specific Protein Modifications

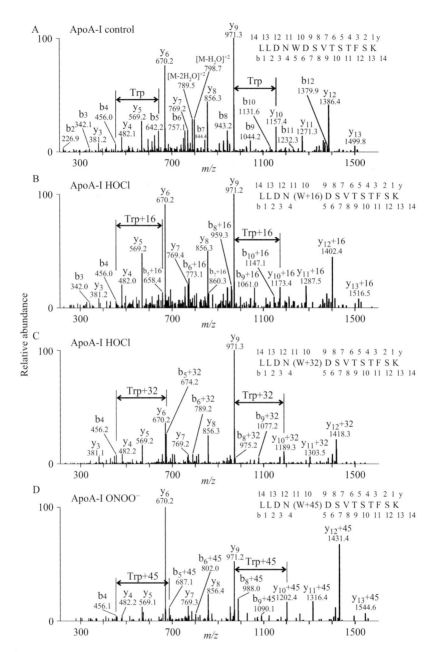

Figure 3.6 MS/MS identification of tryptophan oxidation in lipid-free apoA-I exposed to HOCl or ONOO$^-$. MS/MS analysis of (A) [LLDN*W*DSVTSTFSK + 2H]$^{+2}$ (m/z 807.4), (B) [LLDN(*W*-OH)DSVTSTFSK + 2H]$^{+2}$ (m/z 815.2), (C) [LLDN(*W*-(OH)$_2$)DSVTSTFSK + 2H]$^{+2}$ (m/z 823.2), and (D) [LLDN(*W*-NO$_2$)DSVTSTFSK + 2H]$^{+2}$ (m/z 829.5) in apoA-I oxidized with HOCl or ONOO$^-$. Lipid-free apoA-I (5 μM) was

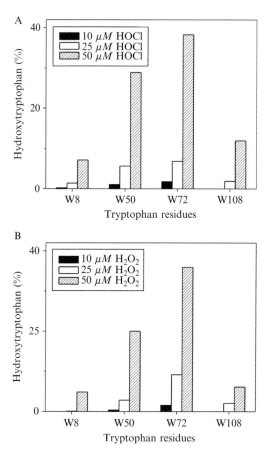

Figure 3.7 Site-specific oxidation of tryptophan residues in lipid-free apoA-I exposed to HOCl or myeloperoxidase-H_2O_2-chloride system. ApoA-I (5 μM) was exposed to (A) HOCl or (B) the myeloperoxidase-H_2O_2-chloride system at the indicated concentrations of oxidant for 60 min at 37 °C in phosphate buffer (pH 7.4). The myeloperoxidase system was supplemented with 0.1 M NaCl. A tryptic digest (for W8 and W50) or Glu-C digest (for W72 and W108) of apoA-I was analyzed by LC-ESI-MS and MS/MS. Oxidized tryptophan (hydroxytryptophan plus dihydroxytryptophan) peptides were detected and quantified as described in the legend to Fig. 3.3. Peptide sequences were confirmed using MS/MS. Results are representative of three independent experiments.

exposed to HOCl (10:1, mol/mol, oxidant:protein) or $ONOO^-$ (25:1, mol/mol, oxidant:protein) for 60 min at 37 °C in phosphate buffer (20 mM sodium phosphate, 100 μM DTPA, pH 7.4). After the reaction was terminated with L-methionine, apoA-I was digested with trypsin, and the peptides were analyzed with LC-ESI-MS/MS.

14. Reactive Nitrogen Species Generates Nitrotryptophan Residues in Lipid-Free ApoA-I

We were unable to identify hydroxytryptophan or dihydroxytryptophan in apoA-I after we exposed the protein to $ONOO^-$ or the MPO-nitrite system. However, we did detect the tryptophan nitration product (W+45). MS/MS analysis demonstrated that $ONOO^-$ nitrated three tryptophan residues (Trp8, Trp50, and Trp72) and that the MPO-H_2O_2-nitrite system nitrated two tryptophan residues (Trp50 and Trp72). Figure 3.6D shows the MS/MS analysis of the nitration product at Trp50. Compared with the same peptide in control apoA-I (Fig. 3.6A), every y ion from y_{10} to y_{13} and every b ion from b_5 of the peptide gained 45 amu (Fig. 3.6D), indicating that Typ50 had been converted to nitrotryptophan. We also measured the amount of nitrotryptophan in the peptide, using the ion current of each precursor and product peptide and reconstructed ion chromatograms. This approach indicated very low product yields (1–3% for Trp50 and Trp72 and <1% for Trp8 at 25:1 ratio of oxidant, mol/mol, oxidant/apoA-I) (data not shown).

15. The YXXK Motif Directs ApoA-I Chlorination

Tyr192, the tyrosine residue in apoA-I that MPO chlorinates most readily, resides two amino acid residues away from a lysine residue. In the α-helix structure that is typical of apoA-I, these Tyr and Lys residues would be adjacent to each other on the same face of the helix. Importantly, the ε-amino group of lysine reacts rapidly with HOCl to form long-lived chloramines, which have been proposed to promote tyrosine chlorination. Based on these observations and studies with synthetic peptides, we propose that the YXXK motif (Y, tyrosine; K, lysine; X, unreactive amino acid) can direct regiospecific (specific to a particular region of apoA-I) tyrosine chlorination in α-helical proteins (Bergt et al., 2004a).

To test this model, we engineered a series of mutations into the cDNA of human apoA-I using site-directed mutagenesis (Shao et al., 2006a). Studies with these altered proteins provided strong evidence that YXXK can direct the regiospecific chlorination of tyrosine residues by reagent HOCl. Virtually identical results were observed with the complete MPO system. These observations strongly support our hypothesis that the YXXK motif directs site-specific chlorination of Tyr192.

Interestingly, we also found that a methionine residue in an MXXK or MXXY motif can inhibit tyrosine chlorination. Thus, Tyr115 resides in a YXXK motif, but it resists chlorination, perhaps because it sits near a methionine residue (Met112) in an MXXY motif. When we mutated

Met112 to alanine, which does not react with HOCl, chlorination of Tyr115 increased markedly. Conversely, introducing a methionine residue into the YXXK motif inhibited the chlorination of Tyr192 (Shao *et al.*, 2006a). Our results support the proposal that protein-bound Met residues inhibit tyrosine chlorination, perhaps by scavenging chlorinating intermediates or disrupting the secondary structure of the apolipoprotein.

16. Quantitative Analysis of Posttranslational Modifications of Proteins

Mass spectrometry, especially tandem mass spectrometry (MS/MS), is a critical tool in the qualitative characterization of posttranslational modifications of proteins (Aebersold and Mann, 2003). There has been growing interest in developing mass spectrometric methods for the quantification of posttranslational modifications (Mann and Jensen, 2003).

We use a straightforward approach to quantify protein oxidation. Product yields for peptide oxidation are monitored using the ion current of each precursor and product peptide within the same sample and reconstructed ion chromatograms. The product yield is defined as the percentage of the peak area of the product ion relative to the sum of the peak areas of the precursor and product ions. Thus, product yield (%) = [peak area of product ion/(peak area of precursor ion + peak areas of product ions)] × 100. This method assumes that the entire precursor consumed is converted into products and that the MS response characteristics of the product ions are similar to those of the precursor ion. It is important to note that precursor and product peptides typically exhibit similar retention times on HPLC, ensuring reproducible results.

Our method generally produces an excellent correlation between the loss of a precursor peptide and the yield of its oxidized product. However, the approach can give incorrect results under two circumstances. The first is if the precursor ion and product ion ionize with very different efficiencies. The second is if the precursor peptide is converted in high yield to product ions that MS fails to detect or to unknown product ions that could not be identified. Despite these potential limitations, our approach is especially suitable for quantifying site-specific modifications when multiple sites can be modified, as is generally the case when proteins are oxidized by HOCl or RNS.

17. ApoA-I Oxidation Impairs Cholesterol Transport by the ABCA1 System

ApoA-I isolated from humans with established cardiovascular disease contains elevated levels of two MPO products: 3-chlorotyrosine and 3-nitrotyrosine (Bergt *et al.*, 2004b; Pennathur *et al.*, 2004; Zheng *et al.*, 2004).

To determine whether oxidation of tyrosine residues interferes with cholesterol efflux, we investigated how chlorination and nitration of apoA-I affected the ability of the protein to promote cholesterol efflux from cells expressing ABCA1 (Bergt et al., 2004b; Shao et al., 2005a). Oxidation by H_2O_2 alone had no effect. In striking contrast, chlorination of apoA-I by MPO or HOCl impaired the ability of the protein to promote cholesterol efflux (Bergt et al., 2004b; Shao et al., 2005a). Zheng et al. (2004) reported similar results. These investigators also found that nitration of apoA-I by MPO impaired the protein's ABCA1-mediated reverse cholesterol transport activity due to the oxidative inhibition of the ability of apoA-I to interact with lipid. In contrast, we found that nitration of apoA-I by either MPO or $ONOO^-$ was much less inhibitory (Shao et al., 2005a). Thus, oxidation of apoA-I by HOCl or the MPO chlorinating system selectively impairs the ability of the protein to remove cholesterol from cells by a pathway requiring ABCA1.

We proposed a model whereby site-specific oxidation alters apoA-I remodeling, impairing the ability of the protein to promote cholesterol efflux by the ABCA1 pathway (Shao et al., 2006b). These observations suggest that HOCl generated by MPO could damage apoA-I and promote the formation of macrophage foam cells in the human artery wall. The accumulation of foam cells could, in turn, promote the development of atherosclerotic lesions.

18. Conclusions

This chapter described LC-ESI-MS/MS approaches for quantifying site-specific modifications in apoA-I protein that has been exposed to HOCl or RNS. Results demonstrated that methionine, tryptophan, and tyrosine residues are targeted for oxidation by HOCl or the MPO-H_2O_2-chloride system. Although Met112 is oxidized more readily than Met86 and Met148, methionine oxidation by HOCl is not highly site specific. Tryptophan residues, especially Trp50 and Trp72, are also oxidized by HOCl. Chlorination of apoA-I's tyrosine residues by HOCl or the MPO-H_2O_2-chloride system is much more site selective: Tyr192 is the major site, whereas Tyr100 and Tyr166 are poorly chlorinated. Our observations also revealed that tyrosine residues are the major targets when apoA-I is exposed to $ONOO^-$ or the MPO-H_2O_2-nitrite system. Although the predominant nitration site for both sources of reactive nitrogen species is Tyr192, reagent $ONOO^-$ is a more selective nitrating agent than the MPO-H_2O_2-nitrite system. Low levels of methionine oxidation have also been observed in apoA-I oxidized by these two agents. Although we identified nitrotryptophan products in apoA-I that had been exposed to $ONOO^-$ or the

MPO-H$_2$O$_2$-nitrite system, the product yields were essentially negligible compared with those seen with tyrosine nitration.

The method described in this chapter is applicable to other proteins modified by HOCl or RNS, as well as to other posttranslational modification systems. Importantly, it can be used to detect protein-bound 3-chlorotyrosine and 3-nitrotyrosine, two widely used biomarkers of inflammation and other diseases involving activated phagocytes. Therefore, it should help elucidate the role of oxidative and nitrative stress in the pathogenesis of disease. The approach is direct and sensitive and does not require isotope labeling.

ACKNOWLEDGMENTS

This work was supported by grants from the National Institutes of Health (HL030086, ES07083, DK017047, HL078527, HL086798). B.S. was supported by a Fellowship Award from the Phillip Morris External Research Program (Philip Morris USA Inc. and Philip Morris International). Mass spectrometry experiments were performed by the Mass Spectrometry Resource, Department of Medicine and the Mass Spectrometry Core, Diabetes and Endocrinology Research Center at the University of Washington.

REFERENCES

Aebersold, R. and Mann, M. (2003). Mass spectrometry-based proteomics. *Nature* **422,** 198–207.

Ajees, A. A., Anantharamaiah, G. M., Mishra, V. K., Hussain, M. M., and Murthy, H. M. (2006). Crystal structure of human apolipoprotein A-I: Insights into its protective effect against cardiovascular diseases. *Proc. Natl. Acad. Sci. USA* **103,** 2126–2131.

Ames, B. N., Shigenaga, M. K., and Hagen, T. M. (1993). Oxidants, antioxidants, and the degenerative diseases of aging. *Proc. Natl. Acad. Sci. USA* **90,** 7915–7922.

Beckman, J. S. Beckman, T. W., Chen, J., Marshall, P. A., and Freeman, B. A. (1990). Apparent hydroxyl radical production by peroxynitrite? Implications for endothelial injury from nitric oxide and superoxide. *Proc. Natl. Acad. Sci. USA* **87,** 1620–1624.

Beckman, J. S., Chen, J., Ischiropoulos, H., and Crow, J. P. (1994). Oxidative chemistry of peroxynitrite. *Methods Enzymol.* **233,** 229–240.

Beckman, J. S. and Koppenol, W. H. (1996). Nitric oxide, superoxide, and peroxynitrite: The good, the bad, and ugly. *Am. J. Physiol.* **271,** C1424–C1437.

Bergt, C., Fu, X., Huq, N. P., Kao, J., and Heinecke, J. W. (2004a). Lysine residues direct the chlorination of tyrosines in YXXK motifs of apolipoprotein A-I when hypochlorous acid oxidizes high density lipoprotein. *J. Biol. Chem.* **279,** 7856–7866.

Bergt, C., Pennathur, S., Fu, X., Byun, J., O'Brien, K., McDonald, T. O., Singh, P., Anantharamaiah, G. M., Chait, A., Brunzell, J., Geary, R. L., Oram, J. F., *et al.* (2004b). The myeloperoxidase product hypochlorous acid oxidizes HDL in the human artery wall and impairs ABCA1-dependent cholesterol transport. *Proc. Natl. Acad. Sci. USA* **101,** 13032–13037.

Berliner, J. A. and Heinecke, J. W. (1996). The role of oxidized lipoproteins in atherogenesis. *Free Radic. Biol. Med.* **20,** 707–727.

Brouillette, C. G. and Anantharamaiah, G. M. (1995). Structural models of human apolipoprotein A-I. *Biochim. Biophys. Acta* **1256**, 103–129.

Crowley, J. R., Yarasheski, K., Leeuwenburgh, C., Turk, J., and Heinecke, J. W. (1998). Isotope dilution mass spectrometric quantification of 3-nitrotyrosine in proteins and tissues is facilitated by reduction to 3-aminotyrosine. *Anal. Biochem.* **259**, 127–135.

Daugherty, A., Dunn, J. L., Rateri, D. L., and Heinecke, J. W. (1994). Myeloperoxidase, a catalyst for lipoprotein oxidation, is expressed in human atherosclerotic lesions. *J. Clin. Invest.* **94**, 437–444.

Davidson, W. S. and Thompson, T. B. (2007). The structure of apolipoprotein a-I in high density lipoproteins. *J. Biol. Chem.* **282**, 22249–22253.

Domigan, N. M., Charlton, T. S., Duncan, M. W., Winterbourn, C. C., and Kettle, A. J. (1995). Chlorination of tyrosyl residues in peptides by myeloperoxidase and human neutrophils. *J. Biol. Chem.* **270**, 16542–16548.

Drapeau, G. R. (1977). Cleavage at glutamic acid with staphylococcal protease. *Methods Enzymol.* **47**, 189–191.

Eiserich, J. P., Hristova, M., Cross, C. E., Jones, A. D., Freeman, B. A., Halliwell, B., and van der Vliet, A. (1998). Formation of nitric oxide-derived inflammatory oxidants by myeloperoxidase in neutrophils. *Nature* **391**, 393–397.

Fenn, J. B., Mann, M., Meng, C. K., Wong, S. F., and Whitehouse, C. M. (1989). Electrospray ionization for mass spectrometry of large biomolecules. *Science* **246**, 64–71.

Finkel, T. and Holbrook, N. J. (2000). Oxidants, oxidative stress and the biology of aging. *Nature* **408**, 239–247.

Foote, C. S., Goyne, T. E., and Lehrer, R. I. (1983). Assessment of chlorination by human neutrophils. *Nature* **301**, 715–716.

Gaut, J. P., Byun, J., Tran, H. D., Lauber, W. M., Carroll, J. A., Hotchkiss, R. S., Belaaouaj, A., and Heinecke, J. W. (2002). Myeloperoxidase produces nitrating oxidants in vivo. *J. Clin. Invest.* **109**, 1311–1319.

Glomset, J. A. (1968). The plasma lecithins:cholesterol acyltransferase reaction. *J. Lipid Res.* **9**, 155–167.

Gordon, D. J., and Rifkind, B. M. (1989). High-density lipoprotein: The clinical implications of recent studies. *N. Engl. J. Med.* **321**, 1311–1316.

Gotto, A. M., Jr., Pownall, H. J., and Havel, R. J. (1986). Introduction to the plasma lipoproteins. *Methods Enzymol.* **128**, 3–41.

Harrison, J. E. and Shultz, J. (1976). Studies on the chlorinating activity of myeloperoxidase. *J. Biol. Chem.* **251**, 1371–1374.

Hawkins, C. L., Pattison, D. I., and Davies, M. J. (2003). Hypochlorite-induced oxidation of amino acids, peptides and proteins. *Amino Acids* **25**, 259–274.

Hazen, S. L., Crowley, J. R., Mueller, D. M., and Heinecke, J. W. (1997). Mass spectrometric quantification of 3-chlorotyrosine in human tissues with attomole sensitivity: A sensitive and specific marker for myeloperoxidase-catalyzed chlorination at sites of inflammation. *Free Radic. Biol. Med.* **23**, 909–916.

Hazen, S. L. and Heinecke, J. W. (1997). 3-Chlorotyrosine, a specific marker of myeloperoxidase-catalyzed oxidation, is markedly elevated in low density lipoprotein isolated from human atherosclerotic intima. *J. Clin. Invest.* **99**, 2075–2081.

Hazen, S. L., Hsu, F. F., Mueller, D. M., Crowley, J. R., and Heinecke, J. W. (1996). Human neutrophils employ chlorine gas as an oxidant during phagocytosis. *J. Clin. Invest.* **98**, 1283–1289.

Heinecke, J. W. (1998). Oxidants and antioxidants in the pathogenesis of atherosclerosis: Implications for the oxidized low density lipoprotein hypothesis. *Atherosclerosis* **141**, 1–15.

Heinecke, J. W., Baker, L., Rosen, H., and Chait, A. (1986). Superoxide-mediated modification of low density lipoprotein by arterial smooth muscle cells. *J. Clin. Invest.* **77**, 757–761.

Heinecke, J. W., Hsu, F. F., Crowley, J. R., Hazen, S. L., Leeuwenburgh, C., Mueller, D. M., Rasmussen, J. E., and Turk, J. (1999). Detecting oxidative modification of biomolecules with isotope dilution mass spectrometry: Sensitive and quantitative assays for oxidized amino acids in proteins and tissues. *Methods Enzymol.* **300,** 124–144.

Heinecke, J. W., Li, W., Daehnke, H. L. D., and Goldstein, J. A. (1993). Dityrosine, a specific marker of oxidation, is synthesized by the myeloperoxidase-hydrogen peroxide system of human neutrophils and macrophages. *J. Biol. Chem.* **268,** 4069–4077.

Hope, H. R., Remsen, E. E., Lewis, C., Jr., Heuvelman, D. M., Walker, M. C., Jennings, M., and Connolly, D. T. (2000). Large-scale purification of myeloperoxidase from HL60 promyelocytic cells: Characterization and comparison to human neutrophil myeloperoxidase. *Protein Expr. Purif.* **18,** 269–276.

Jerlich, A., Hammel, M., Nigon, F., Chapman, M. J., and Schaur, R. J. (2000). Kinetics of tryptophan oxidation in plasma lipoproteins by myeloperoxidase-generated HOCl. *Eur. J. Biochem.* **267,** 4137–4143.

Kelleher, N. L. (2004). Top-down proteomics. *Anal. Chem.* **76,** 197A–203A.

Klebanoff, S. J. (1993). Reactive nitrogen intermediates and antimicrobial activity: Role of nitrite. *Free Radic. Biol. Med.* **14,** 351–360.

Leeuwenburgh, C., Hardy, M. M., Hazen, S. L., Wagner, P., Oh-ishi, S., Steinbrecher, U. P., and Heinecke, J. W. (1997). Reactive nitrogen intermediates promote low density lipoprotein oxidation in human atherosclerotic intima. *J. Biol. Chem.* **272,** 1433–1436.

Levine, R. L., Moskovitz, J., and Stadtman, E. R. (2000). Oxidation of methionine in proteins: Roles in antioxidant defense and cellular regulation. *IUBMB Life* **50,** 301–307.

Lewis, G. F. and Rader, D. J. (2005). New insights into the regulation of HDL metabolism and reverse cholesterol transport. *Circ. Res.* **96,** 1221–1232.

Mann, M. and Jensen, O. N. (2003). Proteomic analysis of post-translational modifications. *Nat. Biotechnol.* **21,** 255–261.

Mendez, A. J., Oram, J. F., and Bierman, E. L. (1991). Protein kinase C as a mediator of high density lipoprotein receptor-dependent efflux of intracellular cholesterol. *J. Biol. Chem.* **266,** 10104–10111.

Morita, Y., Iwamoto, H., Aibara, S., Kobayashi, T., and Hasegawa, E. (1986). Crystallization and properties of myeloperoxidase from normal human leukocytes. *J. Biochem.* **99,** 761–770.

Morris, J. C. (1966). Acid ionization constant of HOCl from 5 to 35 degrees. *J. Phys. Chem.* **70,** 3798–3805.

Navab, M., Anantharamaiah, G. M., Reddy, S. T., Van Lenten, B. J., Ansell, B. J., and Fogelman, A. M. (2006). Mechanisms of disease: Proatherogenic HDL–an evolving field. *Nat. Clin. Pract. Endocrinol. Metab.* **2,** 504–511.

Nelson, D. P., and Kiesow, L. A. (1972). Enthalpy of decomposition of hydrogen peroxide by catalase at 25 degrees C (with molar extinction coefficients of H_2O_2 solutions in the UV). *Anal. Biochem.* **49,** 474–478.

Oram, J. F. and Heinecke, J. W. (2005). ATP-binding cassette transporter A1? A cell cholesterol exporter that protects against cardiovascular disease. *Physiol. Rev.* **85,** 1343–1372.

Panzenbock, U., Kritharides, L., Raftery, M., Rye, K. A., and Stocker, R. (2000). Oxidation of methionine residues to methionine sulfoxides does not decrease potential antiatherogenic properties of apolipoprotein A-I. *J. Biol. Chem.* **275,** 19536–19544.

Pennathur, S., Bergt, C., Shao, B., Byun, J., Kassim, S. Y., Singh, P., Green, P. S., McDonald, T. O., Brunzell, J., Chait, A., Oram, J. F., O'Brien, K., *et al.* (2004). Human atherosclerotic intima and blood of patients with established coronary artery disease contain high density lipoprotein damaged by reactive nitrogen species. *J. Biol. Chem.* **279,** 42977–42983.

Rothblat, G. H., de la Llera-Moya, M., Atger, V., Kellner-Weibel, G., Williams, D. L., and Phillips, M. C. (1999). Cell cholesterol efflux: Integration of old and new observations provides new insights. *J. Lipid Res.* **40,** 781–796.

Ryan, R. O., Forte, T. M., and Oda, M. N. (2003). Optimized bacterial expression of human apolipoprotein A-I. *Protein Expr. Purif.* **27,** 98–103.

Segrest, J. P., Garber, D. W., Brouillette, C. G., Harvey, S. C., and Anantharamaiah, G. M. (1994). The amphipathic alpha helix: A multifunctional structural motif in plasma apolipoproteins. *Adv. Protein Chem.* **45,** 303–369.

Shao, B., Bergt, C., Fu, X., Green, P., Voss, J. C., Oda, M. N., Oram, J. F., and Heinecke, J. W. (2005a). Tyrosine 192 in apolipoprotein A-I is the major site of nitration and chlorination by myeloperoxidase, but only chlorination markedly impairs ABCA1-dependent cholesterol transport. *J. Biol. Chem.* **280,** 5983–5993.

Shao, B., Fu, X., McDonald, T. O., Green, P. S., Uchida, K., O'Brien, K. D., Oram, J. F., and Heinecke, J. W. (2005b). Acrolein impairs ATP binding cassette transporter A1-dependent cholesterol export from cells through site-specific modification of apolipoprotein A-I. *J. Biol. Chem.* **280,** 36386–36396.

Shao, B., Oda, M. N., Bergt, C., Fu, X., Green, P. S., Brot, N., Oram, J. F., and Heinecke, J. W. (2006a). Myeloperoxidase impairs ABCA1-dependent cholesterol efflux through methionine oxidation and site-specific tyrosine chlorination of apolipoprotein A-I. *J. Biol. Chem.* **281,** 9001–9004.

Shao, B., Oda, M. N., Oram, J. F., and Heinecke, J. W. (2006b). Myeloperoxidase: An inflammatory enzyme for generating dysfunctional high density lipoprotein. *Curr. Opin. Cardiol.* **21,** 322–328.

Stadtman, E. R. (1992). Protein oxidation and aging. *Science* **257,** 1220–1224.

Vaisar, T., Pennathur, S., Green, P. S., Gharib, S. A., Hoofnagle, A. N., Cheung, M. C., Byun, J., Vuletic, S., Kassim, S., Singh, P., Chea, H., Knopp, R. H., et al. (2007). Shotgun proteomics implicates protease inhibition and complement activation in the anti-inflammatory properties of HDL. *J. Clin. Invest.* **117,** 746–756.

Vogt, W. (1995). Oxidation of methionyl residues in proteins: Tools, targets, and reversal. *Free Radic. Biol. Med.* **18,** 93–105.

von Eckardstein, A., Walter, M., Holz, H., Benninghoven, A., and Assmann, G. (1991). Site-specific methionine sulfoxide formation is the structural basis of chromatographic heterogeneity of apolipoproteins A-I, C-II, and C-III. *J. Lipid Res.* **32,** 1465–1476.

Whitehouse, C. M., Dreyer, R. N., Yamashita, M., and Fenn, J. B. (1985). Electrospray interface for liquid chromatographs and mass spectrometers. *Anal. Chem.* **57,** 675–679.

Witztum, J. L., and Steinberg, D. (1991). Role of oxidized low density lipoprotein in atherogenesis. *J. Clin. Invest.* **88,** 1785–1792.

Yla-Herttuala, S., Palinski, W., Rosenfeld, M. E., Parthasarathy, S., Carew, T. E., Butler, S., Witztum, J. L., and Steinberg, D. (1989). Evidence for the presence of oxidatively modified low density lipoprotein in atherosclerotic lesions of rabbit, and man. *J. Clin. Invest.* **84,** 1086–1095.

Zheng, L., Nukuna, B., Brennan, M. L., Sun, M., Goormastic, M., Settle, M., Schmitt, D., Fu, X., Thomson, L., Fox, P. L., Ischiropoulos, H., Smith, J. D., et al. (2004). Apolipoprotein A-I is a selective target for myeloperoxidase-catalyzed oxidation and functional impairment in subjects with cardiovascular disease. *J. Clin. Invest.* **114,** 529–541.

CHAPTER FOUR

Influence of Intramolecular Electron Transfer Mechanism in Biological Nitration, Nitrosation, and Oxidation of Redox-Sensitive Amino Acids

Hao Zhang, Yingkai Xu, Joy Joseph, *and* B. Kalyanaraman

Contents

1. Introduction	66
2. Methods	67
2.1. Syntheses and purification of peptides	67
2.2. Syntheses of oxidation, nitration, and nitrosation products of N-acetyl-TyrCys-amide (YC)	68
2.3. Modification of YC peptides by $MPO/H_2O_2/NO_2^-$ system	69
2.4. HPLC analyses of nitration and oxidation products	69
2.5. Electron spin resonance spin-trapping experiments	70
3. Results	70
3.1. Characterization of model tyrosylcysteine peptides for intramolecular electron transfer studies	70
3.2. Intramolecular electron transfer from tyrosine to cysteine induced by $MPO/H_2O_2/NaNO_2$: HPLC products analysis	71
3.3. Identification of radical intermediates from electron transfer reaction using ESR spin-trapping techniques	74
3.4. Intramolecular versus intermolecular electron transfer mechanism between tyrosyl radical and cysteinyl residue in model peptides: ESR spin trapping and HPLC product analysis	79
3.5. Biological implications of intramolecular electron transfer reaction in protein nitration and nitrosation reactions	80
3.6. Proteomics and prediction of electron transfer mechanism	88
3.7. Overall conclusions: Future perspectives	89
Acknowledgments	90
References	90

Department of Biophysics and Free Radical Research Center, Medical College of Wisconsin, Milwaukee, Wisconsin

Abstract

Using both high-performance liquid chromatography (HPLC) and electron spin resonance (ESR) spin-trappng techniques, we developed an analytical methodology for investigating intramolecular electron transfer-mediated tyrosyl nitration and cysteine nitrosation in model peptides. Peptides N-acetyl-TyrCys-amide (YC), N-acetyl-TyrAlaCys-amide, N-acetyl-TyrAlaAlaCys-amide, and N-acetyl-TyrAlaAlaAlaAlaCys-amide were used as models. Product analysis showed that nitration and oxidation products derived from YC and related peptides in the presence of myeloperoxidase (MPO)/H_2O_2/NO_2^- were not detectable. The major product was determined to be the corresponding disulfide (e.g., YCysCysY), suggestive of a rapid electron transfer from the tyrosyl radical to the cysteinyl residue. ESR spin-trapping experiments with 5,5'-dimethyl-1-pyrroline N-oxide (DMPO) demonstrated that thiyl radical intermediates were formed from peptides (e.g., YC) treated with MPO/H_2O_2 and MPO/H_2O_2/NO_2^-. Blocking the thiol group in YC totally abrogated thiyl radical formation. Under similar conditions, we were, however, able to trap the tyrosyl radical using the spin trap dibromonitrosobenzene sulfonic acid (DBNBS). Competition spin-trapping experiments revealed that intramolecular electron transfer is the dominant mechanism for thiyl radical formation in YC peptides. We conclude that a rapid intramolecular electron transfer mechanism between redox-sensitive amino acids could influence both protein nitration and nitrosation reactions. This mechanism brings together nitrative, nitrosative, and oxidative mechanisms in free radical biology.

1. Introduction

Radiation-induced electron transfer reactions between redox-sensitive amino acids have been known for several decades (Bobrowski *et al.*, 1990, 1992; Pruetz *et al.*, 1980, 1985, 1989; Tanner *et al.*, 1998). For example, the electron transfer occurs over a 35-Å distance between a tyrosyl radical and a cysteine via well-defined electron transfer pathways in ribonucleotide reductase catalyzed reactions (Aubert *et al.*, 2000). Protein tyrosine nitration/oxidation and cysteine S-nitrosation/oxidation have long been identified as markers of oxidative and nitrative stress (Rassaf *et al.*, 2002; White *et al.*, 1994; Zhang *et al.*, 1996) and play important roles in several pathological processes (Baldus *et al.*, 2002; Greenacre and Ischiropoulos 2001; Haddad *et al.*, 1994; Kooy *et al.*, 1997; Lanone *et al.*, 2002; Leeuwenburgh *et al.*, 1997; Pavlick *et al.*, 2002; Paxinou *et al.*, 2001; Rubbo *et al.*, 1994; Schopfer *et al.*, 2003; Turko and Murad 2002; Wu *et al.*, 1999). Protein tyrosyl nitration reactions are induced by •NO_2 and the protein tyrosyl radical (Beckman, 1996; Brennan *et al.*, 2002; Ischiropoulos *et al.*, 1992; Jourd'heuil *et al.*, 2001; MacPherson *et al.*, 2001; Pfeiffer *et al.*, 2001; Pryor and Squadrito, 1995). While there is ample evidence for elevated levels of nitrated and nitrosated proteins in diverse pathologies (Jourd'heuil et al., 2001; MacPherson et al., 2001; Pfeiffer et al., 2001; Pryor and Squadrito, 1995; Schopfer et al., 2003), the biophysical basis for selective nitrosation of cysteines in proteins still remains unknown. It has also been suggested that the mechanism of S-nitrosation is likely mediated by "oxidative nitrosylation," a pathway by which reactive nitrogen species ($ONOO^-/^{\bullet}NO_2$) generate reactive intermediates that can oxidize thiols to thiyl radicals, which can react directly with $^{\bullet}NO$ to form the nitrosothiol (Bryan et al., 2004; Espey et al., 2002). Both tyrosine nitration and cysteine S-nitrosation appear to be selective processes in that not all tyrosine or cysteine residues present in a protein are avidly nitrated, S-nitrosylated, or thiolated (Foster and Stamler, 2004; Gao et al., 2005; Newman et al., 2002; Souza et al., 1999). Factors influencing those nitrosation and nitration in proteins are not fully known. As discussed elsewhere (Ischiropoulos, 1998; Ischiropoulos et al., 1992; Schopfer et al., 2003), the local environment of tyrosine or cysteine residues within the secondary and tertiary structure of the protein will likely influence the site of S-nitrosation/nitration (Denicola et al., 1998, 2002; Gow et al., 1996, 2004; Ischiropoulos, 1998; Khairutdinov et al., 2000; Liu et al., 1998; Marla et al., 1997; Radi et al., 1999; Shao et al., 2005; Zhang et al., 2001, 2003). Although no specific amino acid sequence criteria exist for predicting tyrosine nitration or cysteine nitrosation, it has been suggested that protein tyrosyl nitration is decreased when a tyrosine residue is located close to a cysteine or methionine residue (Greenacre and Ischiropoulos, 2001; Ischiropoulos, 1998; Souza et al., 1999).

Our study revealed that intramolecular electron transfer reaction via oxidation-sensitive amino acids may play an important role in selective nitration and S-nitrosation in proteins (Zhang et al., 2005). The presence of a cysteinyl group in the peptide chain modulates or negates tyrosine nitration mediated by myeloperoxidase (MPO)/H_2O_2-derived nitrating species. Alternatively, the presence of a tyrosyl residue may stimulate intramolecular electron transfer-mediated thiol oxidation in the MPO/H_2O_2/NO_2^- system. Moreover, the presence of a Cys residue adjacent to a tyrosyl group in peptides facilitates the intramolecular radical transfer mechanism between the tyrosyl radical and the Cys residue, forming the corresponding thiyl radical and the S-nitrosated product.

2. METHODS

2.1. Syntheses and purification of peptides

Peptides are synthesized chemically using the standard Fmoc solid-phase peptide synthetic procedure supplied by the Advanced Chemtech Model 90 synthesizer (Louisville, KY) (Zhang et al., 2003). Rink Amide MBHA resin (loading 0.72 mmol/g) is used as a solid support. Fmoc-protected amino acids are coupled as HOBt-esters. All amino acids are coupled two at a time using HOBt/DIC. The following steps are performed in the reaction vessel for each double coupling: deprotection of the Fmoc group with 20% piperidine in -(NMP) for 30 min (twice), three NMP washes, two dichloromethane (DCM) washes, first coupling for 1 h with a fivefold excess of Fmoc amino acid in 0.5 M HOBt and 0.5 M DIC, second coupling using a fresh addition of the same reagent for 1 h, followed by three NMP washes and two DCM washes. Final acetylation is performed using an acetic anhydride/HOBt/DIC mixture for 30 min (twice). The resin is washed twice with DCM and three times with methanol and is then dried under vacuum prior to cleavage. The peptide is deprotected and cleaved from the resin with 90% TFA containing (TIS) for 3 h at room temperature. The resin is removed by filtration and washed with TFA, and the combined TFA filtrates are evaporated to dryness under a stream of dry N_2 gas. The oily residue is washed three times with cold ether to remove the scavengers, and the dry crude peptide is dissolved in acetonitrile/H_2O (1:1) and lyophilized. The crude peptides are purified by a semipreparative reversed-phase (RP) HPLC on a RP-C18 (10 × 250 mm) column using a CH_3CN/water gradient (5 to 25% CH_3CN over 60 min) containing 0.1% TFA at a flow rate of 3 ml/min with detection at 280 nm.

2.2. Syntheses of oxidation, nitration, and nitrosation products of *N*-acetyl-TyrCys-amide (YC)

2.2.1. Synthesis of YCys-CysY

The YC peptide (30 mM) is incubated with 30 mM hydrogen peroxide in a phosphate buffer (0.1 M, pH 7.4) containing 1 mM diethylenetriaminepentaacetic acid (DTPA) at room temperature for 1 h. The reaction mixture is injected into HPLC (Agilent 1090) with a RP-C18 semipreparatory column (250 × 10 mm), and the YC peptide disulfide (YCysCysY) is eluted for 26 min with a CH_3CN/water gradient (5 to 25% CH_3CN over 60 min) containing 0.1% TFA at a flow rate of 3 ml/min. Detection is at 280 nm. The structure of the disulfide product is confirmed by ESI-MS analysis ($M+H^+$, 649.3).

2.2.2. Synthesis of CTyr-TyrC

The YC disulfide (15 mM) is incubated with 10 mM H_2O_2 and 100 μg horse radish peroxidase (HRP) in a phosphate buffer (0.1 M, pH 7.4) containing 100 μM DTPA for 20 min. The reaction mixture is then mixed with β-mercaptoethanol (500 mM), and HRP is removed by ultracentrifugation (3K cut-off). The product, YC dityrosine (CYYC), is purified by a preparative HPLC (C-18, 250 × 10 mm) using a fluorescence detector (ex = 294 nm, em = 410 nm). The YC dityrosine is eluted by a linear CH_3CN gradient as described earlier. The product is further confirmed by ESI-MS (Agilent 1100 Series LC-MS).

2.2.3. Synthesis of nitrated YC

The YCysCysY peptide (1 mM) is mixed with peroxynitrite (10 mM) in a phosphate buffer (0.1 M, pH 7.4) containing DTPA (100 μM) for 20 min. The reaction mixture is then mixed with 500 mM β-mecaptoethanol and incubated for 20 min. The resulting nitrated YC peptide [Y(NO$_2$)C] is purified using a preparative HPLC as described earlier. Nitro peptides show a characteristic UV-VIS spectrum. Upon adding NaOH, the 350-nm absorption peak in MeOH is shifted to 430 nm with an extinction coefficient of 4100 M^{-1} m^{-1}. The structure of Y(NO$_2$)C is verified by LC-MS spectrometry on an Agilent 1100 Series LC-MS.

2.2.4. Synthesis of S-nitroso YC (YCysNO)

The YC peptide (20 mM) is mixed with sodium nitrite (20 mM) in 0.03 M HCl for 10 min at room temperature. The solution is then neutralized by 0.1 mM phosphate buffer, pH 7.4, containing 100 μM DTPA. YCysNO (yield >95%) is verified by ESI-MS (M+H$^+$; 355), by UV absorption (λ_{max} = 335 nm), and by Hg^{2+}-induced cleavage.

2.3. Modification of YC peptides by MPO/H$_2$O$_2$/NO$_2^-$ system

2.3.1. Oxidation induced by MPO/H$_2$O$_2$/NO$_2^-$ system

Typically, peptides (0.3 mM) are incubated with NaNO$_2$ (0.5 mM), H$_2$O$_2$ (0.1 M), and MPO (30 nM) in a phosphate buffer (0.1 M, pH 7.4) containing DTPA (100 μM) at room temperature for 30 min. Reactions are stopped by adding the catalase (200 U) enzyme and analyzed by HPLC. Repeat injections in 24 h show no significant oxidation of YC under these experimental conditions.

2.3.2. S-Nitrosation of YC induced by MPO/H$_2$O/•NO system

All reagents are purged with argon gas 60 min before experiments. YC (150 μM) is incubated with H$_2$O$_2$ (50 μM), MPO (30 nM), and DEANO (10–50 μM) in phosphate buffer (50 mM, pH 7.4) containing DTPA (100 μM) at room temperature for 10 min. Reactions are stopped by adding iodoacetamide (20 mM) and are analyzed by HPLC. Repeat injections within 24 h show no significant loss of YCysNO under these experimental conditions.

2.4. HPLC analyses of nitration and oxidation products

Typically, a 20-μl sample is injected into a HPLC system (HP1100) with a C-18 column (250 × 4.6 mm) equilibrated with 5% CH$_3$CN in 0.1% TFA. The peptide and its product are separated by a linear increase of CH$_3$CN concentration to 25% in 60 min at a flow rate of 1 ml/min. Elution is monitored using online UV-VIS and fluorescence detectors. YC and nitrated YC are eluted at 12 and 17.5 min, respectively. Nitration of tyrosine with or without cysteine by $MPO/H_2O_2/NaNO_2$ is performed under the same conditions, and the products are analyzed as reported previously (Goss et al., 1999).

2.5. Electron spin resonance spin-trapping experiments

2.5.1. Thiyl radical trapping

A typical incubation mixture consists of a peptide or tyrosine (1 mM) and cysteine (1 mM), MPO (50 nM), H_2O_2 (1 mM), and 150 mM 5,5'-dimethyl-1-pyrroline N-oxide (DMPO) in a phosphate buffer (0.1 M, pH 7.4) containing DTPA (100 μM). The reaction is initiated by adding H_2O_2. Samples are subsequently transferred to a 100-μl capillary tube, and electron paramagnetic resonance (EPR) spectra are recorded within 30 s after starting the reaction. EPR spectra are recorded at room temperature on a Bruker ER 200 D-SRC spectrometer operating at 9.8 GHz and a cavity equipped with a Bruker Aquax liquid sample cell. Typical spectrometer parameters are scan range, 100 G; field set, 3505 G; time constant, 0.64 ms; scan time, 10 s; modulation amplitude, 1.0 G; modulation frequency, 100 kHz; receiver gain, 5×10^4; and microwave power, 10 mW. Spectra shown are the average of 10 scans.

2.5.2. Tyrosyl radical trapping

Incubations consisting of a peptide or tyrosine (1 mM) and cysteine (1 mM), MPO (100 nM), and 3,5-dibromo-4-nitrosobenzenesulfonic acid (DBNBS) (20 mM) in phosphate buffer (0.1 M, pH 7.4) containing DTPA (0.1 mM) are mixed rapidly with H_2O_2 (1 mM). Samples are subsequently transferred to a 100-μl capillary tube, and ESR spectra are taken within 30 s after starting the reaction. Typical spectrometer parameters are scan range, 100 G; field set, 3505 G; time constant, 0.64 ms; scan time, 20 s; modulation amplitude, 2.0 G; modulation frequency, 100 kHz; receiver gain, 5×10^5; and microwave power, 20 mW. Spectra shown are the average of 30 scans.

3. Results

3.1. Characterization of model tyrosylcysteine peptides for intramolecular electron transfer studies

Table 4.1 summarizes the amino acid sequences and structures of model peptides with mass spectral data. Intramolecular distances between the tyrosyl oxygen atom and the cysteinyl sulfur atom are calculated from the lowest energy conformations in an aqueous environment using the HyperChem 7.1 Package Program (Hypercube Inc.) (von Freyberg and Braun, 1991).

Influence of Intramolecular Electron Transfer Mechanism

Table 4.1 Model peptides

Peptides	Structure	Mass spectra $(M+H^+)$	Distance between O_y and S_c atoms (Angstrom)
N-Acetyl-TyrCys-amide (YC)		326.1	4.45
N-Acetyl-TyrAlaCys-amide (YAC)		397.2	4.42
N-Acetyl-TyrAlaAlaCys-amide (YAAC)		468.2	4.75
N-Acetyl-TyrAlaAlaAlaAlaCys-amide (YAAAAC)		610.2	4.54

3.2. Intramolecular electron transfer from tyrosine to cysteine induced by MPO/H_2O_2/NaNO$_2$: HPLC products analysis

In the presence of H_2O_2, MPO forms higher oxidants (compounds I and II) capable of oxidizing a variety of inorganic ions (Cl^-, Br^-, NO_2^-, etc.) forming one and two electron oxidation products (Brennan et al., 2002; Eiserich et al., 1998; Van Dalen et al, 2000; van der Vliet et al., 1997). Compounds I and II derived from MPO oxidize tyrosine to tyrosyl radical (Tyr•) but not cysteine to the corresponding thiyl radical. In the absence of an electron transfer mechanism, MPO/H_2O_2 or MPO/H_2O_2/NaNO$_2$ should oxidize the YC peptide to form dityrosine, nitrotyrosine, and some disulfide (Fig. 4.1A). However, the reactions should predominantly form disulfide as a product in case of an electron transfer from tyrosine to cysteine (Fig. 4.1A). To understand the electron transfer from the tyrosyl radical to cysteine, we analyzed the products formed from MPO/H_2O_2-dependent

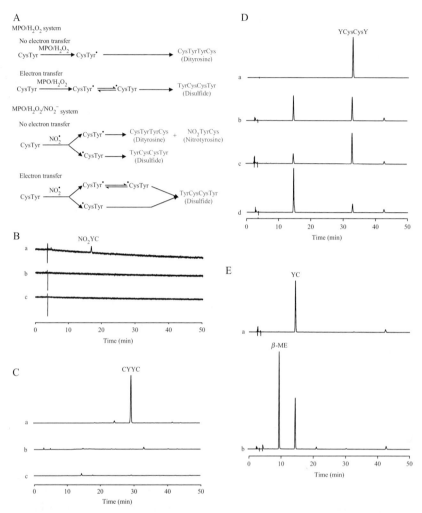

Figure 4.1 HPLC analyses of $MPO/H_2O_2/NO_2^-$-dependent oxidation and nitration products of the YC peptide. The peptide YC (300 μM) was incubated with MPO (30 nM), H_2O_2 (100 μM), and/or $NaNO_2$ (500 μM) in a phosphate buffer (0.1 M, pH 7.4) containing DTPA (100 μM) for 30 min at room temperature, and products were analyzed by HPLC as described in the text. (A) Reaction schemes for MPO system modifying the YC peptide. (B) HPLC traces obtained at 350 nm to monitor nitration products formed from the same incubation mixtures containing (a) the authentic nitrated product [Y(NO_2)C; 0.5 μM]; (b) MPO, YC, and H_2O_2; and (c) MPO, YC, H_2O_2, and $NaNO_2$. (C) HPLC traces obtained using the fluorescence at ex = 290 nm and em = 410 nm for dityrosine-type product. (a) The authentic dityrosyl standard (CYYC; 0.1 μM); (b) MPO, YC, and H_2O_2; and (c) MPO, YC, H_2O_2, and $NaNO_2$. (D) HPLC traces were obtained at 280 nm to monitor the disulfide products formed in the reaction. (a) The authentic disulfide (YCysCysY, 150 μM); (b) MPO, YC, and H_2O_2; (c) MPO, YC, H_2O_2, and $NaNO_2$; and (d) YC and H_2O_2. (E) HPLC traces were obtained at 280 nm to monitor

nitration/oxidation of the YC peptide by a RP-HPLC. The authentic YC peptide (300 μM), YC nitration product [Y(NO$_2$)C; 0.1 μM], dityrosyl product (CYYC; 0.1 μM), and disulfide product (YCysCysY; 150 μM) are eluted at 16, 18, 29, and 32 min, respectively (Figs. 4.1B–4.1E, trace a). The UV-VIS detection at 350 nm is used for nitrated tyrosyl residue, and the fluorescence (ex = 290 nm, em = 410 nm) technique is used for detecting the dityrosyl product of YC (Figs. 4.1B and 4.1C). Incubation of the YC peptide with MPO, H$_2$O$_2$, and NaNO$_2$ in a phosphate buffer (0.1 M, pH 7.4) containing DTPA (100 μM) for 30 min at room temperature failed to yield detectable levels of nitrated tyrosine or the dityrosyl product (Figs. 4.1B and 4.1C). However, upon UV-VIS analysis at 280 nm (for tyrosyl residue), a new product eluting at 32 min was detected (Fig. 4.1D, traces b–d). By comparison with the HPLC profile of the authentic YC disulfide (Fig. 4.1D, trace a), the peak detected at 32 min was attributed to YCysCysY, a disulfide formed from YC. Incubation of YC with MPO/H$_2$O$_2$ also yielded YCys-CysY but no dityrosine peptide (CYYC) (Fig. 4.1C, trace b). In the presence of H$_2$O$_2$ alone, a slight increase in disulfide YCysCysY was detected (Fig. 4.1D, trace d) due to H$_2$O$_2$-dependent oxidation of cysteine to the corresponding disulfide.

Additional confirmation of disulfide formation in this system (i.e., YC/MPO/H$_2$O$_2$/NO$_2^-$) is obtained by incubating the reaction mixture with β-mercaptoethanol. As shown in Fig. 4.1E, the disulfide YCysCysY is reduced back to the parent YC peptide. This result is further confirmed by mass spectrometry. The peak eluting at 32 min in Fig. 4.1D was analyzed by HPLC/ESI-MS. The mass spectral analysis of this peak is identical to the m/z pattern of the authentic YC disulfide (M+H$^+$:649; M+2H$^+$: 325).

To confirm that the inhibition of tyrosine nitration/dimerization of YC induced by MPO/H$_2$O$_2$/NO$_2^-$ is because of the electron transfer to cysteine, we modified the YC peptide with methyl methanethiosulfonate (MMTS), a thiol-blocking reagent (Fig. 4.2A). HPLC analysis of the products from incubation of YCSCH$_3$ with the MPO system showed that a substantial amount of nitrotyrosine and dityrosine type products was formed after thiol in YC was blocked (Fig. 4.2B).

Incubation of other model peptides (YAC, N-acetyl tyrosylalaninylcysteine amide; YAAC, N-acetyl tyrosylalaninylalaninylcysteine amide; and YAAAAC) with MPO/H$_2$O$_2$/NaNO$_2$ also yielded exclusively the corresponding disulfide products with little or no formation of nitrated and dityrosyl products, These data are in contrast to results obtained with

---

the reduction of disulfide by β-mercaptoethanol. (a) The authentic tyrosyl cysteine peptide standard (YC, 300 μM) and (b) MPO, YC, H$_2$O$_2$, and NaNO$_2$, except that β-mercaptoethanol (β-ME) (300 mM) was added to the reaction mixture after 30 min.

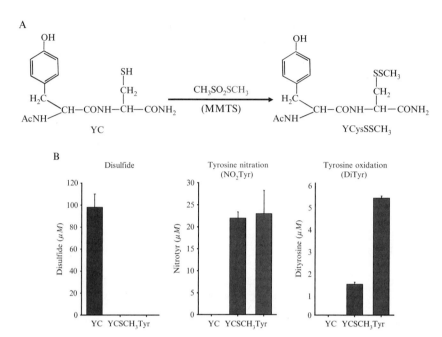

Figure 4.2 HPLC analyses of MPO/H_2O_2/NO_2^--dependent oxidation and nitration products of YC peptide after MMTS modification. The peptide YC, tyrosine, or YCSCH$_3$ (300 μM) was incubated with MPO (30 nM), H_2O_2 (100 μM), and/or NaNO$_2$ (500 μM) in phosphate buffer (0.1 M, pH 7.4) containing DTPA (100 μM) for 30 min at room temperature, and products were analyzed by HPLC as described in the text. (A) The scheme for modification of the YC peptide by MMTS. (B) Graphs showing yields of nitrotyrosine, dityrosine, and disulfide after reaction.

tyrosine alone in the MPO/H_2O_2/NO_2^- system where a substantial amount of nitrotyrosine and dityrosine was formed (Fig. 4.2B).

These results indicate that MPO/H_2O_2/NaNO$_2$-dependent oxidation of YC and related homologs yields the corresponding disulfide as the only major product and that the extent of formation of nitrated tyrosine and dityrosine oxidation products was negligible. The HPLC product analysis clearly revealed that a rapid electron transfer from a tyrosyl radical to cysteine occurs in model peptides and that this reaction leads to form disulfide as a major product.

3.3. Identification of radical intermediates from electron transfer reaction using ESR spin-trapping techniques

To establish a radical transfer mechanism for disulfide formation from the YC, YAC, YAAC, or YAAAAC peptide induced by MPO/H_2O_2, we used a ESR spin-trapping technique to detect the corresponding tyrosyl and thiyl radical intermediates in this reaction (Harman *et al.*, 1984; Saez *et al.*, 1982).

DMPO, a nitrone spin trap, will react with the thiyl radical rapidly ($k = 10^7–10^8\ M^{-1}s^{-1}$) to form DMPO–thiyl radical adducts that exhibit a distinct EPR spectral pattern (Harman et al., 1984; Kalyanaraman et al., 1996; Karoui et al., 1996; Saez et al., 1982). Another spin trap, DBNBS, was used previously to detect a tyrosyl radical generated by metmyoglobin/H_2O_2 (Gunther et al., 2003; Ischiropoulos, 2003). Here, we used both spin traps to establish an electron transfer from the tyrosyl radical to the cysteine forming a thiyl radical, which dimerizes to form a disulfide.

3.3.1. Detection of thiyl radicals formed in MPO/H_2O_2-induced oxidation of tyrosylcysteine model peptides using DMPO

In the presence of DMPO, incubation of YC with MPO and H_2O_2 in a phosphate buffer containing DTPA (100 μM) yields a four-line EPR spectrum (Fig. 4.3A) with the hyperfine coupling constants, $\alpha_N = 15.2$ G and $\alpha_H = 16.2$ G, which are similar to the values reported for the DMPO–glutathionyl adduct ($\alpha_N = 15.2$ G and $\alpha_H = 16.4$ G). Pretreatment of the YC peptide with N-ethylmaleimide, a thiol-blocking reagent, inhibits the DMPO spin trap adduct formation (Fig. 4.3A), indicating that spin adduct formation depends on the thiol group. As superoxide formed during autoxidation of thiols will react with DMPO to yield both DMPO–OOH and DMPO–OH, we performed experiments in the presence of superoxide dismutase (SOD) to exclude the role of oxy radicals. The addition of SOD did not alter the ESR spectrum of the DMPO–SCys adduct (Fig. 4.3A), indicating that there is little or no contribution from both DMPO–OH and DMPO–OOH adducts. Data showed that the ESR spectrum induced by MPO/H_2O_2 is a DMPO–thiyl adduct signal. Oxidation of other peptides (YAC, YAAC, and YAAAAC) in the MPO/H_2O_2 system containing DMPO yields the corresponding DMPO–SCysAY ($\alpha_N = 15.1$ G and $\alpha_H = 16.0$ G), DMPO–SCysAAY ($\alpha_N = 15.1$ G and $\alpha_H = 15.95$ G), and DMPO–SCysAAAAY ($\alpha_N = 15.1$ G and $\alpha_H = 15.75$ G) adducts (Fig. 4.3B). These results indicate that thiyl radicals are formed from these model peptides during MPO/H_2O_2-dependent oxidation reactions.

MPO/H_2O_2 has been shown to effectively oxidize tyrosine to the tyrosyl radical (Eiserich et al., 1998). To verify that the thiyl radical is the product of an electron transfer from tyrosyl radical, we synthesized an YC analog peptide, N-acetyl cysteinylphenylalanine amide (CF), in which phenylalanine replaces tyrosine. It is known that MPO/H_2O_2 cannot oxidize Phe to Phe radical. Oxidation of CF in MPO/H_2O_2 and DMPO does not yield a detectable ESR signal (Fig. 4.4A). This suggests that the tyrosine residue is essential for inducing thiyl radical formation in this system (i.e., MPO/H_2O_2/YC). In the case of including tyrosine in the MPO/H_2O_2/FC reaction, we were able to detect a DMPO–SCysF signal, indicating that the tyrosyl radical formed is responsible for stimulating thiyl radical (Cys•Phe) formation via a hydrogen atom abstraction reaction

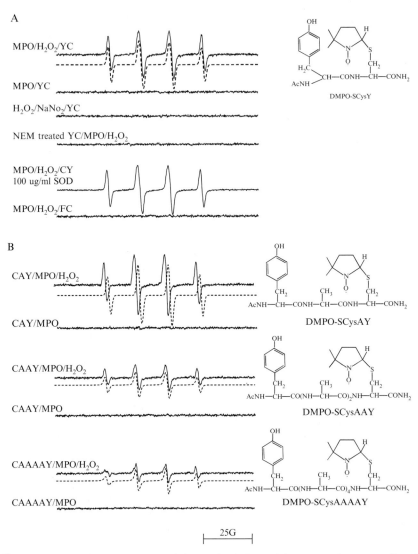

Figure 4.3 DMPO spin trapping of thiyl radicals formed from MPO/H_2O_2-mediated oxidation of model peptides. (A) Incubation mixtures contained the following components as indicated: YC peptide (1 mM), MPO (50 nM), DMPO (150 mM), and H_2O_2 (1 mM) in phosphate buffer (0.1 M, pH 7.4) containing DTPA (100 μM). Immediately after adding all of the components, the reaction mixture was transferred to an ESR capillary tube (100 μl), and spectra were recorded at room temperature. (B) Incubation conditions were the same as just described, except that YAC, YAAC, or YAAAAC was used. Other experimental conditions are as indicated.

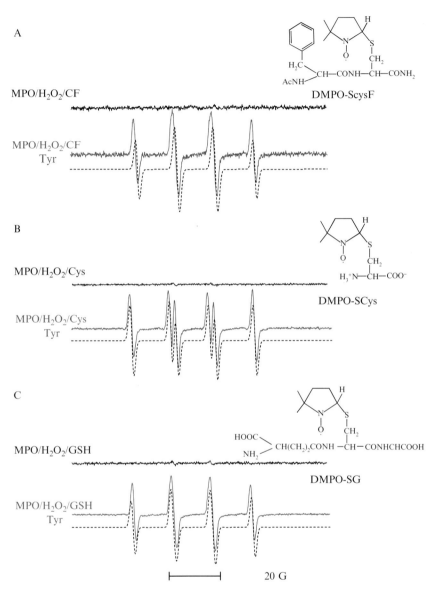

Figure 4.4 Tyrosine enhances the formation of DMPO–thiyl radical adducts formed from MPO/H$_2$O$_2$-dependent oxidation of thiols. (A) Incubation mixtures consisted of CysPhe (CF) (1 mM), MPO (50 nM), DMPO (150 mM), and H$_2$O$_2$ (1 mM) and tyrosine (1 mM) in a phosphate buffer (0.1 M, pH 7.4) containing DTPA (100 μM), and the reaction mixtures were transferred immediately to an ESR capillary tube for analysis at room temperature tube. (B) Same as just described, except that incubation mixtures contained cysteine (1 mM) instead of CF. (C) Same as just described except that GSH (1 mM) was used instead of CF.

from HSCysF (Fig. 4.4A). Incubation of Cys or glutathione (GSH) with MPO/H_2O_2/DMPO in the presence of tyrosine also yields DMPO–SCys ($\alpha_N = 15.05$ G and $\alpha_H = 17.2$ G) (Fig. 4.4B) and DMPO–SGlu adducts (Fig. 4.4B). These results clearly indicate that exogenously added tyrosine or the presence of a tyrosyl residue in the vicinity of a cysteinyl residue stimulates thiyl radical formation in the MPO/H_2O_2-dependent oxidation of cysteine-containing peptides.

3.3.2. Detection of tyrosyl radicals formed in MPO/H_2O_2-induced oxidation of tyrosylcysteine model peptides using DBNBS

To further confirm that the electron transfer mechanism from the tyrosyl radical to cysteine is responsible for thiyl radical formation, we used another spin trap, DBNBS, to investigate the formation of tyrosyl radical in the reaction. The EPR spectrum of the DBNBS–tyrosyl adduct formed from tyrosine oxidation by MPO/H_2O_2 is attributed to a mixture of two spin adducts with different EPR parameters ($\alpha_N = 13.6$ G, $\alpha_N = 13$ G, and $\alpha_H = 6$ G) (not shown). Previous reports indicate that two DBNBS–tyrosyl adducts exist centered at the C1 and C3 or C5 positions of tyrosine (Gunther et al., 2003).

In contrast to tyrosine, incubation of YC with MPO/H_2O_2/DBNBS failed to generate the DBNBS–TyrCys adduct (Fig. 4.5B). However, under these conditions, thiyl radical formation is induced, as shown by DMPO spin trapping (Fig. 4.5A). These results indicate that the presence of a nearby Cys residue dramatically lowers the concentration of tyrosyl radicals formed during MPO/H_2O_2. We surmised that the presence of a sulfhydryl blocking agent will increase the stability of tyrosyl radicals formed from YC and related peptides. The peptide YC was treated with MMTS to block the sulfydryl group. Oxidation of MMTS-treated YC by MPO/H_2O_2 in the presence of DMPO totally abolished formation of the DMPO-SCysY signal, as compared to YC oxidation by MPO/H_2O_2 (Fig. 4.5C). However, oxidation of MMTS-modified YC by MPO/H_2O_2 in the presence of DBNBS spin trap resulted in formation of the corresponding DBNBS–tyrosyl adducts (Fig. 4.5D). These results led us to conclude that there is a rapid electron transfer from the tyrosyl radical to the cysteinyl sulfhydryl group during MPO/H_2O_2-mediated oxidation of YC peptides and that this electron transfer reaction is responsible for the lack of nitration of the tyrosyl residue and enhanced oxidation of the cysteinyl –SH group. Thus, blocking the –SH group enhances tyrosyl nitration of YC and related peptide homologs (Fig. 4.2).

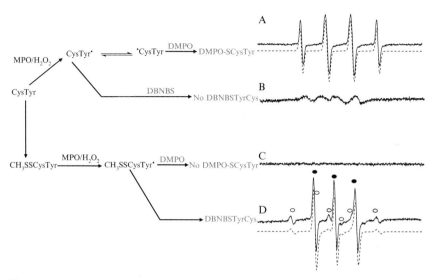

Figure 4.5 The effect of a thiol-blocking agent on DMPO- and DBNBS-adduct formation. (A) YC (1 mM) was mixed with MPO (50 nM), DMPO (150 mM), and H_2O_2 (1 mM), and samples were transferred immediately to an ESR capillary tube for analysis. (B) Same as just described, except that instead of DMPO, the spin trap DBNBS (20 mM) was used for detecting tyrosyl adducts. (C) The dipeptide (YC, 10 mM) was modified using CH_3SSO_3 (MMTS) (20 mM) in phosphate buffer (0.1 M, pH 7.4) containing DTPA (100 μM) at room temperature for 2 h. YCSSCH$_3$ (1 mM) was mixed with MPO (50 nM), DMPO (150 mM), and H_2O_2 (1 mM), and samples were transferred immediately to an ESR capillary tube for analysis. (D) Same as just described, except that instead of DMPO, the spin trap DBNBS (20 mM) was used for detecting tyrosyl adducts. Absorptions of the DBNBS-tyrosyl adduct are marked by solid circles, and open circles denote line positions of an unknown spin adduct.

3.4. Intramolecular versus intermolecular electron transfer mechanism between tyrosyl radical and cysteinyl residue in model peptides: ESR spin trapping and HPLC product analysis

To investigate whether thiyl radicals in YC model peptides are formed from an intermolecular hydrogen abstraction reaction (i.e., TyrO•-HSCys + TyrOH-HSCys → TyrOH-•SCys + TyrOH-HSCys) or from an intramolecular hydrogen abstraction reaction between the tyrosyl radical and the cysteinyl sulfhydryl group (TyrO•-HSCys → TyrOH-•SCysY), we incubated YC with MPO and H_2O_2 in a phosphate buffer containing DTPA and DMPO in the presence of added cysteine. The rationale for this experiment is as follows: If the intermolecular electron transfer is a dominant mechanism, cysteine should react fairly rapidly with the tyrosyl radical of YCs (TyrO•-HSCys) ($k = 10^6$ $M^{-1}s^{-1}$) (Pruetz et al., 1989) forming the cysteinyl radical, which is trapped by DMPO ($k = 10^8$ $M^{-1}s^{-1}$) to form the

DMPO–SCys adduct that exhibits a characteristic EPR spectral pattern (Saez *et al.*, 1982) (Fig. 4.6A). However, if the intramolecular electron transfer is the major pathway, the ESR spectrum should correspond to that of the DMPO–YC thiyl radical adduct (Fig. 4.6A). Figure 4.6B shows that tyrosyl radicals induced thiyl radical formation from exogenously added GSH via an intermolecular hydrogen abstraction reaction in MPO/H_2O_2/Tyr/GSH. Including cysteine in this reaction gave an ESR spectrum from the mixed GS–DMPO adduct (50%) and Cys–DMPO (50%) adduct. However, EPR spin-trapping data showed that there was no change in EPR spectra of the DMPO–SCysY adduct (100%) formed during oxidation of YC by MPO/H_2O_2 in the presence of DMPO and 1 mM cysteine (Fig. 4.6C). Only when the concentration of cysteine exceeded 5 mM was a change detected in the EPR spectrum of the DMPO–SCysY adduct (not shown). Under the same experimental conditions, the addition of cysteine (1 mM) had no effect on DMPO–SCysAY, DMPO–SCysAAY, and DMPO–SCysAAAAY spectra (not shown). These findings suggest that the intramolecular reaction between the tyrosyl radical and the cysteinyl sulfhydryl group (TyrO•-HSCys → TyrOH-•SCys) is quite facile and that this reaction rate must be greater than 10^3–10^4 s^{-1} (Bobrowski *et al.*, 1990, 1992; Pruetz *et al.*, 1980, 1985, 1989; Tanner *et al.*, 1998).

The EPR results were confirmed by HPLC. The reaction mixture was analyzed by HPLC after incubating YC in the presence of MPO/H_2O_2 or MPO/H_2O_2/NO_2^- with and without DMPO for 15 min. A new peak that eluted at 22 min was detected (Figs. 4.7A and 4.7B, trace 2). This particular peak was not detected in the absence of the YC dipeptide from mixtures containing Cys and MPO/H_2O_2/DMPO or MPO/H_2O_2/NO_2^-/DMPO (data not shown). This new peak (Figs. 4.7A and 4.7B, trace 2) was attributed to the DMPO–SCysY nitrone. Consistent with trapping of the •SCysTyr radical, DMPO inhibited the formation of YCysCysY (Fig. 4.7). The structural assignment was further verified by MS (Fig. 4.7C). The addition of 1 mM Cys did not inhibit the signal intensity of the DMPO–SCysY nitrone (Figs. 4.7A and 4.7B, trace 3), suggesting that the intramolecular electron transfer between the tyrosyl radical and Cys in TyrO•-HSCys is faster than the intermolecular reaction between exogenously added Cys and TyrO•-HSCys.

3.5. Biological implications of intramolecular electron transfer reaction in protein nitration and nitrosation reactions

There is increasing evidence for the generation of inflammatory oxidants, including the reactive oxygen/nitrogen species in the progression and pathogenesis of cardiovascular, pulmonary, and neurodegenerative diseases (Baldus *et al.*, 2002; Haddad *et al.*, 1994; Kooy *et al.*, 1997; Leeuwenburgh

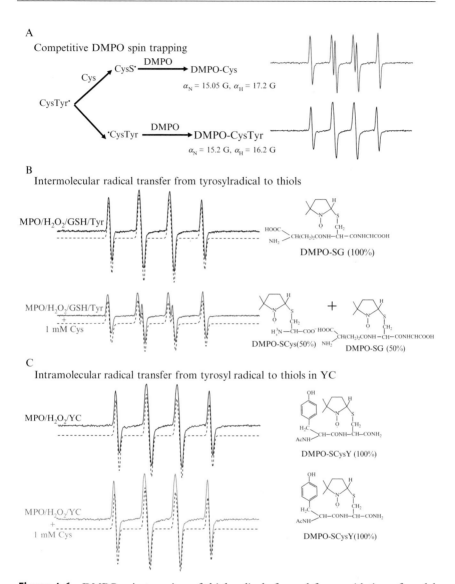

Figure 4.6 DMPO spin trapping of thiyl radicals formed from oxidation of model peptides and tyrosine by MPO/H_2O_2: The effect of added cysteine. (A) The scheme for experimental design for competitive spin trapping. (B) ESR of samples from incubations containing Tyr (1 mM), MPO (50 nM), H_2O_2 (1 mM), DMPO (0.1 M), and GSH (1 mM) in the presence of cysteine (1 mM) in a phosphate buffer (0.1 M, pH 7.4) containing DTPA (100 μM). (C) Incubations contained YC (1 mM), MPO (50 nM), DMPO (150 mM), and H_2O_2 (1 mM) in the presence of cysteine (1 mM) in phosphate buffer (0.1 M, pH 7.4) containing DTPA (100 μM), and samples were transferred immediately to an ESR capillary for analysis.

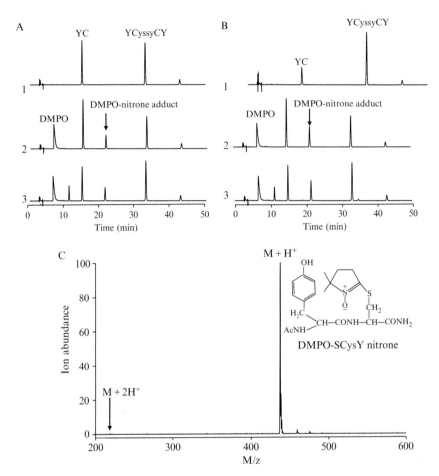

Figure 4.7 HPLC and MS detection and characterization of DMPO–SCysY nitrone. (A) Incubations consist of the following: (trace 1) YC (1 mM), MPO (50 nM), and H$_2$O$_2$ (1 mM) in phosphate buffer (pH 7.4, 0.1 M) containing DTPA (100 μM), (trace 2) same as just described but including DMPO (150 mM), and (trace 3) same as described but including cysteine (1 mM). Following a 15-min incubation period, samples were analyzed by HPLC. (B) All incubations are identical to A, except that they included NaNO$_2$ (1 mM). (C) ESI-MS of DMPO–thiyl adduct isolated at 20 min from A. The molecular ion peak (M+H$^+$) is 437.2.

et al., 1997; Pavlick et al., 2002; Paxinou et al., 2001; Schopfer et al., 2003; Turko and Murad 2002; Wu et al., 1999). Supporting evidence came from identification of the posttranslational modification of protein and lipid oxidation/nitration marker products (Rassaf et al., 2002; White et al., 1994; Zhang et al., 1996). Prominent nitrative, nitrosative, and oxidative reactions in tissues include tyrosine nitration, cysteine and tryptophan nitrosation, tyrosine, tryptophan, histidine, and methionine oxidation and lipid oxidation/nitration (Beckman, 1996; Brennan et al., 2002; Ischiropoulos et al., 1992; Jourd'heuil et al., 2001; MacPherson et al., 2001; Pfeiffer et al., 2001; Pryor and Squadrito, 1995). Factors influencing selective nitration and S-nitrosylation of tyrosyl and cysteinyl residues in protein are not fully known. Several studies have shown that nitration and nitrosation may be controlled by various microenvironmental factors (hydrophobicity, CO_2 levels, membrane oxygen concentration, acidic environment, and amino acid sequence) (Denicola et al., 1998, 2002; Gow et al., 1996, 2004; Ischiropoulos, 1998; Khairutdinov et al., 2000; Liu et al., 1998; Marla et al., 1997; Radi et al., 1999; Shao et al., 2005; Zhang et al., 2001, 2003). Understanding the biophysical/biochemical mechanisms that determine the motif for nitration-sensitive tyrosine residues has physiological and pathophysiological relevance (Ischiropoulos, 2003).

Present data show that a rapid *intramolecular* electron transfer between the tyrosyl radical and the cysteine residue can change the product profile of tyrosine nitration and cysteine oxidation induced by $MPO/H_2O_2/NO_2^-$. Data show that the presence of a cysteinyl group in the peptide chain modulates or negates tyrosine nitration mediated by MPO/H_2O_2-derived nitrating species. Alternatively, the presence of a tyrosyl residue may stimulate intramolecular electron transfer-mediated thiol oxidation in the $MPO/H_2O_2/NO_2^-$ system (Figs. 4.2 and 4.3). However, the presence of a Cys residue adjacent to a tyrosyl group in peptides facilitates the intramolecular radical transfer mechanism between the tyrosyl radical and the Cys residue forming the corresponding thiyl radical and the S-nitrosated product. In addition to forming a disulfide, anaerobic oxidation of YC (300 μM) in the presence of MPO (50 nM), H_2O_2 (50 μM), and NO (1–5 μM/min) for 15 min formed a significant amount of S-nitrosated product (Fig. 4.8). In the absence of MPO, S-nitrosated product was not detected. Therefore, to understand how a protein structure affects nitration and nitrosation profiles, it is imperative to take into consideration the electron transfer mechanism.

In addition to cysteine, intramolecular electron transfer reactions between other redox-sensitive amino acids may also be important in determining the nitration and S-nitrosation profiles of protein. Radiation-induced intramolecular electron transfer reactions between tryptophan and tyrosine were reported many years ago (Bobrowski et al., 1990, 1992; Pruetz et al., 1980, 1985, 1989; Tanner et al., 1998). The rate constants of intramolecular electron transfer reactions for the tryptophan–tyrosine pair (Trp•TyrOH $\rightarrow$ TrpHTyrO•) are in the microsecond timescale ($k = 10^3$–10^6 s^{-1}) (Bobrowski et al., 1990). The insertion of glycines between TrpH and TyrOH slightly changed the intramolecular electron transfer rate but not its efficiency. The carbonate radical or nitrogen dioxide radical derived from peroxynitrite reacts with tryptophan 10 times faster than with tyrosine. This coupled with the rapid electron transfer from the tryptophanyl radical to tyrosine should favor nitrotyrosine formation.

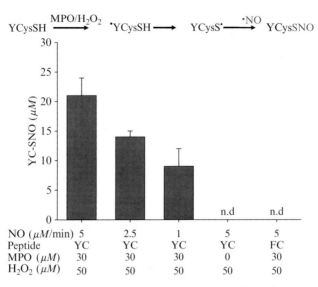

Figure 4.8 MPO/H$_2$O$_2$ induced S-nitrosation of YC peptide in the presence of •NO. Incubation contained YC (150 μM) MPO (20 nM), H$_2$O$_2$ (50 μM), DTPA (100 μM), and DEA-NO (10–50 μM) in 50 mM phosphate (pH 7.4) that had been purged previously with argon for 1 h. After a 10-min incubation, the reaction was stopped by adding iodoacetamide (20 mM). Samples were analyzed by HPLC and eluted by a linear gradient of CH$_3$CN from 10 to 35% in 0.1% TFA for 25 min. YCSNO was eluted around 12 min, and the reaction yield was calculated using the authentic YCSNO as standard.

To confirm this hypothesis, we synthesized a dipeptide, N-acetyl tryptophanyl tyrosine (WY) amide, as a model for investigating the electron transfer mechanism on nitration of the Tryp-Tyr peptide induced by peroxynitrite. The WY peptide (250 μM) is mixed with 0.1 mM peroxynitrite in a phosphate buffer (0.1 M, pH 7.4) containing 100 μM DTPA for 10 min. Products are separated by a RP-HPLC coupled with UV-VIS and fluorescence detectors. The HPLC trace monitored at 280 nm showed that the retention time of the authentic WY peptide was 25 min (Fig. 4.9A). In addition to the peak corresponding to WY, two new peaks appeared at 38 and 41 min in the presence of peroxynitrite. The peak eluting at 38 min showed an absorption peak at 350 nm that shifted to 420 nm after adding NaOH, typical for nitrated tyrosine. The WY(NO$_2$) peptide structure was confirmed by mass spectroscopy. Another peak eluting at 41 min was assigned to a dityosyl type product due to its fluorescence (ex = 290 nm, em = 410 nm) typical for dimeric tyrosine. Figures 4.9A and 4.9B show that the neighboring tryptophan did not inhibit tyrosyl nitration but actually slightly enhanced tyrosine nitration. Although tryptophan reacts with the carbonate radical or nitrogen dioxide radical much faster than tyrosine, a rapid electron transfer from the tryptophanyl radical to tyrosine results in the

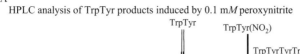
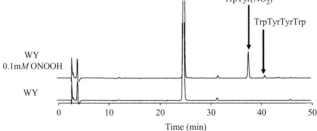
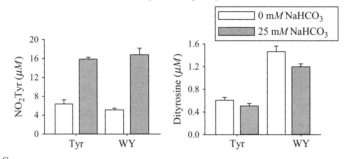
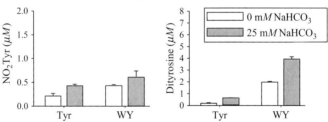

Figure 4.9 WY peptide nitration and oxidation induced by peroxynitrite. (A) HPLC analyses of peroxynitrite-dependent oxidation and nitration products of the WY peptide. The peptide WY (250 μM) was incubated with peroxynitrite (100 μM) in phosphate buffer (0.1 M, pH 7.4) containing DTPA (100 μM) for 30 min at room temperature, and products were analyzed by the same HPLC method as described in the text for the YC peptide. (B) Graphs showing nitration and oxidation yields of the WY peptide. (C) Nitration and oxidation yields induced by SIN-1 (250 μM). The WY peptide was synthesized by a standard Fmoc solid-phase peptide synthesis procedure. The standard WY(NO$_2$) was synthesized using the same procedure as for the nitroYC peptide and purified by HPLC. The standard dityrosine product (WYYW) was synthesized using HRP/H$_2$O$_2$ as described for YCCY synthesis.

formation of nitrated tyrosine and dimeric tyrosine. A similar type of intramolecular electron transfer reaction between methionine radical cation and tyrosine was reported for Met-TyrOH peptide (i.e., Met$^{\bullet+}$-TyrOH → MetH-TyrO$^{\bullet}$) (Bobrowski et al., 1992). Our preliminary study on model peptides indicated that neither tryptophan nor methionine inhibits tyrosine nitration induced by peroxynitrite.

From looking at the peptide sequences of S-nitrosated proteins (Tables 4.2 and 4.3), one can recognize the following trend: S-nitrosation of cysteines occurs more frequently when cysteine residues are present adjacent or closer to a tyrosine residue separated through bond (Table 4.2) (Arstall et al., 1998; Foster and Stamler, 2004; Hao et al., 2004; Ischiropoulos, 2003; Marshall and Stamler, 2002; McVey et al., 1999; Yang et al., 2005; Yao et al., 2004; Yasinska and Sumbayev, 2003) or through space (Table 4.3) (Barrett et al., 2005; Chen et al., 2007; Hao et al., 2006; Huang et al., 2005; Jia et al., 1996; Mallis et al., 2001; Satoh and Lipton, 2007; Stamler, 1992; Tenneti et al., 1997; Uehara et al., 2006). Prominent proteins of relevance to signal transduction (e.g., NF-κB and HIF-1α) share this characteristic at the S-nitrosated target site. Using a proteomic approach, several mitochondrial proteins have been shown to undergo S-nitrosation (Foster and Stamler, 2004). Although the actual S-nitrosation site(s) has not been identified, most S-nitrosated mitochondrial proteins exhibit a Tyr and Cys in close proximity (Table 4.4) (Foster and Stamler, 2004). Several S-nitrosated proteins identified in lipopolysaccharide (LPS)-stimulated RAW 264.7 cells have been found to share a common Tyr and Cys sequence (Gao et al., 2005). These findings

Table 4.2 YC sequences surrounding S-nitrosylated cysteines

Protein	S-Nitrosylation target	Sequence	Reference
Human hypoxia-inducible factor 1α	Cys^{800}	QLTS**Y**^{798}D**C**^{800}EVNAP	57
NF-κB P50	Cys^{62}	FRFR**Y**^{59}V**C**^{62}QGPSH	58
Creatine kinase (B)	Cys^{283}	G**Y**^{279}I LT**C**^{283}PSNLG	59
Creatine kinase (M)	Cys^{283}	G**Y**^{279}VLT**C**^{283}PSNLG	
Parkin	Cys^{263}, Cys^{268}	D**C**^{263}FHL**Y**^{267}**C**^{268}VTRLN	60
Argininosuccinate synthetase	Cys^{132}	RFELS**C**^{132}**Y**^{133}SLAP	61
β-Actin	Not identified	IKEKL**C**^{217}**Y**^{218}VALD	62
Adenylate cyclase (type VI human)	Not identified	RAGGF**C**^{38}TPR**Y**^{42}M GD**C**^{429}**YY C**^{432}VSG A**Y**^{672}VA**C**^{675}ALLVF	63

Table 4.3 Examples of S-nitrosated protein with potential interaction between tyrosine and cysteine

Protein	S-Nitrosated cysteine	Neighboring tyrosine	Distance (Å^3) Cys...Tyr
Caspase 3	Cys^{163}	Tyr^{216}	4.6
NSF	Cys^{91}	Tyr^{83}	10.9
Protein disulfide isomerase	Cys^{39}	Tyr^{33}	9.32
Protein tyrosine phosphatase 1B	Cys^{215}	Tyr^{46}	8.5
Human serum albumin	Cys^{34}	Tyr^{84}	2.7
Protein tyrosine phosphatase SHP-1	Cys^{453}	Tyr^{276}	9.1
Tubulin	Cys$^{295}\alpha$	Tyr$^{319}\alpha$	4.9
	Cys$^{376}\alpha$	Tyr$^{272}\alpha$	7.3
	Cys$^{347}\alpha$	Tyr$^{262}\alpha$	9.8
	Cys$^{12}\beta$	Tyr$^{185}\beta$	8.9
	Cys$^{239}\beta$	Tyr$^{202}\beta$	10.4
	Cys$^{303}\beta$	Tyr$^{202}\beta$	3.5
H Ras-P21 protein	Cys^{118}	Tyr^{141}	8.1
Human hemoglobin	Cys$^{93}\beta$	Tyr$^{42}\alpha$	12

Table 4.4 S-Nitrosated proteins in mitochondria

S-Nitrosylated Protein	CY sequence
Sarcosine dehydrogenase (*Mus musculus*)	CY
Catalase (*M. musculus*)	CxY
Dihydrolipoamide dehydrogenase (*M. musculus*)	CxY
Hydroxymethylglutaryl-CoA synthase (*M. musculus*)	CY
Glutamate oxaloacetate transaminase 2, mitochondrial (*M. musculus*)	CxxxxY
Malate dehydrogenase 2, NAD (mitochondrial)(*M. musculus*)	CxY

raise the possibility that a protein-derived tyrosyl radical could facilitate oxidative S-nitrosation. It has been shown that most sequences contain a cysteine or methionine proximal to nitrotyrosines in neuronal tissues isolated from the mouse brain in an experimental Parkinson's disease (Sacksteder et al., 2006). This was in agreement with our own findings showing that YC nitration in the aqueous phase occurs in the presence of ONOO$^-$ but not in the presence of MPO/H$_2$O$_2$/NO$_2^-$.

3.6. Proteomics and prediction of electron transfer mechanism

Proteomic screening has been used to identify the structural sequence promoting nitration *in vivo* (Sacksteder *et al.*, 2006). These investigators concluded the following. (1) The presence of a positively charged moiety nearer to sites of nitration likely stimulates nitration. (2) Nearby cysteines do not prevent tyrosine nitration *in vivo* in mice treated with MPTP. In that study (Sacksteder *et al.*, 2006), the authors proposed peroxynitrite as the nitrating species. In a previous study describing the electron transfer mechanism, we used $MPO/H_2O_2/NO_2^-$ (not peroxynitrite) where nitration occurs through an interaction between the tyrosyl radical and •NO_2. Unpublished data suggest that authentic peroxynitrite reacts with YC to form the nitrated peptide. At present, we do not know whether nitration with peroxynitrite occurs via a radical or nonradical mechanism. There might be other interpretations as to why previous investigators found intact cysteines adjacent to nitrated tyrosyl residue (Sacksteder *et al.*, 2006). Samples prepared for proteomics screening were treated with Dithiothreitol (DTT) and tributylphosphine (TBP) to reduce disulfides back to thiols. DTT has also been known to transform nitrosothiols to thiols. Scheme 4.1

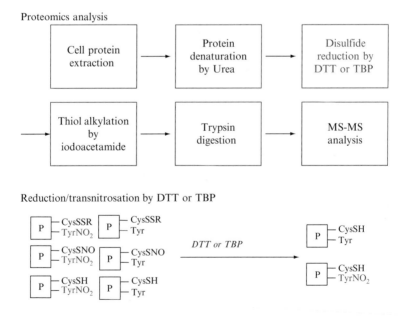

Scheme 4.1 A hypothetical example showing how DTT or TBP modifies disulfide and the nitrosated protein during sample preparation for proteomics screening.

shows a hypothetical example in which DTT and TBP treatment of post-translationally modified protein (nitrated tyrosine and nitrosated/oxidized cysteine) generates a nitrated protein with cysteinyl residues.

3.7. Overall conclusions: Future perspectives

By and large, the approach described in this study provides the experimental strategy for exploring electron transfer reactions in the nitration and nitrosation of peptides. Understanding the role of the electron transfer mechanism in the nitrosation of cysteine and the nitration of tyrosine will greatly facilitate studies using site-directed mutagenesis. These studies highlight a potentially novel possibility: the ultimate site of modification (i.e., buried cysteine residue) may be distant from the initial site of oxidation of a more surface-exposed tyrosine residue and an intramolecular electron transfer mechanism may facilitate this modification. Elucidating the role of structural motifs in electron transfer-dependent nitrosation/nitration reactions of mutated proteins should help define the chemical biology of how specific mutations lead to pathophysiologic processes. Furthermore, acquiring a detailed knowledge of sequence specificity for nitration and nitrosation reactions may assist in mapping important functional regions of proteins that mediate redox signaling. Emerging data support these basic concepts. For example, data consistent with electron transfer between tyrosyl radicals and cysteines have been reported to govern the selectivity of Tyr nitration and Cys oxidation of the sarco/endoplasmic reticulum Ca-ATPase (SERCA) (Sharov *et al.*, 2002). This 110-kDa membrane protein plays a major functional role in muscle contraction and relaxation through the ATP-dependent transport of cytosolic calcium into the sarcoplasmic reticulum. Age-dependent oxidation and nitration profiles of SERCA revealed efficient modification of several Cys residues with low solvent accessibility, supporting the potential importance of these reactions in the pathophysiology (i.e., impaired vessel function) of chronic degenerative diseases. S-nitrosation of Parkin, a ubiquitin E3 ligase that helps target certain proteins for proteasomal degradation, was identified in brains of people with Parkinson's disease and in mice treated with a chemical inducing Parkinson's-like neurodegeneration (Chung *et al.*, 2004; Yao *et al.*, 2004). The peptide sequences of S-nitrosated proteins such as Parkin have a tyrosine residue closer to the S-nitrosated target cysteine. It has been shown that the replacement of tyrosyl residues with fluorotyrosines greatly hinders the electron transport process in biological systems (Seyedsayamdost *et al.*, 2006). It is likely that a similar strategy will enable us to fully understand the significance of an electron transfer mechanism in protein nitration/nitrosation.

ACKNOWLEDGMENTS

This work was supported by National Institutes of Health Grant HL63119. The authors thank Drs. Victor Darley-Usmar and Bruce Freeman for their insightful comments on the role of electron transfer in redox signaling.

REFERENCES

Arstall, M. A., Bailey, C., Gross, W. L., Bak, M., Balligand, J. L., and Kelly, R. A. (1998). Reversible S-nitrosation of creatine kinase by nitric oxide in adult rat ventricular myocytes. *J. Mol. Cell Cardiol.* **30,** 979–988.

Aubert, C., Vos, M. H., Mathis, P., Eker, A. P., and Brettel, K. (2000). Intraprotein radical transfer during photoactivation of DNA photolyase. *Nature* **405,** 586–590.

Baldus, S., Eiserich, J., Brennan, M., Jackson, R., Alexander, C., and Freeman, B. (2002). Spatial mapping of pulmonary and vascular nitrotyrosine reveals the pivotal role of myeloperoxidase as a catalyst for tyrosine nitration in inflammatory diseases. *Free Radic. Biol. Med.* **33,** 1010.

Barrett, D. M., Black, S. M., Todor, H., Schmidt-Ullrich, R. K., Dawson, K. S., and Mikkelsen, R. B. (2005). Inhibition of protein-tyrosine phosphatases by mild oxidative stresses is dependent on S-nitrosylation. *J. Biol. Chem.* **280,** 14453–14461.

Beckman, J. S. (1996). Oxidative damage and tyrosine nitration from peroxynitrite. *Chem. Res. Toxicol.* **9,** 836–844.

Bobrowski, K., Wierzchowski, K. L., Holcman, J., and Ciurak, M. (1990). Intramolecular electron transfer in peptides containing methionine, tryptophan and tyrosine: A pulse radiolysis study. *Int. J. Radiat. Biol.* **57,** 919–932.

Bobrowski, K., Wierzchowski, K. L., Holcman, J., and Ciurak, M. (1992). Pulse radiolysis studies of intramolecular electron transfer in model peptides and proteins. IV. Met/S:. Br→Tyr/O. radical transformation in aqueous solution of H-Tyr-(Pro)n-Met-OH peptides. *Int. J. Radiat. Biol.* **62,** 507–516.

Brennan, M. L., Wu, W., Fu, X., Shen, Z., Song, W., Frost, H., Vadseth, C., Narine, L., Lenkiewicz, E., Borchers, M. T., Lusis, A. J., Lee, J. J., et al. (2002). A tale of two controversies: Defining both the role of peroxidases in nitrotyrosine formation in vivo using eosinophil peroxidase and myeloperoxidase-deficient mice, and the nature of peroxidase-generated reactive nitrogen species. *J. Biol. Chem.* **277,** 17415–17427.

Bryan, N. S., Rassaf, T., Maloney, R. E., Rodriguez, C. M., Saijo, F., Rodriguez, J. R., and Feelisch, M. (2004). Cellular targets and mechanisms of nitros(yl)ation: An insight into their nature and kinetics *in vivo*. *Proc. Natl. Acad. Sci. USA* **101,** 4308–4313.

Chen, Y. Y., Huang, Y. F., Khoo, K. H., and Meng, T. C. (2007). Mass spectrometry-based analyses for identifying and characterizing S-nitrosylation of protein tyrosine phosphatases. *Methods* **42,** 243–249.

Chung, K. K., Thomas, B., Li, X., Pletnikova, O., Troncoso, J. C., Marsh, L., Dawson, V. L., and Dawson, T. M. (2004). S-nitrosylation of parkin regulates ubiquitination and compromises parkin's protective function. *Science* **304,** 1328–13231.

Denicola, A., Batthyany, C., Lissi, E., Freeman, B. A., Rubbo, H., and Radi, R. (2002). Diffusion of nitric oxide into low density lipoprotein. *J. Biol. Chem.* **277,** 932–936.

Denicola, A., Souza, J. M., and Radi, R. (1998). Diffusion of peroxynitrite across erythrocyte membranes. *Proc. Natl. Acad. Sci. USA* **95,** 3566–3571.

Eiserich, J. P., Hristova, M., Cross, C. E., Jones, A. D., Freeman, B. A., Halliwell, B., and van der Vliet, A. (1998). Formation of nitric oxide-derived inflammatory oxidants by myeloperoxidase in neutrophils. *Nature* **391,** 393–397.

Espey, M. G., Thomas, D. D., Miranda, K. M., and Wink, D. A. (2002). Focusing of nitric oxide mediated nitrosation and oxidative nitrosylation as a consequence of reaction with superoxide. *Proc. Natl. Acad. Sci. USA* **99,** 11127–11132.

Foster, M. W., and Stamler, J. S. (2004). New insights into protein S-nitrosylation: Mitochondria as a model system. *J. Biol. Chem.* **279,** 25891–25897.

Gao, C., Guo, H., Wei, J., Mi, Z., Wai, P. Y., and Kuo, P. C. (2005). Identification of S-nitrosylated proteins in endotoxin-stimulated RAW264.7 murine macrophages. *Nitric Oxide* **12,** 121–126.

Goss, S. P., Hogg, N., and Kalyanaraman, B. (1999). The effect of alpha-tocopherol on the nitration of gamma-tocopherol by peroxynitrite. *Arch. Biochem. Biophys.* **363,** 333–340.

Gow, A., Duran, D., Thom, S. R., and Ischiropoulos, H. (1996). Carbon dioxide enhancement of peroxynitrite-mediated protein tyrosine nitration. *Arch. Biochem. Biophys.* **333,** 42–48.

Gow, A. J., Farkouh, C. R., Munson, D. A., Posencheg, M. A., and Ischiropoulos, H. (2004). Biological significance of nitric oxide-mediated protein modifications. *Am. J. Physiol. Lung Cell Mol. Physiol.* **287,** L262–L268.

Greenacre, S. A., and Ischiropoulos, H. (2001). Tyrosine nitration: Localisation, quantification, consequences for protein function and signal transduction. *Free Radic. Res.* **34,** 541–581.

Gunther, M. R., Tschirret-Guth, R. A., Lardinois, O. M., and Ortiz de Montellano, P. R. (2003). Tryptophan-14 is the preferred site of DBNBS spin trapping in the self-peroxidation reaction of sperm whale metmyoglobin with a single equivalent of hydrogen peroxide. *Chem. Res. Toxicol.* **16,** 652–660.

Haddad, I. Y., Pataki, G., Hu, P., Galliani, C., Beckman, J. S., and Matalon, S. (1994). Quantitation of nitrotyrosine levels in lung sections of patients and animals with acute lung injury. *J. Clin. Invest.* **94,** 2407–2413.

Hao, G., Derakhshan, B., Shi, L., Campagne, F., and Gross, S. S. (2006). SNOSID, a proteomic method for identification of cysteine S-nitrosylation sites in complex protein mixtures. *Proc. Natl. Acad. Sci. USA* **103,** 1012–1017.

Hao, G., Xie, L., and Gross, S. S. (2004). Argininosuccinate synthetase is reversibly inactivated by S-nitrosylation *in vitro* and *in vivo*. *J. Biol. Chem.* **279,** 36192–361200.

Harman, L. S., Mottley, C., and Mason, R. P. (1984). Free radical metabolites of L-cysteine oxidation. *J. Biol. Chem.* **259,** 5606–5611.

Huang, Y., Man, H. Y., Sekine-Aizawa, Y., Han, Y., Juluri, K., Luo, H., Cheah, J., Lowenstein, C., Huganir, R. L., and Snyder, S. H. (2005). S-nitrosylation of N-ethylmaleimide sensitive factor mediates surface expression of AMPA receptors. *Neuron* **46,** 533–540.

Ischiropoulos, H. (1998). Biological tyrosine nitration: A pathophysiological function of nitric oxide and reactive oxygen species. *Arch. Biochem. Biophys.* **356,** 1–11.

Ischiropoulos, H. (2003). Biological selectivity and functional aspects of protein tyrosine nitration. *Biochem. Biophys. Res. Commun.* **305,** 776–783.

Ischiropoulos, H., Zhu, L., Chen, J., Tsai, M., Martin, J. C., Smith, C. D., and Beckman, J. S. (1992). Peroxynitrite-mediated tyrosine nitration catalyzed by superoxide dismutase. *Arch. Biochem. Biophys.* **298,** 431–437.

Jia, L., Bonaventura, C., Bonaventura, J., and Stamler, J. S. (1996). S-nitrosohaemoglobin: A dynamic activity of blood involved in vascular control. *Nature* **380,** 221–226.

Jourd'heuil, D., Jourd'heuil, F. L., Kutchukian, P. S., Musah, R. A., Wink, D. A., and Grisham, M. B. (2001). Reaction of superoxide and nitric oxide with peroxynitrite: Implications for peroxynitrite-mediated oxidation reactions *in vivo*. *J. Biol. Chem.* **276,** 28799–28805.

Kalyanaraman, B., Karoui, H., Singh, R. J., and Felix, C. C. (1996). Detection of thiyl radical adducts formed during hydroxyl radical- and peroxynitrite-mediated oxidation of thiols: A high resolution ESR spin-trapping study at Q-band (35 GHz). *Anal. Biochem.* **241**, 75–81.

Karoui, H., Hogg, N., Frejaville, C., Tordo, P., and Kalyanaraman, B. (1996). Characterization of sulfur-centered radical intermediates formed during the oxidation of thiols and sulfite by peroxynitrite: ESR-spin trapping and oxygen uptake studies. *J. Biol. Chem.* **271**, 6000–6009.

Khairutdinov, R. F., Coddington, J. W., and Hurst, J. K. (2000). Permeation of phospholipid membranes by peroxynitrite. *Biochemistry* **39**, 14238–14249.

Kooy, N. W., Lewis, S. J., Royall, J. A., Ye, Y. Z., Kelly, D. R., and Beckman, J. S. (1997). Extensive tyrosine nitration in human myocardial inflammation: Evidence for the presence of peroxynitrite. *Crit. Care Med.* **25**, 812–819.

Lanone, S., Manivet, P., Callebert, J., Launay, J.-M., Payen, D., Aubier, M., Boczkowski, J., and Mebazaa, A. (2002). Inducible nitric oxide synthase (NOS2) expressed in septic patients is nitrated on selected tyrosine residues: Implications for enzymic activity. *Biochem. J.* **366**, 399–404.

Leeuwenburgh, C., Hardy, M. M., Hazen, S. L., Wagner, P., Oh-ishi, S., Steinbrecher, U. P., and Heinecke, J. W. (1997). Reactive nitrogen intermediates promote low density lipoprotein oxidation in human atherosclerotic intima. *J. Biol. Chem.* **272**, 1433–1436.

Liu, X., Miller, M. J., Joshi, M. S., Thomas, D. D., and Lancaster, J. R., Jr. (1998). Accelerated reaction of nitric oxide with O2 within the hydrophobic interior of biological membranes. *Proc. Natl. Acad. Sci. USA* **95**, 2175–2179.

MacPherson, J. C., Comhair, S. A., Erzurum, S. C., Klein, D. F., Lipscomb, M. F., Kavuru, M. S., Samoszuk, M. K., and Hazen, S. L. (2001). Eosinophils are a major source of nitric oxide-derived oxidants in severe asthma: Characterization of pathways available to eosinophils for generating reactive nitrogen species. *J. Immunol.* **166**, 5763–5772.

Mallis, R. J., Buss, J. E., and Thomas, J. A. (2001). Oxidative modification of H-ras: S-thiolation and S-nitrosylation of reactive cysteines. *Biochem. J.* **355**, 145–153.

Marla, S. S., Lee, J., and Groves, J. T. (1997). Peroxynitrite rapidly permeates phospholipid membranes. *Proc. Natl. Acad. Sci. USA* **94**, 14243–14248.

Marshall, H. E., and Stamler, J. S. (2002). Nitrosative stress-induced apoptosis through inhibition of NF-kappa B. *J. Biol. Chem.* **277**, 34223–34228.

McVey, M., Hill, J., Howlett, A., and Klein, C. (1999). Adenylyl cyclase, a coincidence detector for nitric oxide. *J. Biol. Chem.* **274**, 18887–18892.

Newman, D. K., Hoffman, S., Kotamraju, S., Zhao, T., Wakim, B., Kalyanaraman, B., and Newman, P. J. (2002). Nitration of PECAM-1 ITIM tyrosines abrogates phosphorylation and SHP-2 binding. *Biochem. Biophys. Res. Commun.* **296**, 1171–1179.

Pavlick, K. P., Laroux, F. S., Fuseler, J., Wolf, R. E., Gray, L., Hoffman, J., and Grisham, M. B. (2002). Role of reactive metabolites of oxygen and nitrogen in inflammatory bowel disease. *Free Radic. Biol. Med.* **33**, 311–322.

Paxinou, E., Chen, Q., Weisse, M., Giasson, B. I., Norris, E. H., Rueter, S. M., Trojanowski, J. Q., Lee, V. M.-Y., and Ischiropoulos, H. (2001). Induction of alpha-synuclein aggregation by intracellular nitrative insult. *J. Neurosci.* **21**, 8053–8061.

Pfeiffer, S., Lass, A., Schmidt, K., and Mayer, B. (2001). Protein tyrosine nitration in mouse peritoneal macrophages activated *in vitro* and *in vivo*: Evidence against an essential role of peroxynitrite. *FASEB J.* **15**, 2355–2364.

Prutz, W. A., Butler, J., Land, E. J., and Swallow, A. J. (1980). Direct demonstration of electron transfer between tryptophan and tyrosine in proteins. *Biochem. Biophys. Res. Commun.* **96**, 408–411.

Prutz, W. A., Butler, J., Land, E. J., and Swallow, A. J. (1989). The role of sulphur peptide functions in free radical transfer: A pulse radiolysis study. *Int. J. Radiat. Biol.* **55,** 539–556.

Prutz, W. A., Moenig, H., Butler, J., and Land, E. J. (1985). Reactions of nitrogen dioxide in aqueous model systems: Oxidation of tyrosine units in peptides and proteins. *Arch. Biochem. Biophys.* **243,** 125–134.

Pryor, W. A., and Squadrito, G. L. (1995). The chemistry of peroxynitrite: A product from the reaction of nitric oxide with superoxide. *Am. J. Physiol.* **68,** L699–L722.

Radi, R., Denicola, A., and Freeman, B. A. (1999). Peroxynitrite reactions with carbon dioxide-bicarbonate. *Methods Enzymol.* **301,** 353–367.

Rassaf, T., Bryan, N. S., Kelm, M., and Feelisch, M. (2002). Concomitant presence of N-nitroso and S-nitroso proteins in human plasma. *Free Radic. Biol. Med.* **33,** 1590–1596.

Rubbo, H., Radi, R., Trujillo, M., Telleri, R., Kalyanaraman, B., Barnes, S., Kirk, M., and Freeman, B. A. (1994). Nitric oxide regulation of superoxide and peroxynitrite-dependent lipid peroxidation: Formation of novel nitrogen-containing oxidized lipid derivatives. *J. Biol. Chem.* **269,** 26066–26075.

Sacksteder, C. A., Qian, W. J., Knyushko, T. V., Wang, H., Chin, M. H., Lacan, G., Melega, W. P., Camp, D. G., 2nd, Smith, R. D., Smith, D. J., Squier, T. C., and Bigelow, D. J. (2006). Endogenously nitrated proteins in mouse brain: Links to neurodegenerative disease. *Biochemistry* **45,** 8009–8022.

Saez, G., Thornalley, P. J., Hill, H. A., Hems, R., and Bannister, J. V. (1982). The production of free radicals during the autoxidation of cysteine and their effect on isolated rat hepatocytes. *Biochim. Biophys. Acta* **719,** 24–31.

Satoh, T., and Lipton, S. A. (2007). Redox regulation of neuronal survival mediated by electrophilic compounds. *Trends Neurosci.* **30,** 37–45.

Schopfer, F. J., Baker, P. R., and Freeman, B. A. (2003). NO-dependent protein nitration: A cell signaling event or an oxidative inflammatory response? *Trends Biochem. Sci.* **28,** 646–654.

Seyedsayamdost, M. R., Reece, S. Y., Nocera, D. G., and Stubbe, J. (2006). Mono-, di-, tri-, and tetra-substituted fluorotyrosines: New probes for enzymes that use tyrosyl radicals in catalysis. *J. Am. Chem. Soc.* **128,** 1569–1579.

Shao, B., Bergt, C., Fu, X., Green, P., Voss, J. C., Oda, M. N., Oram, J. F., and Heinecke, J. W. (2005). Tyrosine 192 in apolipoprotein A-I is the major site of nitration and chlorination by myeloperoxidase, but only chlorination markedly impairs ABCA1-dependent cholesterol transport. *J. Biol. Chem.* **280,** 5983–5993.

Sharov, V. S., Galeva, N. A., Knyushko, T. V., Bigelow, D. J., Williams, T. D., and Schoneich, C. (2002). Two-dimensional separation of the membrane protein sarcoplasmic reticulum Ca-ATPase for high-performance liquid chromatography-tandem mass spectrometry analysis of posttranslational protein modifications. *Anal. Biochem.* **308,** 328–335.

Souza, J. M., Daikhin, E., Yudkoff, M., Raman, C. S., and Ischiropoulos, H. (1999). Factors determining the selectivity of protein tyrosine nitration. *Arch. Biochem. Biophys.* **371,** 169–178.

Stamler, J. S., Simon, D. I., Osborne, J. A., Mullins, M. E., Jaraki, O., Michel, T., Singel, D. J., and Loscalzo, J. (1992). S-nitrosylation of proteins with nitric oxide: Synthesis and characterization of biologically active compounds. *Proc. Natl. Acad. Sci. USA* **89,** 444–448.

Tanner, C., Navaratnam, S., and Parsons, B. J. (1998). Intramolecular electron transfer in the dipeptide, histidyltyrosine: A pulse radiolysis study. *Free Radic. Biol. Med.* **24,** 671–678.

Tenneti, L., D'Emilia, D. M., and Lipton, S. A. (1997). Suppression of neuronal apoptosis by S-nitrosylation of caspases. *Neurosci. Lett.* **236,** 139–142.

Turko, I. V., and Murad, F. (2002). Protein nitration in cardiovascular diseases. *Pharmacol. Rev.* **54,** 619–634.

Uehara, T., Nakamura, T., Yao, D., Shi, Z. Q., Gu, Z., Ma, Y., Masliah, E., Nomura, Y., and Lipton, S. A. (2006). S-nitrosylated protein-disulphide isomerase links protein misfolding to neurodegeneration. *Nature* **441**, 513–517.

van Dalen, C. J., Winterbourn, C. C., Senthilmohan, R., and Kettle, A. J. (2000). Nitrite as a substrate and inhibitor of myeloperoxidase: Implications for nitration and hypochlorous acid production at sites of inflammation. *J. Biol. Chem.* **275**, 11638–11644.

van der Vliet, A., Eiserich, J. P., Halliwell, B., and Cross, C. E. (1997). Formation of reactive nitrogen species during peroxidase-catalyzed oxidation of nitrite: A potential additional mechanism of nitric oxide-dependent toxicity. *J. Biol. Chem.* **272**, 7617–7625.

von Freyberg, B., and Braun, W. J. (1991). Efficient search for all low energy conformations of polypeptides by Monte Carlo methods. *Comput. Chem.* **12**, 1065.

White, C. R., Brock, T. A., Chang, L. Y., Crapo, J., Briscoe, P., Ku, D., Bradley, W. A., Gianturco, S. H., Gore, J., Freeman, B. A., and Tarpey, M. M. (1994). Superoxide and peroxynitrite in atherosclerosis. *Proc. Natl. Acad. Sci. USA* **91**, 1044–1048.

Wu, W., Chen, Y., and Hazen, S. L. (1999). Eosinophil peroxidase nitrates protein tyrosyl residues: Implications for oxidative damage by nitrating intermediates in eosinophilic inflammatory disorders. *J. Biol. Chem.* **274**, 25933–25944.

Yang, Y., and Loscalzo, J. (2005). S-nitrosoprotein formation and localization in endothelial cells. *Proc. Natl. Acad. Sci. USA* **102**, 117–122.

Yao, D., Gu, Z., Nakamura, T., Shi, Z. Q., Ma, Y., Gaston, B., Palmer, L. A., Rockenstein, E. M., Zhang, Z., Masliah, E., Uehara, T., and Lipton, S. A. (2004). Nitrosative stress linked to sporadic Parkinson's disease: S-nitrosylation of parkin regulates its E3 ubiquitin ligase activity. *Proc. Natl. Acad. Sci. USA* **101**, 10810–10814.

Yasinska, I. M., and Sumbayev, V. V. (2003). S-nitrosation of Cys-800 of HIF-1alpha protein activates its interaction with p300 and stimulates its transcriptional activity. *FEBS Lett.* **549**, 105–109.

Zhang, H., Bhargava, K., Keszler, A., Feix, J., Hogg, N., Joseph, J., and Kalyanaraman, B. (2003). Transmembrane nitration of hydrophobic tyrosyl peptides: Localization, characterization, mechanism of nitration, and biological implications. *J. Biol. Chem.* **278**, 8969–8978.

Zhang, H., Joseph, J., Feix, J., Hogg, N., and Kalyanaraman, B. (2001). Nitration and oxidation of a hydrophobic tyrosine probe by peroxynitrite in membranes: Comparison with nitration and oxidation of tyrosine by peroxynitrite in aqueous solution. *Biochemistry* **40**, 7675–7686.

Zhang, H., Xu, Y., Joseph, J., and Kalyanaraman, B. (2005). Intramolecular electron transfer between tyrosyl radical and cysteine residue inhibits tyrosine nitration and induces thiyl radical formation in model peptides treated with myeloperoxidase, H_2O_2, and NO_2^-: EPR spin trapping studies. *J. Biol. Chem.* **280**, 40684–40698.

Zhang, Y. Y., Xu, A. M., Nomen, M., Walsh, M., Keaney, J. F., Jr., and Loscalzo, J. (1996). Nitrosation of tryptophan residue(s) in serum albumin and model dipeptides: Biochemical characterization and bioactivity. *J. Biol. Chem.* **271**, 14271–14279.

CHAPTER FIVE

Protein Thiol Modification by Peroxynitrite Anion and Nitric Oxide Donors

Lisa M. Landino

Contents

1. Introduction 96
2. Protein Cysteine Oxidation by ONOO⁻ 97
3. Methodology to Detect Protein Thiol Modification 97
4. Iodoacetamide Labeling of Proteins after Oxidant Treatment 98
5. HPLC Separation of Fluorescein-Labeled Peptides 99
6. Detection of Interchain Disulfides by Western Blot 100
7. Repair of Protein Disulfides by Thioredoxin Reductase and Glutaredoxin Systems 101
8. Thioredoxin Reductase Repair of Protein Disulfides 102
9. Quantitation of Protein Disulfides by Measuring NADPH Oxidation 103
10. Quantitation of Total Cysteine Oxidation Using DTNB 103
11. Glutaredoxin/Glutathione (GSH) Reductase Repair of Protein Disulfides 104
12. Detection of Protein *S*-Glutathionylation 105
13. Thiol/disulfide Exchange with Oxidized Glutathione 105
    13.1. Quantitation of GSH following tubulin thiol-disulfide exchange with GSSG 106
14. Protein Thiol Modification by Nitric Oxide Donors 106
    14.1. Detection of tubulin *S*-nitrosation 107
15. Conclusions 108
References 108

Abstract

Oxidation and modification of protein cysteines can have profound effects on protein structure and function. Using tubulin and microtubule-associated proteins (MAP) tau and MAP2 as examples, this chapter summarizes methods employed to characterize total cysteine modification using thiol-specific reagent 5-iodoacetamido-fluorescein labeling. Western blot analysis of

Department of Chemistry, The College of William and Mary, Williamsburg, Virginia peroxynitrite-damaged tubulin under nonreducing conditions reveals the formation of higher molecular weight dimers and tetramers. Disulfides in microtubule proteins are substrates for both the thioredoxin reductase system and the glutaredoxin/glutathione reductase system. The yield of disulfides formed by peroxynitrite anion is quantitated by monitoring the oxidation of NADPH, a cofactor required by the thioredoxin reductase system. Treatment of proteins with *S*-nitrosothiols, including *S*-nitrosoglutathione and *S*-nitroso-*N*-acetyl penicillamine, can yield either disulfides or protein *S*-nitrosation. In the case of tubulin, both types of cysteine modification were detected.

1. INTRODUCTION

Regulation of proteins via oxidation and/or modification of cysteines is a topic of considerable interest (Hess *et al.*, 2001; Hogg, 2003; Lipton *et al.*, 2002). Rather than view such regulation as an "on/off" switch that converts between a thiol (RSH) and a disulfide (RSSR), there is increasing evidence of more subtle responses that depend on the oxidation state and modification by nitric oxide (NO) and glutathione (GSH)(Kim *et al.*, 2002). Like phosphorylation, the covalent modification of cysteines by NO or GSH alters protein structure and, consequently, interactions with downstream signaling targets.

Transcription factors are among the most extensively studied "redox modulators" where protein function depends on the oxidation state of critical cysteines. The *Escherichia coli* transcription factor OxyR is activated by oxidants such as H_2O_2 to transcribe genes involved in the oxidative stress response (Demple, 1998; Kim *et al.*, 2002). A single cysteine of OxyR may be reduced, oxidized to a sulfenic acid (RSOH), *S*-nitrosated (RSNO), or *S*-glutathionylated (RSSG), and each form has distinct transcriptional activation activity (Kim *et al.*, 2002).

Redox regulation of actin, a ubiquitous cytoskeletal protein, has also been studied extensively in recent years. Cysteine oxidation and modification of nonmuscle β/γ actin lead to loss of polymerization activity (Lassing *et al.*, 2007). A recent crystal structure of oxidized β-actin, a protein with six cys residues, reveals a homodimer, composed of two β-actin subunits linked via a disulfide (Lassing *et al.*, 2007). Although a disulfide forms via cys 374 of each subunit, cys 272 has been identified as the most reactive cysteine. The reactivity of protein thiols toward oxidants depends on the protein microenvironment and oxidant accessibility. For example, a thiolate (RS^-) rather than a thiol is more susceptible to oxidation by H_2O_2. Thus knowledge of the three-dimensional structure of a protein allows researchers to predict reactivity. Cys 272 of actin is solvent accessible and adjacent to a glutamic acid that may facilitate deprotonation of cys 272. From the crystal structure, it appears that cys 272 is oxidized to a sulfinic acid (RSO_2^-).

2. Protein Cysteine Oxidation by ONOO−

Our interest in protein thiols stems from recent work with brain-derived cytoskeletal proteins. Tubulin and several microtubule-associated proteins (MAPs), including tau and microtubule-associated protein-2 (MAP2), are oxidized readily by ONOO− *in vitro*, and cysteine thiols of microtubule proteins, rather than other amino acids, are most susceptible to oxidation (Landino *et al.*, 2002, 2004a,d). Indeed, tubulin thiol oxidation correlated with inhibition of microtubule polymerization. Tubulin is unique in that it contains a total of 20 reduced cysteines per protein dimer, 12 in α-tubulin and 8 in β-tubulin (Nogales *et al.*, 1998). ONOO−-induced tubulin disulfides are at least partially responsible for the inhibition of microtubule polymerization because the addition of disulfide-reducing agents, including dithiothreitol (DTT) and tris(2-carboxyethyl)phosphine hydrochloride (TCEP), restored much of the polymerization activity that was lost following ONOO− addition (Landino *et al.*, 2002).

Additional experiments showed that sodium bicarbonate protected the microtubule protein from ONOO−-induced inhibition of polymerization. Carbon dioxide, derived from bicarbonate, reacts rapidly with ONOO− to form the nitrosoperoxocarboxylate adduct (ONOOCO$_2^-$), which decomposes rapidly to carbonate and nitrogen dioxide radicals (Denicola *et al.*, 1996; Lymar and Hurst, 1995). Compared to ONOO− alone, the radical species formed from ONOOCO$_2^-$ induce more tyrosine nitration and, consequently, less thiol oxidation. Tyrosine nitration of porcine tubulin by 1 mM ONOO− increased approximately twofold when 10 mM sodium bicarbonate was present, whereas the extent of cysteine oxidation decreased from 7.5 mol cys/mole tubulin to 6.3 mol cys/mole tubulin (Landino *et al.*, 2002). These results confirm that cysteine oxidation of tubulin by ONOO−, rather than tyrosine nitration, is the primary mechanism of inhibition of microtubule polymerization. Therefore, to confirm a role for cysteine oxidation when using ONOO−, it is important to perform a control using bicarbonate and assess enzyme/protein activity relative to samples without bicarbonate.

3. Methodology to Detect Protein Thiol Modification

Our approach to studying protein thiol modifications requires simple equipment that is available in a typical biochemistry laboratory. In addition, the methodology described herein should be amenable to studies with many proteins.

4. Iodoacetamide Labeling of Proteins after Oxidant Treatment

We routinely monitor cysteine oxidation and reduction by treating protein samples with a thiol-specific fluorescent reagent, 5-iodoacetamidofluorescein (IAF). Oxidized cysteines cannot be labeled; thus, as the oxidant concentration increases, fluorescein incorporation into α- and β-tubulin decreases. In Figure 5.1, cysteines labeled with fluorescein are designated as "S-Fl" and S(ox) represents a higher oxidation state of sulfur.

Tubulin is diluted in either 0.1 M phosphate buffer, pH 7.4 (PB), or 0.1 M PIPES, pH 6.9, 1 mM MgSO$_4$, 1 mM EGTA buffer (PME) and then treated with ONOO$^-$ for 5 min at 37° in a total reaction volume of 30 μl (20 μg protein). ONOO$^-$ stock solutions are diluted with 0.1 M NaOH just prior to use, and the volume of ONOO$^-$ solution added to achieve the indicated concentrations is normalized to avoid variations in pH. IAF in dimethylformamide (DMF) is added to a final concentration of 1.5 mM (2 μl), and samples are incubated at 37° for an additional 30 min. To determine if the volume of ONOO$^-$ in 0.1 M NaOH will change the pH of a particular buffer, the volumes of buffer and 0.1 M NaOH can be scaled up and the pH change detected using a pH meter. In our experience with 0.1 M buffers, 1.5 μl of ONOO$^-$ solution in 0.1 M NaOH added to 20 μl of protein solution in buffer does not affect the reaction pH.

ONOO$^-$ is quite stable in dilute NaOH solution, but ONOOH decomposes rapidly at neutral pH. Thus, a bolus addition of micromolar ONOO$^-$ is misleading because upon addition of a tiny volume of ONOO$^-$ stock solution to a buffered protein solution at physiological pH, ONOO$^-$ either reacts with protein target or isomerizes to form nitrate ion within seconds (Pryor and Squadrito, 1995; Radi, 1998; Radi et al., 1991)(Fig. 5.2).

Of note, it is critical to calculate the molar concentration of cysteines, rather than of protein, and to add at least a fivefold excess of IAF so that IAF does not become a limiting reagent. Proteins are resolved by SDS-PAGE on a 7.5% gel under reducing conditions, and gel images are captured using a

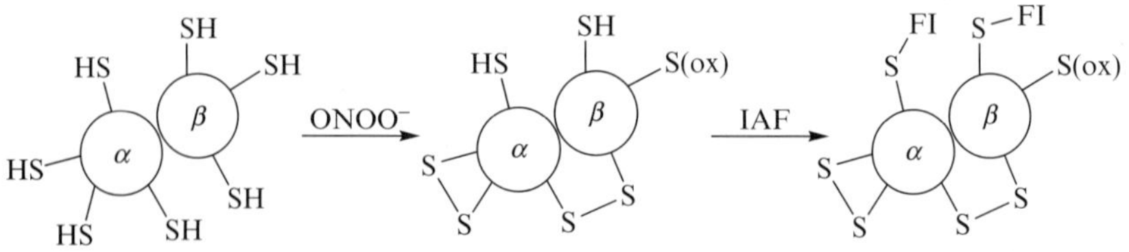

Figure 5.1 IAF labeling of oxidized tubulin.

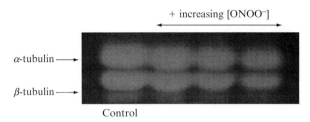

Figure 5.2 Thiol-specific fluorescein labeling of ONOO⁻-treated tubulin. Purified porcine tubulin (1.0 mg/ml, 20 μl) was treated with NaOH (0.1 M, 1.5 μl) or 100, 250, and 500 μM ONOO⁻ (lanes 2–4) for 5 min at 25°. IAF in DMF was added to 1.5 mM (2 μl), and samples were incubated for 30 min at 37°. After reducing SDS-PAGE, the gel was photographed on a UV transilluminator. (See color insert.)

Kodak DC290 system and an ultraviolet (UV) transilluminator. The intensity of the fluorescein-labeled protein bands can be measured using commercial software such as the Kodak 1D Image Analysis package. Alternatively, fluorescein-labeled protein bands can be excised from the gel and placed in 200 μl 0.25 M Tris, pH 8.8. After the protein has diffused from the gel, protein-bound fluorescein is quantitated at 494 nm. The advantage of using IAF is that it is straightforward and changes in cysteine labeling can be assessed immediately following SDS-PAGE.

Another iodoacetamide that can be used to label proteins after oxidant treatment is iodoacetamido-biotin (IAB). To detect biotinylated proteins, they must be transferred from the SDS-PAGE gel to a membrane (nitrocellulose or PVDF) and then detected with an avidin-HRP conjugate in conjunction with a colorimetric or chemiluminescence detection system. These additional steps, although easy to perform, require more time. The advantage of using IAB is that, following oxidant treatment and IAB labeling, a protein is often digested to yield labeled peptides. If one intends to identify the specific cysteines that are labeled, and therefore not oxidized, by mass spectrometry, then this is a useful step. Biotinylated peptides can be captured with immobilized avidin to simplify the peptide mixture and then sequenced.

5. HPLC Separation of Fluorescein-Labeled Peptides

Fluorescein-labeled tubulin samples, prepared as described earlier, are treated with 1% trypsin overnight to generate labeled peptides. The resulting peptides are separated by reversed-phase HPLC on a C8 column using a linear gradient from 20 to 70% acetonitrile in 0.1% trifluoroacetic acid (TFA). Labeled peptides are detected at 440 nm, the wavelength maximum of fluorescein in the presence of acid. By overlaying chromatograms from control and ONOO⁻-treated tubulin, we identified multiple fluorescein-labeled peptides of both α- and β-tubulin for which the incorporation of label decreased as the concentration of ONOO⁻ increased. The cysteines within these peptides are oxidized by ONOO⁻ so that the label cannot be incorporated to the same extent. However, oxidation of tubulin cysteines failed to show any specificity. Multiple fluorescein-labeled peaks decreased in intensity at each ONOO⁻ concentration tested. If a particular cysteine were more reactive toward ONOO⁻, then one would expect labeling of the resulting peptide to selectively decrease at the lowest ONOO⁻ concentration tested. Although our results with tubulin showed no specificity, this methodology is applicable to other proteins.

6. Detection of Interchain Disulfides by Western Blot

Our work with ONOO⁻ showed that interchain disulfides between α- and β-tubulin subunits form readily and can be detected by Western blot. Both tubulin dimers and tetramers of approximately 100 and 200 kDa, respectively, are observed (Landino *et al.*, 2002, 2004a). Likewise, treatment of β-actin with H_2O_2 leads to homodimer formation (Lassing *et al.*, 2007).

Following treatment with ONOO⁻, tubulin species (10 μg total protein per lane) are separated by SDS-PAGE on 7.5% polyacrylamide gels under nonreducing conditions. Proteins are transferred to PVDF membranes, blocked with 3% milk for 30 min, and probed with Tub 2.1, a mouse monoclonal anti-β-tubulin antibody (1:2000) for 2 h. The β-tubulin/antibody complex is visualized by chemiluminescence using the Pierce West Pico system.

Figure 5.3 shows a typical Western blot of 7 μM tubulin (140 μM cys) treated with 50 and 100 μM ONOO⁻. The control in lane 1 shows a major band at 50 kDa corresponding to β-tubulin and a small amount of tubulin dimer. Tubulin used in these experiments is purified in the absence of a reducing agent; thus, some air oxidation does occur. When TCEP, a phosphine-based disulfide-reducing agent, is added, as shown in Fig. 5.3, lane 2, the higher molecular weight dimer disappears. ONOO⁻ treatment yields a substantial increase in tubulin dimmers, and tubulin tetramers are also observed (lanes 3–5 in Fig. 5.3).

In addition to interchain disulfides, intrachain disulfides also form, although they cannot be detected by this method. However, as tubulin becomes oxidized, the main β-tubulin band at 50 kDa broadens due to an increase in intrachain disulfides. Likewise, as tubulin is reduced, the band tightens (compare lanes 1 and 2 in Fig. 5.3). This is attributed to changes in SDS binding as the protein is oxidized or reduced. Following SDS-PAGE

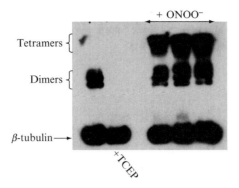

Figure 5.3 Detection of tubulin dimers and tetramers by Western blot A. Tubulin samples (8 μM protein) were treated with TCEP or ONOO$^-$ for 5 min at 25°. Lane 1, control; lane 2, tubulin + TCEP; lane 3, 50 μM ONOO$^-$; and lanes 4 and 5, 100 μM ONOO$^-$. β-Tubulin was detected by Western blot after SDS-PAGE under nonreducing conditions.

separation under nonreducing conditions, the α- and β-tubulin bands are more diffuse, and several dimer and tetramer bands are observed. The protein tubulin, as purified from brain tissue, is composed of multiple individual α- and β-tubulin gene products that vary slightly in molecular weight (Luduena, 1998). Thus, different $\alpha\beta$ dimer combinations yield variations in the mass of interchain disulfide-linked dimers and tetramers. In our experience, the relative amount of higher molecular weight species observed on a Western blot is proportional to total disulfides.

7. Repair of Protein Disulfides by Thioredoxin Reductase and Glutaredoxin Systems

Tubulin disulfides are substrates for the thioredoxin reductase system (TRS) and the glutaredoxin reductase system (GRS) (Landino et al., 2004a,b). Likewise, neuron-specific tau and MAP2 are readily oxidized by both ONOO$^-$ and H_2O_2 to form disulfide-linked species, and the oxidized proteins are substrates for the TRS (Landino et al., 2004d). Notably, the reversible oxidation and reduction of MAPs cysteines have significant effects on microtubule polymerization kinetics. The existence of a repair system for damage to tau and MAP2 suggests that fluctuations in the redox state of neurons may regulate microtubule polymerization. Likewise, an imbalance in tau and MAP2 repair by the TRS may contribute to tau aggregation in Alzheimer's disease. Tau and MAP2 disulfides are repaired by the GRS, and both native (reduced) tau and MAP2 undergo thiol–disulfide exchange with oxidized glutathione (GSSG) in a manner similar to tubulin (Landino et al., 2004c).

8. THIOREDOXIN REDUCTASE REPAIR OF PROTEIN DISULFIDES

Thioredoxin reductase catalyzes the NADPH-dependent reduction of an active site disulfide in oxidized Trx, Trx-S$_2$, to the dithiol, Trx-(SH)$_2$ [Eq. (5.1)]. Reduced Trx undergoes thiol/disulfide exchange with oxidized protein substrates [Eq. (5.2)].

$$\text{NADPH} + \text{H}^+ + \text{Trx-S}_2 \rightarrow \text{NADP}^+ + \text{Trx-(SH)}_2 \quad (5.1)$$

$$\text{Trx-(SH)}_2 + \text{protein-S}_2 \rightarrow \text{Trx-S}_2 + \text{protein-(SH)}_2 \quad (5.2)$$

Trx is a ubiquitous 12-kDa cytosolic protein with multiple known substrates, including ribonucleotide reductase and methionine sulfoxide reductase (Arner et al., 1999). Trx is a highly conserved protein, and the E. coli and mammalian proteins differ by only several amino acids. Trx from E. coli is often used in assays with mammalian TrxR because it does not aggregate as readily as mammalian Trx. We use E. coli Trx in combination with TrxR, purified from rat liver, to characterize the interaction of the TRS with tubulin (Landino et al., 2004a).

In a qualitative assay, assessed by Western blot, the TRS reduces the disulfide bonds that are formed between tubulin subunits following ONOO$^-$ addition. Reduction of the interchain disulfide(s) requires all three components of the TRS, and repair is both time and concentration dependent. A typical reaction contains 20 μM tubulin pretreated with 0.5 mM ONOO$^-$, 5 μM Trx, 100 nM TrxR, and 1.0 mM NADPH.

TrxR is purified from rat liver as described and stored at −80° in TE buffer (50 mM Tris-HCl, 1 mM EDTA, pH 7.5) (Arner et al., 1999). TrxR activity is assayed during purification steps using 5,5′-dithiobis(2-nitrobenzoic acid)(DTNB) as a substrate, and its identity is confirmed by Western blot using a rabbit polyclonal antibody against TrxR (Upstate Biotechnology). The activity of each TrxR preparation is calculated as micromoles NADPH oxidized per minute per microliter of protein solution using DTNB as a substrate. E. coli thioredoxin is from Sigma, and human thioredoxin is from American Diagnostica, Inc.

Because all three components of the TRS are required to reduce tubulin disulfides, Trx must serve as the intermediary that makes contact with tubulin. Indeed, we detected a disulfide-linked complex between β-tubulin and Trx (Landino et al., 2004a). Purified tubulin (3.5 μM) and increasing concentrations of E. coli Trx (3.5–17 μM) are combined and incubated for 15 min at 25°. In these samples, a new protein band of approximately 65 kDa is detected by the anti-β-tubulin antibody. The protein band at

65 kDa is also detected by an antibody against Trx. The 65-kDa band is not observed under reducing conditions.

The *E. coli* Trx that is used contains a mixture of the oxidized and reduced forms, and the oxidized form Trx-S_2 reacts with reduced tubulin, Tub-$(SH)_2$, to form the transient interchain disulfide. In the TRS reaction with $ONOO^-$-oxidized tubulin, reduced Trx, Trx-$(SH)_2$, reacts with oxidized tubulin, Tub-S_2.

Any protein that is oxidized by $ONOO^-$ to form disulfide-linked species, either intermolecularly or intramolecularly, should be treated with the components of the TRS to determine if reduction of the disulfides occurs. Reduction of disulfides can be assessed by IAF labeling. Although many disulfides are substrates for the TRS, it cannot be assumed that all protein disulfides will be reduced.

9. Quantitation of Protein Disulfides by Measuring NADPH Oxidation

To study the kinetics of repair of both inter- and intrasubunit disulfides in tubulin formed by $ONOO^-$, we monitor NADPH oxidation at 340 nm. Samples contain oxidized tubulin, Trx, TrxR, and NADPH. This allows us to quantitate disulfide yield as a function of $ONOO^-$ concentration added and also examine the rate of disulfide reduction. The change in absorbance at 340 nm is converted to a molar quantity using 6220 M^{-1} cm^{-1}, the molar absorptivity of NADPH. TrxR reduces disulfides and sulfenic acids, RSOH, but does not reduce higher oxidation states of sulfur. Thus, TRS repair of an oxidized protein may not restore IAF labeling to control levels.

10. Quantitation of Total Cysteine Oxidation Using DTNB

Ellman's reagent, 5,5'-dithiobis(2-nitrobenzoic acid), has been used extensively to quantitate free thiols in proteins (Riddles *et al.*, 1979). TNB, a product of the reaction of DTNB with free thiols, absorbs strongly at 412 nm, allowing for facile detection of micromolar concentrations of thiols [Eq. (5.3)].

$$DTNB + RSH \rightarrow TNB + RS\text{-}TNB \text{ (mixed disulfide)} \quad (5.3)$$

DTNB assays are used routinely to calculate the decrease in free thiols following oxidant damage.

A typical assay performed in our laboratory is described. Heat-stable porcine MAPs (1.9 µg/µl, 40 µl) are treated with 0.1 M NaOH (control), ONOO$^-$ in 0.1 M NaOH for 5 min, or H_2O_2 in H_2O for 20 min at room temperature. Thirty-five microliters of 8 M guanidine–HCl and 25 µl 1.2 mM DTNB are added to each sample (100 µl total volume), and absorbance values are recorded at 412 nm within 10 min. The addition of guanidine HCl ensures that all cysteines will be accessible. Molar concentrations of free cysteines in the protein samples are calculated from a GSH standard curve (final concentrations from 10 to 100 µM). Absorbance values of the GSH standards are also measured in the presence of guanidine HCl.

Although the TRS will reduce disulfides and sulfenic acids, higher oxidation states of sulfur are not reduced (Holmgren, 1989). By comparing the number of cysteines reduced by the TRS and the number of total cysteines oxidized using DTNB, one can attribute the difference to formation of a higher oxidation state.

11. Glutaredoxin/Glutathione (GSH) Reductase Repair of Protein Disulfides

Glutathione is the primary physiologic redox regulator *in vivo*. GSH can be oxidized by ONOO$^-$ and other cellular oxidants to GSSG and is reduced to GSH by the NADPH-dependent enzyme glutathione reductase (Holmgren and Aslund, 1995). GSH concentrations in most eukaryotic cells are in the 1 to 10 mM range with a typical GSH:GSSG ratio of 100 to 400 (Gilbert, 1995). Under normal physiologic conditions, GSH protects proteins from oxidation by reactive oxygen species. However, under conditions of oxidative stress, the ratio of GSH:GSSG may decrease to 1–10 and thus GSH will not be present in sufficient quantities to prevent protein oxidation. Furthermore, GSSG undergoes thiol/disulfide exchange with protein thiols, which has been observed with tubulin, tau, and MAP2 (Gilbert, 1995; Landino *et al.*, 2004b,c). Protein oxidation is likely when GSSG concentrations increase during oxidative stress.

Although most protein disulfides are substrates for the TRS, the GRS is more specific, requiring a mixed disulfide substrate among oxidized protein, protein-S$_2$, and GSH [Eq. (5.4)] (Dalle-Donne *et al.*, 2003; Holmgren, 1989). Protein-glutathione disulfide (PSSG) represents a mixed disulfide between protein and GSH, and PSH is a reduced protein cysteine.

$$\text{protein-S}_2 + \text{GSH} \rightarrow \text{PSSG} + \text{PSH} \qquad (5.4)$$

Glutaredoxin (Grx) is a small redox-active protein that readily undergoes thiol-disulfide exchange with GSH/GSSG. GSH, Grx, glutathione reductase

(GR), and NADPH comprise the glutaredoxin repair system, which is capable of reducing protein disulfides (Holmgren, 1989).

$$\text{protein-}S_2 + \text{Grx (red)} \rightarrow 2\text{PSH} + \text{Grx (ox)} \quad (5.5)$$

$$\text{Grx (ox)} + 2\text{GSH} \rightarrow \text{Grx (red)} + \text{GSSG} \quad (5.6)$$

$$\text{GSSG} + \text{NADPH} \rightarrow 2\text{GSH} + \text{NADP}^+ \quad (5.7)$$

As shown in equation 5.5, protein-S_2 is reduced by Grx. Oxidized Grx then undergoes thiol-disulfide exchange with GSH to yield GSSG (Eq. 5.6), which is reduced back to GSH by GR. (Eq 5.7). NADPH oxidation supplies the energy for this repair system. Disulfides in tubulin, tau, and MAP2 are substrates for the GRS, although repair is slower and not complete as for the TRS. A typical GRS repair assay contains 20 μM tubulin pretreated with 0.5 mM ONOO$^-$, 1 mM GSH, 5 μM Grx, 1.0 mM NADPH, and 0.2 unit glutathione reductase (Landino *et al.*, 2004b).

12. Detection of Protein *S*-Glutathionylation

Both control and ONOO$^-$-damaged tubulin are treated with GSH or GSSG and assayed for protein *S*-glutathionylation using an anti-GSH antibody (Dalle-Donne *et al.*, 2003). Glutathionylation of tubulin to form the mixed disulfide, PSSG, is detected by dot blot when ONOO$^-$-damaged tubulin is incubated with GSH (Landino *et al.*, 2004b). Given the reactivity of glutaredoxin toward mixed disulfides of protein and GSH, it is not surprising that glutathionylation of tubulin is detected.

13. Thiol/disulfide Exchange with Oxidized Glutathione

To study the extent of reaction of native tubulin with GSSG, we measure the concentration of GSH produced over time by HPLC. Each tubulin disulfide formed yields two molecules of GSH [Eq. (5.8)].

$$\text{Tub-SH}_2 + \text{GSSG} \rightarrow \text{Tub-S}_2 + 2\text{GSH} \quad (5.8)$$

We also analyzed tubulin for total cysteine content by labeling protein with IAF, and the decrease in tubulin cysteines was consistent with the amount of GSH produced by thiol-disulfide exchange. Although the millimolar GGSG concentrations used in these experiments were relatively high, such concentrations were necessary to demonstrate the reactivity of tubulin thiols with GSSG.

13.1. Quantitation of GSH following tubulin thiol-disulfide exchange with GSSG

Purified tubulin (1.9 μg/μl, 20 μl) in PME buffer is reacted with 0–2 mM GSSG in 0.1 M phosphate buffer at pH 6.9, 7.4, or 8.0 at 37° for 30 min (40 μl total volume). IAF in DMF is added to a final concentration of 1.5 mM, and samples are incubated at 37° for 30 min. Cold ethanol is added to 80% to precipitate the fluorescein-labeled tubulin, and the protein is collected by centrifugation at 16,000 g for 20 min. The supernatant containing fluorescein-labeled GSH (GS-AF) is treated with DTT to convert excess IAF to the more stable DTT adduct (DTT-AF). Samples (20 μl supernatant) are analyzed by reversed-phase HPLC on a C8 column (Zorbax). Solvent A contains 20% acetonitrile/0.1% TFA, and solvent B contains 70% acetonitrile/0.1% TFA. A linear gradient from 0 to 50% B over 20 min is used to elute GS-AF (retention time of 7 min) and DTT-AF (retention time of 11 min). Fluorescein-labeled products are detected at 440 nm, the absorbance maximum of fluorescein in solvent A. GS-AF is quantitated by comparison of peak areas to a standard curve created following injection of known amounts of a GS-AF standard prepared by reacting GSH with IAF.

14. Protein Thiol Modification by Nitric Oxide Donors

The general category of nitric oxide donors used routinely include S-nitrosoglutathione (GSNO), S-nitroso-N-acetyl-penicillamine (SNAP), and NONOates such as DEA NONOate, 2-(N,N-diethylamino)-diazenolate 2-oxide. Both GSNO and SNAP are S-nitrosothiols that react with protein cysteines (R'SH) via transnitrosation as shown in Eq. (5.9) (Ji *et al.*, 1999; Park, 1988).

$$\text{RSNO} + \text{R'SH} \rightarrow \text{RSH} + \text{R'SNO} \qquad (5.9)$$

The NONOates are not S-nitroso compounds, but rather release NO-free radical as they decompose in solution.

In a protein such as tubulin with 20 reduced cysteines, it is likely that disulfide bonds will form following S-nitrosation of a tubulin cysteine. According to Eq. (5.10), GSNO reacts with a protein thiol (PSH) to form a nitrosated protein intermediate. The intermediate can be attacked by an

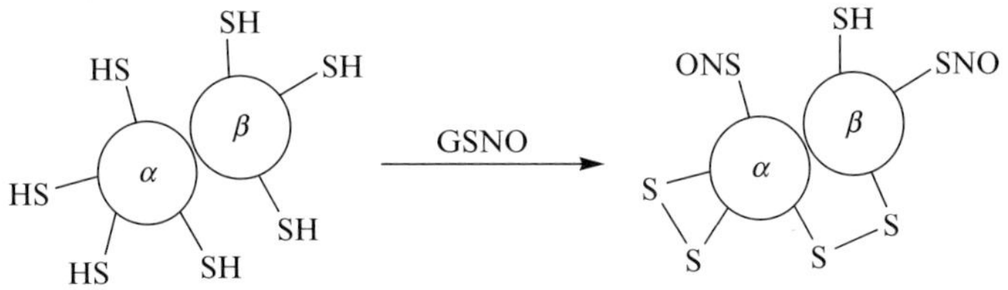

Figure 5.4 GSNO treatment yields disulfides and S-nitrosation.

adjacent protein thiol(ate) to yield a disulfide bond (PSSP′) and nitroxyl (HNO) [Eq. (5.11)]. Indeed, for proteins such as tubulin with multiple cysteines, disulfides would be the expected product (Hogg, 2000; Xian et al., 2000).

$$GSNO + PSH \rightarrow GSH + PSNO \quad (5.10)$$

$$PSNO + P'SH \rightarrow PSSP' + HNO \quad (5.11)$$

For tubulin, both disulfides and S-nitrosation are detected after treatment with GSNO. Interchain disulfides are confirmed by our Western blot method. Tubulin S-nitrosation is quantified by the modified Saville assay described. IAF labeling detects total change in protein cysteines and thus represents the sum of disulfides and S-nitrosation (Fig. 5.4).

14.1. Detection of tubulin S-nitrosation

Purified tubulin (60 μl, 0.75 μg/μl) is diluted with PB, pH 7.4, and treated with GSNO or SNAP for 30 min. Ethanol is added to 80%, samples are left on ice for 15 min to precipitate protein, and the protein pellet is collected by centrifugation at 10,000 g for 10 min. Pellets are washed with 80% ethanol and then resuspended in 60 μl PB, pH 7.4, containing 1% SDS. Samples are heated to 45–50° to aid in dissolving protein pellets. NO bound to protein is detected by the modified Saville assay described by Cook and colleagues (1996). Neutral Greiss reagent (60 μl) and HgCl$_2$ (2 μl, 10 mM) are added to the protein samples, and the product derived from NO$_x$ is detected at 496 nm after 20 min. Concentrations of SNO are calculated from a GSNO standard curve prepared in the presence of 1% SDS. Solutions of SNAP, GSNO, or DEA NONOate are prepared in PB. Neutral Greiss reagent is prepared by mixing equal volumes of 57 mM sulfanilamide in phosphate-buffered saline (PBS) and 1.2 mM N-(1-naphthyl)ethylenediamine dihydrochloride in PBS.

15. Conclusions

The methodology described herein applied to our studies of microtubule protein cysteine oxidation by $ONOO^-$ and NO donors is simple and straightforward. Our work has yielded interesting *in vitro* results that may provide clues as to the role of protein cysteine thiol modification in regulating microtubule dynamics *in vivo*. Many proteins with reduced thiols have the potential to be regulated by oxidation, S-glutathionylation, and S-nitrosation. The experiments described provide a facile starting point to explore this area in more detail.

REFERENCES

Arner, E. S. J., Zhong, L., and Holmgren, A. (1999). Preparation and assay of mammalian thioredoxin and thioredoxin reductase. *Methods Enzymol.* **300,** 226–239.
Cook, J. A., Kim, S. Y., Teague, D., Krishna, M. C., Pacelli, R., Mitchell, J. B., Yodovotz, Y., Nims, R. W., Christodoulou, D., Miles, A. M., Grisham, M. B., and Wink, D. A. (1996). Convenient colorimetric and fluorometric assays for S-nitrosothiols. *Anal. Biochem.* **238,** 150–158.
Dalle-Donne, I., Rossi, R., Giustarini, D., Colombo, R., and Milzani, A. (2003). Actin S-glutathionylation: Evidence against a thiol-disulphide exchange mechanism. *Free Radic. Biol. Med.* **35,** 1185–1193.
Demple, B. (1998). A bridge to control. *Science* **279,** 1655–1656.
Denicola, A., Freeman, B. A., Trujillo, M., and Radi, R. (1996). Peroxynitrite reaction with carbon dioxide/bicarbonate: Kinetics and influence on peroxynitrite-mediated oxidations. *Arch. Biochem. Biophys.* **333,** 49–58.
Gilbert, H. F. (1995). Thiol/disulfide exchange equilibria and disulfide bond stability. *Methods Enzymol.* **251,** 8–28.
Hess, D. T., Matsumoto, A., Nudelman, R., and Stamler, J. S. (2001). S-nitrosylation: Spectrum and specificity. *Nat. Cell Biol.* **3,** E1–E3.
Hogg, N. (2000). Biological chemistry and clinical potential of S-nitrosothiols. *Free Radic. Biol. Med.* **28,** 1478–1486.
Hogg, P. J. (2003). Disulfide bonds as switches for protein function. *Trends Biochem. Sci.* **28,** 210–214.
Holmgren, A. (1989). Thioredoxin and glutaredoxin systems. *J. Biol. Chem.* **264,** 13963–13966.
Holmgren, A., and Aslund, F. (1995). Glutaredoxin. *Methods Enzymol.* **252,** 283–292.
Ji, Y., Akerboom, T. P. M., Sies, H., and Thomas, J. A. (1999). S-nitrosylation and S-glutathiolation of protein sulfhydryls by S-nitrosoglutathione. *Arch. Biochem. Biophys.* **362,** 67–78.
Kim, S. O., Merchant, K., Nudelman, R., Beyer, W. F., Keng, T., DeAngelo, J., Hausladen, A., and Stamler, J. S. (2002). OxyR: A molecular code for redox-related signalling. *Cell* **109,** 383–396.
Landino, L. M., Hasan, R., McGaw, A., Cooley, S., Smith, A. W., Masselam, K., and Kim, G. (2002). Peroxynitrite oxidation of tubulin sulfhydryls inhibits microtubule polymerization. *Arch. Biochem. Biophys.* **398,** 213–220.

Landino, L. M., Iwig, J. S., Kennett, K. L., and Moynihan, K. L. (2004a). Repair of peroxynitrite damage to tubulin by the thioredoxin reductase system. *Free Radic. Biol. Med.* **36,** 497–506.

Landino, L. M., Moynihan, K. L., Todd, J. V., and Kennett, K. L. (2004b). Modulation of the redox state of tubulin by the glutathione/glutaredoxin reductase system. *Biochem. Biophys. Res. Commun.* **314,** 555–560.

Landino, L. M., Robinson, S. H., Skreslet, T. E., and Cabral, D. M. (2004c). Redox modulation of tau and microtubule-associated protein-2 by the glutathione/glutaredoxin reductase system. *Biochem. Biophys. Res. Commun.* **323,** 112–117.

Landino, L. M., Skreslet, T. E., and Alston, J. A. (2004d). Cysteine oxidation of tau and microtubule-associated protein-2 by peroxynitrite: Modulation of microtubule assembly kinetics by the thioredoxin reductase system. *J. Biol. Chem.* **279,** 35101–35105.

Lassing, I., Schmitzberger, F., Bjornstedt, M., Holmgren, A., Nordlund, P., Schutt, C. E., and Lindberg, U. (2007). Molecular and structural basis for redox regulation of β-actin. *J. Mol. Biol.* **370,** 331–348.

Lipton, S. A., Choi, Y.-B., Takahashi, H., Zhang, D., Li, W., Godzik, A., and Bankston, L. A. (2002). Cysteine regulation of protein function as exemplified by NMDA-receptor modulation. *Trends Neurosci.* **25,** 474–480.

Luduena, R. F. (1998). Multiple forms of tubulin: Different gene products and covalent modifications. *Intl. Rev. Cytol.* **178,** 207–275.

Lymar, S. V., and Hurst, J. K. (1995). Rapid reaction between peroxynitrite ion and carbon dioxide: Implications for biological activity. *J. Am. Chem. Soc.* **117,** 8867–8868.

Nogales, E., Wolf, S. G., and Downing, K. H. (1998). Structure of the $\alpha\beta$-tubulin dimer by electron crystallography. *Nature* **391,** 199–203.

Park, J.-W. (1988). Reaction of S-nitrosoglutathione with sulfhydryl groups in proteins. *Biochem. Biophys. Res. Commun.* **152,** 916–920.

Pryor, W. A., and Squadrito, G. L. (1995). The chemistry of peroxynitrite: A product from the reaction of nitric oxide with superoxide. *Am. J. Physiol.* **268,** L699–L722.

Radi, R. (1998). Peroxynitrite reactions and diffusion in biology. *Chem. Res. Toxicol.* **11,** 720–721.

Radi, R., Beckman, J. S., Bush, K. M., and Freeman, B. A. (1991). Peroxynitrite oxidation of sulfhydryls. *J. Biol. Chem.* **266,** 4244–4250.

Riddles, P. W., Blakeley, R. L., and Zerner, B. (1979). Ellman's reagent: 5,5'-dithiobis (2-nitrobenzoic acid): A reexamination. *Anal. Biochem.* **94,** 75–81.

Xian, M., Chen, X., Liu, Z., Wang, K., and Wang, P. G. (2000). Inhibition of papain by S-nitrosothiols: Formation of mixed disulfides. *J. Biol. Chem.* **275,** 20467–20473.

CHAPTER SIX

INDIRECT MECHANISMS OF DNA STRAND SCISSION BY PEROXYNITRITE

Orazio Cantoni *and* Andrea Guidarelli

Contents

1. Introduction	112
2. Materials and Methods	112
2.1. Cell culture and treatments	112
2.2. Measurement of DNA single-strand breakage by the alkaline halo assay	113
2.3. Measurement of DNA single-strand breakage by the comet assay	114
2.4. Microscopic analysis and image processing	114
2.6. Statistical analysis	116
3. Results and Discussion	116
Acknowledgments	119
References	119

Abstract

Peroxynitrite is a highly reactive nitrogen species directly damaging diverse biomolecules in target cells. This is not the case for genomic DNA, as its cleavage is mediated by secondary reactive species produced at the mitochondrial level in a Ca^{2+}-dependent reaction in which ubisemiquinone serves as an electron donor. Under these conditions, superoxide is produced in a time-dependent manner for up to 30 min and dismutates readily to H_2O_2, which can now reach the nucleus and generate DNA strand scission in a reaction of the Fenton type. This chapter provides evidence for this mechanism using the alkaline halo assay, a simple technique for the assessment of DNA strand scission at the single cell level.

Istituto di Farmacologia e Farmacognosia, Università degli Studi di Urbino "Carlo Bo," Urbino, Italy

Methods in Enzymology, Volume 440 © 2008 Elsevier Inc.
ISSN 0076-6879, DOI: 10.1016/S0076-6879(07)00806-3 All rights reserved.

1. Introduction

Peroxynitrite, the coupling product of nitric oxide and superoxide, is a highly reactive nitrogen species with the ability to damage diverse biomolecules, including the DNA (for a review, see Szabó and Ohshima, 1997). This notion, which can be directly established from experiments using isolated DNA, has long been taken as an indication of direct DNA cleavage also in intact cells. While yet unproven, the mechanism of direct strand scission implies that peroxynitrite, regardless of whether generated within the cell or entering the cell from the extracellular milieu, should be able to avoid interactions with other biomolecules in order to reach the DNA. Although the reported half-life of peroxynitrite in a biological buffer would be consistent with this possibility, it nevertheless appears unlikely that such a highly reactive species may escape interactions with reduced glutathione, protein thiols, tyrosine residues, and a variety of additional biological molecules (Beckman and Koppenol, 1996). These considerations led us to challenge the hypothesis that peroxynitrite promotes strand scission of genomic DNA via an indirect mechanism, i.e., by triggering events associated with the formation of secondary reactive species finally leading to DNA strand scission.

This chapter describes some results providing evidence for such an indirect mechanism, in which the DNA-damaging species, superoxide/H_2O_2, are generated at the mitochondrial level via a process requiring both Ca^{2+} and electron transport in the respiratory chain. DNA damage was assessed using a rapid and sensitive assay, recently developed in our laboratory, measuring DNA single-strand breakage at the single cell level.

2. Materials and Methods

2.1. Cell culture and treatments

U937 human myeloid leukemia cells are cultured in suspension in RPMI 1640 medium supplemented with 10% fetal bovine serum, penicillin (100 units/ml), and streptomycin (100 μg/ml), at 37° in T-75 tissue culture flasks gassed with an atmosphere of 95% air–5% CO_2. Peroxynitrite is synthesized by the reaction of nitrite with acidified H_2O_2, as described previously (Radi et al., 1991), with minor modifications (Tommasini et al., 2002). Treatments are performed in prewarmed saline A (8.182 g/liter NaCl, 0.372 g/liter KCl, 0.336 g/liter $NaHCO_3$, and 0.9 g/liter glucose, pH 7.4, at 37°) containing 2.5 × 10^5 cells/ml. The cell suspension (2 ml) is inoculated into 15-ml plastic tubes before the addition of peroxynitrite.

Peroxynitrite is added on the wall of these tubes and immediately mixed to equilibrate its concentration on the cell suspension. To avoid changes in pH due to the high alkalinity of the peroxynitrite stock solution, an appropriate amount of 1.5 N HCl is also added to the wall of the tubes prior to peroxynitrite. Note that this procedure has to be done quickly, as peroxynitrite decomposes rapidly at neutral pH.

Some experiments are performed in permeabilized cells. Permeabilization is achieved by adding digitonin (10 μM) to 2.5×10^5 cells/ml suspended in 0.25 M sucrose, 0.1% bovine serum albumin, 10 mM MgCl$_2$, 10 mM K$^+$-HEPES, 5 mM KH$_2$PO$_4$, pH 7.2, at 37°. Under these conditions, digitonin permeabilizes the plasma membrane but leaves mitochondrial membranes intact (Fiskum et al., 1980). Treatments (10 min) are performed immediately after the addition of digitonin (i.e., in permeabilization buffer), using the same conditions employed with intact cells.

2.2. Measurement of DNA single-strand breakage by the alkaline halo assay

The alkaline halo assay is performed as described by Sestili and Cantoni (1999), with minor modifications (Guidarelli et al., 2001). An aliquot (160 μl) of the cell suspension (4.0×10^4 cells) is centrifuged (300g, 3 min, 4°) and resuspended in 50 μl of ice-cold phosphate-buffered saline (PBS, 8 g/liter NaCl, 1.15 g/liter Na$_2$HPO$_4$, 0.2 g/liter KH$_2$PO$_4$, 0.2 g/liter KCl) supplemented with ethylenediaminetetraacetic acid (EDTA, 5 mM). Fifty microliters of 1.5% low-melting agarose in PBS/EDTA buffer (60°) is then added to the cell suspension and immediately sandwiched between an agarose-coated slide and a coverslip. Note that this last step should be accurate and rapid in order to prevent gelling during pipetting and/or a nonhomogeneous distribution of the cells in the gel. The slide preparations are incubated for 3 to 5 min at ice temperature to allow complete gelling. The coverslips are then removed, and slides are immersed in an alkaline buffer (0.1 M NaOH/ 1 mM EDTA, pH 12.5) for 15 min at room temperature. The slides are subsequently washed with distilled water, stained for 5 min with 70 μl ethidium bromide (10 μg/ml), washed again three times with distilled water, and finally incubated for 30 min in fresh distilled water. Note that the washing steps may cause detachment of the agarose gel. However, this inconvenience is extremely rare when the slides are prepared correctly (see later) and handled delicately.

As a final note, it is important to use appropriate slides. Agarose-coated slides are available commercially. As an alternative, common slides may be prepared as follows: sandpaper is passed 30–40 times in the long edges of the slide prior to the addition of 1 ml of an agarose solution in distilled water (1% low-gelling agarose) kept previously for 15 min at 100°. The agarose solution is spread throughout the surface of the slides and is finally dried by an overnight incubation at room temperature.

2.3. Measurement of DNA single-strand breakage by the comet assay

The comet assay is performed as described by Singh et al. (1988), with minor modifications (Guidarelli et al., 1998b). Briefly, cells are processed and transferred to agarose-coated slides using the same procedure described earlier for the alkaline halo assay. The slides are then immersed (60 min) in ice-cold lysing solution (2.5 M NaCl, 0.1 M EDTA, 10 mM Tris, 1% sarkosyl, 5% dimethyl sulfoxide, and 1% Triton X-100, pH 10.0). The slides are subsequently placed on an electrophoretic tray with an alkaline buffer (0.3 M NaOH/1 mM EDTA) and therein left for 30 min. Electrophoresis is performed at 300 mA, for 20 min, in the same alkaline buffer maintained at 14°. Finally, the slides are washed and stained with ethidium bromide, as described earlier for the alkaline halo assay.

2.4. Microscopic analysis and image processing

The ethidium bromide-labeled DNA is visualized using a BX-51 microscope (Olympus, Tokyo, Japan) equipped with a SPOT-RT camera unit (Diagnostic Instruments, Delta Sistemi, Rome, Italy). The resulting images are digitally acquired and processed at the single-cell level on a personal computer using Scion Image software (Scion, Frederick, MA).

In the alkaline halo assay, the level of DNA strand scission is expressed as a nuclear spreading factor, which represents the ratio between the area of the halo (obtained by subtracting the area of the nucleus from the total area, nucleus + halo) and that of the nucleus. Figure 6.1F shows image analysis in a cell previously treated for 30 min with 200 μM peroxynitrite. Note that this procedure is quite simple and accurate, as both the halo and the nucleus are described by a circle. The average nuclear spreading factor, calculated from four different experiments, corresponds to 7.01 ± 0.57 (Fig. 6.1A). The nuclear spreading factor calculated from untreated cells under identical conditions is 1.86 ± 0.23. Experimental results can also be expressed as relative nuclear spreading factor values, calculated by subtracting the nuclear spreading factor of control cells from that of treated cells. This procedure was used in the calculation of the results reported in Figs. 6.2A and 6.2B.

In the comet assay, the extent of DNA strand scission is expressed as a tail/nucleus ratio, calculated according to the method of Müller et al. (1994). Figure 6.1G shows image analysis in a cell previously treated for 30 min with 200 μM peroxynitrite. Note that this procedure is complicated by the manual drawing of the perimeter of the comet's tail, whereas the head is described by a circle. The tail/nucleus ratio represents the ratio between

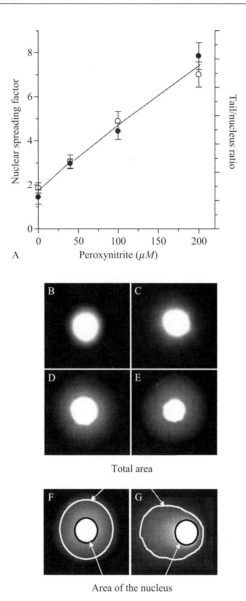

Figure 6.1 DNA single-strand breakage induced by peroxynitrite. (A) Alkaline halo (○) or comet (●) assays were used to assess the level of DNA single-strand breakage induced by a 30-min exposure of U937 cells to increasing concentrations of peroxynitrite. (B–E) Representative micrographs of cells treated with 0 (B), 40 (C), 100 (D), or 200 (E) μM peroxynitrite. (F and G) Representative micrographs of cells treated with 200 μM peroxynitrite, analyzed with alkaline halo (F) or comet (G) assays, and processed for image analysis. The image processing is fast and accurate for F, in which both the halo and the nuclear remnant present a regular shape. The drawing of these circles is computer assisted. The image processing is more complicated for G, as the comet

the fluorescence specifically associated to the area of the tail (obtained by subtracting the area of the nucleus from the total area, nucleus + comet's tail) and the area of the head (nucleus).

In each experiment, the tail/nucleus ratio and the nuclear spreading factor values are calculated from 50 to 75 randomly selected cells. Images are stained digitally with a computer-generated pseudo-color scale ("Fire-1" option from the "Color Tables" menu of the Scion Image software), which stains low-fluorescence regions (i.e., halos or comet tails) red and high-fluorescence regions (i.e., nuclear remnants) bright orange to yellow.

2.6. Statistical analysis

Statistical analysis of data for multiple comparison is performed by ANOVA followed by a Dunnett's test.

3. Results and Discussion

Figure 6.1A shows that a 30-min exposure to 40 to 200 μM peroxynitrite promotes DNA single-strand breakage in U937 cells. Identical results were obtained using two different techniques that measure DNA damage at the single cell level, the comet assay and a simplified version of the alkaline halo assay, originally developed in 1999 in our laboratory (Sestili and Cantoni, 1999). Our method is as sensitive as the comet assay but presents a number of advantages: very simple, rapid, and inexpensive. The alkaline halo assay involves a few steps in which the cells are first embedded in melted agarose and then lysed in a hypotonic alkaline solution prior to DNA staining with ethidium bromide and image analysis with a fluorescent microscope. Under these conditions, single-stranded DNA fragments spread radially from the nuclear cage and generate a fluorescent image that resembles a halo concentric to the nuclear remnants. The area of the halos increases at increasing levels of DNA fragmentation; this process is associated with a progressive reduction of areas of the nuclear remnants. Typical images of cells treated with peroxynitrite as indicated previously, and processed with the alkaline halo assay, are shown in Figs. 6.1B–6.1E.

Thus, the alkaline halo assay presents some analogies with the comet assay but is based on a different principle. Indeed, while the early steps are the same in both assays, our method does not use electrophoresis to separate

---

presents an irregular shape that needs to be drawn manually. As a consequence, calculation of the nuclear spreading factor after the alkaline halo assay is more rapid, accurate, and reliable than calculation of the tail/nucleus ratio after the comet assay. Results illustrated in A represent the means ± SEM calculated from four separate experiments.

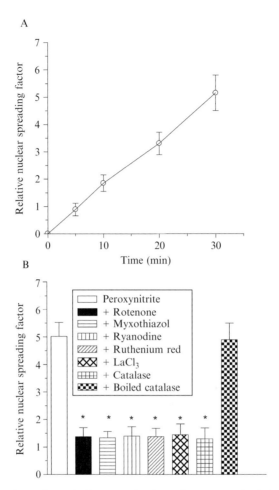

Figure 6.2 Evidence for an indirect mechanism of DNA strand scission by peroxynitrite based on studies performed in intact and permeabilized cells. (A) A bolus of peroxynitrite (200 μM) was added to the culture medium of U937 cells, and the extent of DNA strand scission was assessed using the alkaline halo assay in cells harvested at increasing time intervals. (B) Cells were first permeabilized and then exposed for 10 min to 0 or 200 μM peroxynitrite in the absence or presence of rotenone (0.5 μM), myxothiazol (5 μM), ryanodine (20 μM), ruthenium red (200 nM), LaCl$_3$ (100 μM), or 10 U/ml catalase (active enzymatically or heat inactivated). The level of DNA strand scission is expressed as relative nuclear spreading factor values, calculated as detailed in the text. Data represent the means ± SEM calculated from three to five separate experiments. $*P < 0.01$ compared with cells exposed to peroxynitrite alone (one-way ANOVA followed by Dunnett's test).

damaged from undamaged DNA, as does the comet assay, but rather a short-term incubation in a hypotonic buffer. An additional advantage is at the image processing level, as the area of the halos (circle, Fig. 6.1F) can be calculated more conveniently and accurately than the area of the comets, characterized by an irregular shape resembling that of a pear (Fig. 6.1G).

These results indicate that peroxynitrite promotes DNA single-strand breakage detected conveniently by the alkaline halo assay. The question of whether this response is the result of direct or indirect effects of peroxynitrite was addressed with a very simple experiment (Guidarelli et al., 2000). We reasoned that if DNA is indeed its immediate target, then DNA damage should be maximal and readily detected soon after the addition of peroxynitrite, an agent with a very short half-life. A bolus of peroxynitrite was therefore added to the cultures, and the extent of DNA cleavage was measured in cells harvested after increasing time intervals of incubation. As shown in Fig. 6.2A, DNA strand scission was barely detectable 5 min after the addition of peroxynitrite and increased linearly over time of incubation. These results are therefore in contrast with the hypothesis of direct DNA strand scission and imply that the very short exposure to peroxynitrite (due to its very short half-life) triggers events leading to the time-dependent formation of species causing strand scission of genomic DNA.

We surmised that mitochondria are likely targets of peroxynitrite, as these organelles actively release reactive oxygen species in response to a variety of toxins. In addition, the DNA cleavage generated by short-chain lipid hydroperoxides, as *tert*-butylhydroperoxide, is also mediated by mitochondrially generated reactive oxygen species (Guidarelli et al., 1997a,b,c, 1998a, 2001).

It was found (Guidarelli et al., 2000, 2007) that the DNA-damaging response evoked by peroxynitrite is blunted by the respiration-deficient phenotype, as well as by the inhibition of electron entry in the respiratory chain through complex I, e.g., by rotenone, or inhibition of the electron flow from the reduced coenzyme Q to cytochrome c_1, e.g., by myxothiazol, thereby resulting in the prevention of ubisemiquinone formation (Brand and Hermfisse, 1997). Consistently, bona fide inhibitors of complex III (as antimycin A) enhanced DNA strand scission induced by peroxynitrite via a rotenone/myxothiazol-sensitive mechanism (Guidarelli et al., 2000, 2007).

In short, previous results indicate that the DNA single-strand breakage induced by peroxynitrite is in fact mediated by superoxide generated in a reaction in which ubisemiquinone serves as an electron donor. Superoxide, however, was not the final DNA-damaging species but rather its dismutation product H_2O_2, which, via its interaction with divalent iron, promotes the formation of the highly reactive hydroxyl radical (Guidarelli et al., 2000).

Further analyses revealed that the mitochondrial formation of superoxide/H_2O_2 is Ca^{2+}-dependent (Guidarelli et al., 2006, 2007). As a consequence, very low concentrations of peroxynitrite, which would not cause DNA cleavage in the absence of additional manipulations, were found to promote maximal damage in the presence of agents causing mitochondrial accumulation of the cation (Guidarelli et al., 2006, 2007). However, at levels causing maximal DNA cleavage, peroxynitrite mobilized Ca^{2+} from the ryanodine receptor, and a significant fraction of the cation was taken up by the mitochondria. This process appears to be critical for DNA strand scission, as this event is abolished under conditions in which ryanodine suppresses Ca^{2+} mobilization and the ensuing mitochondrial accumulation of the cation (Guidarelli et al., 2006, 2007).

Experiments performed in permeabilized cells provide results consistent with the notion that the DNA-damaging response evoked by peroxynitrite indeed requires both electron transport and mitochondrial Ca^{2+} accumulation.

Figure 6.2B shows that the DNA strand scission induced by 200 μM peroxynitrite (10 min) in permeabilized cells is sensitive to rotenone, myxothiazol, and ryanodine, as observed previously in intact cells (see earlier discussion). In addition, DNA cleavage was suppressed by a very low concentration of ruthenium red, which, under these conditions, specifically prevents mitochondrial Ca^{2+} uptake (Carafoli, 1987), as well as by lanthanium ions, known to competitively inhibit mitochondrial Ca^{2+} uptake (Thomas and Reed, 1988). Finally, DNA strand scission was prevented by enzymatically active catalase.

Thus, the information obtained with intact cells is both confirmed and implemented by the use of the permeabilized cell system. Hence, our method for the rapid assessment of DNA single-strand breakage proved to be reliable, versatile, and very effective for establishing the indirect induction of these lesions by peroxynitrite. Our studies provide compelling evidence against the commonly held notion that peroxynitrite promotes DNA strand scission directly. Future studies may lead to the development of novel pharmacological strategies preventing DNA damage mediated by peroxynitrite in inflamed tissues, an event of importance in carcinogenesis (Ohshima et al., 2003).

ACKNOWLEDGMENTS

This work was supported by grants from the Associazione Italiana per la Ricerca sul Cancro and from Ministero dell'Università e della Ricerca Scientifica e Tecnologica, Progetti di Interesse Nazionale (OC).

REFERENCES

Beckman, J. S., and Koppenol, W. H. (1996). Nitric oxide, superoxide and peroxynitrite: The good, the bad, and the ugly. *Am. J. Physiol.* **271**, C1424–C1437.

Brand, K. A., and Hermfisse, U. (1997). Aerobic glycolysis by proliferating cells: A protective strategy against reactive oxygen species. *FASEB J.* **11**, 388–395.

Carafoli, E. (1987). Intracellular calcium homeostasis. *Annu. Rev. Biochem.* **56**, 395–433.

Fiskum, G., Craig, S. W., Decker, G. L., and Lenhinger, A. L. (1980). The cytoskeleton of digitonin-treated rat hepatocytes. *Proc. Natl. Acad. Sci. USA* **77**, 3430–3434.

Guidarelli, A., Brambilla, L., Clementi, E., Sciorati, C., and Cantoni, O. (1997a). Stimulation of oxygen consumption promotes mitochondrial calcium formation of *tert*-butylhydroperoxide-induced DNA single-strand breakage. *Exp. Cell Res.* **237,** 176–185.

Guidarelli, A., Cerioni, L., and Cantoni, O. (2007). Inhibition of complex III promotes loss of Ca^{2+} dependence for mitochondrial superoxide formation and permeability transition evoked by peroxynitrite. *J. Cell Sci.* **120,** 1908–1914.

Guidarelli, A., Clementi, E., Brambilla, L., and Cantoni, O. (1997b). Mechanism of antimycin A-mediated enhancement of *tert*-butylhydroperoxide-induced single-strand breakage in DNA. *Biochem. J.* **328,** 801–806.

Guidarelli, A., Clementi, E., De Nadai, C., Barsacchi, R., and Cantoni, O. (2001). TNFα enhances the DNA single-strand breakage induced by the short-chain hydroperoxide analogue *tert*-butylhydroperoxide via ceramide-dependent inhibition of complex III followed by enforced superoxide and hydrogen peroxide formation. *Exp. Cell Res.* **270,** 56–65.

Guidarelli, A., Clementi, E., Sciorati, C., and Cantoni, O. (1998a). The mechanism of nitric oxide-mediated enhancement of *tert*-butylhydroperoxide-induced DNA single-strand breakage. *Br. J. Pharmacol.* **125,** 1074–1080.

Guidarelli, A., Clementi, E., Sciorati, C., Cattabeni, F., and Cantoni, O. (1997c). Calciumdependent mitochondrial formation of species mediating DNA single strand breakage in U937 cells exposed to sublethal concentrations of *tert*-butylhydroperoxide. *J. Pharmacol. Exp. Ther.* **283,** 66–74.

Guidarelli, A., Sciorati, C., Clementi, E., and Cantoni, O. (2006). Peroxynitrite mobilizes calcium ions from ryanodine-sensitive stores, a process associated with the mitochondrial accumulation of the cation and the enforced formation of species mediating cleavage of genomic DNA. *Free Radic. Biol. Med.* **41,** 154–164.

Guidarelli, A., Sestili, P., and Cantoni, O. (1998b). Opposite effects of nitric oxide on DNA strand scission and toxicity caused by tert-butylhydroperoxide in U937 cells. *Br. J. Pharmacol.* **123,** 1311–1316.

Guidarelli, A., Tommasini, I., Fiorani, M., and Cantoni, O. (2000). Essential role of the mitochondrial respiratory chain in peroxynitrite-induced strand scission of genomic DNA. *IUBMB Life* **50,** 195–201.

Müller, W. U., Bauch, T., Streffer, C., Niedereichlz, F., and Bocker, W. (1994). Comet assay studies of radiation-induced DNA damage and repair in various tumour cell lines. *Int. J. Radiat. Biol.* **65,** 315–319.

Ohshima, H., Tatemichi, M., and Sawa, T. (2003). Chemical basis of inflammation-induced carcinogenesis. *Arch. Biochem. Biophys.* **417,** 3–11.

Radi, R., Beckman, J. S., Bush, K. M., and Freeman, B. A. (1991). Peroxynitrite oxidation of sulfhydryls: The cytotoxic potential of superoxide and nitric oxide. *J. Biol. Chem.* **266,** 4244–4250.

Sestili, P., and Cantoni, O. (1999). Osmotically driven radial diffusion of single-stranded DNA fragments on an agarose bed as a convenient measure of DNA strand scission. *Free Radic. Biol. Med.* **26,** 1019–1026.

Singh, N. P., McCoy, M. T., Tice, R. R., and Schneider, E. L. (1988). A simple technique for quantitation of low levels of DNA damage in individual cells. *Exp. Cell Res.* **175,** 184–191.

Szabó, C., and Ohshima, H. (1997). DNA damage induced by peroxynitrite: Subsequent biological effects. *Nitric Oxide* **1,** 373–385.

Thomas, C. E., and Reed, D. J. (1988). Effect of extracellular Ca^{++} omission on isolated hepatocytes. II. Loss of mitochondrial membrane potential and protection by inhibitors of uniport Ca^{++} transduction. *J. Pharmacol. Exp. Ther.* **245,** 501–507.

Tommasini, I., Sestili, P., and Cantoni, O. (2002). Delayed formation of hydrogen peroxide mediates the lethal response evoked by peroxynitrite in U937 cells. *Mol. Pharmacol.* **61,** 870–878.

CHAPTER SEVEN

NITRIC OXIDE: INTERACTION WITH THE AMMONIA MONOOXYGENASE AND REGULATION OF METABOLIC ACTIVITIES IN AMMONIA OXIDIZERS

Ingo Schmidt

Contents

1. Ammonia-Oxidizing Bacteria	122
2. Aerobic Ammonia Oxidation	122
3. Anaerobic Ammonia Oxidation with Nitrogen Dioxide as Oxidant Releases NO as Product	123
4. Aerobic Ammonia Oxidation with Nitrogen Dioxide as Oxidant	125
5. Regulation of Metabolic Activities in Ammonia Oxidizers	127
6. Nitric Oxide Induces Denitrification in *N. europaea*	128
7. Nitric Oxide Induces the Biofilm Formation of *N. europaea*	129
8. Nitric Oxide Is Required in *N. europaea* to Restore Ammonia Oxidation after Chemoorganotrophic Denitrification	132
References	133

Abstract

The biological nitrogen cycle is a complex interplay of many microorganisms. In the past, oxidation of the inorganic nitrogen compound ammonia by the ammonia oxidizing bacteria was thought to be restricted to oxic environments, and the metabolic flexibility of these organisms seemed to be limited. The discovery of an anaerobic metabolism in the late 1990s showed that these assumptions are no longer valid. NO and NO_2 are essential intermediates in ammonia oxidation. Both gases have wide-ranging regulatory effects on ammonia oxidation, denitrification, and biofilm formation of the ammonia oxidizing bacteria.

Mikrobiologie, Universität Bayreuth, Bayreuth, Germany

1. Ammonia-Oxidizing Bacteria

Microorganisms participating in nitrification have been characterized as obligatory chemolithoautotrophic ammonia- or nitrite-oxidizing bacteria. Lithotrophic nitrifiers belong to the family Nitrobacteriaceae (Watson et al., 1989), although they are not necessarily related phylogenetically. Although they were thought to be restricted in their metabolic potentialities, they have been found in many different ecosystems, such as fresh water, salt water, sewage systems, soils, and on/in rocks, as well as in masonry. Additionally, they are found in extreme habitats such as in Antarctic soils and in environments with pH values of about 4 (e.g., acid tea and forest soils), as well as with pH values of about 10 (e.g., soda lakes). Growth under suboptimal conditions seems to be possible by ureolytic activity, aggregate formation, or in biofilms on the surfaces of substrata. It is even more interesting to note that aerobic nitrifiers were also found in anoxic environments (Abeliovich and Vonhak, 1992). For many years there was no consistent explanation for this observation, as ammonia oxidizers were thought to be obligatory aerobic bacteria. The proposed idea of a drift from oxic layers was not convincing, as the cell numbers of ammonia oxidizers are (too) high even deep in the anoxic layers. Furthermore, bacteria responded almost immediately with increasing ammonia oxidation activities to the exposure to ammonia and oxygen (Abeliovich and Vonhak, 1992), indicating an active metabolism in these anoxic ecosystems as well. First evidence for a metabolic activity of ammonia oxidizers in anoxic environment was given by identifying their anaerobic denitrification capability. *Nitrosomonas* can grow with organic compounds or hydrogen as the electron donor and nitrite as the electron acceptor (Bock et al., 1995; Schmidt et al., 2004b). *Nitrosomonas* was further shown to oxidize ammonia in an oxygen-independent process (Schmidt and Bock, 1997).

Ammonia is a central compound of the global nitrogen cycle and the major source of nitrogen for most organisms. Key to the nitrification activity of ammonia oxidizers such as *Nitrosomonas europaea* are the enzymes ammonia monooxygenase (AMO) and hydroxylamine oxidoreductase (HAO) catalyzing the sequential oxidation of ammonia via hydroxylamine (AMO) to nitrite (HAO). This chapter focuses on *N. europaea* as the most studied ammonia oxidizer and stresses the importance of nitrogen oxides (NO, NO_2) for the bacteria's metabolism and its regulation.

2. Aerobic Ammonia Oxidation

Ammonia oxidizers first oxidize ammonia to hydroxylamine [Eq. (7.1)] with AMO. Second in this metabolism is the oxidation of hydroxylamine to nitrite [Eq. (7.2)] by HAO. Two of the four electrons released from the ammonia oxidation [Eq. (7.2)] are required for the AMO reaction [Eq. (7.1)], whereas the other two are employed in generation of a proton motive force in order to regenerate ATP and, via a reverse electron transport, NADH (Hooper *et al.*, 1997).

$$NH_3 + O_2 + 2H^+ + 2e^- \rightarrow NH_2OH + H_2O \qquad (7.1)$$

$$NH_2OH + H_2O \rightarrow HNO_2 + 4H^+ + 4e^- \qquad (7.2)$$

3. Anaerobic Ammonia Oxidation with Nitrogen Dioxide as Oxidant Releases NO as Product

To study the anaerobic ammonia oxidation, a Clark-type oxygen electrode (Schmidt and Bock, 1997) is used as the reaction vessel. The electrode is equipped with a gassing system that permanently flushes the vessel with oxygen-free helium gas. NO_2 is added from a gas cylinder containing 2000 ppm NO_2 in oxygen-free helium using a gas diluter. The gas stream through the system is adjusted at 150 ml min^{-1}. The chemical NO_2 consumption by nitrite and nitrate formation is measured in sterile control experiments, and chemical NO_2 consumption, as well as the formation of nitrite and nitrate, is taken into account for determining the stoichiometry of the anaerobic ammonia oxidation [Eqs. (7.3–7.5)]. Biological NO_2 consumption is evaluated from the amount of NO_2 added to a cell suspension minus the NO_2 emitted minus the NO_2 converted into nitrite or nitrate. The formation of NO is followed by two independent methods: First, the NO concentration is analyzed online in the gas headspace by a chemiluminescence-based assay (Brien *et al.*, 1996) using an NO/NO$_x$ analyzer (Eco physics). The analyzer is equipped with an additional valve that allows reducing the gas flow through the analyzer to below 100 ml min^{-1}. The experiments are gassed with 150 ml min^{-1} of an air/NO$_2$/NO mixture, and the low gas demand of the NO/NO$_x$ analyzer allows one to measure undiluted gas from the gas headspace. The analyzer is calibrated with gas standards between 0 and 200 ppm NO. Second, the Clark-type oxygen electrode is applied to analyze the NO concentration online in the cell suspension. The electrode is polarized with a potential of 800 mV. A mixture of NaCl and HCl (1 *M*) is used as the electrolyte. Because ammonia oxidation is a slow process in chemical terms, an experiment takes 5 to 15 min, and the relatively slow response of the electrode does not negatively influence the analysis of NO production during the experiments. The electrode is calibrated by flushing the headspace of the electrode chamber with helium gas with different NO partial pressures. Then, the NO concentration in the medium (water) is calculated (solubility of NO: $1.7 \cdot 10^{-3}$ M at 25°C, P_{NO} of 1 atm). Total NO formation during anaerobic ammonia oxidation is determined on the basis of NO released into the headspace plus the NO solved in the medium.

In the absence of oxygen, ammonia oxidizers are able to oxidize ammonia with nitrogen dioxide or dinitrogen tetroxide (dimeric form of NO_2) as the oxidant, respectively (Schmidt and Bock, 1997). This anaerobic ammonia oxidation [Eq. (7.3)] is also catalyzed by AMO (Schmidt and Bock, 1998; Schmidt et al., 2001a), replacing molecular oxygen by NO_2/N_2O_4 [Eq. (7.3)].

$$NH_3 + N_2O_4 + 2H^+ + 2e^- \rightarrow NH_2OH + H_2O + 2NO \quad (7.3)$$

$$NH_2OH + H_2O \rightarrow HNO_2 + 4H^+ + 4e^- \quad (7.4)$$

Ammonia and N_2O_4 are consumed by *Nitrosomonas* in a ratio of about one to one [Eq. (7.3)]. In the course of the reaction, N_2O_4 is reduced to NO. Per 1 mol N_2O_4, 2 mol NO is formed and released from the cells. Hydroxylamine, which is formed in a 1:1 stoichiometry to the ammonia consumed, is further oxidized to nitrite [HAO; Eq. (7.4)]. The oxidation of hydroxylamine does not depend on oxygen and is under both oxic and anoxic conditions catalyzed by the HAO. While under oxic conditions, oxygen is used as the terminal electron acceptor; under anoxic conditions, the nitrite produced during ammonia oxidation is used instead. This leads to the formation of dinitrogen [Eq. (7.5)]. About two-thirds of the produced nitrite is converted to N_2. The amount of N-nitrite plus N-dinitrogen is almost equivalent to the amount of consumed N-ammonium. A model illustrating the anaerobic ammonia oxidation of *N. europaea* is shown in Fig. 7.1A.

$$HNO_2 + 3H^+ + 3e^- \rightarrow 0.5N_2 + 2H_2O \quad (7.5)$$

Under anoxic conditions with 25 ppm NO_2 in the gas headspace, a specific anaerobic ammonia oxidation activity of about 130 μmol ammonia (g protein)$^{-1}$ h^{-1} can be measured. In the same experimental setup, but in an air atmosphere (21% oxygen), about 1400 μmol ammonia (g protein)$^{-1}$ h^{-1} is oxidized (Schmidt and Bock, 1997). The specific anaerobic ammonia oxidation activity is low (10-fold lower) due to the fact that only low NO_2 concentrations can be added to a culture of *Nitrosomonas*, as NO_2 concentrations of more than 75 ppm already have an inhibitory effect on the ammonia oxidation activity of these bacteria. When an atmosphere with 25 ppm O_2 is applied, ammonia oxidation is hardly detectable and the specific activity is below 10 μmol ammonia (g protein)$^{-1}$ h^{-1}. These findings indicate that the affinity of the AMO for NO_2/N_2O_4 [$K_s(NO_2)$ 52 ppm in the gas headspace] is higher than for O_2 [$K_s(NO_2)$ 55,000 ppm in the gas headspace]. This paved the way for the assumption that NO_2 might be a suitable oxidant for aerobic ammonia oxidation as well (see later).

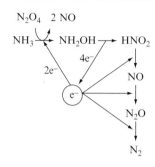 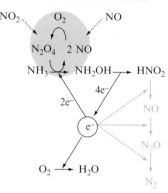

Figure 7.1 Hypothetical models of anaerobic (A) and aerobic (B) ammonia oxidation by *Nitrosomonas* (adapted from Schmidt and Jetten, 2004). The NO_x cycle proposed for aerobic ammonia oxidation at AMO is shown on the gray background (B). (A) N_2O_4 (NO_2)-dependent ammonia oxidation under anoxic conditions and (B) aerobic ammonia oxidation. Dotted lines: When the gas headspace is supplemented with NO or NO_2, small amounts of both compounds are consumed and ammonia oxidation activity is increased significantly. Under both oxic and anoxic conditions, the same enzyme, AMO, is responsible for the oxidation of ammonia to hydroxylamine. NO and NO_2 are assumed to act as additional oxidants, filling up the "NO_x pool" at the AMO. NO is an end product of anaerobic ammonia oxidation. It is released and can be measured in the medium and the gas headspace (A). In the presence of oxygen, only minor amounts of NO are detectable. Evidence shows that NO is reoxidized to NO_2 in *N. europaea* cells, most likely at AMO. NO_2/N_2O_4 can serve as the oxidant for the ammonia oxidation again [NO_x cycle of *Nitrosomonas* (B)]. The oxidation of hydroxylamine is the same under oxic and anoxic conditions. From the four electrons derived from the oxidation of hydroxylamine, two are required in ammonia oxidation. Under anoxic conditions, the other two electrons are transferred to nitrite as terminal electron acceptors and, under oxic conditions, to oxygen or to a lesser extent to nitrite [(B) Schmidt and Jetten, 2004].

4. Aerobic Ammonia Oxidation with Nitrogen Dioxide as Oxidant

Aerobic ammonia oxidation is studied as described earlier, with the only difference being that the Clark-type oxygen electrode is gassed with an air/NO or NO_2 mixture. NO concentrations in the medium are monitored with the Clark-type electrode; NO and NO_2 concentrations in the gas headspace are measured with the NO/NO_x analyzer.

Results show that aerobic ammonia oxidation also uses NO_2/N_2O_4, but not oxygen as the oxidizing agent at the AMO (Fig. 7.1B). NO is formed [Eq. (7.3)] and then oxygen is required to reoxidize NO to NO_2. NO_2/N_2O_4 is again used as the oxidant for ammonia oxidation. This NO/NO_2

conversion is called the "NO_x cycle" (Fig. 7.1B; Schmidt et al., 2001a). Evidence for this mechanism is described in this chapter.

When NO or NO_2 is added to the gas headspace of a *Nitrosomonas* cell suspension under oxic growth conditions, the ammonia oxidation activity is increased significantly (Table 7.1), and both NO and NO_2 are consumed.

In contrast to anaerobic ammonia oxidation, a stoichiometry between the ammonia and the NO or NO_2 consumption is not detectable for aerobic ammonia oxidation. Because only very small amounts of NO or NO_2 are consumed, both gases cannot directly affect this increased ammonia oxidation activity by acting as a substrate. In such a case, NO, NO_2, and ammonia consumption in the same order of magnitude has to be expected, but in comparison with the ammonia oxidation rate, the NO and NO_2 consumption rates are 100 to 1000 times lower, and a fixed stoichiometry between the consumption rates was never observed (Schmidt et al., 2001c; Zart et al., 2000). Therefore, it was suggested that NO and NO_2 "fill up" the NO_x pool at the AMO, resulting in a higher availability of the oxidant NO_2/N_2O_4 and consequently in an increased ammonia oxidation activity (Fig. 7.1B). Further experiments provide additional evidence of a function of NO and NO_2 in ammonia oxidation: First, a rapid inactivation of ammonia oxidation, e. g., by adding the specific inhibitor acetylene or by rapidly reducing the temperature to below 4 °C, leads to a release of NO. Per AMO, 0.5 to 2.8 NO molecules are released. Second, when inactive *N. europaea* cells (starved for ammonia for 1 day or longer) are transferred into a buffer containing ammonia, the length of time of the lag phase is dependent of the availability of NO and/or NO_2. Removing NO and NO_2 by strong aeration with NO_x-free air, the lag phase is prolonged significantly, whereas the addition of NO or NO_2 results in a shorter lag phase (Schmidt et al., 2001c). As a consequence, ammonia oxidizers should be able to regulate their ammonia oxidation activity by controlling their endogenous NO production. In cells with increased denitrification activity (increased NO production), ammonia oxidation activity is significantly increased as well (Zart and Bock, 1998). Therefore, NO is assumed to be the regulatory signal controlling ammonia oxidation activity.

Table 7.1 Effect of NO and NO_2 on the specific ammonia oxidation activity of *N. eutropha* (adapted from Zart and Bock, 1998)

NO_x concentration in the gas headspace	Specific ammonia oxidation activity [μmol (mg protein)$^{-1}$ h^{-1}]
Without NO_x	2.0
25 ppm NO	12.4
50 ppm NO	11.4
25 ppm NO_2	13.2
50 ppm NO_2	17.4

The new model of aerobic ammonia oxidation (Fig. 7.1B) resulting from the aforementioned findings is not in conflict with the previous model in which oxygen was thought to act directly as an oxidant in ammonia oxidation (Hooper *et al.*, 1997). According to the new model, the oxygen in hydroxylamine still originates from O_2, but O_2 is not used to oxidize ammonia directly, but rather to (re-)oxidize NO to NO_2, which then serves as an oxidant introducing the oxygen atom into ammonia, thereby forming hydroxylamine.

Summarizing, the following findings have been the basis in developing the model of anaerobic and aerobic ammonia oxidation (Fig. 7.1). First, NO_2/N_2O_4 is used as an oxidant and NO is released during anaerobic ammonia oxidation. Second, NO_2 is not available in natural environments under anoxic conditions. Anaerobic ammonia oxidation is therefore dependent on the transport of NO_2 from oxic layers. Third, under anoxic conditions, nitrite is the only suitable terminal electron acceptor available (Fig. 7.1A). Two of the four electrons released during the oxidation of hydroxylamine are required for ammonia oxidation. The other two electrons, used to conserve energy, are transferred to the nitrite as a terminal acceptor. To reduce nitrite to molecular dinitrogen (N_2), three electrons per nitrogen atom are necessary. As a consequence, up to two-thirds of the produced nitrite is converted to N_2 (Schmidt and Bock, 1998). Fourth, under oxic growth conditions, the produced NO can be (re)oxidized to NO_2 (N_2O_4) (Fig. 7.1B; Schmidt *et al.*, 2001a). Because of this (re)oxidation of NO, only small amounts of this gas are detectable in the gas headspace of *Nitrosomonas* cell suspensions. Fifth, ammonia oxidation with N_2O_4 as the oxidant is more favorable from an energetic point of view [Eq. (7.3); $\Delta G^{0,}$ -140 kJ mol^{-1}] compared to a reaction directly with oxygen [Eq. (7.1), $\Delta G^{0,}$ -120 kJ mol^{-1}]. The first step of ammonia oxidation is not the energy-generating step of the process, but the lower reaction enthalpy might accelerate the oxidation of ammonia, providing more hydroxylamine for energy generation. This fact might explain the increased ammonia oxidation activity and the higher growth rate of *Nitrosomonas* with NO_2/N_2O_4 as the oxidant (Schmidt *et al.*, 2002; Zart and Bock, 1998). Sixth, the addition of NO_2 or NO increases the specific ammonia oxidation activity under oxic conditions. The addition of NO or NO_2 may supply the NO_x cycle with additional substrate, increasing the amount of oxidant available for the AMO and thus finally increasing the specific ammonia oxidation activity.

5. REGULATION OF METABOLIC ACTIVITIES IN AMMONIA OXIDIZERS

When NO and NO_2/N_2O_4 were discovered as key elements of ammonia oxidation of *N. europaea* and other ammonia oxidizers (Fig. 7.1) in the 1990s, interest naturally focused on the regulatory functions of NO and NO_2. This section discusses the regulatory effects of NO on the metabolism of ammonia oxidizers.

6. NITRIC OXIDE INDUCES DENITRIFICATION IN *N. EUROPAEA*

The genome of *N. europaea* encodes for a nitrite reductase and a NO reductase. Under oxic growth conditions, both proteins are expressed (Beaumont et al., 2002, 2004a). Nevertheless, the aerobic denitrification activity of *N. europaea* remains fairly low and nitrite consumption activity hardly exceeds 5% of ammonia consumption activity. It is interesting to note that an N_2O reductase has not been identified in the genome of *N. europaea*, although N_2 has been reported as a product of the denitrification pathway (Bock et al., 1995; Poth, 1985). The denitrification pathway is of great importance for *N. europaea* when grown under anoxic conditions by chemoorganotrophic denitrification. Nitrite, NO, and N_2O are suitable electron acceptors supporting the growth of *N. europaea* (Schmidt et al., 2004b). The function of the denitrification pathway under oxic conditions is not understood yet. It was proposed that aerobic denitrification might be important for ammonia oxidation with NO (NO_x cycle; Schmidt and Jetten, 2004): First, denitrification can provide NO, which is required for the NO_x cycle. The ammonia oxidation activity correlates with the availability of NO (see earlier discussion). Second, the simultaneous utilization of nitrite and oxygen as terminal electron acceptors might accelerate the energy metabolism (energy conservation) of *Nitrosomonas*, resulting in significantly increased growth rates (Zart and Bock, 1998). Another assumption is that aerobic denitrification might reduce the concentration of nitrite, which is toxic for *N. europaea* at high concentrations (Beaumont et al., 2002). Whether the usually low denitrification activity is sufficient to keep the nitrite concentration below a toxic level is questionable, though.

Anyhow, when NO is added to the gas headspace of an *N. europaea* cell suspension (chemostat) or NO-donating compounds are added, a significantly increased aerobic denitrification activity is detectable (Zart and Bock, 1998). Temporarily, more than 50% of the nitrite produced by ammonia oxidation is denitrified [N loss >50%; N loss is the loss of dissolved N compounds (ammonia, nitrite) by the formation of gaseous N compounds (NO, N_2O, N_2)]. The NO concentration and the N loss of *Nitrosomonas* are correlated.

The regulatory mechanisms controlling the denitrification activity in *N. europaea* are not fully understood yet. Very recent results indicate that transcription of the nitric oxide reductase in *N. europaea* is up to six times upregulated at NO concentrations above 200 ppm in the gas headspace

(Beyer and Schmidt, unpublished result). In these experiments, the transcription of the nitrite reductase (*nirK*) was hardly influenced by NO. A variety of regulatory systems that control the expression of denitrification enzymes or is involved in the defense against reactive oxygen species such as SoxR, FNR, NorR, Fur, NnrR, NsnR, and H-NOX have been described in bacteria (Spiro, 2006). So far, in *N. europaea*, only the FNR and NsrR systems have been investigated (Beaumont *et al.*, 2002, 2004a,b). The FNR system is not essential for the expression of nitrite reductase and NO reductase. In contrast, NsrR represses transcription from the promotor upstream of the *nirK*. Anyhow, the control of the expression by NsrR is obviously dependent on the concentration of nitrite and the pH value of the medium, but not on the NO or O_2 concentration (Beaumont *et al.*, 2004a). Because the NO concentration was shown to influence the denitrification activity of *N. europaea* (Zart and Bock, 1998), further regulatory systems with NO as the signal are an obvious presumption.

7. NITRIC OXIDE INDUCES THE BIOFILM FORMATION OF *N. EUROPAEA*

A regulatory function of NO was observed in *N. europaea* when biofilm formation was studied (Schmidt *et al.*, 2004a). Changes in the growth mode (planktonic/motile or biofilm) of *N. europaea* were examined depending on different parameters, such as NO, ammonium, nitrite, and oxygen concentrations, as well as the pH value and temperature. Studies showed that the NO concentration is the only parameter that significantly influences the growth mode. When an *N. europaea* population growing in a chemostat is supplemented with NO (aeration gas: air–NO mixture), the cell number of planktonic/motile cells decreases while biofilm formation is detectable (Fig. 7.2). About 65% of the planktonic/motile cells actively attach to the substratum, and the number of biofilm cells increases rapidly during NO supplementation (assembly of the biofilm; Fig. 7.2B, days 3 to 8). The change in growth mode is accompanied by a change in the protein pattern. Interestingly, the NO concentration stabilizes at a high level (about 40 ppm) when NO supplementation is stopped (Fig. 7.2, days 8 to 17). Obviously, the exposure of *Nitrosomonas* to NO not only influences the growth mode, but also the metabolic activity by inducing increased denitrification and NO production (see earlier discussion). As a consequence of a high NO concentration, cells remain in the biofilm growth mode. When NO is actively removed from the system by increasing the aeration rate, and the NO concentration is reduced to less than 5 ppm, the cell number of the planktonic/motile cells increases again (Fig. 7.2B, days 17 to 22). It is interesting to note that NO added to a *Nitrosomonas* culture leads to a net

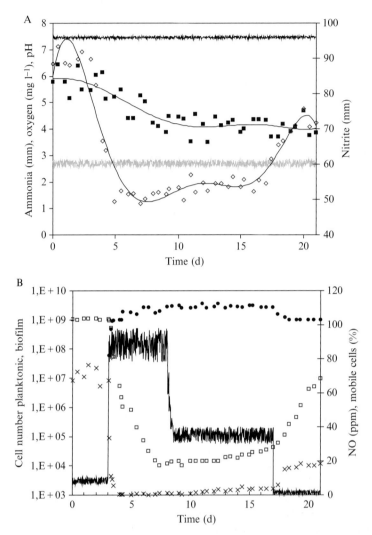

Figure 7.2 *N. europaea* grown in chemostat (from Schmidt *et al.*, 2004a). Depending on the NO concentration, cells switch between planktonic/motile or biofilm mode. The 5-liter laboratory scale reactor was operated as a chemostat with 3.5 liters mineral medium (Schmidt and Bock, 1997). (A) Ammonium (■), oxygen (gray line), nitrite (◊) concentration, and pH value (black line). (B) NO concentration in the gas headspace (black line), cell number of motile/planktonic cells (□) and biofilm cells (●), and the ratio of motile to motile/planktonic cells (x). The reactor was aerated with about 0.5 liter air min^{-1}; the oxygen concentration was at 5 ± 0.2 mg liter^{-1}. During days 3 to 8, the aeration gas was supplemented with NO (110 ppm) to maintain an NO concentration of about 90 ppm in the gas headspace (B). At day 8, NO supplementation was stopped. The NO detectable in the headspace (about 40 ppm) was now produced by *N. europaea*. On day 17, aeration was increased to 5 liters air/N$_2$ min^{-1} in order to reduce the NO concentration in the chemostat. An air/N$_2$ mixture was used to maintain the oxygen

consumption of NO (NO$_x$ cycle and Fig. 7.2B, days 3–8), whereas without added NO, a net production of NO is detectable (Fig. 7.2B, days 0–3 and 8–22). On the one hand, an excess of NO seems to be consumed by the NO$_x$ cycle, whereas on the other hand, the cells have an endogenous overproduction of NO, which is released to the gas headspace.

Several proteins of *N. europaea* are highly significantly up- or downregulated, depending on the growth mode (Schmidt *et al.*, 2004a). Flagella and flagella assembly proteins are expressed in planktonic/motile cells, but are not detectable in immotile biofilm cells. An increased NO concentration (above 30 ppm) induces the change in the growth mode from planktonic/motile toward biofilm cells, and the expression of flagella protein is downregulated. The reverse process occurs when NO supplementation is stopped and the NO concentration is adjusted below 5 ppm; the flagella protein is upregulated and the number of planktonic/motile cells increases (Fig. 7.2B; Schmidt *et al.*, 2004a). Typical of biofilm cells is the expression of two regulatory proteins, a CheY-like protein (response regulator consisting of a CheY-like receiver domain and a winged-helix DNA-binding domain like that of OmpR) and cheZ (chemotaxis protein), whereas in planktonic/motile cells, cheW (chemotaxis protein) is detectable. It can be speculated that these regulatory proteins might be involved in NO sensing and/or signal transduction. Interestingly, expression of the proteins nitrosocyanin and succinyl-CoA synthetase is also modulated. The concentration of both proteins is significantly higher in biofilm cells (approximately by a factor of 10). The succinyl-CoA synthetase catalyzes the reversible conversion of succinyl-CoA to succinate. Although the citric acid cycle is central to the energy yielding metabolism in many microorganisms, the C4 and C5 carbon intermediates serve as biosynthetic precursors for a variety of products. In view of the higher content of succinyl-CoA synthetase in biofilm cells, two pathways might be discussed: (1) the gluconeogenesis and subsequently the production of extracellular polymer substances via oxaloacetate as the basis of biofilm formation and (2) the synthesis of porphyrin and heme starting with succinyl-CoA, which might explain the high concentration of nitrosocyanin. Nitrosocyanin is a periplasmic protein with homology to plastocyanin, the electron donor copper region of the N$_2$O reductase (Whittaker *et al.*, 2000). The function of nitrosocyanin has not been determined yet. Interestingly, when a high nitrosocyanin concentration is detectable in the cells, the specific denitrification activity increases about fivefold (N loss from 9 to 50%, Fig. 7.2A). Because plastocyanin is an electron shuttle involved in N$_2$O reduction, it can be assumed that

---

concentration at 5 ± 0.2 mg liter^{-1}, although the gas flow was increased 10 times. The dilution rate of the chemostat was 0.1 h^{-1}. Throughout the experiment, the temperature was 28°C. The pH value was kept at 7.4 [20% (w/v) Na$_2$CO$_3$].

nitrosocyanin has a similar function in the denitrification (N$_2$O reduction) activity of *N. europaea*.

8. NITRIC OXIDE IS REQUIRED IN *N. EUROPAEA* TO RESTORE AMMONIA OXIDATION AFTER CHEMOORGANOTROPHIC DENITRIFICATION

When *Nitrosomonas* cells are grown with pyruvate as the electron donor and nitrite as the electron acceptor under anoxic conditions, the amount of AMO in the cells decreases. After about 2 weeks, AMO is no longer detectable in the denitrifying cells (Schmidt et al., 2001b). Consequently, when these cells are transferred to an ammonium-containing medium and to oxic conditions, they are not able to oxidize ammonia. Evidence shows that the recovery of ammonia oxidation activity strongly depends on the NO and NO$_2$ concentration (Table 7.2). Upon addition of NO or NO$_2$, ammonia consumption starts after 1 to 4 days. Most notably, the concentration of AMO, as well as the concentrations of a- and c-type cytochromes, increases with increasing ammonia oxidation activity (Schmidt et al., 2001c). When removing NO and NO$_2$ produced by the cells by intensive aeration (about 1 liter air min^{-1} to aerate a 400-ml *N. europaea* cell suspension), a recovery of ammonia oxidation is not detectable within 4 weeks.

In denitrifying *Nitrosomonas* cells, only small amounts of cytochrome are detectable, but with reactivation of ammonia oxidation activity, the cytochrome concentration of both a- and c-type cytochromes increases simultaneously (Schmidt et al., 2001b). Therefore, a- and c-type cytochromes seem to be related to the ammonia oxidation of *Nitrosomonas*.

Table 7.2 Recovery of ammonia oxidation of denitrifying *N. eutropha* cells incubated in different gas headspaces[a]

Gas headspace	Time required for the recovery of ammonia oxidation[b] (hours)
Air without NO or NO$_2$	214
Air with 25 ppm NO	88
Air with 50 ppm NO	74
Air with 200 ppm NO	49
Air with 25 ppm NO$_2$	77
Air with 50 ppm NO$_2$	65
Air with 100 ppm NO$_2$[c]	34

[a] Adapted from Schmidt et al. (2001c).
[b] The time required to recover ammonia oxidation is defined as the period of time until the first 200 μM ammonium is oxidized.
[c] A NO$_2$ concentration of 200 ppm is lethal for *N. europaea* in the experiments.

Because hydroxylamine oxidation activity also increases during the recovery of ammonia oxidation, it can be assumed that the HAO (a heme protein) and electron transport components containing a- and c-type cytochromes are synthesized during this period. Addition of the translation inhibitor chloramphenicol renders cells unable to recover their ammonia oxidation activity. These observations indicate that a recovery of ammonia oxidation corresponds with *de novo* synthesis of AMO, HAO, and electron transport components.

It might be speculated that the nitrogen oxide couple $NO-NO_2$ might function as an oxygen-sensing element responsible for switching between anaerobic denitrification and aerobic ammonia oxidation. Under anoxic conditions, NO is a stable compound—oxidation to NO_2 is not possible— and *Nitrosomonas* cells remain in the "denitrification mode." As soon as oxygen enters the system (environment), NO_2 becomes available by the oxidation of NO. This might be the signal that switches *Nitrosomonas* into the "ammonia oxidation mode." Hence, the $NO-NO_2$ concentrations are the regulatory signal for a shift between pyruvate-dependent denitrification and ammonia oxidation of *Nitrosomonas*.

Metabolism of the ammonia oxidizers, family Nitrobacteriaceae, is highly flexible and allows this group of organisms not only to survive, but also to grow in diverse habitats. Scientific research has just started to uncover the regulation of their metabolism and the niche differentiation of different species, and next to nothing is known about the evolutionary development of these organisms. It had been speculated that microorganisms with intracytoplasmic membrane systems might have the same origin and might be closely related to phototrophic microorganisms (Watson *et al.*, 1989). The recent discoveries of the metabolic flexibility of the ammonia oxidizers allows one to speculate about the origin of the ammonia oxidizers as denitrifying microorganisms: According to this hypothesis, the exposure to oxygen put an exerted selective pressure on these bacteria to develop a detoxification pathway for compounds such as O_2, NO, and NO_2. In ammonium-rich environments, ammonia might have been a reasonable electron donor for this detoxification. Later evolution may have established ammonia oxidation as a pathway to generate energy.

Future research may help uncover the background of the still undeciphered development of the ammonia oxidizers.

REFERENCES

Abeliovich, A., and Vonhak, A. (1992). Anaerobic metabolism of *Nitrosomonas europaea*. *Arch. Microbiol.* **158,** 267–270.

Beaumont, H. J., Hommes, N. G., Sayavedra-Soto, L. A., Arp, D. J., Arciero, D. M., Hooper, A. B., Westerhoff, H. V., and van Spanning, R. J. (2002). Nitrite reductase of

Nitrosomonas europaea is not essential for production of gaseous nitrogen oxides and confers tolerance to nitrite. *J. Bacteriol.* **184**, 2557–2560.

Beaumont, H. J., Lens, S. I., Reijnders, W. N., Westerhoff, H. V., and van Spanning, R. J. (2004a). Expression of nitrite reductase in *Nitrosomonas europaea* involves NsrR, a novel nitrite-sensitive transcription repressor. *Mol. Microbiol.* **54**, 148–158.

Beaumont, H. J., van Schooten, B., Lens, S. I., Westerhoff, H. V., and van Spanning, R. J. (2004b). *Nitrosomonas europaea* expresses a nitric oxide reductase during nitrification. *J. Bacteriol.* **186**, 4417–4421.

Bock, E., Schmidt, I., Stüven, R., and Zart, D. (1995). Nitrogen loss caused by denitrifying *Nitrosomonas* cells using ammonia or hydrogen as electron donors and nitrite as electron acceptor. *Arch. Microbiol.* **163**, 16–20.

Brien, J. F., McLaughlin, B. E., Nakatsu, K., and Marks, G. (1996). Chemiluminescence headspace-gas analysis for determination of nitric oxide formation in biological systems. *Methods Enzymol.* **268**, 83–92.

Hooper, A. B., Vannelli, T., Bergmann, D. J., and Arciero, D. M. (1997). Enzymology of the oxidation of ammonia to nitrite by bacteria. *Antonie van Leeuwenhoek* **71**, 59–67.

Poth, M. (1986). Dinitrogen production from nitrite by a *Nitrosomonas* isolate. *Appl. Environ. Microbiol.* **52**, 957–959.

Schmidt, I., and Bock, E. (1997). Anaerobic ammonia oxidation with nitrogen dioxide by *Nitrosomonas eutropha*. *Arch. Microbiol.* **167**, 106–111.

Schmidt, I., and Bock, E. (1998). Anaerobic ammonia oxidation by cell free extracts of *Nitrosomonas eutropha*. *Antonie van Leeuwenhoek* **73**, 271–278.

Schmidt, I., Bock, E., and Jetten, M. S. M. (2001a). Ammonia oxidation by *Nitrosomonas eutropha* with NO_2 as oxidant is not inhibited by acetylene. *Microbiology* **147**, 2247–2253.

Schmidt, I., Hermelink, C., van de Pas-Schoonen, K., Strous, M., op den Camp, H. J., Kuenen, J. G., and Jetten, M. S. M. (2002). Anaerobic ammonia oxidation in the presence of nitrogen oxides (NOx) by two different lithotrophs. *Appl. Environ. Microbiol.* **68**, 5351–5357.

Schmidt, I., and Jetten, M. S. M. (2004). Anaerobic oxidation of inorganic nitrogen compounds. *In* "Strict and Facultative Anaerobes: Medical and Environmental Aspects" (M. M. Nakano and P. Zuber, eds.), pp. 283–303. Horizon Scientific Press, Springer-Verlag.

Schmidt, I., Steenbakkers, P. J. M., op den Camp, H. J. M., Schmidt, K., and Jetten, M. S. M. (2004a). Physiologic and proteomic evidence for a role of nitric oxide in biofilm formation by *Nitrosomonas europaea* and other ammonia oxidizers. *J. Bacteriol.* **186**, 2781–2788.

Schmidt, I., van Spanning, R. J. M., and Jetten, M. S. M. (2004b). Denitrification and ammonia oxidation by *Nitrosomonas europaea* wild-type, and NirK and NorB-deficient mutants. *Microbiology* **150**, 4107–4114.

Schmidt, I., Zart, D., and Bock, E. (2001b). Effects of gaseous NO_2 on cells of *Nitrosomonas eutropha* previously incapable of using ammonia as an energy source. *Antonie van Leeuwenhoek* **79**, 39–47.

Schmidt, I., Zart, D., and Bock, E. (2001c). Gaseous NO_2 as a regulator for ammonia oxidation of *Nitrosomonas eutropha*. *Antonie van Leeuwenhoek* **79**, 311–318.

Spiro, S. (2006). Regulators of bacterial responses to nitric oxide. *FEMS Microbiol. Rev.* **31**, 193–211.

Watson, S. W., Bock, E., Harms, H., Koops, H.-P., and Hooper, A. B. (1989). Genera of ammonia-oxidizing bacteria. *In* "Bergey's Manual of Systematic Bacteriology" (J. T. Staley, M. P. Bryant, N. Pfennig, and J. G. Holt, eds.) pp. 1822–1834. Williams & Wilkins, Baltimore, MD.

Whittaker, M., Bergmann, D., Arciero, D., and Hooper, A. B. (2000). Electron transfer during the oxidation of ammonia by the chemolithotrophic bacterium *Nitrosomonas europaea*. *Biochim. Biophys. Acta* **1459,** 346–355.

Zart, D., and Bock, E. (1998). High rate of aerobic nitrification and denitrification by *Nitrosomonas eutropha* grown in a fermentor with complete biomass retention in the presence of gaseous NO_2 or NO. *Arch. Microbiol.* **169,** 282–286.

Zart, D., Schmidt, I., and Bock, E. (2000). Significance of gaseous NO for ammonia oxidation by *Nitrosomonas eutropha*. *Antonie van Leeuwenhoek* **77,** 49–55.

CHAPTER EIGHT

Chemiluminescent Detection of S-Nitrosated Proteins: Comparison of Tri-iodide, Copper/CO/Cysteine, and Modified Copper/Cysteine Methods

Swati Basu,[*] Xunde Wang,[†] Mark T. Gladwin,[†,‡] and Daniel B. Kim-Shapiro[*]

Contents

1. Introduction	138
2. Methods for Detection of S-Nitrosothiols	139
3. Chemiluminescent-Based Detection of S-Nitrosothiols	140
3.1. Tri-iodide method	140
3.2. Cu/CO/cysteine method	144
3.3. Modified 2C method	145
4. Advantages and Disadvantages	146
4.1. Tri-iodide method	146
4.2. Cu/CO/cysteine method	146
4.3. Modified 2C method	147
5. Comparisons and Validations	147
6. Conclusions	151
References	153

Abstract

The precise quantification of high and low molecular weight S-nitrosothiols (RSNOs) in biological samples is necessary for the study of nitric oxide-dependent posttranslational signal transduction. Several chemiluminescence-based methods are used for the detection of S-nitrosothiols using the nitric oxide analyzer, including the tri-iodide method and the Cu/CO/cysteine (3C) method.

[*] Department of Physics, Wake Forest University, Winston-Salem, North Carolina
[†] Pulmonary and Vascular Medicine Branch, National Heart Lung and Blood Institute, National Institutes of Health, Bethesda, Maryland
[‡] Critical Care Medicine Department, Clinical Center, National Institutes of Health, Bethesda, Maryland

Despite the fact that the tri-idodide method is the most widely used and validated methodology, the levels of *S*-nitrosated hemoglobin (SNO-Hb) and *S*-nitrosated albumin have been lower than those reported using photolysis coupled to chemiluminescence. This chapter demonstrates that the tri-iodide method and a newly developed modified copper/cysteine (2C) method compare favorably with the 3C method. Our comparisons include physiologically relevant conditions where the ratio of SNO to heme is low and the frequency of nitrosation of a given Hb tetramer is less than 1. In our studies, the tri-iodide method, the 3C method, and the modified 2C method give consistent and reproducible results. These studies suggest that the proper use of any of these methods can be effective in the accurate measurement of *S*-nitrosothiols in biological samples. Using more than one in combination has the potential to resolve controversies related to the role of hemoglobin in the generation of RSNOs or the role of RSNOs in biology, whether the RSNOs derive from nitrite or other nitrosative pathways.

1. Introduction

S-Nitrosothiols (RSNOs) are derived from covalent bonding of the oxidized form of the vasodilator nitric oxide (NO) with thiols such as glutathione (GSH), cysteine, hemoglobin, and albumin. Bioactive RSNOs are found in many tissues and are proposed to have a wide range of biological activities, which include vasoregulation, activation of sGC, inhibition of mitochondrial respiration, and regulation of platelet activation and hepatocyte and neuronal function (Arnelle and Stamler, 1995; Clementi *et al.*, 1998; Hothersall and Noronha-Dutra, 2002; Liu *et al.*, 2004; Mathews and Kerr, 1993; Miersch and Mutus, 2005). Low and high molecular weight S-nitrosothiols in the plasma are proposed to play a role in the regulation of bioavailability of NO and blood flow in humans (Foster *et al.*, 2003). In general, RSNOs have been proposed to participate in intracellular signal transduction (Gow *et al.*, 2004; Guikema *et al.*, 2005; Hogg, 2002; Miersch and Mutus, 2005) and have been considered as candidate therapeutic NO donors (Folkerts and Nijkamp, 2006; Gaston *et al.*, 1994; Janero *et al.*, 2004; Snyder *et al.*, 2002). Additionally, changes in RSNOs levels in specific tissues and blood have been associated with disease conditions (Chung *et al.*, 2005a, b; Elahi *et al.*, 2007; Feelisch *et al.*, 2002).

After the first researchers synthesized RSNO (Tasker and Jones, 1909) in 1974, Incze *et al.* (1974) showed that RSNO possessed antibacterial efficacy. This was followed by the discovery that RSNOs can activate soluble guanylyl cyclase, the target enzyme of NO (Ignarro, 1980). After this groundbreaking discovery, investigators developed sensitive methods to measure RSNOs *in vivo* (Stamler *et al.*, 1992). Since then many laboratories have used various methodologies to determine the levels of RSNOs in plasma, red blood cells, and blood (MacArthur et al., 2007).

2. METHODS FOR DETECTION OF S-NITROSOTHIOLS

Among the methods, absorption is the simplest method of detection of low molecular weight S-nitrosothiols such as S-glutathione (GSNO). The nitrosothiol covalent bond generally absorbs light around 340 nm. There is a weaker band around 545 nm. Some reported extinction coefficients for GSNO include 980 $M^{-1}cm^{-1}$ at 338 nm and 767 to 908 $M^{-1}cm^{-1}$ at 334 nm. Nitrosothiols need to be protected against light, as the S–NO bond is light sensitive and undergoes photolytic decomposition to NO and the thiyl radical (Singh et al., 1996). Shortcomings of this method include interference from other colored species (so it is very difficult to use absorbance spectroscopy with hemoglobin) and the detection limit, which is ≈5 μM, depending on the spectrophotometer and the path length of the cuvette used. Also, the presence of nitrite and low pH conditions can result in artifactual SNO measurements.

Other methods of detection of RSNOs include the use of colorimetry and fluorimetry (Tarpey et al., 2004). The Greiss/Saville colorimetric method allows for the detection of S-nitrosated proteins, as well as low molecular weight S-nitrosothiols (Saville, 1958). The nitrosonium ion (NO^+) is cleaved from RSNO by mercury chloride and then reacts with the Greiss/Saville reagents to form an Azo dye that can be detected by absorbance at 540 nm. A major limitation is the sensitivity, which is 0.1 μM for low molecular weight S-nitrosothiols and higher for high molecular weight S-nitrosothiols. In blood, levels of RSNO measured are small and the presence of hemoglobin lowers the sensitivity further. Therefore, hemoglobin should be filtered out from the sample to obtain a more accurate measurement of RSNO by this method (Basu et al., 2006). This method may be used for the detection of higher concentrations of RSNO, although at the same time other methods could also be used for verification, such as the neutral pH method and fluorescence techniques, which use similar principles and are more sensitive (Wink et al., 1999). Combining HPLC with the Griess method [as employed by the commercially available ENO instrument (EiCom Corp.)] may be used to measure nitrite and increase the sensitivity of the Greiss/Saville method (Kleinbongard et al., 2002; Yamada et al., 1998).

Fluorescence techniques for the detection of RSNO include labeling by the dye 4,5-diaminofluorescein. The nitrosonium ion is released from RSNOs by treatment with mercury chloride in the presence of the dye. This results in the formation of the fluorescent triazolofluorescein. Use of the thiol-reactive fluorescent dye, Cy5 maleimide, the biotin-switch assay, and use of antibodies against S-nitrocysteine are other methods that have been used for the identification of RSNOs and these have been reviewed elsewhere (Gladwin *et al.*, 2006; Gow *et al.*, 2007; Jaffrey and Snyder, 2001; Jaffrey *et al.*, 2001; Kettenhofen *et al.*, 2007).

3. Chemiluminescent-Based Detection of S-Nitrosothiols

Chemiluminescence-based detection of S-nitrosothiols can be carried out using a nitric oxide analyzer (Sievers NOA 280i, GE Analytical Instruments). The specificity of the detection of S-nitrosothiols relies on the use of specific chemical reagents to selectively cleave the NO from the S–NO bond to release free NO into the gas phase, which is detected by the NOA. The NOA has a reaction chamber where ozone is produced from oxygen. The NO, liberated by the reductive release of S-nitrosothiols in the purge vessel, reacts with ozone and results in the production of nitrogen dioxide. A fraction of the NO_2 is produced in an electronically excited state ($NO_2^\star$), which, upon decay to its ground state, emits light in the near-infrared region and can be detected and quantified by a photomultiplier tube. Provided the ozone is in excess, and reaction conditions are kept constant, the intensity of light emitted is directly proportional to the NO concentration. Analysis of data can be accomplished using software that integrates the area under the curve (AUC) of peaks obtained for the signal of NO released in millivolts with time. Appropriate standard curves need to be made depending on the range of RSNO to be measured. The NOA can detect levels of NO gas from as low as 0.3 pmol. With the system closed and the reagent of choice in the reaction chamber of the purge vessel, the cell pressure as indicated on the NOA should be slightly higher than the cell pressure when the system is opened (purge vessel uncapped).

Several different procedures are used to specifically and accurately detect RSNO in the background of other compounds that could release or be converted to NO. Three of these methods—tri-iodide, Cu/CO/cysteine (3C), and modified copper/cysteine (2C)—are described in detail here. Other methods, such as ascorbic acid/CuCl (Nagababu *et al.*, 2006) and ultraviolet photolysis (Alpert *et al.*, 1997; Gow *et al.*, 2007; Keaney *et al.*, 1993), have also been used to release NO from RSNOs followed by the detection of NO by ozone chemiluminescence.

3.1. Tri-iodide method

3.1.1. Principle
The tri-iodide (I_3^-) method was developed by Samouilov and Zweiler (1998) and was later modified for the measurement of S-nitrosothiols on albumin and hemoglobin (Gladwin *et al.*, 2002; Marley *et al.*, 2000). Helium

gas is bubbled through 9 ml of the I_3^- reagent in the glass purge vessel of the NOA at room temperature and constant cell pressure. The vessel is linked to a trap containing 15 ml of 1 N NaOH, which traps traces of iodine or acid before transfer to the chemiluminescence nitric oxide analyzer. The I_3^- reagent is made of 67 mM potassium iodide (KI), 28.5 mM iodine in acetic acid, and deionized water. The I_3^- reagent reacts with S-nitrosothiols, iron-nitrosyls (Fe^{II}-NO) and nitrite (NO_2^-), to stoichiometrically release NO gas, which is detected by the NOA as described earlier. The KI and acetic acid (without iodine) form HI and selectively reduce nitrite to NO. This reaction produces I_2, and thus (together with iodine) I_3^-, so that S-nitrosothiols can be reduced as well. However, inclusion of crystalline iodine is necessary to obtain reproducible measurements of S-nitrosothiols in the presence or absence of nitrite. S-Nitrosothiols are distinguished from nitrite by treating samples with 5% acidified sulfanilamide. The acidified sulfanilamide reacts with nitrite to form a diazonium complex that does not have a signal in the I_3^- chemiluminescence assay. S-Nitrosothiols can be further distinguished from iron–nitrosyl complexes by reaction of the acidified sulfanilamide sample with and without 5 mM mercuric chloride (which reduces S–NO to nitrite). The mercury-stable signals may also result from species such as nitrosoamines. Therefore, this method allows for the detection of nitrite, iron–nitrosyl hemoglobin, and S-nitrosohemoglobin.

3.1.2. Procedure

Measurement of samples with S-nitrosated complexes such as S-nitrosated hemoglobin requires treatment of the sample with a SNO stabilization solution (Yang *et al.*, 2003) containing 4 mM potassium ferricyanide (K_3FeCN_6), 10 mM N-ethylmaleimide (NEM), and 100 μM diethylenetriaminepentaacetic acid (DTPA) for 10 min at room temperature prior to injection into the I_3^- solution in the purge vessel. Ferricyanide is required to remove NO from the heme while preserving the SNO bond. Oxidation of heme by ferricyanide prevents ferrous heme from reacting with S–NO and prevents interference by heme autocapture during the assay. If ferricyanide is not used, the hemoglobin S–NO bond could be reduced by reactions with reduced heme (Spencer *et al.*, 2000). The NEM binds all free thiols to eliminate artifactual S-nitrosation, and DTPA is a metal chelator that prevents the destabilization of S-nitrosothiols. After incubation, the sample is passed through two G-25 Sephadex columns (PD10, Amersham Biosciences) to remove all molecules with low molecular weights, including unreacted K_3FeCN_6, NEM, and DTPA. Note that this is an important step if working with samples that have a high nitrite level to reduce the background nitrite concentration (MacArthur *et al.*, 2007). The sample is injected into the purge vessel directly. Part of the sample is treated with acid sulfanilamide for 3 min [9:1 v/v, sample:5% (w/v) acid sulfanilamide] and then injected, while another part of the sample is treated with 5 mM

mercuric chloride ($HgCl_2$) for 3 min followed by treatment with acid sulfanilamide for 3 min and then injected into the purge vessel.

The stock I_3^- reagent is prepared as follows.

1. Dissolve 400 mg of potassium iodide and 260 mg iodine in 8 ml of deionized water.
2. Add 28 ml of 100% acetic acid and mix for 30 min. (The solution should be dark brown in color. Fresh solution should be made prior to use daily.)
3. Pipette 9 ml of reagent into the purge vessel of the NOA, maintain purge vessel at room temperature (keep water bath at 25°), and allow the system to reach a steady baseline.
4. Inject samples after treatment with the SNO stabilization solution described next.

The SNO stabilization solution is prepared as follows.

    a. Prepare the following stock solutions: 200 mM potassium ferricyanide (K_3FeCN_6), 250 mM NEM, and 10 mM diethylenetriaminepentaacetic acid (DTPA).

    b. Make SNO stabilization solution and mix with sample as follows:

Volume (μl)	Reagent	Final concentration
60 μl	200 mM K_3FeCN_6	4 mM
120 μl	250 mM NEM	10 mM
30 μl	10 mM DTPA	100 μM
1290 μl	Phosphate-buffered saline	
1500 μl	Sample	

Note: Samples may be diluted in SNO stabilization solution up to 10-fold; the concentrations of the chemicals in the final volume should remain as indicated earlier. For small signals, up to 1 ml of sample can be injected.

    c. After incubation for 10 min at room temperature, follow the flowchart shown in Fig. 8.1.

       Note: Injection volumes may vary depending on the background S–NO concentration.

5. Allow the millivolt peaks to return to baseline before every subsequent injection.
6. Change the reagent in the purge vessel after a few injections, especially when injecting high protein concentrations, which may cause foaming.
7. Prepare a calibration curve using known concentrations of nitrite or GSNO. Known concentrations of GSNO can be made using absorption spectroscopy.

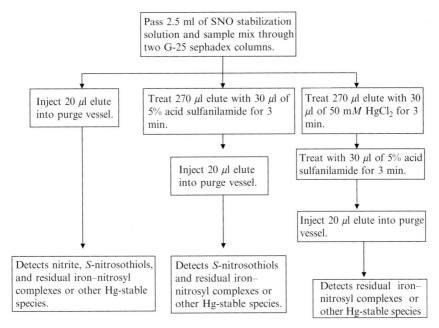

Figure 8.1 Flowchart.

8. For calculations, use the Liquid Program to calculate AUC and concentration of NO released based on the calibration curve. Some laboratories have found it better to use other programs to integrate the peaks and get the AUC, such as Origin (OriginLab Corp.)
9. Account for dilutions by treatment with the SNO stabilization solution (multiply the detected concentration by 2), dilution due to columns (take absorbance of sample before and after running through the columns and calculate dilution factor), acid sulfanilamide and mercuric chloride (multiply the concentration by 1.11 and 1.22, respectively). The concentration of protein determined by absorption is best obtained by performing a least-squares fit to basis absorption spectrum of potential species present in the mixture, including oxygenated and deoxygenated Hb, methemoglobin, nitrosyl hemoglobin, and nitrite-bound methemoglobin.
10. Subtract calculated concentrations of sample treated with $HgCl_2$ followed with acid sulfanilamide from the sample treated with acid sulfanilamide only to get S-nitrosothiol levels.

This method, described for particular application to the measurement of SNO-Hb, can be modified to detect S-nitrosothiols in standards, cell media, and cell lysates and in samples with isolated proteins. If red blood cells are in samples, the SNO stabilization solution needs to contain 1% Nonidet P-40 (NP-40) to solubilize the membranes. All procedures are carried out in the dark, as S-nitrosothiols are sensitive to light.

3.2. Cu/CO/cysteine method

3.2.1. Principle

In the copper chloride (Cu), carbon monoxide (CO), and cysteine (Cys) method (Doctor *et al.*, 2005), NO is formed by the conversion of nitrosothiols in a solution of saturated cuprous chloride or Cu(I)Cl and 1 mM L-cysteine in 50 mM phosphate buffer, pH 6.5, at 50° (Fang *et al.*, 1998). The NO released is then detected by chemiluminescence (NOA). Excess Cys favors the formation of CSNO, which is initially formed by transnitrosation from other nitrosothiols. Cu(I) then reduces the NO$^+$ equivalent in CSNO and forms NO [and Cu(II)]. Cysteine also functions to reduce Cu(II) and regenerate Cu(I). Carbon monoxide added to the gas flow of helium in the purge vessel prevents autocapture of NO by heme Fe(II). Metal carbonyls in CO gas ($\approx$0.7 ppm) need to be removed, as carbonyls of Ni and Fe chemiluminescence in the presence of O_3. Therefore, the CO gas is passed through iodine crystals and activated charcoal (in series) and is then mixed with helium gas in a gas proportioner (Aalborg, Orangeburg, NY). In this method, after every injection of sample, the oxidized Cu needs to be replaced and the residual Hb needs to be removed by changing the solution in the purge vessel. Detergents such as Triton X-100 or NP-40 produce artifactual signals. Samples need to be treated with 10-fold excess mercuric chloride to verify specificity.

3.2.2. Procedure

Measurement of S-nitrosated proteins by the Cu/CO/Cys method does not require any treatment of the sample prior to injections into the purge vessel.

1. Prepare the following stock solutions.
   a. Solution of 50 mM phosphate buffer, pH to 6.5.
   b. Two milliliters of 1 mg/1 ml saturated solution of Cu(I)Cl in deionized water. Vortex every time before adding to the purge vessel.
   c. Two milliliters of 0.1 M L-cysteine in deionized water.
2. After turning on the NOA, pour 10 ml of 50 mM phosphate buffer in the purge vessel and maintain temperature of the purge vessel at 50°. Fill trap with 15 ml deionized water.
3. Right before injection of sample, add 100 μl each of 1 mg/ml CuCl (100 μM final concentration) and 0.1 M cysteine (1 mM final concentration).
4. Start the CO and helium (maintain 1:1 CO:He gas flow by the gas proportioner) gas mixture into the purge vessel and allow the baseline to stabilize.
5. Inject 5 μl of sample and let the millivolt peak return to baseline.
6. Change phosphate buffer solution in purge vessel and repeat steps 2 to 5 after treatment of 270-μl sample with 30 μl of 50 mM HgCl$_2$ for 3 min.

7. Monitor any CO leaks by a CO detector (preferably one that displays CO in ppm) and excessive heating of the hopcolite filter due to the flow of CO gas through it.
8. Monitor foaming caused by high protein concentrations, which can overflow into the NOA via the trap.
9. Prepare a calibration curve using known concentrations of GSNO and calculate AUC and concentration of NO released based on the calibration curve for each injection.
10. Account for dilutions due to $HgCl_2$ treatment and subtract calculated concentrations of sample treated with $HgCl_2$ from the calculated concentration of the same sample injected directly.

This method is specific for the measurement of S-nitrosothiols only if $HgCl_2$ treatment is included (see Section 4). It does not detect any nitrite. The sensitivity is linear to 0.00062 SNO/Hb (Doctor et al., 2005).

3.3. Modified 2C method

3.3.1. Principle

The novel modified 2C (Cu/Cysteine) method (see Section 5) is based on the 2C method (Fang et al., 1998) with pretreatment of samples to prevent NO autocapture. In the modified 2C method, therefore, the sample is treated with the SNO stabilization solution (as in the I_3^- method), which includes 4 mM K_3FeCN_6, resulting in the conversion of any Fe(II)heme to Fe(III)heme and thereby preventing any autocapture of NO by Fe(II)heme. As mentioned earlier, the SNO stabilization solution also contains 10 mM NEM, which prevents artifactual S-nitrosation by free thiols, and 100 μM DTPA, which prevents destabilization of the SNO bond by metals. Part of the sample is also treated with 5 mM $HgCl_2$ prior to injection into the purge vessel for verification of the presence of SNO.

3.3.2. Procedure

Samples containing heme-based proteins are treated with the SNO stabilization solution (for preparation, see procedure for I_3^- method) for 10 min at room temperature. After incubation, samples are passed through two G-25 Sephadex columns. The sample is injected directly, and part of the sample is treated with 5 mM $HgCl_2$ for 3 min prior to injection into the purge vessel.

1. Prepare the following stock solutions.
   a. Solution of 50 mM phosphate buffer, pH to 6.5.
   b. Two milliliters of 1 mg/1 ml saturated solution of Cu(I)Cl in deionized water. Vortex every time before adding to the purge vessel.
   c. Two milliliters of 0.1 M cysteine in deionized water.

2. After turning on the NOA, pour 10 ml of 50 mM phosphate buffer in the purge vessel and maintain the temperature of the purge vessel at 50°. Fill trap with 15 ml deionized water.
3. Right before injection of sample, add 100 μl each of 1 mg/ml CuCl (100 μM final concentration) and 0.1 M cysteine (1 mM final concentration) and allow baseline to stabilize.
4. Inject 5 μl (or higher depending on strength of signal) of sample and let the millivolt peak return to baseline.
5. Change phosphate buffer solution in the purge vessel and repeat steps 2 to 5 after treatment of the 270-μl sample with 30 μl of 50 mM HgCl$_2$ for 3 min.
6. Prepare a calibration curve using known concentrations of GSNO and calculate AUC and concentration of NO released based on the calibration curve for each injection.
7. Account for dilutions due to treatment with the SNO stabilization solution, due to columns, and due to HgCl$_2$ treatment and subtract calculated concentrations of sample treated with HgCl$_2$ from calculated concentration of the same sample injected directly.

4. Advantages and Disadvantages

4.1. Tri-iodide method

Among the advantages of this method, the tri-iodide reagent releases NO from nitrosamines, iron–nitrosyls, and nitrite. Therefore, it allows for the accurate measurement of S-nitrosothiols in samples with nitrite, which is often present as a contaminant in solutions and laboratory glassware. Smaller injection volumes are required and, in samples with high protein amounts, there is minimal foaming. This is a significant advantage of the tri-iodide method. If foaming occurs, a silica-based antifoam agent, antifoam B emulsion (Sigma; 0.1% vol), may be used. Pretreatment of samples with the SNO stabilization solution (in particular ferricyanide) prevents autocapture of NO by heme in samples with hemoglobin, artificial S-nitrosation of free thiols, and destabilization due to metal contaminants (MacArthur et al., 2007).

4.2. Cu/CO/cysteine method

A major advantage of the 3C method is that it requires no pretreatment of samples prior to injection into the purge vessel. It is nitrite silent and therefore contamination of nitrite in samples does not cause artifactual signals. Because this method detects iron–nitrosyl hemoglobin, samples with hemoglobin need to be treated with $HgCl_2$ for verification (see Section 5). Autocapture of NO in samples with heme is prevented by the flow of carbon monoxide gas. The drawback of using carbon monoxide is destruction of the hopcolite chemical filter attached to the NOA (which filters ozone) due to excessive heating. Removal of the filter has been suggested to solve the problem (Gow *et al.*, 2007), but because the hopcolite filter eliminates excess ozone from the NOA from entering the pump, this could be detrimental to the pump. Instead, application of ice bags over the hopcolite filter (see Section 5) may be used. Heating of the filter and damage still need to be monitored. Samples with high proteins cause excessive foaming and are thereby difficult to use with this method. The changing of the purge vessel solution after every injection is tedious, as there is an increase in time taken for return of the millivolt signal to baseline.

4.3. Modified 2C method

This method combines the advantages of the 3C method with that of the tri-iodide method. Pretreatment of the sample with the SNO stabilization solution as in the tri-iodide method prevents autocapture of NO by heme and artifactual *S*-nitrosation by free thiols. Therefore, unlike the 3C method, CO flow is not required, eliminating damage to the hopcolite filter and increased time required for return of the millivolt signal to baseline after each injection. Like the 3C method, it is nitrite silent, but does require verification of the SNO bond when measuring SNO-Hb by treatment with $HgCl_2$, as it does have some sensitivity to residual iron–nitrosyl hemoglobin or other species that produce Hg-stable peaks (see Section 5).

5. Comparisons and Validations

A wide range of levels of RSNO differing by three orders of magnitude have been reported (Giustarini *et al.*, 2004; MacArthur *et al.*, 2007). These variations are probably because of the use of methods that differ in their sample treatment, reductive purge vessel reagents, and reaction conditions. Extensive validation of the tri-iodide method has been done elsewhere (Gladwin *et al.*, 2000, 2002; MacArthur *et al.*, 2007; Wang *et al.*, 2006). This method has been used by various laboratories to report levels of *S*-nitrosated proteins in blood.

To compare and verify methods, simultaneous measurements of identical serial dilutions of stock solutions of SNO-Hb were performed by both the 3C method and the tri-iodide method as shown in Fig. 8.2 (Huang *et al.*, 2006). The two methods had a correlation coefficient of $r = 0.999258$, $p < 0.001$, showing close correlation between these methods. This published comparison

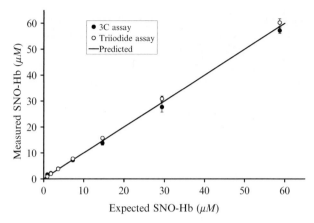

Figure 8.2 Validation of the tri-iodide method. A SNO-Hb standard sample that was 125 μM in heme and 59 μM in nitrosated thiol was serially diluted 1:1, reaching a final concentration of 0.9 μM (SNO). For each concentration, the sample was split in two and assayed by the tri-iodide method (○) or the 3C method (●). The measured concentration of SNO was plotted against the expected amount based on the amount that the initial sample was diluted. The solid line shows the predicted amount. The initial amount of SNO-Hb was taken as the average of that using the 3C and the tri-iodide methods (57 and 60 μM, respectively). This research was originally published in Huang et al. (2006). © American Society of Hematology.

has led to some criticism of the tri-iodide method, where it was pointed out that the values measured by tri-iodide are significantly less than that measured by 3C for the lowest concentration of SNO-Hb measured (Doctor and Gaston, 2006). The expected amount of SNO-Hb measured for this point was 0.9 μM (based on serial dilutions from starting material that gave virtually identical determinations from 3C and tri-iodide). Analysis by tri-iodide of this single sample gave the expected 0.9 μM, whereas 3C gave 1.7 μM, so the proposition that the discrepancy at 0.9 μM (Fig. 8.2) is due to a problem in tri-iodide is invalid. Further comparisons of 3C and tri-iodide described later support the notion that both of these methods yield equivalent results.

Rogers and colleagues (2005) reported that sensitivity of the tri-iodide method was affected by autocapture of NO by cell-free deoxygenated hemes in the reaction chamber. They proposed adding ferricyanide to the reaction chamber, thereby reducing heme autocapture (Rogers et al., 2005). However, it has been demonstrated (Wang et al., 2006) that treatment with SNO stabilization solutions (containing ferricyanide), as described earlier, gives the same results as the modified tri-iodide method of Rogers et al. (2005). The question of autocapture is a red herring, as the SNO stabilization solution oxidizes all the hemes to the ferric form, which does not capture NO effectively.

Another "false" problem that has sometimes been brought up involves the stability of SNO-Hb and other nitrosothiols in acid sulfanilamide.

By monitoring absorption spectroscopy in conjunction with tri-iodide methods, Wang et al. (2006) clearly showed that pretreatment of samples before injection for measurements by acid sulfanilamide does not result in destabilization of S-nitrosothiols.

Another group (Nagababu et al., 2006) used the ascorbic acid/CuCl method to measure much higher SNO-Hb when deoxygenated Hb was reacted with nitrite than that measured by the modified 2C assay (Kim-Shapiro group, unpublished result) for the same reaction conditions. The ascorbic acid/CuCl method uses ascorbic acid to reduce cupric chloride to cuprous chloride, thereby releasing NO from nitrosothiols [purge vessel has 1 mM cupric(II) chloride (or Cu (II)Cl$_2$) and 1.25 mM ascorbic acid in 50 mM sodium phosphate buffer, pH 7.4, at 37°]. Samples are oxidized to prevent the binding of released NO to Fe(II) hemes by treatment with 50-fold excess potassium ferricyanide (K$_3$FeCN$_6$) for 10 min and then passed through a Sephadex G-25 column to remove K$_3$FeCN$_6$ before injections are made into the purge vessel. As this method does not involve pretreatment with NEM, it does not get rid of signals generated from free thiols, resulting in artifactual SNO measurements after treatment with ferricyanide (Bryan et al., 2004). In the absence of NEM in the SNO stabilization solution, higher levels of SNO-Hb are measured than in the presence of NEM. As described previously (Marley et al., 2000), pretreatment with NEM is essential when measuring S-nitrosothiols using most methodologies.

Gow and co-workers (2007) reported that unlike the 3C method, sample modifications made for the tri-iodide method between the time of protein extraction from cells and that of measurements affects the results. In addition, Hausladen et al. (2007) compared SNO-Hb measured by photolysis/chemiluminescence with that measured by the tri-iodide method and concluded that there is a reduction in the sensitivity of measurements by the tri-iodide method. They also were critical of the tri-iodine reagent, which, according to their analysis, could generate NOI (nitrosyl iodide) from nitrite or GSNO and cause nitrosation and artifactual increased SNO measurements. However, Wang et al. (2006) showed that S-nitroso species such as CSNO, SNO-Hb, and SNO-albumin are stable under the conditions employed by the tri-iodide assay. They also reported that the formation of iodides potentially could form due to the conditions of the modified assay used by Hausladen et al. (2007), causing chemical instability of the purge vessel reagent. Measurements of SNO-Hb [prepared according to procedures used by Hausladen et al. (2007)] by the 3C and tri-iodide methods by the Kim-Shapiro group are reported in Table 8.1. It is important to point out that in many of these preparations, the amount of SNO per tetramer was essentially always 0 or 1, mimicking likely conditions *in vivo*. Each trial in Table 8.1 is from a separate SNO-Hb preparation.

To further verify the tri-iodide method, serial dilutions of a SNO-Hb preparation (in which individual Hb tetramers were essentially all singly nitrosated or not nitrosated at all) into excess (SNO-free) Hb were subsequently measured by the tri-iodide method (Fig. 8.3). These showed a very close correlation between the expected values based on the 3C and tri-iodide methods and values obtained by the tri-iodide method, with a correlation coefficient $r = 0.999$ and concentrations reaching 300 nM SNO in 440 μM Hb. Previous validation studies have shown that the tri-iodide method is sensitive down to 5 nM SNO-Hb in whole blood or 0.00005% SNO/heme (Gladwin et al., 2002). Taken together, data in Table 8.1 (showing that preparations of SNO-Hb give similar yields in 3C and tri-iodide measurements) and Fig. 8.3 (showing that the tri-iodide method gives expected results in serial dilutions down to very low SNO/heme ratios) lend strong support to the tri-iodide method.

Table 8.1 SNO-Hb measured in μM by Cu/CO/cysteine (3C) and tri-iodide (I_3^-) methods ($\pm$SD)

Trial	3C method	I_3^- method	% Difference	Total heme (μM)
1	43.1 $\pm$ 1.4	44.7 $\pm$ 2.1	3.8	209
2	97.9 $\pm$ 2.5	86.1 $\pm$ 1.7	12.9	1760
3	5.5 $\pm$ 0.4	3.9 $\pm$ 0.4	32.5	262
4	4.9 $\pm$ 0.2	4.9 $\pm$ 0.5	2.1	423
5	57.1 $\pm$ 1.1	60.2 $\pm$ 1.5	5.3	125

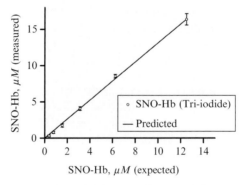

Figure 8.3 Measurement of serial dilutions of SNO-Hb by tri-iodide method. The SNO-Hb sample prepared according to Hausladen et al. (2007) was measured by both the 3C method and the tri-iodide method. This standard SNO-Hb sample (1760 μM in heme and 92 μM in SNO-Hb) was diluted fourfold, and further serial dilutions (1:1, using 440 μM heme), resulting in a range of concentrations from 16.4 to 0.32 μM with the range of SNO/heme ratio from 0.038 to 0.001, were measured by the tri-iodide method. The measured range of concentrations was plotted against the expected amount based on the amount the initial sample (measured by both methods) was diluted. Results are an average of three repeats $\pm$ standard deviation.

Although pretreatment of a sample for removal of nitrite is not required in the 3C method, iron–nitrosyl hemoglobin can be detected. A large signal of pure iron–nitrosyl hemoglobin measured by the 3C assay (by the Kim-Shapiro group) is shown in Fig. 8.4. Therefore, treatment of all samples with $HgCl_2$ is recommended to verify levels of S-nitrosothiols by this method. Under physiological conditions, the amount of iron–nitrosyl Hb is likely to be below the detection limit for 3C, but the use of Hg for verification is recommended (Doctor et al., 2005). In addition, hopcolite filter heating due to the flow of CO was reduced greatly by the use of ice bags during the course of measurements by the 3C method.

Measurements of previously prepared SNO-Hb and SNO-albumin (SNO-HSA) were each done by modified 2C and tri-iodide methods (Fig. 8.5) by the Gladwin group, showing a close correlation between the two methods with a correlation coefficient $r = 0.993$ for SNO-albumin (Figure 8.5A) and $r = 0.9979$ for SNO-Hb (Figure 8.5B).

The aforementioned studies show that both the tri-iodide method and the modified 2C method compare well with the 3C method.

6. Conclusions

The aforementioned experiments, taken together with the validation studies performed previously by many independent laboratories (Crawford et al., 2004; Feelisch et al., 2002; Huang et al., 2006; Marley et al., 2000; Samouilov and Zweier, 1998; Wang et al., 2006), clearly demonstrate that the tri-iodide method is a sensitive, reliable, and reproducible method for the measurement of S-nitrosothiols.

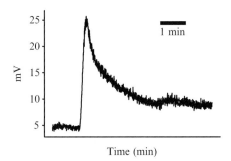

Figure 8.4 Representative raw chemiluminescence data of iron–nitrosyl Hb measured by the 3C method; 62.9 μM iron–nitrosyl Hb (HbNO) was made by treating 120 μM deoxyhemoglobin (95.8%) with 2 mM NO buffer. HbNO (100 μl) was treated with 1 mM $HgCl_2$ for 5 min. Of this, 5 μl was injected into the purge vessel for measurement by the 3C method. Results are raw chemiluminescence signal of millivolts with time in minutes.

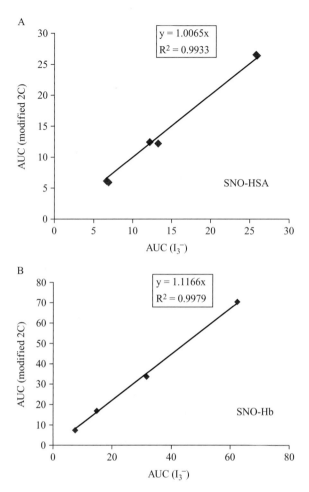

Figure 8.5 Comparison of measurements of SNO-HSA and SNO-Hb by the modified 2C method and the tri-iodide method. The copper/cysteine assay (2C assay) was performed as reported (Fang et al., 1998). To compare the AUC from tri-iodide and copper/cysteine assays, SNO-HSA and SNO-Hb were measured. Both SNO-Hb and SNO-HSA were made by incubation of hemoglobin with CSNO or GSNO (Xu et al., 2002). SNO-Hb (60, 120, 240, and 480 pmol SNO) was first treated with 10 mM K$_3$Fe(CN)$_6$ and cyanide in the dark for 10 min before measuring by the 2C assay. SNO-HSA (55, 110, and 220 pmol SNO) was injected without any treatment. The same amount of SNO-Hb and SNO-HSA was treated with sulfanilamide as described previously (MacArthur et al., 2007), and SNO was measured by the tri-iodide method. The program Origin was used to process data. AUC was calculated and compared.

For measurement of the low levels of S-nitrosothiols in biological samples, it is necessary to use techniques that are sensitive and highly specific and with detection levels that are low and at the same time include sample treatment that limits degradation and contamination of the species being studied. It is important that multiple methods are used in conjunction for verification purposes. For example, when samples contain iron–nitrosyl Hb, electron paramagnetic resonance spectroscopy may be used to verify the levels of this compound. Proper control experiments should be performed, such as NEM blocked samples to prevent the formation of artifactual thiols and appropriate treatments for conditions where an excess of other potential constituents may occur, such as nitrogen oxides or heme. For all methods it is important to use appropriate standards and calibration curves prepared in frequent intervals for reaction conditions being used, such as temperature, pH, and the presence of destabilizing agents such as metals or GSH, which occur in high concentrations in blood.

As shown in this chapter, ozone-based chemiluminescence, coupled with reductive methods such as tri-iodide, modified 2C, and 3C, can be used effectively in the measurement of S-nitrosothiols. Use of these methods may resolve the controversies surrounding the levels of S-nitrosothiols in biological samples and related conclusions regarding the role of hemoglobin in the generation of RSNOs or the role of RSNOs in pathophysiological conditions.

REFERENCES

Alpert, C., Ramdev, N., George, D., and Loscalzo, J. (1997). Detection of S-nitrosothiols and other nitric oxide derivatives by photolysis-chemiluminescence spectrometry. *Anal. Biochem.* **245,** 1–7.

Arnelle, D. R., and Stamler, J. S. (1995). No+, No(Center-Dot), and No- donation by S-nitrosothiols: Implications for regulation of physiological functions by S-nitrosylation and acceleration of disulfide formation. *Arch. Biochem. Biophys.* **318,** 279–285.

Basu, S., Hill, J. D., Shields, H., Huang, J., Bruce King, S., and Kim-Shapiro, D. B. (2006). Hemoglobin effects in the Saville assay. *Nitric Oxide* **15,** 1–4.

Bryan, N. S., Rassaf, T., Rodriguez, J., and Feelisch, M. (2004). Bound NO in human red blood cells: Fact or artifact? *Nitric Oxide Biol. Chem.* **10,** 221–228.

Chung, K. K. K., Dawson, T. M., and Dawson, V. L. (2005a). Nitric oxide, S-nitrosylation and neurodegeneration. *Cell. Mol. Biol.* **51,** 247–254.

Chung, K. K. K., Dawson, V. L., and Dawson, T. M. (2005b). S-nitrosylation in Parkinson's disease and related neurodegenerative disorders. *In* "Nitric Oxide, Pt E," pp. 139–150.

Clementi, E., Brown, G. C., Feelisch, M., and Moncada, S. (1998). Persistent inhibition of cell respiration by nitric oxide: Crucial role of S-nitrosylation of mitochondrial complex I and protective action of glutathione. *Proc. Natl. Acad. Sci. USA* **95,** 7631–7636.

Crawford, J. H., Chacko, B. K., Pruitt, H. M., Piknova, B., Hogg, N., and Patel, R. P. (2004). Transduction of NO-bioactivity by the red blood cell in sepsis: Novel mechanisms of vasodilation during acute inflammatory disease. *Blood* **104,** 1375–1382.

Doctor, A., and Gaston, B. (2006). Detecting physiologic fluctuations in the S-nitrosohemoglobin micropopulation: Triiodide versus 3C. *Blood* **108,** 3225–3226.

Doctor, A., Platt, R., Sheram, M. L., Eischeid, A., McMahon, T., Maxey, T., Doherty, J., Axelrod, M., Kline, J., Gurka, M., Gow, A., and Gaston, B. (2005). Hemoglobin conformation couples erythrocyte S-nitrosothiol content to O-2 gradients. *Proc. Natl. Acad. Sci. USA* **102,** 5709–5714.

Elahi, M. M., Naseem, K. M., and Matata, B. M. (2007). Nitric oxide in blood: The nitrosative-oxidative disequilibrium hypothesis on the pathogenesis of cardiovascular disease. *FEBS J.* **274,** 906–923.

Fang, K., Ragsdale, N. V., Carey, R. M., MacDonald, T., and Gaston, B. (1998). Reductive assays for S-nitrosothiols: Implications for measurements in biological systems. *Biochem. Biophys. Res. Commun.* **252,** 535–540.

Feelisch, M., Rassaf, T., Mnaimneh, S., Singh, N., Bryan, N. S., Jourd'Heuil, D., and Kelm, M. (2002). Concomitant S-, N-, and heme-nitros(yl)ation in biological tissues and fluids: Implications for the fate of NO *in vivo. FASEB J.* **16,** 1775–1785.

Folkerts, G., and Nijkamp, F. P. (2006). Nitric oxide in asthma therapy. *Curr. Pharm. Design* **12,** 3221–3232.

Foster, M. W., McMahon, T. J., and Stamler, J. S. (2003). S-nitrosylation in health and disease. *Trends Mol. Med.* **9,** 160–168.

Gaston, B., Drazen, J. M., Loscalzo, J., and Stamler, J. S. (1994). The biology of nitrogen-oxides in the airways. *Am. J. Resp. Crit. Care* **149,** 538–551.

Giustarini, D., Milzani, A., Colombo, R., Dalle-Donne, I., and Rossi, R. (2004). Nitric oxide, S-nitrosothiols and hemoglobin: Is methodology the key? *Trends Pharmacol. Sci.* **25,** 312–316.

Gladwin, M. T., Shelhamer, J. H., Schechter, A. N., Pease-Fye, M. E., Waclawiw, M. A., Panza, J. A., Ognibene, F. P., and Cannon, R. O. (2000). Role of circulating nitrite and S-nitrosohemoglobin in the regulation of regional blood flow in humans. *Proc. Natl. Acad. Sci. USA* **97,** 11482–11487.

Gladwin, M. T., Wang, X., and Hogg, N. (2006). Methodological vexation about thiol oxidation versus S-nitrosation: A commentary on "An ascorbate-dependent artifact that interferes with the interpretation of the biotin-switch assay." *Free Radic. Biol. Med.* **41,** 557–561.

Gladwin, M. T., Wang, X. D., Reiter, C. D., Yang, B. K., Vivas, E. X., Bonaventura, C., and Schechter, A. N. (2002). S-nitrosohemoglobin is unstable in the reductive erythrocyte environment and lacks O-2/NO-linked allosteric function. *J. Biol. Chem.* **277,** 27818–27828.

Gow, A., Doctor, A., Mannick, J., and Gaston, B. (2007). S-Nitrosothiol measurements in biological systems. *J. Chromatogr. B* **851,** 140–151.

Gow, A. J., Farkouh, C. R., Munson, D. A., Posencheg, M. A., and Ischiropoulos, H. (2004). Biological significance of nitric oxide-mediated protein modifications. *Am. J. Physiol. Lung C* **287,** L262–L268.

Guikema, B., Lu, Q., and Jourd'heuil, D. (2005). Chemical considerations and biological selectivity of protein nitrosation: Implications for NO-mediated signal transduction. *Antioxid. Redox Sign.* **7,** 593–606.

Hausladen, A., Rafikov, R., Angelo, M., Singel, D. J., Nudler, E., and Stamler, J. S. (2007). Assessment of nitric oxide signals by triiodide chemiluminescence. *Proc. Natl. Acad. Sci. USA* **104,** 2157–2162.

Hogg, N. (2002). The biochemistry and physiology of S-nitrosothiols. *Annu. Rev. Pharmacol. Toxicol.* **42,** 585–600.

Hothersall, J. S., and Noronha-Dutra, A. A. (2002). Nitrosothiol processing by platelets. *Methods Enzymol.* **359,** 238–244.

Huang, K. T., Azarov, I., Basu, S., Huang, J., and Kim-Shapiro, D. B. (2006). Lack of allosterically controlled intramolecular transfer of nitric oxide from the heme to cyseine in the beta subunit of hemoglobin. *Blood* **107,** 2602–2604.

Ignarro, L., and Gruetter, C. A. (1980). Requirement of thiols for activation of coronary arterial guanylate cyclase by glyceryl trinitrate and sodium nitrite: Possible involvement of S-nitrosothiols. *Biochim. Biophys. Acta* **631,** 221–231.

Incze, K., Parkes, J., Mihalyi, V., and Zukal, E. (1974). Antibacterial effect of cysteine-nitrosothiol and possible precursors thereof. *Appl. Microbiol.* **27,** 202–205.

Jaffrey, S. R., Erdjument-Bromage, H., Ferris, C. D., Tempst, P., and Snyder, S. H. (2001). Protein S-nitrosylation: A physiological signal for neuronal nitric oxide. *Nat. Cell Biol.* **3,** 193–197.

Jaffrey, S. R., and Snyder, S. H. (2001). The biotin switch method for detection of S-nitrosylayted proteins. *Sci. STKE* **86,** 1–10.

Janero, D. R., Bryan, N. S., Saijo, F., Dhawan, V., Schwalb, D. J., Warren, M. C., and Feelisch, M. (2004). Differential nitros(yl)ation of blood and tissue constituents during glyceryl trinitrate biotransformation *in vivo*. *Proc. Natl. Acad. Sci. USA* **101,** 16958–16963.

Keaney, J. F., Simon, D. I., Stamler, J. S., Jaraki, O., Scharfstein, J., Vita, J. A., and Loscalzo, J. (1993). NO forms an adduct with serum albumin that has endothelium-derived relaxing factor-like properties. *J. Clin. Invest.* **91,** 1582–1589.

Kettenhofen, N. J., Broniowska, K. A., Keszler, A., Zhang, Y., and Hogg, N. (2007). Proteomic methods for analysis of S-nitrosation. *J. Chromatogr. B Analyt. Technol. Biomed. Life Sci.* **851,** 152–159.

Kleinbongard, P., Rassaf, T., Dejam, A., Kerber, S., and Kelm, M. (2002). Griess method for nitrite measurement of aqueous and protein-containing samples. *Methods Enzymol.* **359,** 158–168.

Liu, L., Yan, Y., Zeng, M., Zhang, J., Hanes, M. A., Ahearn, G., McMahon, T. J., Dickfeld, T., Marshall, H. E., Que, L. G., and Stamler, J. S. (2004). Essential roles of S-nitrosothiols in vascular horneostasis and endotoxic shock. *Cell* **116,** 617–628.

MacArthur, P., Shiva, S., and Gladwin, M. T. (2007). Measurement of circulating nitrite and S-nitrosothiols by reductive chemiluminescence. *J. Chromatogr. B Analyt. Technol. Biomed. Life Sci.* **851,** 93–105.

Marley, R., Feelisch, M., Holt, S., and Moore, K. (2000). A chemiluminescense-based assay for S-nitrosoalbumin and other plasma S-nitrosothiols. *Free Radic. Res.* **32,** 1–9.

Mathews, W. R., and Kerr, S. W. (1993). Biological activity of S-nitrosothiols: The role of nitric oxide. *J. Pharmacol. Exp. Ther.* **267,** 1529–1537.

Miersch, S., and Mutus, B. (2005). Protein S-nitrosation: Biochemistry and characterization of protein thiol-NO interactions as cellular signals. *Clin. Biochem.* **38,** 777–791.

Nagababu, E., Ramasamy, S., and Rifkind, J. M. (2006). S-Nitrosohemoglobin: A mechanism for its formation in conjunction with nitrite reduction by deoxyhemoglobin. *Nitric Oxide* **15,** 20–29.

Rogers, S. C., Khalatbari, A., Gapper, P. W., Frenneaux, M. P., and James, P. E. (2005). Detection of human red blood cell-bound nitric oxide. *J. Biol. Chem.* **280,** 26720–26728.

Samouilov, A., and Zweier, J. L. (1998). Development of chemiluminescence-based methods for specific quantitation of nitrosylated thiols. *Anal. Biochem.* **258,** 322–330.

Saville, B. (1958). A scheme for colorimetric determination of microgram amounts of thiols. *Analyst* **83,** 670–672.

Singh, R. J., Hogg, N., Joseph, J., and Kalyanaraman, J. (1996). Mechanism of nitric oxide release from S-nitrosothiols. *J. Biochem.* **271,** 18596–18603.

Snyder, A. H., McPherson, M. E., Hunt, J. F., Johnson, M., Stamler, J. S., and Gaston, B. (2002). Acute effects of aerosolized S-nitrosoglutathione in cystic fibrosis. *Am. J. Respir. Crit. Care Med.* **165,** 922–926.

Spencer, N. Y., Zeng, H., Patel, R. P., and Hogg, N. (2000). Reaction of S-nitrosoglutathione with the heme group of deoxyhemoglobin. *J. Biol. Chem.* **275,** 36562–36567.

Stamler, J. S., Jaraki, O., Osborne, J., Simon, D. I., Keaney, J., Vita, J., Singel, D., Valeri, C. R., and Loscalzo, J. (1992). Nitric oxide circulates in mammalian plasma primarily as an S-nitroso adduct of serum albumin. *Proc. Natl. Acad. Sci. USA* **89,** 7674–7677.

Tarpey, M. M., Wink, D. A., and Grisham, M. B. (2004). Methods for detection of reactive metabolites of oxygen and nitrogen: *In vitro* and *in vivo* considerations. *Am. J. Physiol. Regul. Integr. Comp. Physiol.* **286,** R431–R444.

Tasker, H. S., and Jones, H. Q. (1909). The action of mercaptans on acid chlorides. II. The acid chlorides of phosphorus, sulphur and nitrogen. *J. Chem. Soc.* **95,** 1910.

Wang, X. D., Bryan, N. S., MacArthur, P. H., Rodriguez, J., Gladwin, M. T., and Feelisch, M. (2006). Measurement of nitric oxide levels in the red cell: Validation of tri-iodide-based chemiluminescence with acid-sulfanilamide pretreatment. *J. Biol. Chem.* **281,** 26994–27002.

Wink, D. A., Kim, S., Coffin, D., Cook, J. C., Vodovotz, Y., Chistodoulou, D., Jourd'heuil, D., and Grisham, M. B. (1999). Detection of S-nitrosothiols by fluorometric and colorimetric methods. *In* "Nitric Oxide, Pt C," pp. 201–211.

Xu, X. L., Lockamy, V. L., Chen, K. J., Huang, Z., Shields, H., King, S. B., Ballas, S. K., Nichols, J. S., Gladwin, M. T., Noguchi, C. T., Schechter, A. N., and Kim-Shapiro, D. B. (2002). Effects of iron nitrosylation on sickle cell hemoglobin solubility. *J. Biol. Chem.* **277,** 36787–36792.

Yamada, T., Hisanaga, M., Nakajima, Y., Kanehiro, h., Watanabe, A., Takao, O., Nishio, K., Sho, M., Nagao, M., Harada, A., Matsushima, K., and Nakano, H. (1998). Serum interlukin-6, interlukin-8, hepatocyte, growth factor, and nitric oxide changes during thoracic surgery. *World J. Surg.* **22,** 783–790.

Yang, B. K., Vivas, E. X., Reiter, C. D., and Gladwin, M. T. (2003). Methodologies for the sensitive and specific measurement of S-nitrosothiols, iron-nitrosyls, and nitrite in biological samples. *Free Radic. Res.* **37,** 1–10.

CHAPTER NINE

S-NITROSOTHIOL ASSAYS THAT AVOID THE USE OF IODINE

Lisa A. Palmer *and* Benjamin Gaston

Contents

1. Introduction	158
2. *S*-Nitrosothiol Synthesis	158
3. *S*-Nitrosothiol Assays	160
3.1. Chemiluminescence	160
3.2. Photolysis–chemiluminescence	161
3.3. Reduction–chemiluminescence in the presence of CuCl and cysteine	161
3.4. Reductive chemiluminescence in the presence of carbon monoxide (CO)	164
3.5. Reductive chemiluminescence in the presence of potassium iodine and iodide	165
3.6. 4,5-Diaminofluorescein fluorescence assay (DAF assay)	166
3.7. Saville assay	167
3.8. Mass spectrometry	168
3.9. Biotin substitution	170
3.10. Anti-*S*-nitrosocysteine antibodies	170
4. *S*-Nitrosothiols in Health and Disease: The Importance of Getting the Assay Right	171
5. Summary	171
References	171

Abstract

S-Nitrosylation is a ubiquitous signaling process in biological systems. Research regarding this signaling has been hampered, however, by assays that lack sensitivity and specificity. In particular, iodine-based assays for *S*-nitrosothiols (1) produce nitrosyliodide, a potent nitrosating agent that can be lost to reactions in the biological sample being studied; (2) require pretreatment of biological samples with several reagents that react with proteins, artifactually forming or breaking S–NO bonds before the assay; and (3) are not sensitive or

Department of Pediatrics, University of Virginia School of Medicine, Charlottesville, Virginia

Methods in Enzymology, Volume 440
ISSN 0076-6879, DOI: 10.1016/S0076-6879(07)00809-9

© 2008 Elsevier Inc.
All rights reserved.

specific for nitrogen oxides in biological samples, reporting a wide range of different concentrations and falsely reporting NO-modified proteins, to be nitrite. These data, therefore, suggest that iodine-based assays should never be used for biological S-nitrosothiols. There are other assays that provide reasonably sensitive and accurate data regarding biological S-nitrosothiols, including assays based on mass spectrometry, spectrophotometry, chemiluminescence, fluorescence, and immunostaining. Each assay, however, has limitations and should be quantitatively complemented by separate assays. Continued improvement in assays will facilitate improved understanding of S-nitrosylation signaling.

1. Introduction

Nitric oxide (NO) is a cell-signaling molecule involved in a number of physiological and pathophysiological processes. NO is synthesized by a family of enzymes known as NO synthases (NOS). At least three isoforms exist: neuronal NOS (nNOS, or type I NOS), inducible NOS (iNOS, or type II NOS), and endothelial NOS (eNOS, or type III NOS) (Stuehr, 1999). Each enzyme is a homodimer requiring three cosubstrates (L-arginine, NADPH, and O_2) and five cofactors (FAD, FMN, Ca^{2+}–calmodulin, heme, and tetrahydrobiopterin) for activity (Gorren and Mayer, 1998).

Nitric oxide has a complex redox chemistry involving three redox species: NO^+ (nitrosonium cation), NO^- (nitroxyl anion), and $NO^\bullet$ (free radical) (Arnelle et al., 1995; Gaston et al., 2006; Stamler et al., 1997). $NO^\bullet$ is a free radical gas with a short half-life under physiological conditions. NO^- is formed by the reduction of $NO^\bullet$ and is highly labile and reactive. NO^+ is the oxidized form of $NO^\bullet$; it is extremely reactive and is always in complex with another species, such as a thiolate anion (Gaston et al., 2006; Hess et al., 2005). S-Nitrosylation is a redox-associated modification of a cysteine thiol by NO^+ that transduces NOS- or exogenous nitrogen oxide-induced bioactivity. S-Nitrosothiols (RSNOs) are products of S-nitrosylation that are synthesized, stored, transported, and degraded in cell systems (Gaston et al., 2006; Hess et al., 2005).

2. S-Nitrosothiol Synthesis

S-Nitrosothiols can be formed by a number of enzymatic and nonenzymatic reactions in biological systems as recently reviewed (Carver, et. al., 2005). Increased production of endogenous RSNOs is associated with the activation of each NOS form (Gow et al., 2002; Mayer et al., 1998). S-Nitrosylation can result from the direct interaction of NO with thiyl radicals to form S-nitrosothiols. For example, NO can react with thiols, producing the S-nitrosothiol radical; these radicals can be stabilized through the loss of an electron to form S-nitrosothiol or through protonation. Additionally, NO can be oxidized enzymatically or inorganically to NO^+ by a variety of cellular electron acceptors, including oxygen and both protein– and nonprotein–metal ion complexes (Gaston et al., 2003, 2006; Hess et al., 2005). Many of these reactions may be compartmentalized in biological membranes or hydrophobic pockets in proteins (Mannick et al., 2001; Whalen et al., 2007). These pockets promote the formation of S-nitrosothiols by either oxidation and/or because of increased solubility of NO in hydrophobic areas (Hess et al., 2005; Nedospasov et al., 2000). The position of acidic and basic residues surrounding the target cysteine residue—such as the S-nitrosylation motif of X (H,R,K) C(D,E)—has been proposed to be an important determinant of protein S-nitrosylation (Stamler et al., 1997). These proton donating and receiving groups can also be in close proximity to the S γ atom in the three-dimensional structure of the protein (Ascenzi et al., 2000; Perez-Mato et al., 1999).

Specific proteins other than NOS catalyze the formation of S-nitrosothiols. Ceruloplasmin catalyzes S-nitrosoglutathione (GSNO) formation (Inoue et al., 1999). In addition, hemoglobin (Hb) binds NO at cysteine (cys) β93 (Chan et al., 1998). Hemoglobin has been shown to act as nitrite reductase, generating RSNOs (Angelo et al., 2006; Nagababu et al., 2006). In the case of hemoglobin, loss of O_2-heme binding changes the conformation of Hb from the oxygenated state to the deoxygenated state. This permits the transfer of NO^+ equivalents from the thiol cysβ93 (Chan et al., 1998; Jia et al., 1996; Luchsinger et al., 2003) to a recipient cys in the cytosolic amino terminus of anion-exchange protein 1 (AE1) on the red blood cell (RBC) cell membrane (Pawloski et al., 2001). The NO^+ equivalents on AE1 are then exported from the RBC by a mechanism that is not well defined, but may occur directly through cell–cell transfer or indirectly through the formation—through transnitrosation—of intermediates such as GSNO or S-nitrosocysteine (CSNO) (Doctor et al., 2005; Lipton et al., 2001; Liu et al., 2004; Palmer et al., 2007). Subsequent entry of GSNO and CSNO into cells can be mediated by γ-glutamyl transpeptidase (γGT), the L-AT transport system, protein disulfide isomerase, and/or other proteins (Askew et al., 1995; Hogg et al., 1997; Lipton et al., 2001; Zhang and Hogg, 2004).

Cellular catabolism

The catabolism of S-nitrosothiols is also regulated. One enzyme that plays a physiological role is S-nitrosoglutathione reductase (Liu et al., 2001, 2004). This ubiquitously expressed enzyme is responsible for the breakdown of endogenously produced GSNO to oxidized glutathione and ammonia.

The enzyme does not appear to directly affect S-nitrosothiol–protein substrates; however, proteins may be affected through altered transnitrosation equilibria with GSNO. Other enzymes systems have been implicated in the catabolism of S-nitrosothiols *in vivo* (Gaston *et al.*, 2006). These include thioredoxin/ thioredoxin reductase (Trujillo *et al.*, 1998), Cu/Zn superoxide dismutase (Johnson *et al.*, 2001; Jourd'heuil *et al.*, 1999; Schonhoff *et al.*, 2006), and γGT (Askew *et al.*, 1995; Hogg *et al.*, 1997; Lipton *et al.*, 2001).

There are many aspects of signaling through S-nitrosylation that have parallels to protein phosphorylation. Both are posttranslational modifications that alter the function or activity of a protein. S-Nitrosylation modifies an array of receptors, including the ryanodine receptor (Xu *et al.*, 1998) and the N-methyl-D-aspartate receptor (Choi *et al.*, 2000). It modifies the expression and/or function of many transcription factors, including HIF-1 (Cho *et al.*, 2007; Li *et al.*, 2007; Palmer *et al.*, 2007), specificity proteins 1 and 3 (Zaman *et al.*, 2004), NFκB (della Torre *et al.*, 1997, 1998; Mathews *et al.*, 1996; Park *et al.*, 1997), and oxy R (Kim *et al.*, 2002; Hausladen *et al.*, 1996). It modifies ubiquitin ligases such as Parkin (Yao *et al.*, 2004) and protein von Hippel Lindau (Palmer *et al.*, 2007). It modifies caspases (Mannick *et al.*, 1999, 2001), cyclooxygenase (Kim *et al.*, 2005), phosphatases (Carver *et al.*, 2007), G-proteins (Whalen *et al.*, 2007), ion channels (Chen *et al.*, 2004; Joksovic *et al.*, 2007), and a broad array of other targets.

3. S-Nitrosothiol Assays

Because S-nitrosylation signaling is a novel and relatively unexplored area of biology, assays that can accurately measure RSNOs in cells and organisms are in demand. Unfortunately, each of the currently available assays has limitations, an issue that has led to a good deal of controversy in the field (Gaston *et al.*, 2006; Gow *et al.*, 2007). Biological S-nitrosothiols can be assayed using a variety of techniques. These include indirect methods, which involve cleaving in the S–NO bond, followed by measurement of the liberated nitrogen oxide. There are also methods to detect the presence of S-nitrosothiols directly, including mass spectrometry (MS) and the use of antinitrosocysteine antibody. Most of these assays can be used for the detection of levels of S-nitrosothiols in or near the biological range.

3.1. Chemiluminescence

One of the most widely used and versatile methods to detect S-nitrosothiols involves homolytic cleavage of the S–NO bond, followed by an assay for NO using chemiluminescence. Chemiluminescent detectors use the principle that NO reacts with ozone to generate NO_2^*, which, in turn, decays to NO_2, releasing light (Gow et al., 2007). This reaction occurs in the gas phase and is dependent on the rate of mixing of NO and ozone (Sexton et al., 1994). The release of NO from S-nitrosothiols into the gas phase can be achieved anaerobically by either photolysis or chemical reduction.

3.2. Photolysis–chemiluminescence

Exposure of RSNOs to light (300–350 nm) causes homolytic cleavage of the S–N bond through light $n_O \rightarrow \pi^*$ transition (Sexton et al., 1994; Shishido and de Oliveira, 2000). Activated S-nitrosothiols formed after light exposure decay to free NO and the corresponding RS radical. Following photolysis, NO is ordinarily removed in an inert gas stream (argon) and is analyzed by chemiluminescence (Fig. 9.1). In this assay, nitrite contamination can cause NO release if the pH is allowed to fall, particularly to <6 (Chen et al., 2004; Mannick et al., 1999). To achieve optimal S–N cleavage, it is best to use a broad-spectrum lamp that does not span lower than 300 nm. In addition, lamp wattage should not exceed 100 W, as the heat generated could result in aberrant signals. This assay is highly sensitive and highly reproducible, with a detection limit around 5 pmol (Arnelle and Stamler, 1995). At extremely high intensity, it is possible to form S-nitrosothiols from nitrate (NO_3^-) and reduced thiols (Dejam et al., 2003); however, these are not conditions used in conventional assays, and control experiments in which samples are spiked with $NaNO_3$ have shown that there is no significant S-nitrosothiol formation in biological samples using these systems.

3.3. Reduction–chemiluminescence in the presence of CuCl and cysteine

This assay uses excess cysteine to drive the transnitrosation equilibrium to form CSNO from a biological S-nitrosothiol, followed by CuCl-mediated reduction of CSNO to NO (Fang et al., 1998). Typically, this assay is carried out in helium or argon in an anaerobic purge vessel containing 1 mM cysteine and saturated (100 μM) CuCl at pH 7.0, 50°. The detection limit is $\approx$5 to 10 pmol (Fang et al., 1998; Gow et al., 2002). Standard curves should be generated using the specific S-nitrosothiols being assayed. Negative controls should include water, samples spiked with 10 μM nitrite and nitrate, and samples pretreated with excess $HgCl_2$ (which breaks the S-nitrosothiol bond, leading to NO_2^- formation before the assay is performed). While performing this assay, the pH of the solutions must be monitored as (1) the pH of the stock solution of CuCl/cysteine can fall with time and (2) $HgCl_2$ added to the cysteine can change the pH. This assay is versatile, as it can be used in intact cells (Doctor et al., 2005) in tandem with high-performance liquid chromatography (HPLC)

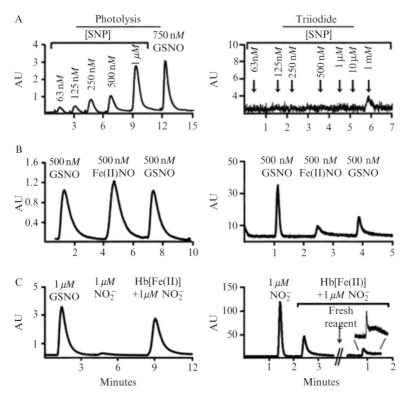

Figure 9.1 Comparison of the sensitivity and specificity of photolysis and tri-iodide assays for paired samples of Fe_{NO} compounds. (A) SNP, a model Fe(III)NO compound, at the indicated concentrations, with GSNO shown as a standard. AU, arbitrary units. Note that in this and subsequent figures, the magnitudes of signals generated by photolysis and by tri-iodide, expressed as arbitrary units, cannot be compared directly (the two methods exhibit equivalent sensitivity for NO). (B) A sequence of injections of a GSNO standard (500 nM), a Fe(II)NO Hb solution (1 mM Hb[Fe(II)] containing 500 nM Hb[Fe(II)NO]), and a repetition of the GSNO standard after Hb. The repeat injection of GSNO in the tri-iodide gave a distorted, diminished signal. (C) Hb Fe(III)NO/Fe(II)NO$^+$ equivalent (SNO precursor). Photolysis accurately measures transient formation of a Fe(III)NO/Fe(II)NO$^+$ equivalent generated from 1 μM nitrite/1 mM deoxyHb (x5–10 s), with scant response from nitrite alone (1 μM GSNO shown for comparison). In contrast, nitrite produces a prominent signal in the tri-iodide assay, whereas its signal in the presence of 250 μM deoxyHb is attenuated markedly. Furthermore, the signal generated by such samples can be variable and difficult to quantify (the line shape of a second sample, which is magnified for clarity, hampers reliable integration). From Hausladen *et al.* (2007), with permission.

(Fang *et al.*, 1998) and/or following protein immunoprecipitation procedures (Fig. 9.2; Mannick *et al.*, 2001).

In the case of immunoprecipitation experiments, several additional controls need to be performed. First, equal concentrations of an isotype-matched

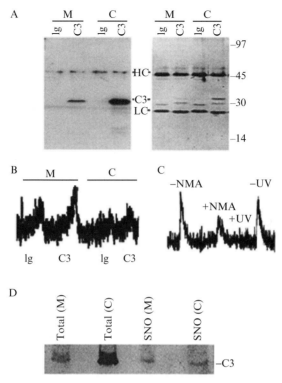

Figure 9.2 S-Nitrosylation of mitochondrial and cytoplasmic caspase-3. (A) Caspase-3 immunoprecipitation. Proteins were immunoprecipitated from mitochondrial (M) and cytoplasmic (C) cellular fractions using a caspase-3-specific monoclonal antibody (C3) or equal concentrations of an isotype-matched control antibody (Ig). Immunoprecipitated proteins were visualized on silver-stained gels (right) or caspase-3 Western blot analysis (left). Molecular weight markers, immunoglobulin heavy chain (HC), light chain (LC), and caspase-3 (C3) are shown. (B) S-Nitrosylation of caspase-3. The SNO-derived chemiluminescence signal of Ig control (Ig) and caspase-3 (C3) immunoprecipitations obtained from mitochondrial (M) and cytoplasmic (C) fractions of 10C9 cells are shown. NO chemiluminescence in arbitrary units is plotted on the y axis, and time is plotted on the x axis. The NO released from each sample is proportional to the area under the curve. Data are representative of 1 of 10 separate experiments. (C) Caspase-3 is S-nitrosylated endogenously. The SNO-derived chemiluminescence signal of mitochondrial caspase-3 immunoprecipitates from CEM cells after they had been grown for 48 h in the presence (+NMA) or absence (-NMA) of 4.5 mM L-NMA is shown. Data are representative of one of two separate experiments. Mitochondrial caspase-3 immunoprecipitates from control cells were divided into two samples, one of which was exposed to UV light for 10 min (+UV) and the other was left untreated in the dark at room temperature for the same period (-UV). SNO-derived chemiluminescence signals from UV-treated and untreated samples are shown. Data are representative of one of four separate experiments. (D) A higher percentage of mitochondrial than cytoplasmic caspase-3 is S-nitrosylated. S-Nitrosylated proteins in cytoplasmic (C) and mitochondrial (M) fractions were labeled selectively with biotin and then purified over

antibody should be used as a negative control (Mannick *et al.*, 2001; Zaman *et al.*, 2006). Second, SDS-PAGE, followed by silver staining and/or immunoblotting, should be used to confirm that the only protein brought down in detectable amounts is the protein of interest. Third, injections of the immunoprecipitate should be paired with an injection of an aliquot of the sample treated with $HgCl_2$ or ultraviolet (UV) light to remove the NO^+/NO group. It should be noted that in some instances, removal of the protein from the immunoprecipitation beads may alter the pH or disturb S–NO bonds. In this case, it is possible to inject the washed immunoprecipitation bead preparation directly into the reflux chamber. Control injections should be performed with the beads alone to ensure that there is no baseline signal (Mannick *et al.*, 2001). More importantly, for all immunoprecipitation studies, it is critical to study the protein in which the cysteine of interest has been mutated; and in cells in which NOS has been inactivated/inhibited (Lancaster and Gaston, 2004).

Low mass S-nitrosothiols can be separated by reverse-phase HPLC [e.g., C-18 column, mobile phase: 20% methanol, 80% phosphate-buffered saline (PBS) with 1 mM 1-octanesulfoic acid, pH 2.2, at 1 ml/min] (Fang *et al.*, 1998). Peaks are monitored spectrophotometrically at 220 or 340 nm; additionally (or alternatively), the eluate is collected at 1-min intervals for the CuCl–cysteine reduction assay as described earlier. Using HPLC, it is critical to show that the signal can be eliminated (photolysis or $HgCl_2$). In addition, it is important to perform control experiments which the injection is spiked with reduced thiols to be certain that S-nitrosothiol is not formed artifactually from NO_2^- on the column.

3.4. Reductive chemiluminescence in the presence of carbon monoxide (CO)

The analysis of S–NO bonds in heme-containing proteins presents a challenge because the NO is autocaptured by heme groups after reduction (Doctor *et al.*, 2005). Reductive chemiluminescence can be modified by adding CO to the inert gas flow through the reflux chamber to prevent the autocapture of NO (Doctor *et al.*, 2005; Palmer *et al.*, 2007). Unfortunately, research grade CO contains metal carbonyls, which chemiluminescence in the presence of ozone. These must be removed by passage through an iodide crystal and activated charcoal (Burgard *et al.*, 2006; Doctor *et al.*, 2005).

---

streptavidin-agarose as described previously (Jaffrey *et al.*, 2001). The purified S-nitrosylated proteins were then analyzed by caspase-3 Western blot analysis. One out of 100 of the total protein in the mitochondrial or cytoplasmic starting sample (16 μg) was loaded in the lanes marked total. Purified S-nitrosylated proteins obtained from each fraction were loaded in the lanes labeled SNO. Caspase-3 (C3) is indicated. From Mannick *et al.* (2001), with permission.

In addition, headspace pressures should be maintained at 100 to 200 torr greater than atmospheric as this pressure determines CO partial pressure in the aqueous reactant matrix. Oxidized cysteine should be replaced and residual heme proteins removed by refreshing the reflux chamber after each sample injection. It should be noted that several chemiluminescent NO analyzers use hopcalite in their ozone-scavenging system. The reaction between hopcalite and CO is exothermic, and the heat produced may melt or ignite the ozone-scavenging cassette. Thus, hopcalite traps should not be used with reductive chemiluminescence in the presence of CO, but the ozone should be vented to a well-functioning fume hood. Moreover, additional safety precautions must be set in place due to the use of CO. This includes the use of an environmental CO alarm to warn investigators of a CO leak in the system, a complete check to ensure tight coupling at all points within the operating circuit, and the cessation of CO gas flow—followed by a flush with inert gas—before opening any part of the system to the atmosphere. Note also that this assay relies on an optimal-sized reflux chamber and careful regulation of each parameter. To be safe and successful, it requires substantial investigator experience.

3.5. Reductive chemiluminescence in the presence of potassium iodine and iodide

This process has been validated in comparison with itself under different conditions in many studies, but should never be used as an assay for biological S–NO modifications in proteins. Results obtained using this method are in disagreement with those obtained by several other methods (Chan et al., 1998; Doctor and Gaston, 2006; Doctor et al., 2005; Gaston et al., 2006; Hausladen et al., 2007; Luchsinger et al., 2003; McMahon and Doctor, 2005; McMahon et al., 2002), particularly in physiologically relevant concentration ranges ($<900\ \mu M$). Further, the reduction process identifies a variety of species other than S-nitrosothiols, with various yields requiring that samples be pretreated with multiple reagents before the assay can be carried out (Fig. 9.1; Hausladen et al., 2007). These pretreatments modify the concentration of S-nitrosothiol directly and/or by altering the tertiary or quaternary structure of proteins. For example, several papers have been written using this assay to identify a vascular role for endogenous NO_2^- based on blood pretreatment with sulfanilamide. More recently, it has been appreciated that this "nitrite" signal is actually composed of NO-modified proteins, most likely RSNOs (Fig. 9.3; Rogers et al., 2007).

Moreover, using reduction with iodine, there is the potential to produce the potent nitrosative agent nitrosyliodide, which is a highly reactive nitrosating agent; it is consumed by reactions with oxygen, amines, and other nucleophiles that compete with its homolytic decomposition to NO (which is assayed) and I_2 (Hausladen et al., 2007). This is especially

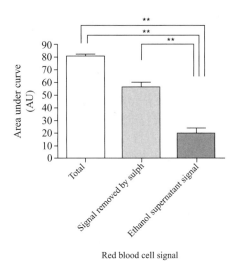

Figure 9.3 Origin of the red blood cell signal removed by acidified sulfanilamide (sulf). Arterial red blood cell fractions were assessed for total red blood cell signal; sulfanilamide labile signal (total minus sulfanilamide stable signal); and red blood cell signal minus protein (ethanol precipitation) ($n = 5$). All samples were measured in the modified tri-iodide reagent. **$p < 0.01$. From Rogers et al. (2007), with permission.

problematic when measuring hemoglobin and may account for the many discrepancies between iodine-based S-nitrosothiol hemoglobin assays and all other S-nitrosothiol hemoglobin assays (Doctor and Gaston, 2006; Doctor et al., 2005; Gaston et al., 2006).

3.6. 4,5-Diaminofluorescein fluorescence assay (DAF assay)

An additional indirect way to measure biological S-nitrosothiol content is through the DAF fluorescence assay. This assay is based on a modification of fluorescein, which permits reactions with NO in the presence of dioxygen to yield a green, fluorescent triazole (Itoh et al., 2000). One commonly used fluorescent probe is 4,5-diaminofluorescein (DAF-2), which reacts with an oxidation product of NO to form the highly fluorescent triazolofluorescein, DAF-2T. In this assay, the transfer of NO^+ equivalents to DAF, forming DAF-2T, is measured (Doctor et al., 2005; King et al., 2005; McMahon et al., 2002). To perform this assay, proteins are solubilized in 10 mM PBS (pH 7.4) and are incubated in the presence or absence of $HgCl_2$ (10 min) to break the S–NO bond, forming NO_2^-. The proteins are then removed by Microcon centrifugation/ultrafiltration (10-kDa filter, 10,600 g, 20 min) or by acetone precipitation to reduce background fluorescence. The filtrate is reduced with 0.4 M HCl, protonating NO_2^- to form an NO donor, and is then incubated with 150 μM DAF2 (made in 10 mM PBS, pH 7.4, at an optimal ratio of DAF:S-nitrosothiol = 10:1). Samples are placed in black microplates, and fluorescence is determined using a fluorometer (such as Fl_x-800). Optimal fluorescence occurs after titration back to pH 8 with NaOH. Readings are compared to a standard curve made of a (synthetic) S-nitrosothiol protein of interest.

A number of factors need to be considered when using this assay to measure biological S-nitrosothiol levels. First, if $HgCl_2$ is used, it must be in significant molar excess, as some chelation of Hg^{2+} can occur on non-S-nitrosothiol sites on proteins. In addition, $HgCl_2$ can alter the background DAF fluorescence; thus, Hg^{2+} must be constant between samples and must be added to controls following filtration. Furthermore, complex protein mixtures or samples are often prepared in buffer containing ion chelators; pilot experiments may be needed to determine the optimum concentrations of Hg^{2+} required. This protocol has been modified to be performed after gel electrophoresis to identify S-nitrosylated proteins (King et al., 2005).

3.7. Saville assay

This is a colorimetric assay to RSNOs. It is reproducible, but lacks sensitivity. In this assay, sulfanilamide [3.4% (w/v)] and N-(1-naphthyl)ethylenediamine (0.1% w/v) are often prepared in 0.4 M HCl in the presence or absence of 1% (w/v) $HgCl_2$ (Gow et al., 2007; Saville, 1958). Standards/samples are mixed with sulfanilamide in the absence or presence of $HgCl_2$; these are then reacted with N-(1-naphyl)ethylenediamine. The reaction product is monitored spectrophotometrically at 540 nm. The difference in the NO_2^- concentration measured in the absence or presence of $HgCl_2$ represents the concentration of S-nitrosothiol. Although this assay is easy inexpensive to perform, it is not very useful for the detection of biological levels of S-nitrosothiol-modified proteins because the limit of detection (500 nM) is too close to biological concentrations (Gow et al., 2007). Note, however, that recent data (unpublished) suggest that novel spectroscopic methods can improve the sensitivity of the Saville reaction.

Improved sensitivity, coupled with chromatographic selectivity for specific low mass S-nitrosothiols, has been reported by Akaike and co-workers (1997), who coupled a C-18 reversed-phase (4.6 × 250 μM; Tskgel DS-T_s; Tosh, Tokyo) column (mobile phase 10 μM Na acetate, pH 5.5, with 0–7% methanol; 5.5 ml/min) with in series with a pump introducing either 1.75 mM $HgCl_2$ or 1.75 mM $CuSO_4$ to degrade the RSNO (pumped at 0.2 ml/min) followed by the Griess reagent (as described earlier, 0.23 ml/min) for analysis at 540 nM (Fig. 9.4). The yield was 100% with the $HgCl_2$ and 30% with $CuSO_4$; the sensitivity was as low as 3 nM. By substituting an 8 × 30-mm gel filtration column with mobile phase 0.15 M NaCl

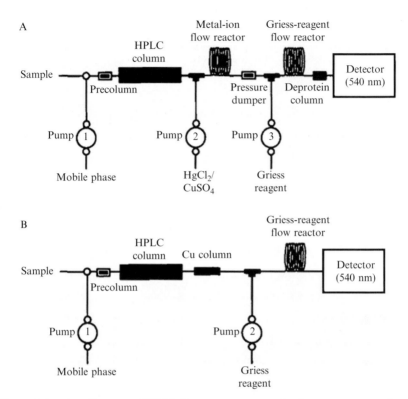

Figure 9.4 Flow diagrams of HPLC flow reactor systems for the measurement of various RSNOs using either a Hg^{2+}/Cu^{2+} flow reactor (A) or a copper metal-loaded column (B) for RSNO decomposition. In both systems, the Griess reagent flow reactor was employed for the colorimetric detection (540 nm) of NO_2^-; the RSNO eluted from the HPLC column was detected as NO_2^- after metal-induced decomposition. (A) A reversed-phase HPLC column (TSK gel ODS-80Ts) and a gel nitration column (YMC Diol-120) were used for low-molecular weight and high-molecular weight RSNOs, respectively. A deproteinization column was included in the system for the S-NO-protein measurement before access to the detector. See text for details. From Aiaike et al. (1997), with permission.

(0.55 ml/min), coupled with $HgCl_2$ and Griess reagent pumps (as described earlier), S-nitrosylated proteins can be separated.

3.8. Mass spectrometry

Mass spectrometry can be used both to measure S-nitrosothiol levels in a sample and to identify S-nitrosothiol-modified proteins. Currently, this is the gold standard for the measurement and identification of RSNOs in biological samples. Identification is through selective monitoring of the appropriate mass/charge (m/z) ratio. Liquid chromatography (LC)-MS can be used if the LC phase is relatively rapid, the sample is cool, and the solvent is inert. Particularly if the assay is performed in acid, several controls must be performed, including (1) injection of the corresponding reduced thiol to ensure that there is not artifactual formation of S-nitrosothiols from nitrite on the column; (2) matched S-nitrosothiol standards using ^{15}NO; and (3) photolysis and/or HgCl$_2$ displacement before assay (Fig. 9.5; Lancaster and Gaston, 2004; Palmer et al., 2007). Samples can be analyzed by electrospray ionization MS with ion monitoring at an appropriate mass m/z ratio. Alternatively, MS/MS fragmentation studies can be performed for proteins MS has been successfully performed for endogenously produced S-nitrosothiols such as GSNO (Lipton et al., 2001). In general, there is a limit of detection of 200 nM (Lancaster and Gaston, 2004; Palmer et al., 2007), although this is instrument specific. Lower concentrations (pM) of GSNO have been identified in human plasma using the LC-MS system (Taubert et al., 2007), but these methods have not yet been replicated.

Gas chromatography (GC)-MS can also be used to measure S-nitrosothiols, both as plasma proteins (such as S-nitrosoalbumin) or—coupled to HPLC—in low mass fractions. The S-nitrosothiol is reacted with HgCl$_2$ (as in the Saville reaction), which, in turn, is reacted to form a pentafluorobenzyl derivative; this derivative is measured by GC-MS with a sensitivity as low as 150 to 200 nM and recovery in excess of 93%. ^{15}N-labeled nitrite is used to create standards. This method has been validated extensively (Tsikas et al., 1999, 2001, 2002).

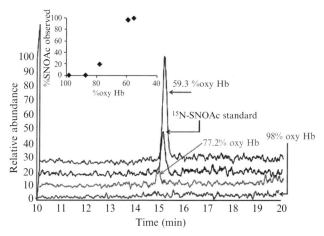

Figure 9.5 Oxygenated erythrocytes were deoxygenated *ex vivo* (argon) in the presence of 100 μM NAC; supernatant SNOAc was measured by MS (see text). The SNOAc concentration increased with oxyHb desaturation (co-oximetry: inset), being maximal at 59% saturation, less at 77% saturation, and undetectable at 98% saturation. From Palmer et al. (2007), with permission. (See color insert.)

3.9. Biotin substitution

One method used to identify and detect S-nitrosylated proteins involves biotin substitution (Jaffrey et al., 2001). This assay is ordinarily selective for S-nitrosylated proteins and does not recognize S-oxidized (sulfenic acid and disulfide) groups in proteins (Forrester et al., 2007). Moreover, this is an inexpensive, relatively easy assay to perform in cellular extracts. Protocols vary. In our procedure, proteins (from 10 μg extract) are precipitated in 2 volumes of acetone (in the dark, 4°) and resuspended in 100 μl of 250 mM HEPES, 0.1 mM EDTA 0.1 μM neocuproine (HEN; buffer pH 7.7), and thiols present on proteins are blocked by adding 20 mM (in 9:1 HEN buffer:25% SDS) methyl methanethiosulfonate. After reprecipitation with acetone, cell pellets are resuspended in 75 μl HEN buffer with 25 μl of 4 mM (N-[6-biotinamido)hexyl]-1′-(2′[pyridyldithiopropionamide biotin-HPDP) in dimethyl sulfoxide (fresh) and 1 μl of 0.1 M ascorbate. This ascorbate reduces the S–NO bond; the resulting thiol is biotinylated. The degree of biotinylation can be determined through the use of antibiotin immunoblotting. Alternatively, biotinylated proteins can be purified using strepatavin-agarose beads, eluted with β-mercaptoethanol, and identified by Western blot analysis (Fig. 9.2; Mannick et al., 2001) or used in conjunction with MS. Various blocking agents, such as NEM and reducing agents, have been used successfully in this protocol (Kettenhofen et al., 2007). A control sample that has not been treated with ascorbate is often reported, and ascorbate controls are likely critically important (Huang et al., 2006; Palmer et al., 2007). As a cautionary note, this assay should be performed in the dark, as indirect sunlight can result in false-positive results (Forrester et al., 2007). Specificity should also be shown by pretreatment of the samples with UV light, HgCl$_2$, and/or analysis of proteins in which the reactive cysteine has undergone site-directed mutagenesis.

3.10. Anti-S-nitrosocysteine antibodies

Polyclonal and monoclonal antibodies have been isolated that react with the S-nitrosothiol moiety (Gow et al., 2002; Munson et al., 2005). Typically, these antibodies have been used successfully in the detection of S-nitrosothiol–proteins by immunohistochemistry (Gow et al., 2004). Positive and negative controls are required and should be conducted on serial sections in which the test section is flanked by both a positive and a negative controls. The most effective positive control is chemical nitrosation using added nitrite in 0.4 M HCl. Negative controls are conventionally slides that have been pretreated with HgCl$_2$. These antibodies are not ordinarily useful for Western blotting because the S–NO bond can be broken during gel electrophoresis.

4. S-Nitrosothiols in Health and Disease: The Importance of Getting the Assay Right

Current evidence suggests that S-nitrosothiols play a role in both health and disease. In health, they have been associated with a broad range of physiological functions, from gene regulation to neuronal signaling. However, abnormal S-nitrosylation is associated with numerous disease states. These include arthritis, diabetes, multiple sclerosis, sepsis, asthma, cystic fibrosis, pulmonary arterial hypertension, systemic hypertension, Parkinson's disease, and heart failure (Foster *et al.*, 2003; Gaston *et al.*, 2006). Thus, the ability to measure RSNOs levels in tissues, biological fluids, and subcellular compartments in health and disease states accurately and precisely is critical for understanding the mechanisms by which S-nitrosothiol signaling is involved. The availability of more precise measurements will enable characterization of (1) normal S-nitrosothiol levels in individual tissues and cell compartments; (2) changes associated with each disease state; and (3) the consequences of altering abnormal S-nitrosothiol metabolism and S-nitrosylation signaling. Accurate identification of S-nitrosothiol levels could lead to the development of physiologically relevant model systems in which to study mechanisms of the disease and to develop novel therapeutics.

5. Summary

Many obstacles, both technical and biological, need to be overcome when measuring physiological levels of RSNOs. Many of the assays are used near the limit of detection. The S–NO bond is often labile, and the potential for artifactual nitrosylation or denitrosylation produced by manipulating samples is high. Indeed, endogenous S-nitrosothiols can be lost during protein isolation. To overcome these limitations, S-nitrosylation of a protein should be verified using several methods. Standard curves for these assays must use the relevant standard protein concentration range expected in the biological sample. Manipulation of the proteins should be minimized: the most reliable assays are those that modify the cell or tissue samples the least.

REFERENCES

Akaike, T., Inoue, K., Ikamoto, T., Nishino, H., Otagiri, M., Fujii, S., and Maeda, H. (1997). Nanomolar quantification and identification of various nitrosothiols by high performance liquid chromatography coupled with flow reactors of metals and Griess reagent. *J. Biochem.* **122**, 459–466.

Angelo, M., Singel, D. J., and Stamler, J. S. (2006). An S-nitrosothiol (SNO) synthase function of hemoglobin that utilizes nitrite as a substrate. *Proc. Natl. Acad. Sci. USA* **103**, 8366–8371.

Arnelle, D. R., and Stamler, J. S. (1995). NO^+, NO, and NO- donation by S-nitrosothiols: Implication for regulation of physiological functions by S-nitrosylation and acceleration of disulfide formation. *Arch. Biochem. Biophys.* **318**, 279–285.

Ascenzi, P., Colasanti, M., Persichini, T., Muolo, M., Poltcelli, F., Venturini, G., Bordo, D., and Bolognesi, M. (2000). Re-evaluation of amino acid sequence and structural consensus rules for cysteine-nitric oxide reactivity. *Biol. Chem.* **381**, 623–627.

Askew, S., Bulter, A., Flitney, F., Kemp, G., and Megson, I. (1995). Chemical mechanism underlying the vasodilator and platelet anti-aggregating properties of S-nitroso-N-acetyl-DL penicillamidne and S-nitrosoglutathione. *Bioorg. Med. Chem.* **3**, 1–9.

Burgard, D. A., Abaham, J., Allen, A., Craft, J., Foley, J., Robinson, J., Wells, B., Zu, C., and Stedman, D. H. (2006). Chemiluminescent reactions of nickel iron, and cobalt carbonyls with ozone. *Appl. Spectrosc.* **60**, 99–102.

Carver, D. J., Gaston, B., deRonde, K., and Palmer, L. A. (2007). Akt-mediated activation of HIF-1 in pulmonary vascular endothelial cells by S-nitrosoglutathione. *Am. J. Respir. Cell Mol. Biol.* **37**, 255–263.

Carver, J., Palmer, L. A., Doctor, A., and Gaston, B. (2005). S-Nitrosothiol formation. *Methods Enzymol.* **396**, 95–105.

Chan, N., Rogers, P., and Arnone, A. (1998). Crystal structure of the S-nitroso form of liganded human hemoglobin. *Biochemistry* **37**, 16459–16464.

Chen, E. Y., Bartlett, M. C., Loo, T. W., and Clarke, D. M. (2004). The DeltaF508 mutation disrupts packing of the transmembrane segments of the cystic fibrosis transmembrane conductance regulator. *J. Biol. Chem.* **279**, 39620–39627.

Cho, H., Ahn, D., Park, H., and Yang, E. G. (2007). Modulation of p300 binding by post translational modification of the C-terminal activation domain of hypoxia inducible factor 1 alpha. *FEBS Lett.* **581**, 1542–1548.

Choi, Y. B., Tenneti, L., Le, D. A., Ortiz, J., Bai, G., Chen, H. S., and Lipton, S. A. (2000). Molecular basis of NMDA receptor-coupled ion channel modulation by S-nitrosylation. *Nat. Neurosci.* **3**, 15–21.

Dejam, A., Kleinbongard, P., Tienush, R., Hamada, S., Gharini, P., Rodriguez, J., Feelisch, M., and Kelm, M. (2003). Thiols enhance NO formation from nitrate photolysis. *Free Radic. Biol. Med.* **35**, 1151–1159.

Della Torre, A., Schroeder, R. A., Bartlett, S. T., and Kuo, P. C. (1998). Differential effects of nitric oxide-mediated S-nitrosylation on p50 and cjun DNA binding. *Surgery* **124**, 137–141.

Della Torre, A., Schroeder, R. A., and Kuo, P. C. (1997). Alteration of NF-κB p50 DNA binding kinetics by S-nitrosylation. *Biochem. Biophys. Res. Commun.* **238**, 703–706.

Doctor, A., Gaston, B., and Kim-Shapiro, D. B. (2006). Detecting physiologic fluctuations in the S-nitrosohemoglobin micropopulation: Triiodide versus 3C. *Blood* **108**, 3225–3226.

Doctor, A., Platt, R., Sheram, M. L., Eischeid, A., McMahon, T., Maxey, T., Doherty, J., Axelrod, M., Kline, J., Gurka, M., Gow, A., and Gaston, B. (2005). Hemoglobin conformation couples S-nitrosothiol content in erythrocytes to oxygen gradients. *Proc. Natl. Acad. Sci. USA* **102**, 5709–5714.

Fang, K., Ragsdale, N. V., Carey, R. M., Macdonald, T., and Gaston, B. (1998). Reductive assays for S-nitrosothiols: Implication for measurements in biological system. *Biochem. Biophys. Res. Commun.* **252**, 535–540.

Forrester, M. T., Foster, M. W., and Stamler, J. S. (2007). Assessment and application of the biotin switch technique for examining protein S-nitrosylation under conditions of pharmacologically induced oxidative stress. *J. Biol. Chem.* **282**, 13977–13983.

Foster, M. W., McMahon, T. J., and Stamler, J. S. (2003). S-nitrosylation in health and disease. *Trends Mol. Med.* **9,** 160–168.

Gaston, B., Singel, D., Doctor, A., and Stamler, J. S. (2006). S-nitrosothiol signaling in respiratory biology. *Am. J. Respir. Crit. Care Med.* **173,** 1186–1193.

Gaston, B. M., Carver, J., Doctor, A., and Palmer, L. A. (2003). S-nitrosylation signaling in cell biology. *Mol. Interven.* **3,** 253–262.

Gorren, A. C., and Mayer, B. (1998). The versatile and complex enzymology of nitric oxide synthase. *Biochemistry* **63,** 734–743.

Gow, A., Doctor, A., Mannick, J., and Gaston, B. (2007). S-nitrosothiol measurements in biological systems. *J. Chromatogr. B Analyt. Technol. Biomed. Life Sci.* **851,** 140–151.

Gow, A. J., Chen, Q., Hess, B. J., Day, H., Ischiropoulos, H., and Stamler, J. S. (2002). Basal and stimulated protein S-nitrosylation in multiple cell types and tissues. *J. Biol. Chem.* **277,** 9637–9640.

Gow, A. J., Davis, C. W., Munson, D., and Ischiropoulos, H. (2004). Immunohistochemical detection of S-nitrosylated proteins. *Methods Mol. Biol.* **279,** 167–172.

Hausladen, A., Privalle, C., Keng, T., DeAngelo, J., and Stamler, J. (1996). Nitrosative stress: Activation of the transcription factor OxyR. *Cell* **86,** 719–729.

Hausladen, A., Rafikov, R., Angelo, M., Singel, D. J., and Stamler, J. S. (2007). Assessment of nitric oxide signals by triiodide chemiluminescence. *Proc. Natl. Acad. Sci. USA* **104,** 2157–2162.

Hess, D. T., Matsumoto, A., Kim, S. O., Marshall, H. E., and Stamler, J. S. (2005). Protein S-nitrosylation: Purview and parameters. *Nat. Rev. Mol. Cell Biol.* **6,** 150–166.

Hogg, H., Singh, R., Konorev, E., Joseph, H., and Kalyanaraman, B. S. (1997). S-nitrosoglutathione as a substrate for γ-glutamyl transpeptidase. *Biochem. J.* **323,** 477–481.

Huang, B., and Chen, C. (2006). An ascorbate-dependent artifact that interferes with the interpretation of the biotin switch assay. *Free Radic. Biol. Med.* **41,** 557–561.

Inoue, K., Akaike, T., Miyamoto, Y., Okamoto, T., Sawa, T., Otagiri, M., Suzuki, S., Yoshimura, T., and Maeda, H. (1999). Nitrosothiol formation catalyzed by ceruloplasmin: Implication for cytoprotective mechanism *in vivo*. *J. Biol. Chem.* **274,** 27069–27075.

Itoh, Y., Ma, F. H., Hoshi, M., Oka, M., Noda, K., Ukai, Y., Kohima, H., Nagano, T., and Toda, N. (2000). Determination and bioimaging method for nitric oxide in biological specimens by diaminofluorescein fluorometry. *Anal. Biochem.* **15,** 203–209.

Jaffrey, S. R., Erdjument-Bromage, H., Gerris, C. D., Tempst, P., and Snyder, S. H. (2001). Protein S-nitrosylation: A physiological signal for neuronal nitric oxide. *Nat. Cell Biol.* **3,** 193–197.

Jia, L., Bonaventura, C., Bonaventura, J., and Stamler, J. S. (1996). Nitrosohaemoglobin: A dynamic activity of blood involved in vascular control. *Nature* **380,** 221–226.

Johnson, M. A., Macdonald, T. L., Mannick, J. B., Conaway, M. R., and Gaston, B. (2001). Accelerated S-nitrosothiol breakdown by amyotrophic lateral sclerosis mutant copper, zinc superoxide dismutase. *J. Biol. Chem.* **276,** 39872–39878.

Joksovic, P. M., Doctor, A., Gaston, B., and Todorovic, S. M. (2007). Functional modulation of T-type calcium channels by S-nitrosothiols in the rat thalamus. *J. Neurophysiol.* **97,** 2712–2721.

Jourd'heuil, D., Laroux, F. S., Miles, A. M., Wink, D. A., and Grisham, M. B. (1999). Effect of superoxide dismutase on the stability of S-nitrosothiols. *Arch. Biochem. Biophys.* **361,** 323–330.

Kettenhofen, N. J., Broniowska, K. A., Keszler, A., Zhang, Y., and Hogg, N. (2007). Proteomic methods for analysis of S-nitrosation. *J. Chromatogr. B Analyt. Technol. Biomed. Life Sci.* **851,** 152–159.

Kim, S. F., Huri, D. A., and Snyder, S. H. (2005). Inducible nitric oxide synthase binds, S-nitrosylates, and activates cyclooxygenase-2. *Science* **310,** 1966–1970.

Kim, S. O., Merchange, K., Nudelman, R., Beyer, W. F., Jr., Keng, T., DeAngelo, J., Hausladen, A., and Stamler, J. S. (2002). Oxy R: A molecular code for redox-related signaling. *Cell* **109,** 383–396.

King, M., Gildemeister, O., Gaston, B., and Mannick, J. B. (2005). Assessment of S-nitrosothiols on diaminofluorescein gels. *Anal. Biochem.* **346,** 69–76.

Lancaster, J., and Gaston, B. (2004). NO and nitrosothiols: Spatial confinement and free diffusion. *Am. J. Physiol. Lung Cell Mol. Physiol.* **287,** 465–466.

Li, F., Sonveaux, P., Tabbani, Z. N., Liu, S., Yan, B., Huang, Q., Vujaskovic, Z., Dewhirst, M. W., and Ci, C. Y. (2007). Regulation of HIF-1alpha stabilitiy through S-nitrosylation. *Mol. Cell* **26,** 63–74.

Lipton, A., Johnson, M., Macdonald, T., Lieberman, M., Gozal, D., and Gaston, B. (2001). S-nitrosothiols signal the ventilatory response to hypoxia. *Nature* **413,** 171–174.

Liu, L., Hausladen, A., Zeng, M., Que, L., Heitman, J., and Stamler, J. S. (2001). A metabolic enzyme for S-nitrosothiol conserved from bacteria to humans. *Nature* **410,** 490–494.

Liu, L., Yan, Y., Zeng, M., Zhang, J., Hanes, M. A., Aheam, G., McMahon, T. J., Dickfeld, T., Marshall, H. E., Que, L. G., and Stamler, J. S. (2004). Essential role of S-nitrosothiols in vascular homeostasis and endotoxic shock. *Cell* **116,** 617–628.

Luchsinger, B. P., Rich, E. N., Gow, A. J., Williams, E. M., Stamler, J. S., and Singel, D. J. (2003). NO interaction with oxidized hemes in human hemoglobin: Routes to SNO-hemoglobin function with preferential reactivity within the β subunits. *Proc. Natl. Acad. Sci. USA* **100,** 461–466.

Mannick, J., Schonhoff, C., Papeta, N., Ghafourifar, P., Szibor, M., Fang, K., and Gaston, B. (2001). S-Nitrosylation of mitochondrial caspases. *J. Cell Biol.* **154,** 1111–1116.

Mannick, J. B., Hausladen, A., Liu, L., Hess, D. T., Zeng, M., Miao, Q. X., Kane, L. S., Gow, A. J., and Stamler, J. S. (1999). Fas-induced caspase denitrosylation. *Science* **284,** 651–654.

Mathews, J. R., Botting, C. H., Panico, M., Morris, H. R., and Hay, R. T. (1996). Inhibition of NF-κB binding by nitric oxide. *Nucleic Acids Res.* **24,** 2236–2242.

Mayer, B., Pfeiffer, S., Schrammel, A., Koesling, D., Schmidt, K., and Brunner, F. (1998). A new pathway of nitric oxide/cyclic GMP signaling involving S-nitrosoglutathione. *J. Biol. Chem.* **273,** 3264–3270.

McMahon, T. J., Ahearn, G. S., Moya, M. P., Gow, A. J., Huang, Y. C., Luchsinger, B. P., Nudelman, R., Yan, Y., Krichman, A. D., Bashore, T. M., Califf, R. M., Singel, D. J., et al. (2005). A nitric oxide processing defect of red blood cells created by hypoxia: Deficiency of S-nitrosohemoglobin in pulmonary hypertension. *Proc. Natl. Acad. Sci. USA* **102,** 14801–14806.

McMahon, T. J., Moon, R. E., Luschinger, B. P., Carraway, M. S., Stone, A. E., Stolp, B. W., Gow, A., Pawloski, J. R., Watke, P., Singel, D. J., Piantadosi, C. A., and Stamler, J. S. (2002). Nitric oxide in the human respiratory cycle. *Nat. Med.* **8,** 711–717.

Munson, D. A., Grubb, P. H., Kerecman, J. D., McCurnin, D. C., Yoder, B. A., Hazen, S. L., Shaul, P. W., and Ishiropoulos, H. (2005). Pulmonary and systemic nitric oxide metabolites in a baboon model of neonatal chronic lung disease. *Am. J. Respir. Cell Mol. Biol.* **33,** 582–588.

Nagababu, E., Ramasamy, S., and Rifkind, J. M. (2006). S-nitrosohemoglobin: A mechanism for its formation in conjunction with nitrite reduction by deoxyhemoglobin. *Nitric Oxide* **6,** 657–666.

Nedospasov, A., Rafifov, R., Beda, N., and Nudler, E. (2000). An autocatalytic mechanism of protein nitrosylation. *Proc. Natl. Acad. Sci. USA* **97,** 13543–13548.

Palmer, L. A., Doctor, A., Chhabra, P., Sheram, M. L., Laubach, V. E., Karlinsey, M. Z., Forbes, M. S., Macdonald, T., and Gaston, B. (2007). S-Nitrosothiols signal hypoxia-mimetic vascular pathology. *J. Clin. Invest.* **117,** 2592–2601.

Park, S. K., Lin, J. L., and Murphy, S. (1997). Nitric oxide regulates nitric oxide synthase-2 gene expression by inhibiting NF-κB binding to DNA. *Biochem. J.* **322**, 609–613.

Pawloski, J. R., Hess, D. T., and Stamler, J. S. (2001). Export by red blood cell of nitric oxide bioactivity. *Nature* **409**, 622–626.

Perez-Mato, I., Castro, C., Ruiz, F. A., Corrales, F. J., and Mato, J. M. (1999). Methionine adenosyltransferase S-nitrosylation is regulated by the basic and acidic amino acids surrounding the target thiol. *J. Biol. Chem.* **274**, 17075–17079.

Rogers, S. C., Khalatbari, A., Datta, B. N., Ellery, S., Paul, V., Frenneaux, M. P., and James, P. E. (2007). NO metabolite flux across the human coronary circulation. *Cardiovasc. Res.* **75**, 434–441.

Saville, B. (1958). A scheme for the colorimetric determination of microgram amounts of thiols. *Analyst* **83**, 670–672.

Schonhoff, C. M., Matsuoka, M., Tummala, H., Johnson, M. A., Estevez, A. G., Kamaid, A., Ricart, K. C., Hashimoto, Y., Gaston, B., Macdonald, T. L., Xu, Z., and Mannick, J. B. (2006). S-nitrosothiol depletion in amyotrophic lateral sclerosis. *Proc. Natl. Acad. Sci. USA* **103**, 2404–2409.

Sexton, D. J., Muruganandam, A., McKenny, S. J., and Mutus, B. (1994). Visible light photochemical release of nitric oxide from S-nitrosoglutathione: Potential photochemotherapeutic applications. *Photochem. Photobiol.* **59**, 463–467.

Shishido, S. M., and de Oliveira, M. G. (2000). Polyethylene glycol matrix reduces the rates of photochemical and thermal release of nitric oxide from S-nitroso-N-acetylcysteine. *Photochem. Photobiol.* **71**, 273–280.

Stamler, J. S., Toone, E., Lipton, S., and Sucher, N. (1997). (S) NO signals: Translocation, regulation, and a consensus motif. *Neuron* **18**, 691–696.

Stuehr, D. J. (1999). Mammalian nitric oxide synthases. *Biochim. Biophys. Acta* **1411**, 217–230.

Taubert, D., Roesen, R., Lehmann, C., Jung, N., and Schömig, E. (2007). Effects of low habitual cocoa intake on blood pressure and bioactive nitric oxide: A randomized controlled trial. *JAMA* **298**, 49–60.

Trujillo, M., Alvarez, M., Peluffo, G., Freeman, B., and Radi, R. (1998). Xanthine oxidase-mediated decomposition of S-nitrosothiols. *J. Biol. Chem.* **273**, 7929–7934.

Tsikas, D., Denker, K., and Frölich, J. C. (2001). Artifactual-free analysis of S-nitrosoglutathione and S-nitroglutathione by neutral-pH, anion-pairing, high-performance liquid chromatography: Study on peroxynitrite-mediated S-nitration of glutathione to S-nitroglutathione under physiological conditions. *J. Chromatogr. A* **915**, 107–116.

Tsikas, D., Sandmann, J., Gutzki, F. M., Stichtenoth, D. O., and Frölich, J. C. (1999). Measurement of S-nitrosoalbumin by gas chromatography–mass spectrometry. II. Quantitative determination of S-nitrosoalbumin in human plasma using S-[15N]nitrosoalbumin as internal standard. *J. Chromatogr. B Biomed. Sci. Appl.* **726**, 13–24.

Tsikas, D., Sandmann, J., Rossa, S., and Frölich, J. C. (2002). Measurement of S-nitrosoalbumin by gas chromatography–mass spectrometry. III. Quantitative determination in human plasma after specific conversion of the S-nitroso group to nitrite by cysteine and Cu(2+) via intermediate formation of S-nitrosocysteine and nitric oxide. *J. Chromatogr. B Biomed. Sci. Appl.* **772**, 335–346.

Whalen, E. J., Foster, M. W., Matsumoto, A., Ozawa, K., Violin, J. D., Que, L. G., Nelson, C. D., Benhar, M., Keys, J. R., Rockman, H. A., Koch, W. J., Daaka, Y., et al. (2007). Regulation of beta-adrenergic receptor signaling by S-nitrosylation of G-protein-coupled receptor kinase 2. *Cell* **129**, 511–522.

Xu, L., Eu, J. P., Meissner, G., and Stamler, J. S. (1998). Activation of the cardiac calcium release channel (ryanodine receptor) by poly-S-nitrosylation. *Science* **279**, 234–237.

Yao, D., Gu, Z., Nakamura, T., Shi, Z., Ma, Y., Gaston, B., Palmer, L., Rockenstein, E., Zhang, Z., Masliah, E., Uehara, T., and Lipton, S. (2004). Nitrosative stress linked to sporadic Parkinson's disease: S-Nitrosylation of Parkin regulates its E3 ligase activity. *Proc. Nat. Acad. Sci. USA* **101,** 10810–10814.

Zaman, K., Carraro, S., Doherty, J., Henderson, E. M., Lendermon, E., Liu, L., Verghese, G., Zigler, M., Ross, M., Park, E., Palmer, L., Doctor, L., *et al.* (2006). S-nitrosylating agents: A novel class of compounds that increase cystic fibrosis transmembrane conductance regulator expression and maturation in epithelial cells. *J. Mol. Pharmacol.* **70,** 1435–1442.

Zaman, K., Palmer, L. A., Doctor, A., Hunt, J. F., and Gaston, B. (2004). Concentration dependent effects of endogenous S-nitrosoglutathione on gene reglation by specificity proteins SP3 and SP1. *Biochem. J.* **380,** 67–74.

Zhang, Y., and Hogg, N. (2004). The mechanism of transmembrane S-nitrosothiol transport. *Proc. Natl. Acad. Sci. USA* **101,** 7891–7896.

CHAPTER TEN

Analysis of Citrulline, Arginine, and Methylarginines Using High-Performance Liquid Chromatography

Guoyao Wu[*,†] and Cynthia J. Meininger[†]

Contents

1. Introduction	178
2. The HPLC Apparatus	179
3. Chemicals and Materials	179
4. General Precautions in Sample Preparation and HPLC Analysis	180
5. Analysis of Citrulline and Arginine in Physiological Samples	181
6. Analysis of Methylarginines in Physiological Samples	185
7. Conclusion	188
Acknowledgments	188
References	188

Abstract

Citrulline is a product of arginine degradation by nitric oxide synthase and is a precursor for arginine synthesis in animal cells. After arginine is incorporated into proteins, it may undergo methylation to form N^G-monomethylarginine, which may be converted to asymmetric dimethylarginine and symmetric dimethylarginine. The degradation of these methylated proteins produces free methylarginines. This chapter focuses on the analysis of these amino acids in biological samples (including plasma/serum, urine, cell culture medium, and tissues) using high-performance liquid chromatography that involves precolumn derivatization with o-phthaldialdehyde. Fluorescence is monitored at excitation and emission wavelengths of 340 and 455 nm, respectively. Detection limits are 5 nM for amino acids. The assays are linear between 1 and 100 μM for citrulline and arginine and between 0.1 and 10 μM for methylarginines. These chromatographic methods are highly sensitive, specific, accurate, and easily automated and provide a useful tool to study the regulation of the arginine–nitric oxide pathway.

---

[*] Department of Animal Science, Texas A&M University, College Station, Texas
[†] Cardiovascular Research Institute and Department of Systems Biology and Translational Medicine, Texas A&M Health Science Center, College Station, Texas

Methods in Enzymology, Volume 440
ISSN 0076-6879, DOI: 10.1016/S0076-6879(07)00810-5

© 2008 Elsevier Inc.
All rights reserved.

1. Introduction

L-Arginine is the nitrogenous substrate for the synthesis of nitric oxide (NO) and L-citrulline by NO synthase (NOS) in animal cells. Interestingly, citrulline is converted readily into arginine virtually via argininosuccinate synthase and lyase in all cell types (Wu and Morris, 1998). After arginine is incorporated into proteins, it may undergo methylation to form N^G-monomethylarginine (NMMA) (McBride and Silver, 2001). Protein-bound NMMA may be converted to asymmetric dimethylarginine (ADMA) and symmetric dimethylarginine (SDMA). The degradation of these proteins produces free methylarginines. Both NMMA and ADMA are competitive inhibitors of NOS, whereas elevated levels of SDMA may inhibit arginine transport by cells (Tsikas et al., 2000). Concentrations of citrulline, arginine, NMMA, ADMA, and SDMA in plasma of postabsorptive healthy adults are approximately 30, 100, 0.2, 0.5, and 0.5 μM, respectively (Marliss et al., 2006; Teerlink, 2007; Wu and Morris, 1998). Because of the presence of more than 30 amino acids and low concentrations of methylarginines in physiological fluids, the separation and quantification of these substances are technically challenging.

A number of methods have been developed for amino acid analysis using high-performance liquid chromatography (HPLC). They include precolumn derivatization with 4-chloro-7-nitrobenzofurazan, 9-fluorenyl methylchloroformate, phenylisothiocyanate, naphthalene-2,3-dicarboxaldehyde, 6-aminoquinolyl-N-hydroxysuccinimidyl carbamate, and o-phthaldialdehyde (OPA) (Jones and Gilligan, 1983; Teerlink, 2007; Valtonen et al., 2005). OPA reacts with primary amino acids (e.g., citrulline, arginine, and methylarginines) and amines (e.g., agmatine and spermine) in the presence of 2-mercaptoethanol or 3-mercaptopropionic acid to form a highly fluorescent adduct (Fig. 10.1). This method is used most widely in research laboratories because of the following advantages: simple procedures for the preparation of samples, reagents, and mobile phase solutions; rapid formation of OPA derivatives and their efficient separation at room temperature; high sensitivity

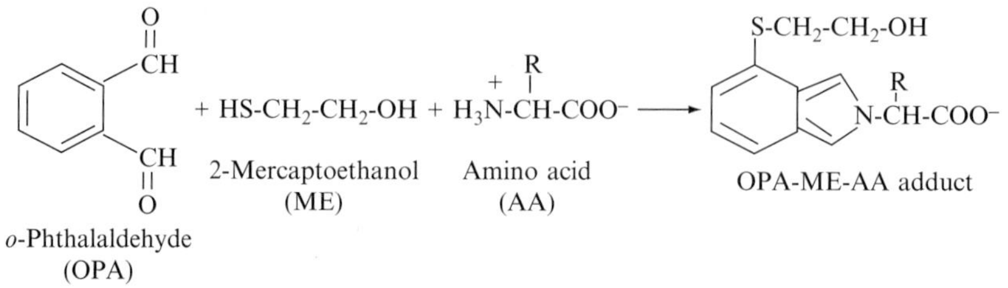

Figure 10.1 Reaction of amino acids with o-phthaldialdehyde in the presence of 2-mercaptoethanol to form highly fluorescent adducts.

of detection at picomole levels; easy automation on the HPLC apparatus; few interfering side reactions; a stable chromatography baseline and accurate integration of peak areas; and rapid regeneration of guard and analytical columns (Jones and Gilligan, 1983; Wu et al., 1996). Because concentrations of citrulline and arginine in plasma and cells are generally 60 to 300 times those of methylarginines (Kohli et al., 2004), separate protocols are more desirable for their accurate analysis by HPLC.

2. The HPLC Apparatus

The Waters HPLC apparatus consists of the following: A Model 600E Powerline multisolvent delivery system with 100-μl heads, a Model 600E system controller, a Model 717plus autosampler, an analytical column protected by a guard column, a Model 2475 Multi λ fluorescence detector, and a Millenium-32 Workstation (Waters Inc., Milford, MA). Fluorescence is monitored at excitation and emission wavelengths of 340 and 455 nm, respectively. All mobile phase solutions should be degassed with helium for at least 30 min before use. To prevent accumulation of salts in the HPLC system, both columns and pumps should be washed thoroughly with water before and after each run. Additionally, the autosampler should be purged with water twice immediately before starting a sample set.

3. Chemicals and Materials

All chemicals of highest purity are obtained from Sigma (St. Louis, MO), and all of the prepared solutions are stored at room temperature, unless specified.

1. HPLC-grade methanol and H_2O: These are obtained from Fisher Scientific (Houston, TX) and used for preparing all reagents and mobile phase solutions, as well as making dilutions for standards and samples. Methanol and H_2O should be checked regularly for possible contamination with amino acids, particularly glycine.
2. Benzoic acid (1.2%): Dissolve 8.4 g benzoic acid (an antibiotic) in 525 ml H_2O, followed by the addition of 175 ml of saturated $K_2B_4O_7$ (prepared in H_2O). This solution is stable for 2 years.
3. Sodium borate (40 mM, pH 9.5): Dissolve 30.51 g $Na_2B_4O_7 \cdot 10H_2O$ in 2 liters of H_2O. This solution is stable for 2 years.
4. OPA reagent solution (30 mM OPA, 50 mM 2-mercaptoethanol, 40 mM sodium borate, and 3.1% Brij-35, pH 9.5): Add 50 mg OPA

(stored at 4°) and 1.25 ml methanol to a brown bottle, followed by the addition of 11.2 ml of 40 mM sodium borate buffer (pH 9.5), 50 μl of 2-mercaptoethanol, and 0.4 ml of Brij-35. The solution is mixed thoroughly, stored at 4°, and used within 36 h after preparation. Brij-35 is used for optimum separation of citrulline from threonine and of ornithine from lysine.

5. Glass vials (4-ml volume) for HPLC analysis: These reusable vials are obtained from Waters Inc. Before use, they are washed three times with distilled/deionized water, dried at 60°, and placed in a 400° furnace overnight to eliminate organic matter.

4. General Precautions in Sample Preparation and HPLC Analysis

Care should be taken in the collection, processing, and storage of biological samples, as well as in the preparation of all reagents and mobile phase solutions (Jobgen et al., 2007). Particularly, glassware, pipettes, tubes, and a researcher's hands should be washed thoroughly with distilled/deionized water and should have no contamination with any amino acid. Freshly obtained whole blood, plasma, serum, urine, amniotic fluid, allantoic fluid, milk, other physiological fluids, cell culture medium, and cells should be immediately deproteinized and neutralized (Kohli et al., 2004). Perchloric acid and potassium carbonate are used as an acid to deproteinize samples and neutralize the acidified solution, respectively, because these two reagents do not contain any substances that interfere with the derivatization of amino acids with OPA or the chromatographic separation of their derivatives. Heparin- or EDTA-containing tubes can be used to withdraw blood without affecting amino acid analysis, but hemolysis should be avoided to prevent the release of arginase and other enzymes from erythrocytes. Tissues other than blood (e.g., skeletal muscle, liver, small intestine, kidney, and brain) should be rapidly dissected and frozen in liquid nitrogen to minimize the catabolism of amino acids. In addition, protein degradation, a potential source of amino acids, should be prevented in sample collection and processing. If possible, frozen tissue samples should be homogenized immediately in an acid solution and neutralized. All neutralized solutions should be stored at $-80°$ (if possible) or at least $-20°$.

Deproteinized and neutralized biological samples are stable at $-80°$ for 6 months without any detectable loss. When unprocessed plasma or serum samples from pigs are stored at $-80°$, losses of citrulline at the end of 0.5, 1, 2, 3, 4, and 5 years are 0.8, 1.4, 2.1, 2.6, 3.3, and 4.5%, respectively, and losses of arginine at the end of 0.5, 1, 2, 3, 4, and 5 years are 1.5, 2.9, 4.0, 6.7, 8.5, and 9.3%, respectively ($n = 20$). Similar results are obtained for human plasma and serum.

Recoveries of amino acids from a biological matrix of interest should be determined using known amounts of standards and corrected for data calculation. Importantly, blank chromatograms should be checked on the day of analysis using reagent solutions (e.g., 0.1 ml of 1.2% benzoic acid, 0.1 ml of neutralized 1.5 M HClO$_4$/2 M K$_2$CO$_3$ extracts, and 1.4 ml H$_2$O for citrulline and arginine analysis) to ensure that they do not contain substances that have the same retention times as amino acids in samples. Furthermore, the linearity of the assays should be established using known amounts of amino acid standards.

5. Analysis of Citrulline and Arginine in Physiological Samples

1. Preparation of amino acid standards not included in regular protein hydrolysate standards. These amino acids are glutamine, asparagine, β-alanine, tryptophan, citrulline, taurine, and ornithine and are referred to as GATCOTA (2.5 mM for each amino acid). To prepare this solution, one needs to dissolve 73 mg glutamine, 66 mg asparagine, 102 mg tryptophan, 87.6 mg citrulline, 62.6 mg taurine, 84.3 mg ornithine, and 44.5 mg β-alanine in 200 ml H$_2$O. This stock solution can be stored in 0.5-ml aliquots at $-80°$ for 6 months without any detectable loss.
2. Protein hydrolysate amino acid standard solution (2.5 mM for each amino acid): This solution (Sigma) contains all protein amino acids (including arginine, lysine, and histidine), except glutamine, asparagine, and tryptophan. It is important that arginine, histidine, and lysine standards be prepared fresh from powder forms on the day of analysis to verify their concentrations in a commercial source of amino acid standard solution, as these basic amino acids may adhere to the surface of their glass container during shipping and storage.
3. Working amino acid standards (10 μM for each amino acid):
    a. Mix 50 μl of GATCOTA (2.5 mM each) and 50 μl of protein hydrolysate amino acid standard solution (2.5 mM each) with 400 μl H$_2$O in a 15-ml polypropylene tube.
    b. Add 500 μl of 1.5 M HClO$_4$ to the solution and vortex the tube. After 2 min, add 11.25 ml H$_2$O and 250 μl of 2 M K$_2$CO$_3$. The tube is vortexed and centrifuged at 3000 g for 5 min. This solution is stored in 0.5-ml aliquots and is stable at $-80°$ for 6 months.
4. Processing of biological samples.
    a. General notes. After biological samples are deproteinized and neutralized, the supernatant fluid is analyzed immediately for amino acid analysis or stored at $-80°$ until analysis. If there is a limited

availability of a sample, the following protocols can be scaled down proportionately.

b. Protocol for the preparation of physiological fluids. Mix 50 μl of plasma (or other physiological fluid) with 50 μl of 1.5 M HClO$_4$ in a 1.5-ml microcentrifuge tube. After 2 min, add 1.125 ml H$_2$O and 25 μl of 2 M K$_2$CO$_3$. The tubes are vortexed and centrifuged (10,000 g for 1 min) to obtain the supernatant fluid for analysis. Note that the dilution factor for a sample is 25. Recovery rates of citrulline and arginine from plasma, milk, amniotic fluid, allantoic fluid, and urine are 98 to 100% ($n = 10$).

c. Protocol for the preparation of tissue extracts. A frozen tissue sample (≈100 mg) is homogenized, with use of a glass homogenizer, in 1 ml of 1.5 M HClO$_4$. The homogenizer is rinsed twice with H$_2$O (3 ml each). The combined solution is transferred to a 15-ml polypropylene tube, followed by slow addition of 0.5 ml of 2 M K$_2$CO$_3$. Tubes are vortexed and centrifuged (3000 g for 5 min) to obtain the supernatant fluid for analysis. Recovery rates of citrulline and arginine from skeletal muscle, small intestine, liver, kidney, placenta, brain, adipose tissue, and mammary tissue are 97 to 100% ($n = 10$).

d. Protocol for the preparation of cell extracts. Cells (e.g., 1.5×10^6 endothelial cells) are obtained by centrifugation (10,000 g for 1 min) or by passing through a layer of silicon oil (Wu and Meininger, 1993). Cells are then immediately acidified with 0.2 ml of 1.5 M HClO$_4$ in a 1.5-ml microcentrifuge tube. After 1 min of vortexing, 0.1 ml of 2 M K$_2$CO$_3$ is added. The tubes are vortexed and centrifuged (10,000 g for 1 min) to obtain the supernatant fluid for analysis. Recovery rates of citrulline and arginine from endothelial cells, hepatocytes, adipocytes, muscle cells, enterocytes, macrophages, and lymphocytes are 98 to 100% ($n = 10$).

6. HPLC analysis.

a. Mobile phase solutions. Mobile phase A (0.1 M sodium acetate, pH 7.2): Add 27.3 g sodium acetate (trihydrate) to 1.6 liters H$_2$O in a glass bottle, followed by sequential addition of 96 μl of 6 N HCl, 180 ml methanol, and 10 ml tetrahydrofuran. Mixing is performed after each addition. Adjust the final volume of the solution to 2 liters with H$_2$O. Mobile phase B: 100% methanol in a brown glass bottle.

b. Preparation of standard or sample vials. To a 4-ml glass vial, add 0.1 ml of 1.2% benzoic acid, 0.1 ml of an amino acid standard (10 μM each) or sample solution, and 1.4 ml H$_2$O. If there is a limited availability of a sample, a microinsert tube placed in a spring holder within the glass vial (all from Waters Inc.) can be used to scale down all the solutions 10 times (a final volume of 160 μl). Vortex all tubes before placing them onto an autosampler (one standard vial at the start and the end, with samples in the middle). For a 49- and 35-min running program, the number of samples should not exceed 12 and 17, respectively, in a sample set to ensure the stability of both OPA and amino acids.

c. Preparation of an analytical column (C_{18}; 4.6 mm × 15 cm, 3 μm) and a guard column (C_{18}; 4.6 mm × 5 cm, 20–40 μm): These two columns are obtained from Supelco (Bellefonte, PA; now a Sigma division). Before running a sample set, the columns should be washed sequentially with water and methanol (at least 15 min with each solvent) and then equilibrated with mobile phase A for at least 20 min at a flow rate of 1.1 ml/min.

d. The autosampler is programmed to mix 25 μl of a standard or sample solution with 25 μl of the OPA reagent solution for 1 min in a reaction loop. The derivatized solution is immediately delivered into the HPLC column without any delay time.

e. Mobile phase gradients (a total running time of 49 min; flow rate, 1.1 ml/min): Because ornithine is a major product of arginine catabolism in animal cells, it is desirable to simultaneously determine citrulline, ornithine, and arginine in a sample (Table 10.1). Satisfactory separation of these three amino acids requires a 49-min program, including the time for column regeneration. If ornithine analysis is not needed, a shorter program with a total running time of 35 min can be used. After each sample set is completed, the columns should be washed sequentially with water, water/methanol (50:50; vol/vol), and methanol (at least 20 min with each solvent) at a flow rate of 1.1 ml/min.

A representative HPLC chromatogram for 25 amino acids is illustrated in Fig. 10.2. All amino acids are well separated. This 49-min program can be used for the analysis of argininosuccinate (6.5 min), phosphoarginine (7.0 min), glucosamine-6-phosphate (8.3 min), glucosamine (13.6 min), N^w-hydroxy-nor-L-arginine (17.9 min), methionine sulfoxide (between glycine and histidine peaks), γ-aminobutyric acid (between alanine and tyrosine peaks), S-adenosylmethionine (21.4 min), carnosine (immediately after

Table 10.1 HPLC gradient program for separation of citrulline, arginine, and ornithine[a]

Mobile phase (%)	Time (min)										
	0	15	20	24	26	34	38	40	42	42.1	49
A	86	86	70	65	53	50	30	0	0	86	86
B	14	14	30	35	47	50	70	100	100	14	14

[a] Flow rate: 1.1 ml/min.

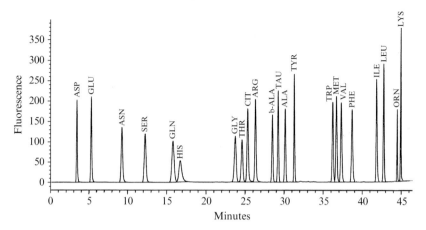

Figure 10.2 An HPLC chromatogram of amino acid analysis. Concentrations of each amino acid were 10 μM.

arginine), 3-methylhistidine (immediately before β-alanine), S-adenosylhomocysteine (32.5 min), dimethylarginine (immediately after taurine), agmatine (immediately before phenylalanine), and hydroxylysine (immediately after leucine). The arginine peak should be verified by treatment of samples with arginase (Wu and Brosnan, 1992).

A gradient program with a total running time of 35 min (including the time for column regeneration) at a flow rate of 1.1 ml/min can be used for the separation of citrulline and arginine. The proportion of mobile phase B is as follows: 0 min, 14%; 15 min, 14%; 20 min, 30%; 24 min, 35%; 24.1 min, 100%; 26 min, 100%; 26.1 min, 14%; and 35 min, 14%. The separation of amino acids before arginine is the same as described earlier for the 49-min program.

Using amino acid standards in the range of 1 nM to 100 μM, we have established that detection limits (defined as a signal-to-noise ratio of 3) are 5 nM and that the assays are linear between 1 and 100 μM for citrulline and arginine ($r^2 = 0.996$). The precision (agreement between replicate measurements) of the HPLC method, as evaluated by the relative deviation (mean of absolute deviation/mean of replicate measurements × 100%; Li et al., 2000), is 1.3 to 1.7% for plasma, milk, urine, amniotic fluid, allantoic fluid, endothelial cells, placenta, small intestine, brain, liver, kidney, mammary tissue, and skeletal muscle ($n = 10$). The accuracy (the nearness of an experimental value to the true value) of the HPLC method, as determined with known amounts of amino acid standards and expressed as relative errors [(measurement value − true value)/(true value × 100%); Li et al., 2000], is 1.6 to 2.2% for the aforementioned biological samples ($n = 10$).

6. ANALYSIS OF METHYLARGININES IN PHYSIOLOGICAL SAMPLES

We have modified the methods of Bode-Böger *et al.* (1996) and Teerlink *et al.* (2002) for methylarginine analysis to improve sample cleanup, eliminate the need for column heating, and facilitate automation on the HPLC apparatus (Fu *et al.*, 2005; Kohli *et al.*, 2004). Conditions for the derivatization of methylarginines with OPA are the same as described previously for citrulline and arginine analysis.

1. Preparation of methylarginine standards: They are stored in 0.5-ml aliquots and stable at $-80°$ for 6 months.
    a. 2 mM NMMA: Dissolve 5.06 mg NMMA-acetate salt in 10 ml H_2O. 2 mM ADMA: Dissolve 4.04 mg ADMA-HCl salt in 10 ml H_2O. 2 mM SDMA: Dissolve 3.04 mg SDMA-di(p-hydroxyazobenzene-p'-sulfonate) salt in 2 ml H_2O.
    b. 25 μM methylarginine standards: Add 50 μl of 2 mM ADMA, 50 μl of 2 mM SDMA, and 50 μl of 2 mM NMMA to 3.85 ml H_2O. Vortex the tube.
    c. 1 μM methylarginine standards: Mix 0.4 ml of 25 μM methylarginine standards with 9.6 ml H_2O. Vortex the tube.
2. Materials for sample cleanup.
    a. Oasis MCX SPE column (1 ml, 30 mg): This cation-exchange, solid-phase extraction column is obtained from Waters Inc. The column is washed with 5 ml of 0.1 M HCl immediately before sample loading, using a vacuum manifold (Waters Inc.) with a capacity for 20 columns.
    b. Cleanup solution: 29.5% ammonium hydroxide/H_2O/methanol (10/40/50; by volume).
3. Processing of biological samples.
    a. General notes. Methylarginine standard solutions should be processed in parallel with biological samples. The pH of extracts from biological samples should be around 7.0.
    b. Protocol for the preparation of physiological fluids. A physiological fluid (e.g., plasma; 0.2 ml) or 0.2 ml of 1 μM methylarginine standards is mixed with 0.1 ml of 1.5 M $HClO_4$, followed by the addition of 0.05 ml of 2 M K_2CO_3. This whole neutralized extract is mixed with 0.7 ml of phosphate-buffered saline (pH 7.0), and the entire solution is loaded into an Oasis MCX SPE column. The column is washed sequentially with 1 ml of 0.1 M HCl, 1 ml methanol, and 1 ml of the cleanup solution, using a vacuum manifold. The last 1-ml

fraction that contains methylarginines is collected, and the solvent is removed using a Model RC10.10 centrifugal evaporator (Jouan Inc., Winchester, VA) at 70° for approximately 3 h. The residues are suspended in 0.2 ml H_2O. Recovery rates of NMMA, ADMA, and SDMA from the solid-phase extraction column are 98, 86, and 97%, respectively. Their recovery rates from the whole process (including deproteinization, neutralization, and column cleanup) are 92–94, 81–84, and 92–94% for physiological fluids (plasma, urine, amniotic fluid, and allantoic fluid), respectively.

c. Protocol for the preparation of tissue extracts. A frozen tissue sample ($\approx$200 mg) is homogenized, with use of a glass homogenizer, in 1 ml of 1.5 M $HClO_4$. The homogenizer is rinsed with 1 ml H_2O. The combined solution is transferred to a 10-ml polypropylene tube, followed by slow addition of 0.5 ml of 2 M K_2CO_3. Similarly, 0.1 ml of a 25 μM methylarginine standard solution is mixed with 1.0 ml of 1.5 M $HClO_4$, 0.5 ml of 2 M K_2CO_3, and 0.9 ml H_2O to obtain 1 μM methylarginine. The tubes are vortexed and centrifuged (3000 g for 5 min). An aliquot (0.2 ml) of the supernatant fluid is used for cleanup, as described earlier. Recovery rates of NMMA, ADMA, and SDMA from the whole process (including deproteinization, neutralization, and column cleanup) are 88–92, 80–82, and 90–93% for tissues (skeletal muscle, placenta, liver, brain, kidney, small intestine, heart, and mammary tissue), respectively.

d. Protocol for the preparation of cell extracts. Cells (e.g., 3×10^6 endothelial cells) are acidified with 0.2 ml of 1.5 M $HClO_4$ in a 1.5-ml microcentrifuge tube. After 1 min of vortexing, 0.1 ml of 2 M K_2CO_3 is added. The tubes are vortexed and centrifuged (10,000 g for 1 min). Methylarginine standards (1 μM) are prepared as described earlier. An aliquot (0.2 ml) of the supernatant fluid is used for cleanup, as described previously. Recovery rates of NMMA, ADMA, and SDMA from the whole process (including deproteinization, neutralization, and column cleanup) are 91–94, 82–84, and 91–94% for cells (endothelial cells, adipocytes, hepatocytes, macrophages, lymphocytes, and enterocytes), respectively.

5. HPLC analysis.

   a. Mobile phase (33.3% methanol/0.64% sodium citrate, pH 6.8): Dissolve 14.4 g sodium citrate in 1.5 liters H_2O. Adjust the solution pH to 6.8 with 0.6 ml of 6 N HCl, followed by the addition of 750 ml methanol.

   b. Preparation of standard or sample vials. To a microinsert tube placed in a spring holder within a 4-ml glass vial, add 10 μl of 1.2% benzoic acid, 10 μl of 1 μM methylarginine standard or sample solution after column cleanup, and 140 μl H_2O. Vortex all tubes before placing them onto an autosampler (one standard vial at the start and the end, with samples in the middle). The number of samples should not exceed 17 in a sample set to ensure the stability of both OPA and methylarginines.

c. Nucleosil 100–5 C6H5 column (4.6 × 250 mm; Macherey Nagel, Easton, PA): Before running a sample set, the columns should be washed sequentially with water and methanol (at least 20 min with each solvent) and then equilibrated with mobile phase A for 30 min at a flow rate of 0.7 ml/min.

d. The autosampler is programmed to mix 15 μl of a standard or sample solution with 15 μl of the OPA reagent solution for 1 min in a reaction loop. The derivatized solution is immediately delivered into the HPLC column without any delay time.

e. Isocratic elution with the mobile phase (33.3% methanol/0.64% sodium citrate, pH 6.8) at a flow rate of 0.7 ml/min. Satisfactory separation of NMMA, SDMA, and ADMA is achieved in a 35-min program, including the time for column regeneration. After each sample set is completed, the column should be washed sequentially with water, water/methanol (50:50; vol/vol), and methanol (30 min with each solvent) at a flow rate of 0.7 ml/min.

A representative HPLC chromatogram is shown in Fig. 10.3. NMMA, SDMA, and ADMA are well separated, with retention times of 20.7, 25.9, and 29.1 min, respectively. Amino acids other than methylarginines also

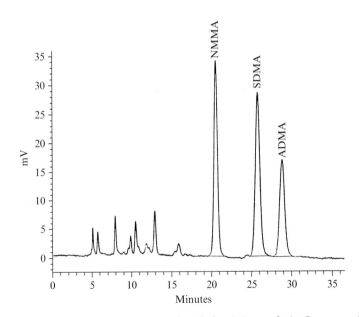

Figure 10.3 An HPLC chromatogram of methylarginine analysis. Concentrations of each amino acid were 1 μM.

react with OPA to yield fluorescent derivatives, and they are all eluted before NMMA. This characteristic is very useful for designing a strategy of fluorescence detection. Because concentrations of protein amino acids, ornithine, citrulline, and taurine are much greater than those of methylarginines in biological samples, suppressing the high fluorescence from the OPA derivatives of amino acids other than methylarginines will aid in setting gains in the detector and accurate quantification of methylarginines. This can be achieved simply by using an excitation wavelength at 440 nm, rather than 340 nm, between 0.0 and 18 min.

Using methylarginine standards in the range of 1 nM to 10 μM, we have established that detection limits (defined as a signal-to-noise ratio of 3) are 5 nM and that the assays are linear between 0.1 and 10 μM for an amino acid ($r^2 = 0.994$). The precision of the HPLC method is 2.3 to 3.5% for various biological samples, including plasma, milk, urine, amniotic fluid, allantoic fluid, endothelial cells, placenta, liver, small intestine, brain, kidney, and skeletal muscle ($n = 10$). The accuracy of the HPLC method is 2.5 to 3.8% for these samples ($n = 10$).

7. Conclusion

Because of its versatility and relatively low costs, HPLC with fluorescence detection is widely used in research laboratories for the accurate analysis of amino acids, including citrulline, arginine, and methylarginines. They react rapidly with OPA, and their derivatives are efficiently separated on a reversed-phase C_{18} column at room temperature. This method is applicable to all biological samples and provides a useful tool to study the regulation of the arginine–NO pathway in cells, tissues, and the whole body.

ACKNOWLEDGMENTS

This work was supported, in part, by grants from USDA CSREES National Research Initiative Competitive Program (Nos. 2001-35203-11247, 2003-35206-13694, and 2008-35206-18764), National Institutes of Health (No. 1R21 HD049449), American Heart Association (Nos. 0655109Y and 0755024Y), Juvenile Diabetes Research Foundation (No. 1-2002-228), and Texas Agricultural Experiment Station (No. H-8200).

REFERENCES

Bode-Böger, S. M., Böger, R. H., Kienke, S., Junker, W., and Frölich, J. C. (1996). L-Arginine/dimethylarginine ratio contributes to enhanced systemic NO production by dietary L-arginine in hypercholesterolemic rabbits. *Biochem. Biophys. Res. Commun.* **219,** 598–603.

Fu, W. J., Haynes, T. E., Kohli, R., Hu, J., Shi, W., Spencer, T. E., Carroll, R. J., Meininger, C. J., and Wu, G. (2005). Dietary L-arginine supplementation reduces fat mass in Zucker diabetic fatty rats. *J. Nutr.* **135,** 714–721.

Jobgen, W. S., Jobgen, S. C., Li, H., Meininger, C. J., and Wu, G. (2007). Analysis of nitrite and nitrate in biological samples using high-performance liquid chromatography. *J. Chromatogr. B Analyt. Technol. Biomed. Life Sci.* **851,** 71–82.

Jones, B. N., and Gilligan, J. P. (1983). o-Phthaldialdehyde precolumn derivatization and reversed-phase high-performance liquid chromatography of polypeptide hydrolysates and physiological fluids. *J. Chromatogr.* **266,** 471–482.

Kohli, R., Meininger, C. J., Haynes, T. E., Yan, W., Self, J. T., and Wu, G. (2004). Dietary L-arginine supplementation enhances endothelial nitric oxide synthesis in streptozotocin-induced diabetic rats. *J. Nutr.* **134,** 600–608.

Li, H., Meininger, C. J., and Wu, G. (2000). Rapid determination of nitrite by reversed-phase high-performance liquid chromatography with fluorescence detection. *J. Chromatogr. B Biomed. Sci. Appl.* **746,** 199–207.

Marliss, E. B., Chevalier, S., Gougeon, R., Morais, J. A., Lamarche, M., Adegoke, O. A. J., and Wu, G. (2006). Elevations of plasma methylarginines in obesity and ageing are related to insulin sensitivity and rates of protein turnover. *Diabetologia* **49,** 351–359.

McBride, A. E., and Silver, P. A. (2001). State of the arg:protein methylation at arginine comes of age. *Cell* **106,** 5–8.

Teerlink, T. (2007). HPLC analysis of ADMA and other methylated L-arginine analogs in biological fluids. *J. Chromatogr. B Analyt. Technol. Biomed. Life Sci.* **851,** 21–29.

Teerlink, T., Nijveldt, R. J., de Jong, S., and van Leeuwen, P. A. M. (2002). Determination of arginine, asymmetric dimethylarginines, and symmetric dimethylarginine in human plasma and other biological samples by high-performance liquid chromatography. *Anal. Biochem.* **303,** 131–137.

Tsikas, D., Böger, R. H., Sandmann, J., Bode-Böger, S. M., and Frölich, J. C. (2000). Endogenous nitric oxide synthase inhibitors are responsible for the L-arginine paradox. *FEBS Lett.* **478,** 1–3.

Valtonen, P., Karppi, J., Nyyssonen, K., Valkonen, V. P., Halonen, T., and Punnonen, K. (2005). Comparison of HPLC method and commercial ELISA assay for asymmetric dimethylarginine (ADMA) determination in human serum. *J. Chromatogr. B Analyt. Technol. Biomed. Life Sci.* **828,** 97–102.

Wu, G., Bazer, F. W., Tuo, W., and Flynn, S. P. (1996). Unusual abundance of arginine and ornithine in porcine allantoic fluid. *Biol. Reprod.* **54,** 1261–1265.

Wu, G., and Brosnan, J. T. (1992). Macrophages can convert citrulline into arginine. *Biochem. J.* **281,** 45–48.

Wu, G., and Meininger, C. J. (1993). Regulation of L-arginine synthesis from L-citrulline by L-glutamine in endothelial cells. *Am. J. Physiol. Heart Circ. Physiol.* **265,** H1965–H1971.

Wu, G., and Morris, S. M., Jr. (1998). Arginine metabolism: Nitric oxide and beyond. *Biochem. J.* **336,** 1–17.

CHAPTER ELEVEN

Quantitative Proteome Mapping of Nitrotyrosines

Diana J. Bigelow* *and* Wei-Jun Qian[†]

Contents

1. Introduction	192
2. Multidimensional LC-MS/MS Provides Large Data Sets for Identification of Nitrotyrosine-Modified Proteins	193
3. Retention of Complexity in Samples Prepared from Global Proteomic Analysis	195
4. Confident Identification of Nitrotyrosine-Containing Peptides	197
5. Comparative Quantitation of Nitrotyrosine-Modified Peptide/Proteins	198
6. Summary	203
References	203

Abstract

An essential first step in the understanding disease and environmental perturbations is the early and quantitative detection of the increased levels of the inflammatory marker nitrotyrosine, as compared with its endogenous levels within the tissue or cellular proteome. Thus, methods that successfully address a proteome-wide quantitation of nitrotyrosine and related oxidative modifications can provide early biomarkers of risk and progression of disease, as well as effective strategies for therapy. Multidimensional separations LC coupled with tandem mass spectrometry (LC-MS/MS) has, in recent years, significantly expanded our knowledge of human (and mammalian model system) proteomes, including some nascent work in identification of posttranslational modifications. This chapter discusses the application of LC-MS/MS for quantitation and identification of nitrotyrosine-modified proteins within the context of complex protein mixtures presented in mammalian proteomes.

---

* Cell Biology and Biochemistry Group, Division of Biological Sciences, Pacific Northwest National Laboratory, Richland, Washington
[†] Division of Biological Sciences, Environmental Molecular Sciences Laboratory, Pacific Northwest National Laboratory, Richland, Washington

1. Introduction

Recent advances in the sensitivity of mass spectrometry instrumentation, coupled with improved separations methods, now offer increasing opportunities for understanding cell and tissue responses at a whole proteome level. Current technologies allow the abundance of thousands of proteins to be monitored simultaneously. Thus, the entire proteome of simple organisms can be resolved, and for larger mammalian proteomes, these capabilities offer significant access into the changing protein landscape of cells and tissues in response to stress and disease, as well as a means to monitor the efficacy of therapeutic interventions. Moreover, sensitive identification of changes in both abundance and posttranslational modification of proteins can provide early and valuable biomarkers of disease, as well as insights into molecular mechanisms of disease progression.

Despite the important role of posttranslational modifications in the activation of signal transduction pathways or as markers of inflammatory events, proteomics strategies to address this dynamic aspect of the proteome have not been fully exploited. In particular, in view of the frequent contribution of inflammation and oxidative stress in pathological tissue changes, 3-nitrotyrosine modifications are of particular interest in monitoring initiation and progression of disease. Indeed, since an antinitrotyrosine antibody became available over a decade ago, increased protein nitration has been reported in over 100 human pathologies and their animal or cell models (Pacher *et al.*, 2007; Ye *et al.*, 1996). However, associated information has lagged regarding the identity of nitrated proteins in specific pathophysiological conditions, the extent of their nitration, and how their modification alters protein and cellular function. Thus, the picture of pathways that explain the correlations between nitrotyrosine and disease is substantially incomplete. As an additional confounding factor, the cell death that often accompanies chronic pathology is likely to contribute to a limited understanding of mechanisms related to disease progression. Thus, detection of proteome changes at early stages in disease progression is essential in understanding disease and in the development of effective therapeutic strategies; an initial step requires the establishment of a baseline of endogenous oxidative stress from a quantitative identification of nitrotyrosine-modified proteins in tissues in the absence of pathology. From such analyses, additional information can be gained regarding the cellular distribution of nitrated proteins as a signature of reactive nitrogen chemistries in the cell, as well as potential vulnerabilities to inflammation and redox regulation of specific proteins and pathways; moreover, examination of nitrotyrosine sites suggests protein structural features that enhance nitration *in vivo*.

2. Multidimensional LC-MS/MS Provides Large Data Sets for Identification of Nitrotyrosine-Modified Proteins

In order to establish a quantitative baseline level of protein nitration, several criteria must be met, i.e., (i) samples must be sufficiently complex so that their composition reflects that of the original cell or tissue; (ii) analytical techniques must be sensitive enough to provide the large data sets necessary to identify relatively low abundance modifications of multiple proteins; and (iii) methods must include a means to identify precise sites of modification sites. In large part, multidimensional LC or other high resolution methods for separation of peptides coupled with tandem mass spectrometry (LC/LC-MS/MS) have provided the best solution to these requirements. Typically, these methods from global proteomic analyses utilize complex protein samples of interest, e.g., cellular, tissue, or organelle homogenates, subjected to proteolytic digestion. The resulting peptides are separated according to physical properties, e.g., charge, size, or hydrophobicity, followed by a second dimension of separation, often inline with very sensitive mass spectrometry (MS). For each MS run (corresponding to one fraction from second-dimension separation), the five most abundant parent masses are selected for further fragmentation for MS/MS spectra from which peptide sequence information is obtained. Because most separation schemes provide more high-quality parent masses per fraction than the MS/MS capacity of the associated spectrometer, undersampling remains a technical limitation to this analytic approach (Qian *et al.*, 2006). Nevertheless, the size of the data sets provided by current LC/LC-MS/MS capabilities are sufficiently large to challenge the overall throughput; for example, a recent global proteomic analysis of brain resulted in ≈751,000 individual MS/MS spectra (Wang *et al.*, 2006). Subsequently, these experimental MS/MS spectra are matched with theoretical mass spectra calculated based on sequences from a given protein database for an organism of choice by means of automated database-searching algorithms (e.g., SEQUEST, MASCOT, X!Tandem) that provide peptide sequence identification (Craig and Beavis, 2004; Perkins *et al.*, 1999; Yates *et al.*, 1995).

An alternative method, which has been commonly used for the identification of nitrotyrosine-modified proteins, consists of MS identification of peptides extracted from proteins bands/spots on one- or two-dimensional electrophoresis gels after in-gel proteolysis. This approach has several shortcomings for quantitative global analysis of nitrated proteins, related primarily to high losses of protein during peptide extraction (Aulak *et al.*, 2001; Castegna *et al.*, 2003; Kanski *et al.*, 2005a). Typically, nitrated proteins identified from gel-based approaches have been limited to abundant and soluble proteins, which are detectable by the nitrotyrosine antibody and for which constituent peptides can be eluted from two-dimensional gels in sufficient amounts for MS identification (Aulak *et al.*, 2001; Castegna *et al.*, 2003; Kanski *et al.*, 2005a,b; Turko *et al.*, 2003). Thus, in view of the common comigration of multiple proteins on electrophoresis gels and the incomplete sequence coverage of extracted proteins, considerable uncertainty exists regarding the identity of a nitrated peptide if, as is frequently the case, the nitrated sequence is not among the extracted peptides from a gel spot. Moreover, LC-MS/MS approaches avoid biases introduced through the use of antibodies for either enrichment or detection of nitrated proteins, as LC-MS/MS identification of nitrated peptides is based on the appearance of nitrotyrosine sites within a specific peptide sequence. Of note, several studies have shown substantial differential sensitivity to various proteins of antinitrotyrosine antibodies from different sources, differences that could lead to failure to detect major nitrated proteins (Barreiro *et al.*, 2002; Sacksteder *et al.*, 2006). Moreover, membrane proteins, which are typically lost by aggregation during the isoelectric focusing step of two-dimensional gels, are identified more abundantly when separations involve peptides rather than proteins (Wu and Yates, 2003). For example, membrane proteins comprised 26% of the proteins identified in a global screen of mouse brain, consistent with predictions that 20 to 30% of all open reading frames encode for membrane proteins (Ahram and Springer, 2004; Krogh *et al.*, 2001; Wallin and von Heijne, 1998). However, it should be noted that transmembrane sequences themselves are in relatively low abundance in most LC-MS/MS proteomics screens (Blonder *et al.*, 2002). In view of model studies that have shown more frequent nitration of intramembrane tyrosine probes as compared with tyrosines in aqueous solution, it might be expected that numerous nitrotyrosines present *in vivo* are being missed by current proteomics screens (Zhang *et al.*, 2001). Thus, development of methods that enhance the recovery of hydrophobic membrane-spanning sequences will be important improvements for a more complete understanding of protein nitration *in vivo*. Other advantages of this analytical approach include its sensitivity and dynamic range, which provides identification of both high and low abundance cellular peptides and proteins. Finally, utilizing the same complex protein mixture to identify both nitrated peptides and their nonnitrated analogs makes a semiquantitative estimate of nitrotyrosine stoichiometry possible. Thus, this chapter focuses on the discussion of current applications of LC-MS/MS for quantitative mapping of nitrotyrosine sites within cells and tissues. In view of several excellent reviews available regarding developments in high-resolution mass spectrometric instrumentation and separation capabilities for proteomics, this chapter focuses on specific experimental strategies related to sample requirements, confident peptide identification, and quantitation of nitrotyrosine-modified peptide/proteins

(Domon and Aebersold, 2006; Ong and Mann, 2005; Qian et al., 2006; Sadygov et al., 2004; Shen and Smith, 2005; Smith et al., 2006).

3. Retention of Complexity in Samples Prepared from Global Proteomic Analysis

As in any cost- and labor-intensive analyses, such as proteomics, sample quality is the most important factor. In order that nitroproteomic analyses are informative of the tissue, organelle, or cell of interest, samples should reflect the original protein composition. Thus, although several chemical and immunoaffinity methods have been developed to selectively enrich nitrated proteins for proteomic analysis, these approaches lack information regarding the extent of nitration within the original cell or tissue (Zhan and Desiderio, 2004; Zhang et al., 2007). A sample preparation that retains as much native complexity as is practical and is still compatible with its application to separation matrices is an essential first step in any proteomic analysis; in particular, this principle applies to the quantitative analysis of nitrotyrosines for which quantitation in context of the total protein is sought. Tissue homogenates that include the removal of connective tissue and other large particles ("cell debris") are commonly used as a source of peptides for global analyses. However, recent proteomics studies have included a limited number of additional fractionation steps prior to high-sensitivity separations that provide subproteomes to further reduce sample complexity, thus enhancing the detection of low abundance proteins and overall proteomic coverage.

For example, Wang and co-workers (2006) subjected a portion of tryptic peptides from brain homogenate to cysteine-peptidyl enrichment (thiopropyl Sepharose) prior to cysteine alkylation; the remaining portion was analyzed without cysteine alkylation, essentially depleting this latter portion of cysteinyl-peptides to provide complementary proteomic samples. From the combined analysis of these two samples, from 1564 to 1859 additional proteins were identified as compared with the yield from either one of these samples. Similarly, the proteome coverage of human mammary epithelial cells was increased from 10 to 34% with cysteine-peptidyl enrichment (Liu et al., 2005).

Other simple fractionation steps prior to proteolysis have been employed that permit the identification of more proteins from the combined results of each subproteome as compared with a single global fraction, such as a tissue homogenate (Foster et al., 2006; Wu et al., 2007). Figure 11.1 shows a comparison of results from two different protein fractionation strategies applied to mouse heart homogenates prior to LC-MS/MS analyses. One of these involves a single centrifugation step of the tissue homogenate that

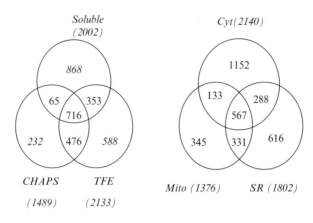

Figure 11.1 Fractionation of mouse heart homogenates reduces sample complexity and increases proteome coverage. Venn diagrams indicate protein identifications resulting from different fractionation strategies applied to LC-MS/MS proteomic analyses of mouse hearts. (Left) Heart homogenates were centrifuged at 48,000 g; the resulting membrane pellets were solubilized with either 2% CHAPS or 30% TFE prior to processing for LC-MS/MS analysis in parallel to that for the supernatant fraction (Soluble). Combined analysis resulted in the identification of 3298 unique proteins; the total protein identified for each fraction is indicated in parentheses. Numbers inside the circles indicate identified proteins that are unique or in common with one or two other fractions. (Right) Heart homogenates were differentially centrifuged at 15,000 g, resulting in a mitochondrially enriched pellet (Mito), and at 48,000 g, resulting in an sarcoplasmic reticulum-enriched pellet (SR); and the 48,000 g supernatant was designated as a cytosolic-enriched (Cyt) fraction. Samples were digested and processed for LC-MS/MS analysis, resulting in the identification of 3432 unique proteins from the combined results; proteins unique or common with one or two other fractions are indicated within the circles as indicated for each fraction. For proteomic analysis, strong cation exchange was used for the first dimension of peptide separation, followed by separation of the resulting 30 fractions on a reversed-phase (C18) capillary HPLC system coupled online with an LTQ ion trap mass spectrometer (ThermoFinnigan San Jose, CA) using an in-house manufactured electrospray ionization interface. MS/MS data were searched against the mouse International Protein Index (IPI) database (version 1.25, available online at http://www.ebi.ac.uk/IPI) and a sequence-reversed IPI database (to assess false positives) using SEQUEST (ThermoFinnigan).

largely separates membrane vesicles from soluble supernatant proteins; the resulting membrane pellet is solubilized with either the detergent 3-[(3-cholamidopropyl)dimethylammonio]-1-propanesulfonate (CHAPS) or the organic solvent trifluoroethanol (TFE). In particular, the use of two different solubilization agents permits enhanced numbers of unique proteins identified in the membrane fractions so that their total number approximates that of the soluble fraction. As an alternative method, differential centrifugation was used to prepare fractions enriched in mitochondria, sarcoplasmic reticulum membranes, and cytosolic proteins (Fig. 11.1). This approach, too, increased the total proteins identified relative to either fraction alone or as compared with a single homogenate fraction, in which approximately 900 proteins were initially identified (data not shown). Sequence coverage was also improved. Similar to overall protein identification, nitrotyrosine-modified protein identification was also enhanced by these initial fractionation strategies. Any number of fractionation schemes might be devised with several caveats, i.e, (i) all subproteomes should be analyzed, avoiding the discarding of fractions, so that the combined results reflect the original proteome of interest and that (ii) within the LC-MS capabilities, fractionation/separations resolution increases have to be balanced with their concomitant decreases in overall sample throughput. In particular, the number of subproteome fractionations at the front end of global analysis increases total MS runs rapidly and therefore decreases throughput. For example, in the case when each experimental sample includes approximately 30 first-dimension separation fractions applied to the second-dimension separation online with the mass spectrometer, the three subproteome samples used in Fig. 11.1 require 90 MS runs per experimental sample, an analysis that requires significant instrumental and computational time under current system capabilities.

4. CONFIDENT IDENTIFICATION OF NITROTYROSINE-CONTAINING PEPTIDES

The automated database searching used to identify native peptides can be adjusted to accommodate searches for specific posttranslational modifications of interest by including the addition of an appropriate mass to the corresponding amino acid within peptide sequences. For example, in the case of nitrotyrosine, searches are performed for the addition of 44.98 Da applied to tyrosines within peptide sequences. Searches for additional modifications may be desired to widen the search for nitrated peptides, as, for example, cysteine carboxamidomethylation is a usual step in sample processing for MS analysis, requiring searches for 57.02 Da applied to cysteines. In a recent LC-MS proteomic analysis, almost two-thirds of nitrated peptides also contained at least one Cys residue; these are peptides that would elude identification without consideration of appropriate modifications on cysteines. Additionally, methionine sulfoxides (with their additional mass of 15.99 Da) should be included in searches for nitrated peptides, as these methionine modifications commonly result from nitrating chemistries, such as peroxynitrite, and are frequently present within nitrated peptides. In biological samples where oxidative stress and inflammation are expected, other stable oxidative modifications may be informative and necessary for the efficient identification of nitrotyrosine modifications, as, for example, sulfinic (RSO_2H) and sulfonic (RSO_3H) acid derivatives of cysteines, methionine sulfones ($MetSO_2$), nitrotryptophans, and bromo- and chlorotyrosines. However, the substantial computational times required for simultaneous searching of multiple modifications within large data sets place practical limits on most global proteomic searches to no more than three or four modifications at a time.

The confidence of peptide identifications by automated database searching is based on the use of stringent criteria for filtering data. To address possible false-positive peptide identifications, sequence-reversed protein database searches have been used in order to establish filtering criteria that ensure low levels (e.g., <2%) of false-positive peptide identifications. Moreover, especially in the case of low abundance modifications such as nitrotyrosines, nitrated peptide identifications can be validated by manual inspection of the corresponding MS/MS spectra to ensure that (i) spectral peaks are clearly defined above the spectral noise; (ii) that no major unidentified peaks are present, and that (iii) major peak assignments are consistent with predicted patterns as illustrated by MS/MS spectra for the nitrated and unmodified versions of the peptide, DSYVAIANACCAPR (Fig. 11.2). As observed by comparison of these two spectra, all detected y fragment ions match for the two spectra, but all b ions containing the nitrotyrosine from the nitrated peptide (b_3–b_{12}) have masses that are 44.98 Da higher than the corresponding b ions from the nonnitrated peptide.

5. Comparative Quantitation of Nitrotyrosine-Modified Peptide/Proteins

Pathophysiological or environmental perturbations often alter the protein abundances, including posttranslational modifications such as nitrotyrosines in cells, tissues, or biological fluids. The ability to quantitatively measure relative protein abundance and their modification differences between different conditions is essential for identifying key molecular targets or candidate biomarkers involved in different diseases. Current quantitative characterization of relative protein abundance differences can be categorized as two general approaches: (1) MS peptide ion intensity-based quantitation, often in combination with stable isotope labeling, and (2) MS/MS spectrum count (or peptide hit)-based quantitation. In the second approach, quantitation information is derived from the number of MS/MS sampling identifying a given protein within a complex mixture. Spectrum counts correspond to the number of MS/MS spectra observed for each distinct peptide, which may include repetitive identification or the

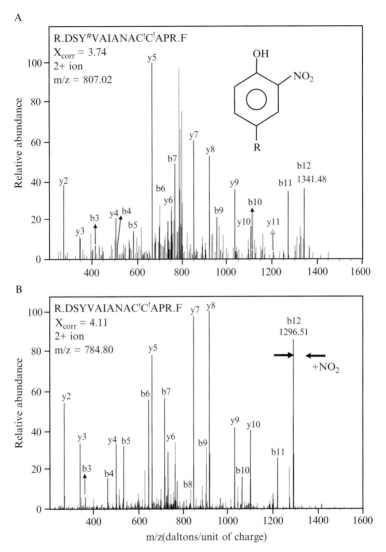

Figure 11.2 MS/MS spectra of the nitrated (A) and nonnitrated (B) peptide, DSYVAIANACCAPR, from the vesicular inhibitory amino acid transporter. Y# indicates nitrotyrosine modification, and C! indicates a cysteinyl residue modified by carboxamidomethylation. The two spectra match well on all detected b and y fragment ions, with the exception that b ions containing nitrotyrosine, i.e., b_3–b_{12}, from the nitrated peptide have masses 44.98 Da higher than the corresponding b ions from the nonnitrated peptide, as illustrated by arrows for b12.

same peptide in one or multiple first-dimension fractions. These values have been demonstrated to provide a reproducible and semiquantitative measure of peptide concentration in complex samples (Qian *et al.*, 2005).

Because nitrotyrosine is often measured within the context of the global proteome, both approaches can theoretically be applied for quantitative measurements of differences in nitration levels. However, the use of stable isotope labeling for the quantitation of nitrotyrosine modifications has not been reported to date. As discussed here, MS/MS spectrum counts derived from current LC-MS/MS capabilities provide the most reliable quantitation of nitrotyrosine modification extent.

For MS ion intensity-based quantitation, most approaches have involved the use of stable isotope labeling to achieve good accuracy in quantifying the relative abundance differences. Many stable isotope-labeling methods have been introduced successfully for quantitative proteomics and are reviewed elsewhere (Ong and Mann, 2005; Qian et al., 2004). In general, stable isotope labels can be introduced into proteins or peptides by metabolic labeling, chemical labeling on specific functional groups, and enzymatic transfer of ^{18}O from water to the C terminus of peptides. All of these methods incorporate either a light or a heavy version of the stable isotope into chemically identical peptides from two different samples, and the resulting peptide pairs are differentiated by MS due to the mass difference between the light and the heavy isotope-coded pair. The ratio of MS ion intensities for the peptide pair can then accurately indicate the abundance ratios of the peptide/protein from the two different samples. More recently, there has been a significant interest in applying "label-free" direct quantitation due to the greater flexibility for comparative analyses and simpler sample processing procedures compared to labeling approaches (Qian et al., 2006; Wang et al., 2003). The isotope labeling and label-free approaches are complementary, and each approach has different sources of variations. While the stable isotope-labeling approach is generally more accurate in quantitation as compared with the label-free approach as a consequence of the unbiased measurement of peptide pairs by MS and the minimum variations from sample processing postlabeling, isotopic ratios often have a limited dynamic range for quantitating large changes as a consequence of the overlap of isotopic envelopes between light and heavy members of the pair. However, the label-free approach is often more ideal for the quantitation of more dramatic changes in protein abundances or in estimating the stoichiometry of posttranslational modification of low abundance modifications as nitrotyrosines. With the proper use of replicate analyses and data normalization to mitigate changes in instrumental performance between analyses, accurate comparative quantitation can be achieved for biological and clinical applications.

The MS/MS spectrum counting approach has been reported as an alternative approach for relative quantitation based on the correlation of protein abundances with the number of MS/MS spectra (or peptide hits) identifying a given proteins (Liu et al., 2004; Qian et al., 2005; Zybailov et al., 2005). It should be noted that the spectrum count approach is a semiquantitative method for comparative analyses because the spectrum counts may not have a linear correlation with protein abundances. Additionally, the spectrum count approach will become problematic when proteins are observed with only a few counts in both conditions. Despite these potential disadvantages, this approach has been increasingly applied for several reasons, including (1) the experiments can be performed on a common ion-trap instrument without the need of high-resolution MS measurements, (2) the approach can be easily coupled with multiple dimensional LC separations, and (3) the reproducibility in terms of spectrum counts is observed for both one-dimensional LC (Liu *et al.*, 2004, 2006) and two-dimensional LC-MS/MS (Qian *et al.*, 2005). Moreover, the spectrum count information provides an estimation of the relative abundances of different proteins within the same sample (Wang *et al.*, 2006).

Specifically, when protein modifications such as nitrotyrosine are of interest, the extent of modification or the stoichiometry of modifications can be estimated by comparing spectrum counts of the modified peptides to total peptides from the global proteome for a given proteins. Here, the large dynamic range of spectrum counts is valuable, as typical levels of nitrated peptides relative to their nonnitrated analogs are quite low, $<1/1000$, levels that would disallow reliable quantitation from isotopic ratios. Another limitation of using stable isotope ratios for the quantitation of posttranslational modifications is that the number of peptide pairs resolved by this method is generally considerably smaller than the peptides identified by label-free approaches, which would further reduce the number of detected nitrotyrosine-modified peptides that are quantifiable. A more common problem for spectrum count-associated quantitation of nitrotyrosines is the detection either of only a few counts of nitrated peptides or of differences derived from different experimental conditions. These low levels may result from incomplete sampling because of stringent criteria for identification and their low abundance; guidelines for the evaluation and statistical significance of spectrum counts at low levels have been discussed previously (Qian *et al.*, 2005).

Reproducibility for LC-MS/MS proteomic analyses can be excellent; a recent study comparing results of proteomic analyses of nine individually processed technical replicates of human plasma samples showed very low variation in peptide intensities with an overall Pearson's correlation coefficient of 0.94 ± 0.02 (Qian *et al.*, 2006). In comparison, plasma samples from nine individual human patients exhibited significantly more variation with a correlation coefficient of 0.85 ± 0.06; however, analyses from nine individual mouse plasma samples showed only a slightly reduced correlation (0.92 ± 0.05) as compared with technical replicates. Studies such as these highlight the importance of minimizing any instrumental contribution to the variability in proteomics results in order to enhance the detection of experimental differences.

As an illustration of spectrum counts used to quantitate nitrotyrosine levels in several tissues, we compared the results of three multidimensional LC-MS/MS proteomic screens of mouse brain, heart, and skeletal muscle, identifying endogenous nitrotyrosine-modified peptides and associated proteins from SEQUEST searches that consider methionine sulfoxides and carboxymethylated cysteines as well as nitrotyrosines. Calculating the ratio of the sum of spectrum counts associated with all nitrotyrosine-containing peptides relative to the sum of all tyrosine-containing peptides, we found significantly different levels of endogenous protein nitration in these three tissues that correlate well with the differential staining intensities exhibited by nitrotyrosine immunoblots of these samples where the greatest abundance of nitrated proteins is found in skeletal muscle followed by heart followed by brain (Fig. 11.3). Notably, the nitrotyrosine immunoblot of

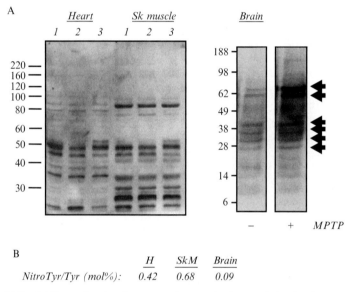

Figure 11.3 Different levels of protein nitration in heart, skeletal (Sk) muscle, and brain. Homogenate proteins from heart, hind limb skeletal muscle, and brain were separated on SDS-PAGE prior to immunoblotting with the antinitrotyrosine antibody (A) for comparison with relative levels of nitrotyrosine-containing peptides identified from multidimensional LC-MS/MS analyses of parallel samples (B). Immunoblots show heart and skeletal muscle homogenates from three individual animals; shown are immunoblots for normal ($-$) and for ($+$) MPTP-lesioned brains, prepared as described previously (Sacksteder *et al.*, 2006). Proteomic-derived values for nitrotyrosine levels were determined from the ratios of (i) the sum of spectrum counts associated with identified nitrotyrosines in peptides to (ii) the sum of total spectrum counts for identified tyrosines in peptides. Proteomic analyses were performed essentially as described in Fig. 11.1 and detailed by Sacksteder and co-workers (2006), with the exception that muscle homogenates were prefractionated according to the scheme outlined in the left side of Fig. 11.1.

homogenate from the 1-methyl-4-phenyl-1,2,3,6-tetrahydropyridine (MPTP)-lesioned brain, a neurotoxin-induced model of Parkinson's disease, exhibits more intense staining relative to normal brain of the same major nitrated bands, highlighting the importance of identifying nitration-sensitive proteins as potential indicators of neurodegeneration and pathology. Indeed, from LC-MS/MS identification of endogenously nitrated proteins in normal brain, more than half of the identified nitrated proteins have been previously reported to be associated with neurodegeneration, further supporting the link between nitration-sensitive proteins and their vulnerability to disease.

6. Summary

Multidimensional LC MS/MS offers expanding possibilities in the quantitative mapping of the nitroproteome as a signature of inflammatory events in normal and disease-progressing tissues. Recent global proteomics have begun to identify significantly greater numbers of nitrated proteins within the context of the complex mixture, as well as to define endogenous levels of inflammation in different tissues. The detection of multiple nitrotyrosine-modified proteins in tissues is just the tip of the iceberg with regard to posttranslational modifications of the proteome. Future work will be aimed toward greater coverage of these modified proteins.

REFERENCES

Ahram, M., and Springer, D. L. (2004). Large-scale proteomic analysis of membrane proteins. *Expert Rev. Proteomics* **1,** 293–302.

Aulak, K. S., Miyagi, M., Yan, L., West, K. A., Massillon, D., Crabb, J. W., and Stuehr, D. J. (2001). Proteomic method identifies proteins nitrated *in vivo* during inflammatory challenge. *Proc. Natl. Acad. Sci. USA* **98,** 12056–12061.

Barreiro, E., Comtois, A. S., Gea, J., Laubach, V. E., and Hussain, S. N. (2002). Protein tyrosine nitration in the ventilatory muscles: Role of nitric oxide synthases. *Am. J. Respir. Cell. Mol. Biol.* **26,** 438–446.

Blonder, J., Goshe, M. B., Moore, R. J., Pasa-Tolic, L., Masselon, C. D., Lipton, M. S., and Smith, R. D. (2002). Enrichment of integral membrane proteins for proteomic analysis using liquid chromatography-tandem mass spectrometry. *J. Proteome Res.* **1,** 351–360.

Castegna, A., Thongboonkerd, V., Klein, J. B., Lynn, B., Markesbery, W. R., and Butterfield, D. A. (2003). Proteomic identification of nitrated proteins in Alzheimer's disease brain. *J. Neurochem.* **85,** 1394–1401.

Craig, R., and Beavis, R. C. (2004). TANDEM: Matching proteins with tandem mass spectra. *Bioinformatics* **20,** 1466–1467.

Domon, B., and Aebersold, R. (2006). Mass spectrometry and protein analysis. *Science* **312,** 212–217.

Foster, L. J., de Hoog, C. L., Zhang, Y., Zhang, Y., Xie, X., Mootha, V. K., and Mann, M. (2006). A mammalian organelle map by protein correlation profiling. *Cell* **125,** 187–199.

Kanski, J., Behring, A., Pelling, J., and Schoneich, C. (2005a). Proteomic identification of 3-nitrotyrosine-containing rat cardiac proteins: Effects of biological aging. *Am. J. Physiol. Heart Circ. Physiol.* **288,** H371–H381.

Kanski, J., Hong, S. J., and Schoneich, C. (2005b). Proteomic analysis of protein nitration in aging skeletal muscle and identification of nitrotyrosine-containing sequences *in vivo* by nanoelectrospray ionization tandem mass spectrometry. *J. Biol. Chem.* **280,** 24261–24266.

Krogh, A., Larsson, B., von Heijne, G., and Sonnhammer, E. L. (2001). Predicting transmembrane protein topology with a hidden Markov model: Application to complete genomes. *J. Mol. Biol.* **305,** 567–580.

Liu, H., Sadygov, R. G., and Yates, J. R., 3rd (2004). A model for random sampling and estimation of relative protein abundance in shotgun proteomics. *Anal. Chem.* **76,** 4193–4201.

Liu, T., Qian, W. J., Chen, W. N., Jacobs, J. M., Moore, R. J., Anderson, D. J., Gritsenko, M. A., Monroe, M. E., Thrall, B. D., Camp, D. G., 2nd, and Smith, R. D. (2005). Improved proteome coverage by using high efficiency cysteinyl peptide enrichment: The human mammary epithelial cell proteome. *Proteomics* **5,** 1263–1273.

Liu, T., Qian, W. J., Mottaz, H. M., Gritsenko, M. A., Norbeck, A. D., Moore, R. J., Purvine, S. O., Camp, D. G., 2nd, and Smith, R. D. (2006). Evaluation of multiprotein immunoaffinity subtraction for plasma proteomics and candidate biomarker discovery using mass spectrometry. *Mol. Cell Proteomics* **5,** 2167–2174.

Ong, S. E., and Mann, M. (2005). Mass spectrometry-based proteomics turns quantitative. *Nat. Chem. Biol.* **1,** 252–262.

Pacher, P., Beckman, J. S., and Liaudet, L. (2007). Nitric oxide and peroxynitrite in health and disease. *Physiol. Rev.* **87,** 315–424.

Perkins, D. N., Pappin, D. J., Creasy, D. M., and Cottrell, J. S. (1999). Probability-based protein identification by searching sequence databases using mass spectrometry data. *Electrophoresis* **20,** 3551–3567.

Qian, W. J., Camp, D. G., 2nd, and Smith, R. D. (2004). High-throughput proteomics using Fourier transform ion cyclotron resonance mass spectrometry. *Expert Rev. Proteomics* **1,** 87–95.

Qian, W. J., Jacobs, J. M., Camp, D. G., 2nd, Monroe, M. E., Moore, R. J., Gritsenko, M. A., Calvano, S. E., Lowry, S. F., Xiao, W., Moldawer, L. L., Davis, R. W., Tompkins, R. G., *et al.* (2005). Comparative proteome analyses of human plasma following *in vivo* lipopolysaccharide administration using multidimensional separations coupled with tandem mass spectrometry. *Proteomics* **5,** 572–584.

Qian, W. J., Jacobs, J. M., Liu, T., Camp, D. G., 2nd, and Smith, R. D. (2006). Advances and challenges in liquid chromatography-mass spectrometry-based proteomics profiling for clinical applications. *Mol. Cell Proteomics* **5,** 1727–1744.

Sacksteder, C. A., Qian, W. J., Knyushko, T. V., Wang, H., Chin, M. H., Lacan, G., Melega, W. P., Camp, D. G., 2nd, Smith, R. D., Smith, D. J., Squier, T.C, and Bigelow, D. J. (2006). Endogenously nitrated proteins in mouse brain: Links to neurodegenerative disease. *Biochemistry* **45,** 8009–8022.

Sadygov, R. G., Cociorva, D., and Yates, J. R., 3rd (2004). Large-scale database searching using tandem mass spectra: Looking up the answer in the back of the book. *Nat. Methods* **1,** 195–202.

Shen, Y., and Smith, R. D. (2005). Advanced nanoscale separations and mass spectrometry for sensitive high-throughput proteomics. *Expert Rev. Proteomics* **2,** 431–447.

Smith, R. D., Tang, K., and Shen, Y. (2006). Ultra-sensitive and quantitative characterization of proteomes. *Mol. Biosyst.* **2,** 221–230.

Turko, I. V., Li, L., Aulak, K. S., Stuehr, D. J., Chang, J. Y., and Murad, F. (2003). Protein tyrosine nitration in the mitochondria from diabetic mouse heart: Implications to dysfunctional mitochondria in diabetes. *J. Biol. Chem.* **278,** 33972–33977.

Wallin, E., and von Heijne, G. (1998). Genome-wide analysis of integral membrane proteins from eubacterial, archaean, and eukaryotic organisms. *Protein Sci.* **7,** 1029–1038.

Wang, H., Qian, W. J., Chin, M. H., Petyuk, V. A., Barry, R. C., Liu, T., Gritsenko, M. A., Mottaz, H. M., Moore, R. J., Camp Ii, D. G., Khan, A. H., Smith, D. J., *et al.* (2006). Characterization of the mouse brain proteome using global proteomic analysis complemented with cysteinyl-peptide enrichment. *J. Proteome Res.* **5,** 361–369.

Wang, W., Zhou, H., Lin, H., Roy, S., Shaler, T. A., Hill, L. R., Norton, S., Kumar, P., Anderle, M., and Becker, C. H. (2003). Quantification of proteins and metabolites by mass spectrometry without isotope labeling or spiked standards. *Anal. Chem.* **75,** 4818–4826.

Wu, C. C., and Yates, J. R., 3rd (2003). The application of mass spectrometry to membrane proteomics. *Nat. Biotechnol.* **21,** 262–267.

Wu, W. W., Wang, G., Yu, M. J., Knepper, M. A., and Shen, R. F. (2007). Identification and quantification of basic and acidic proteins using solution-based two-dimensional protein fractionation and label-free or ^{18}O-labeling mass spectrometry. *J. Proteome Res.* **6,** 2447–2459.

Yates, J. R., 3rd, Eng, J. K., and McCormack, A. L. (1995). Mining genomes: Correlating tandem mass spectra of modified and unmodified peptides to sequences in nucleotide databases. *Anal. Chem.* **67,** 3202–3210.

Ye, Y. Z., Strong, M., Huang, Z. Q., and Beckman, J. S. (1996). Antibodies that recognize nitrotyrosine. *Methods Enzymol.* **269,** 201–209.

Zhan, X., and Desiderio, D. M. (2004). The human pituitary nitroproteome: Detection of nitrotyrosyl-proteins with two-dimensional Western blotting, and amino acid sequence determination with mass spectrometry. *Biochem. Biophys. Res. Commun.* **325,** 1180–1186.

Zhang, H., Joseph, J., Feix, J., Hogg, N., and Kalyanaraman, B. (2001). Nitration and oxidation of a hydrophobic tyrosine probe by peroxynitrite in membranes: Comparison with nitration and oxidation of tyrosine by peroxynitrite in aqueous solution. *Biochemistry* **40,** 7675–7686.

Zhang, Q., Qian, W. J., Knyushko, T. V., Clauss, T. R., Purvine, S. O., Moore, R. J., Sacksteder, C. A., Chin, M. H., Smith, D. J., Camp, D. G., 2nd, Bigelow, D. J., and Smith, R. D. (2007). A method for selective enrichment and analysis of nitrotyrosine-containing peptides in complex proteome samples. *J. Proteome Res.* **6,** 2257–2268.

Zybailov, B., Coleman, M. K., Florens, L., and Washburn, M. P. (2005). Correlation of relative abundance ratios derived from peptide ion chromatograms and spectrum counting for quantitative proteomic analysis using stable isotope labeling. *Anal. Chem.* **77,** 6218–6224.

SECTION TWO

CELLULAR METHODS

CHAPTER TWELVE

Protein S-Nitrosation in Signal Transduction: Assays for Specific Qualitative and Quantitative Analysis

Vadim V. Sumbayev *and* Inna M. Yasinska

Contents

1. Introduction	210
2. Examples Demonstrating the Contribution of Protein S-Nitrosation to Intracellular Signal Transduction	210
3. Assays for Analysis of S-Nitrosation of Specific Cellular Proteins	214
3.1. Jaffrey's method and its modifications	214
3.2. ELISA assay for detection of S-nitrosation of overexpressed proteins	216
4. Conclusions	217
Acknowledgments	218
References	218

Abstract

S-Nitrosation is a type of protein posttranslational modification that has now been found comparable with phosphorylation and acetylation in terms of its contribution to the intracellular signaling networks associated with pathological cell reactions such as host–pathogen interactions, low oxygen availability, cell cycle arrest, and programmed cell death. Therefore, elegant approaches are required to analyze endogenous S-nitrosation of intracellular proteins employed in the regulation of intracellular signal transduction. This chapter describes and discusses recently developed methods that allow both qualitative and quantitative analyses of S-nitrosation of intracellular proteins based on Western blot and enzyme-linked immunosorbent assay.

Medway School of Pharmacy, University of Kent, United Kingdom

1. Introduction

Protein S-nitrosation is a type of posttranslational modification that has been found to be regulated in time and space. This kind of posttranslational modification has now been found to be comparable with phosphorylation and acetylation in terms of its contribution to the intracellular signaling networks (Hess et al., 2005). However, protein S-nitrosation is often reported for cases when the experimental cell lines, proteins, or even sometimes their modified peptide fragments are exposed to NO donors releasing different forms of nitric oxide that could also differ according to their charge and reactivity.

It is therefore very important to understand the physiological relevance of modifications reported and whether they exist upon changes in the intracellular concentrations of enzymatically produced reactive nitrogen species (RNS) under normal and pathological cell reactions associated with impact on the NO production (e.g. inflammatory/innate immune responses; Sumbayev et al., 2007). In this regard, this chapter discusses currently developed assays that allow analysis of physiologically relevant S-nitrosation of proteins that participate in intracellular signaling networks. These methods have been used in current research devoted to S-nitrosation to mitogen-activated protein (MAP) kinase-dependent apoptotic and hypoxic signal transduction.

2. Examples Demonstrating the Contribution of Protein S-Nitrosation to Intracellular Signal Transduction

In recent years it has become increasingly evident that S-nitrosation contributes to the regulation of intracellular signaling networks responsible for cell cycle regulation and that also contribute to programmed cell death (apoptosis; Hess et al., 2005). The contribution of S-nitrosation is reviewed by Hess et al. (2005). This chapter describes two findings that underline the importance of S-nitrosation in the regulation of pathological cell reactions associated with programmed death and low oxygen availability.

S-Nitrosation was found to be one of the mechanisms that could be employed in regulation of the MAP kinase-dependent apoptotic pathway. Nitric oxide was reported to induce apoptosis in neurons and rat pheochromocytoma PC12 cells (Han et al., 2001; Sarker et al., 2003). It was found that apoptosis is mediated via activation of the MAP kinase-dependent pathway. It has also been shown that nitric oxide could S-nitrosate Trx in human embryonic kidney (HEK293) cells, which leads to an activation of apoptosis signal-regulating kinase 1 (ASK1; Sumbayev, 2003). Stimulation of HEK293 cells with S-nitrosoglutathione (GSNO) for 2, 4, 8, and 16 h also caused Trx S-nitrosation, which showed straight correlation with ASK1 activation. It was further investigated that S-nitrosation of Trx induces ASK1 activation (Sumbayev, 2003). Treatment of cells with N-acetyl-cysteine for 2 h after 8 h of pretreatment with GSNO caused an increase in glutathione and nullified ASK1 activation. However, ASK1 is inhibited by S-nitrosation of its Cys869 residue (Park et al., 2004). The appearance of superoxide and NO in equal concentrations leads to peroxynitrite formation; however, peroxynitrite actions abolished by a two- to threefold excess of NO resulted in further preformation of peroxynitrite into N_2O_3-like species, which nitrosate protein thiol groups (Daiber et al., 2002). We have also observed S-nitrosation of reactive Trx thiol groups in the nitric oxide/superoxide system. It was found that Trx thiol groups are targets for S-nitrosation by N_2O_3-like species generated in the system containing xanthine/xanthine oxidase (superoxide producing system) and DEA/NO—the ★NO donating compound; however, they have shown low sensitivity to the NO radical derived from DEA/NO. N_2O_3-dependent S-nitrosation of Trx at approximately twofold of NO excess was comparable to the superoxide amount resulting in the dissociation and activation of ASK1. However, approximately a fourfold of NO excess compared to a superoxide production preserved the level of dissociated ASK1 but decreased its activity due to S-nitrosation of the enzyme, more likely in the position of Cys869 (Yasinska et al., 2004).

Other findings have suggested that Cys69 of Trx is a potential target for S-nitrosation; however, this posttranslational modification has an antiapoptotic impact in endothelial cells (Haendeler et al., 2002). These findings have been supported by another group, who found the same effect in the cardiovascular system using cell culture assays and animal models. Those data also reflect the situation in the cardiovascular system and do not provide any evidence about other cell types (Haendeler et al., 2002; Tao et al., 2004). Taken together, data reporting the pro- and antiapoptotic impacts of Trx S-nitrosation suggest a cell type-specific Trx reactivity to the NO-dependent modifications of thiol groups. In this respect, one could speculate that S-nitrosation of Trx has an apoptotic impact in neurons and HEK293 cells (Sumbayev, 2003; Yasinska et al., 2004); however, in the cardiovascular system, it protects cells against programmed death (Haendeler et al., 2002; Park et al., 2004; Tao et al., 2004).

Evidence suggests that S-nitrosation is one of the posttranslational modifications employed in the regulation of hypoxic signal transduction under normal oxygen availability (Sumbayev et al., 2007). Cells exposed to low oxygen conditions respond by initiating defense mechanisms, including the accumulation of hypoxia-inducible factor (HIF) 1α, a general hypoxia-inducible transcription factor, which plays a pivotal role in mediating cellular responses to the hypoxia upregulating expression of genes such as those involved in angiogenesis, glycolysis, cell adhesion, and other processes, and that is also important for the regulation of cellular oxygen utilization (Fig. 12.1) (Semenza, 2001, 2002; Oda et al., 2006; Walmsley et al., 2005). By administrating the aforementioned physiological reactions, HIF-1α therefore regulates cellular and tissue adaptation to hypoxia, promoting mechanisms of cell survival during inflammatory reactions and innate immune responses, as well as malignant tumor growth (Semenza, 2001, 2002; Oda et al., 2006; Walmsley et al., 2005). Considering that some thiol groups found in HIF-1α are particularly reactive, the question came

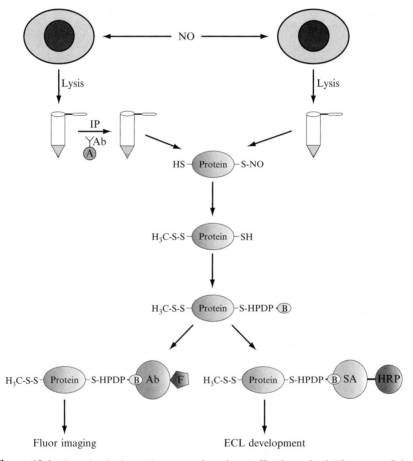

Figure 12.1 Protein S-nitrosation assays based on Jaffrey's method. The steps of the assays starting from cell lysis followed by chemical modifications of S-nitrosated proteins and Western blot analysis are presented. IP, immunoprecipitation; Ab, antibody; B, biotin; SA, streptavidin; F, FITC. (See color insert.)

along of whether HIF-1α is a target for S-nitrosation. Taking into account that nitrogen species are recognized for posttranslation protein modifications, among others S-nitrosation, we studied whether HIF-1α is a target for S-nitrosation (Sumbayev et al., 2003). In vitro NO^+ donating NO donors such as GSNO and S-nitroso-N-acetlypenicillamine (SNAP) provoked massive S-nitrosation of purified HIF-1α. All 15 free thiol groups found in human HIF-1α are subjected to S-nitrosation. Thiol modification is not shared by spermine-NONOate, a NO radical donating compound. However, spermine-NONOate, in the presence of O_2^-, generated by xanthine/xanthine oxidase, regained S-nitrosation, most likely via formation of a N_2O_3-like species. In vitro, S-nitrosation of HIF-1α was attenuated by the addition of GSH or ascorbate. In renal carcinoma cells 4 (RCC4) and human embryonic kidney (HEK293) cells, GSNO or SNAP produced S-nitrosation of HIF-1α, but with a significantly reduced potency that amounted to the modification of three to four thiols only. Importantly, endogenous formation of NO in RCC4 cells via inducible NO synthase activated by cytokines (interleukin-1β, interferon-γ, and lipopolysaccharide) elicited S-nitrosation of HIF-1α that was sensitive to inhibition of inducible NO synthase activity with N-monomethyl-L-arginine. NO-stabilized HIF-1α was susceptible to the addition of N-acetyl-cysteine that destabilized HIF-1α in close correlation to the disappearance of S-nitrosated HIF-1α. Therefore, HIF-1α is a target for S-nitrosation by exogenously and endogenously produced NO (Sumbayev et al., 2003).

Another group has reported that normoxic HIF-1 activity can be upregulated through NO-mediated S-nitrosation and stabilization of HIF-1α (Li et al., 2007). In murine tumors, exposure to ionizing radiation stimulates the generation of nitric oxide in tumor-associated macrophages. As a result, the HIF-1α protein is S-nitrosylated at Cys533 in the oxygen-dependent degradation domain, which prevents its destruction. Importantly, this mechanism appears to be independent of the prolyl hydroxylase-based pathway that is involved in the oxygen-dependent degradation of HIF-1α. This indicates that S-nitrosation could be an alternative approach to get HIF-1α stabilized without impacting prolyl hydroxylase activity. Selective disruption of this S-nitrosation significantly attenuated both radiation- and macrophage-induced activation of HIF-1α. This interaction between NO and HIF-1 was suggested to shed new light on their involvement in the tumor response to treatment as well as the mammalian inflammation process in general (Li et al., 2007).

The examples just discussed indicate that elegant methods are required to analyze S-nitrosation of the proteins as the posttranslational modification employed in the regulation of intracellular signaling. The next section describes recently developed methods that allow one to perform such types of analysis.

3. Assays for Analysis of *S*-Nitrosation of Specific Cellular Proteins

3.1. Jaffrey's method and its modifications

This approach allows qualitative and quantitative analyses of *S*-nitrosation. The method was first designed by Jeffrey *et al.* (2001) and is based on the biotinylation of S–NO groups followed by the detection of biotin by Western blot analysis.

3.1.1. Reagents

1. HEN buffer: 250 mM HEPES, pH 7.7, 1 mM EDTA, and 0.1 mM neocuproine
2. *S*-Methyl methanethiosulfonate (MMTS): 2 M solution in dimethylformamide (DMF)
3. Blocking buffer: 9 volumes of HEN buffer, 1 volume of 25% SDS, and 20 mM MMTS
4. Acetone (with highest grade of purity)
5. *N*-(6-[Biotinamido])hexyl-3′-(2′-pyridyldithio)propionamide (HPDP-biotin): the 50 mM suspension in DMF is normally prepared and then diluted by dimethyl sulfoxide (DMSO) up to a 4 mM concentration
6. Sodium ascorbate: 10 M water solution (prepare fresh, not earlier then 5 min before use and keep on ice)
7. Horseradish peroxidase (HRP)-labeled streptavidin (for qualitative or semiquantitative detection) or FITC-labeled antibiotin antibody (for quantitative analysis)
8. Agarose beads coated with specific protein-capture antibody and all other reagents for immunoprecipitation (in the case that immunoprecipitation of the protein of interest is planned)
9. All reagents and materials for SDS-PAGE electrophoresis and Western analysis

3.1.2. Practical approach

We will now describe the *S*-nitrosation assay for both purified and intracellular proteins. The purified *S*-nitrosated protein should first of all be subjected to the blockage of free SH groups—4 volumes of blocking buffer [9 volumes of HEN buffer plus 1 volume 25% SDS, 20 mM MMTS (from a 2 M stock prepared in DMF)] should be added to the 1 volume of the protein solution and incubated for 20 min at 50° with frequent vortexing. MMTS is then removed by protein precipitation with acetone. Biotin-HPDP (final concentration of 2 mM) and sodium ascorbate (1 μl to reach a final concentration of 1 mM) are then supplied. After incubation for 1 h at 25°,

SDS-PAGE sample buffer is added, and the samples are resolved by SDS-PAGE and transferred for immunoblotting. Considering that cysteine biotinylation is reversible, SDS-PAGE sample buffer must be prepared without reducing agents. Furthermore, to prevent nonspecific reactions of biotin-HPDP, samples should not be boiled prior to electrophoresis. It is better to perform all steps preceding SDS-PAGE in the dark. Immunoblots are then washed twice with TBS (140 mM NaCl, 50 mM Tris-HCl, pH 7.3) containing 0.1% Tween 20, blocked to avoid unspecific binding with TBS plus 5% skim milk for 1 h, and incubated with enzyme HRP-labeled streptavidin (1:500 in TBS plus 5% milk) for 1 h at room temperature. The nitrocellulose membrane should be washed five times for 5 min each with TBS containing 0.1% Tween 20 prior to the detection of S-nitrosated proteins by electrochemiluminescent development. Western blot analysis of protein S-nitrosothiols in the cells are performed as described earlier. However, cells should be washed twice with ice-cold phosphate-buffered saline (PBS). Lysis buffer (50 mM Tris, pH 7.5, 150 mM NaCl, 5 mM EDTA, and 0.5% NP-40) should be used. Four volumes of blocking buffer [9 volumes of HEN buffer plus 1 volume 25% SDS, 20 mM MMTS (from a 2 M stock prepared in DMF)] must be added to the 1 volume of the cell extract solution and incubated for 20 min at 50° with frequent vortexing, followed by the performance of the procedures described earlier in this section.

It is strongly recommended to load the same samples simultaneously onto the other part of the gel followed by blotting. This second part of the gel/membrane should be analyzed using the antibodies against the protein of interest, and the running distance passed by the protein of interest should be compared in both parts of the membrane— the one analyzed with the help of the S-nitrosation assay and the other part subjected to analysis with the anti-"protein of interest" antibody (Jeffrey et al., 2001).

To make the assay even more specific, it is recommended to immunoprecipitate the protein (when its expression level allows this) using agarose beads coated with the specific protein-capture antibody. Upon complete of immunoprecipitation, the protein of interest has to be released (all reagents should not contain the reducing agents to protect S–NO groups from degradation). This approach is the best for tagged proteins overexpressed in the cells upon transfection. It is easy to immunoprecipitate them using agarose beads coated with the antitag capture antibody.

To make a quantitative analysis, the FITC-labeled antibiotin antibody should be used. Immunoblots must be washed twice with TBS (140 mM NaCl, 50 mM Tris-HCl, pH 7.3) containing 0.1% Tween 20, blocked to avoid unspecific binding with TBS plus 5% skim milk for 1 h, and incubated with the FITC-labeled biotin antibody (1:500 in TBS plus 5% milk) for 1 h at room temperature. The nitrocellulose membrane is washed five times for 5 min each with TBS containing 0.1% Tween 20 prior to detection of S-nitrosated proteins. For this purpose, the membrane is scanned with a fluor imager [e.g., FluorImager 595 (Molecular Dynamics)]. The amount of S-NO groups in the protein can be calculated using the program Image-Quant (all the approaches described are summarized in Fig. 12.1). As a standard, it is recommended to use bovine serum albumin (BSA), which has one reactive SH group per molecule, S-nitrosated in the presence of the NO donor GSNO for 60 min (Sumbayev et al., 2003).

3.2. ELISA assay for detection of S-nitrosation of overexpressed proteins

We have developed another quantitative approach to analyze S-nitrosation of intracellular proteins. This approach has only been tested for overexpressed intracellular proteins. We have studied S-nitrosation of apoptosis signal-regulating kinase 1 (ASK1, the plasmid was generously provided by Professor Hidenori Ichijo, University of Tokyo, Japan) transfected into THP-1 human myeloid macrophages. This method is based on immobilization of the overexpressed protein in the wells of a MaxiSorp plate coated with capture antibody and then the colorimetric assay can be applied. This might include simple $HgCl_2$-dependent degradation of S–NO groups followed by nitrite by Griess assay (Cook et al., 1996). In case higher sensitivity is needed, biotinylation of S–NO groups should be applied, followed by detection with HRP-labeled streptavidin (Jeffrey et al., 2001). The last approach has to be tested further in terms of application for the analysis of S-nitrosation of intracellular (not overexpressed) proteins.

3.2.1. Reagents

1. Coating buffer (0.1 M Na_2HPO_4/50 mM citric acid, pH 5.0)
2. Saturation solution (50 mM Tris-HCl, pH 7.0; 150 mM NaCl; 0.05% sodium azide, 0.2% BSA, and 6% D-sorbitol).
3. Capture antibody
4. TBST buffer (10 mM Tris-HCl, pH 8.0, 150 mM NaCl, and 0.05% Tween 20)
5. $HgCl_2$ (0.18 mM solution if quantification based on the Griess assay is used)
6. Reagents for Griess assay (self-made or Griess assay kit)
7. Blocking buffer: 9 volumes of HEN buffer (see earlier discussion), 1 volume of 25% SDS, and 20 mM MMTS
8. HPDP-biotin: the 50 mM suspension in DMF is prepared normally and is then diluted by DMSO up to a 4 mM concentration
9. Sodium ascorbate: 10 M water solution (prepare fresh, not earlier then 5 min before use and keep on ice)
10. HRP-labeled streptavidin
11. All reagents for ortho-phenylenediamine/H_2O_2 analysis

Reagents 7–11 are used if the biotinylation approach is applied.

3.2.2. Practical approach

First of all, the capture antibody has to be diluted in the coating buffer to a final concentration of 5.0 μg/ml. Two hundred microliters of obtained coating solution is dispensed into each well of a MaxiSorp 96-well plate (we use Nunc plates) followed by overnight incubation at 37°. Then each exposed well has to be washed at least three times with TBST buffer followed by the dispensing of 250 μl of saturation solution into each coated well. A minimum 2-h incubation with saturation solution at room temperature is recommended. The saturation solution is then aspirated from the wells, and the plate is dried for 2 h at 37°. The plates can be stored dry at 4° with moisture adsorbent (Sumbayev et al., 2005; Välimaa et al., 2003).

Cells are washed twice with ice-cold PBS. Lysis buffer (50 mM Tris, pH 7.5, 150 mM NaCl, 5 mM EDTA, and 0.5% NP-40) is used to lyse the cells. Fifty microliters of cell lysate (approximately 30 μg of protein) is dispensed into each well (coated with capture antibody) and incubated for 1 h at room temperature. Plates are washed five times with TBST followed by dispensing 100 μl of 0.18 mM HgCl$_2$. After 15 min of incubation with HgCl$_2$, nitrite is detected by the Griess assay (Cook et al., 1996).

When using the biotinylation approach (Jeffrey et al., 2001), S-nitrosated protein immobilized on the plate through the capture antibody should first of all be subjected to the blockage of free SH groups—4 volumes of blocking buffer [9 volumes of HEN buffer plus 1 volume 25% SDS, 20 mM MMTS (from a 2 M stock prepared in DMF)] should be added (total volume should not exceed 200 μl) into each well and incubated for 20 min at 50° with frequent vortexing. MMTS is then removed by at least five to six washings with TBST buffer. Biotin-HPDP (final concentration of 2 mM) and sodium ascorbate (final concentration of 1 mM) should be supplied (total volume should not exceed 200 μl). Plates have to be washed five times with TBST, followed by dispensing of 100 μl/well HRP-labeled streptavidin (1:500–1:1000). After 1 h of incubation at room temperature, the plate is washed at least five times with TBST buffer and then at least twice with Tween 20-free TBST. A peroxidase reaction (ortho-phenylenediamine/H$_2$O$_2$) is then applied. The reactions are quenched after 10 min with an equal volume of 1 M H$_2$SO$_4$, and color development is measured in a microplate reader (any model is acceptable) as the absorbance at 492 nm (Sumbayev et al., 2005).

4. Conclusions

Methods described in this chapter allow investigation of endogenous S-nitrosation of specific intracellular proteins performing qualitative, semiquantitative, and quantitative analyses. However, it is clear from the description that all the methods presented here could be modified and modernized further. The recently developed ELISA approach could be designed for intracellular proteins with normal or even low expression levels, and one of our goals is to contribute to this in the near future.

ACKNOWLEDGMENTS

We thank Professor Hidenori Ichijo (University of Tokyo, Japan) who has generously provided us with the constructs encoding wild-type human ASK1 and its dominant negative form. This work was supported by the start-up grant provided to Dr. V. Sumbayev by Medway School of Pharmacy, University of Kent (United Kingdom).

REFERENCES

Cook, J. A., Kim, S. Y., Teague, D., Krishna, M. C., Pacelli, R., Mitchell, J. B., Vodovotz, Y., Nims, R. W., Christodoulou, D., Miles, A. M., Grisham, M. B., and Wink, D. A. (1996). Convenient colorimetric and fluorometric assays for S-nitrosothiols. *Anal. Biochem.* **238,** 150–158.

Daiber, A., Frein, D., Namgaladze, D., and Ullrich, V. (2002). Oxidation and nitrosation in the nitrogen monoxide/superoxide system. *J. Biol. Chem.* **277,** 11882–11888.

Haendeler, J., Hoffmann, J., Tischler, V., Berk, B. C., Zeiher, A. M., and Dimmeler, S. (2002). Redox regulatory and anti-apoptotic functions of thioredoxin depend on S-nitrosylation at cysteine 69. *Nat. Cell Biol.* **4,** 743–749.

Han, O. J., Joe, K. H., Kim, S. W., Lee, H. S., Kwon, N. S., Baek, K. J., and Yun, H. Y. (2001). Involvement of p38 mitogen-activated protein kinase and apoptosis signal-regulating kinase-1 in nitric oxide-induced cell death in PC12 cells. *Neurochem. Res.* **26,** 525–532.

Hess, D. T., Matsumoto, A., Kim, S. O., Marshall, H. E., and Stamler, J. S. (2005). Protein S-nitrosylation: Purview and parameters. *Nat. Rev. Mol. Cell Biol.* **6,** 150–166.

Jaffrey, S. R., Erdjument-Bromage, H., Ferris, C. D., Tempst, P., and Snyder, S. H. (2001). Protein S-nitrosylation: A physiological signal for neuronal nitric oxide. *Nat. Cell Biol.* **3,** 193–197.

Li, F., Sonveaux, P., Rabbani, Z. N., Liu, S., Yan, B., Huang, Q., Vujaskovic, Z., Dewhirst, M. W., and Li, C. Y. (2007). Regulation of HIF-1alpha stability through S-nitrosylation. *Mol. Cell.* **26,** 63–74.

Oda, T., Hirota, K., Nishi, K., Takabuchi, S., Oda, S., Yamada, H., Arai, T., Fukuda, K., Kita, T., Adachi, T., Semenza, G. L., and Nohara, R. (2006). Activation of hypoxia-inducible factor 1 during macrophage differentiation. *Am. J. Physiol. Cell Physiol.* **291,** C104–C113.

Park, H. S., Yu, J. W., Cho, J. H., Kim, M. S., Huh, S. H., Ryoo, K., and Choi, E. J. (2004). Inhibition of apoptosis signal-regulating kinase 1 by nitric oxide through a thiol redox mechanism. *J. Biol. Chem.* **279,** 7584–7590.

Sarker, K. P., Biswas, K. K., Rosales, J. L., Yamaji, K., Hashiguchi, T., Lee, K. Y., and Maruyama, I. (2003). Ebselen inhibits NO-induced apoptosis of differentiated PC12 cells via inhibition of ASK1-p38 MAPK-p53 and JNK signaling and activation of p44/42 MAPK and Bcl-2. *J. Neurochem.* **87,** 1345–1353.

Semenza, G. L. (2001). HIF-1 and mechanisms of hypoxia sensing. *Curr. Opin. Cell Biol.* **13,** 167–171.

Semenza, G. L. (2002). HIF-1 and tumor progression: Pathophysiology and therapeutics. *Trends Mol. Med.* **8,** S62–S67.
Sumbayev, V. V. (2003). S-nitrosylation of thioredoxin mediates activation of apoptosis signal-regulating kinase 1. *Arch. Biochem. Biophys.* **415,** 133–136.
Sumbayev, V. V., Bonefeld-Jørgensen, E. C., Wind, T., and Andreasen, P. A. (2005). A novel pesticide-induced conformational state of the oestrogen receptor ligand-binding domain, detected by conformation-specific peptide binding. *FEBS Lett.* **579,** 541–548.
Sumbayev, V. V., Budde, A., Zhou, J., and Brüne, B. (2003). HIF-1 alpha protein as a target for S-nitrosation. *FEBS Lett.* **535,** 106–112.
Sumbayev, V. V., and Yasinska, I. M. (2007). Mechanisms of hypoxic signal transduction regulated by reactive nitrogen species. *Scand. J. Immunol.* **65,** 399–406.
Tao, L., Gao, E., Bryan, N. S., Qu, Y., Liu, H. R., Hu, A., Christopher, T. A., Lopez, B. L., Yodoi, J., Koch, W. J., Feelisch, M., and Ma, X. L. (2004). Cardioprotective effects of thioredoxin in myocardial ischemia and reperfusion: Role of S-nitrosation. *Proc. Natl. Acad. Sci. USA* **101,** 11471–11476.
Välimaa, L., Pettersson, K., Vehniäinen, M., Karp, M., and Lövgren, T. (2003). A high-capacity streptavidin-coated microtitration plate. *Bioconjug. Chem.* **14,** 103–111.
Walmsley, S. R., Cadwallader, K. A., and Chilvers, E. R. (2005). The role of HIF-1alpha in myeloid cell inflammation. *Trends Immunol.* **26,** 434–439.
Yasinska, I. M., Kozhukhar, A. V., and Sumbayev, V. V. (2004). S-nitrosation of thioredoxin in the nitrogen monoxide/superoxide system activates apoptosis signal-regulating kinase 1. *Arch. Biochem. Biophys.* **428,** 198–203.

CHAPTER THIRTEEN

DETERMINATION OF MAMMALIAN ARGINASE ACTIVITY

Diane Kepka-Lenhart,* David E. Ash,[†] and Sidney M. Morris*

Contents

1. Introduction	222
2. Principle of Assay	222
3. Preparation of Cell and Tissue Extracts	223
4. Buffers, Reagents and Other Materials for Assay Protocol I	224
5. Assay Protocol I	225
6. Buffers, Reagents, and Other Materials for Assay Protocol II	226
7. Assay Protocol II	226
8. Limitations of Assay	227
9. Determination of Arginase Activity in Cultured Cells	228
Acknowledgment	229
References	229

Abstract

Of all arginine catabolic enzymes, the arginases and nitric oxide (NO) synthases are the ones that are of greatest interest to many investigators. Mammalian arginases catalyze the hydrolysis of arginine to ornithine and urea and are composed of two distinct isozymes: arginase I, located within the cytosol, and arginase II, located within mitochondria. The arginases not only can inhibit NO synthesis by reducing arginine availability, but also can promote the synthesis of polyamines or proline via production of the common precursor ornithine. Because of their inducibility in many cell types and to their potential impact on multiple biochemical pathways in health and disease, there is growing interest in assays of arginase activity. Although arginase activity may be determined by either spectrophotometric or radiochemical assays, radiochemical assays afford greater sensitivity and do not require correction for any ornithine or urea that may be present in the samples. Part of the arginase assay protocol described in this chapter also can be used for radiochemical assays of enzymes

* Department of Microbiology and Molecular Genetics, University of Pittsburgh School of Medicine, Pittsburgh, Pennsylvania
[†] Department of Chemistry, Central Michigan University, Mt. Pleasant, Michigan

that catalyze decarboxylation reactions. No activity assay currently available is capable of distinguishing the arginase isozymes.

1. Introduction

The arginases (L-arginine amidinohydrolase, EC 3.5.3.1) catalyze the hydrolysis of L-arginine to L-ornithine and urea. Mammals express two isozymes of arginase that are encoded by distinct nuclear genes: arginase I, a cytosolic enzyme, and arginase II, a mitochondrial matrix enzyme (Cederbaum *et al.*, 2004; Morris, 2002). Arginase I is highly expressed in the liver as a component of the urea cycle, whereas arginase II is expressed at the highest levels in kidney and prostate (Gotoh *et al.*, 1996; Morris *et al.*, 1997; Vockley *et al.*, 1996). However, both enzymes also can be induced in multiple cell types and in various diseases (Bansal and Ochoa, 2003; Mori, 2007; Morris, 2004). Consequences of arginase induction can include impaired nitric oxide (NO) synthesis as a result of reduced arginine bioavailability, as well as increased synthesis of polyamines and/or proline as a result of increased ornithine production (Morris, 2004, 2007). Increased arginase activity in plasma also can occur as a result of liver damage (Ikemoto *et al.*, 2001; Langle *et al.*, 1995) or—in the case of humans and certain primates—hemolysis due to the presence of arginase in erythrocytes (Morris *et al.*, 2005; Spector *et al.*, 1985). Disease also results as a consequence of inherited arginase I deficiency (Crombez and Cederbaum, 2005). Growing evidence that alterations in arginase activity or expression may be involved in disease has increased interest in methods for assaying arginase activity. Although both spectrophotometric and radiochemical assays of arginase activity have been described in the literature, the radiochemical assay is preferred not only because of its greater sensitivity, but also because it is suitable for analyzing tissue homogenates or plasma without requiring correction for ornithine or urea that may be present in the sample.

2. Principle of Assay

With L-[guanidino-^{14}C]arginine as the substrate, the arginase reaction generates [^{14}C]urea as the only labeled product. In assay protocol I, [^{14}C] urea is separated from L-[guanidino-^{14}C]arginine by a subsequent incubation with urease to generate $^{14}CO_2$, which can be volatilized by acidification of the reaction and trapped on NaOH-soaked filters as $Na_2{}^{14}CO_3$ for quantification by scintillation counting. The arginase and urease reactions are shown later. This protocol is adapted from previously described methods (Carulli *et al.*, 1968; Klein and Morris, 1978; Schimke, 1964; Spector *et al.*,

1980). In assay protocol II, [^{14}C]urea is separated from L-[guanidino-^{14}C] arginine by using an ion-exchange resin. This protocol is adapted from Ruegg and Russell (1980). Essential features and differences between the two protocols are outlined in Fig. 13.1. Although both assays have similar sensitivity and specificity, both protocols are presented here because a specific protocol may be better suited to a particular laboratory because of differences in requirements for specific reagents and equipment.

Arginase reaction: L-[guanidino-^{14}C]arginine $\rightarrow$ L-ornithine + [^{14}C]urea
Urease reaction: [^{14}C]urea + H_2O + 2 H^+ $\rightarrow$ $^{14}CO_2$ + 2 NH_4^+

Because arginases I and II are so similar in enzymatic properties (Ash, 2004), they cannot be distinguished by any activity assays currently available. Thus, the assay will represent the sum of the activities when both arginase isozymes are present.

3. Preparation of Cell and Tissue Extracts

Extracts may be prepared from freshly isolated cells or tissues or from tissues that have been snap frozen and stored at $-76°$. Cells or tissues are homogenized in 10 to 20 volumes of ice-cold 0.1% Triton X-100 containing protease inhibitors, followed by centrifugation to remove insoluble material. Suitable mixtures of protease inhibitors can be prepared using

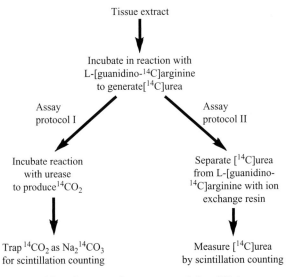

Figure 13.1 Outline of key features of assay protocols I and II. Assay protocol I includes a second enzymatic reaction step that is not required in assay protocol II. Assay protocol II includes a centrifugation step that is not required for assay protocol I.

complete protease inhibitor cocktail tablets (Roche Applied Science) or by combining individual inhibitors to give desired final concentrations (e.g., 2 mM Pefabloc SC-10 μg/ml leupeptin–2 μg/ml pepstatin) (Morris et al., 1998). Samples of human plasma and serum—fresh or stored at $-76°$—may be used in the arginase assay without further treatment. Because human erythrocytes contain arginase, however, care must be taken to avoid hemolysis when drawing blood or during the preparation of plasma or serum.

4. Buffers, Reagents and Other Materials for Assay Protocol I

250 mM L-arginine hydrochloride; store aliquots at $-20°$

L-[Guanidino-^{14}C]arginine; from American Radiolabeled Chemicals, Inc., or Moravec Biochemicals and Radiochemicals. (Because different lots of commercial L-[guanidino-^{14}C]arginine contain varying amounts of [^{14}C] urea, it is essential to first set up a control reaction without any experimental sample and to run the assay from step 4 in order to determine the amount of labeled urea present. If the labeled arginine contains sufficient [^{14}C]urea to provide unacceptably high background values in the control reaction, the arginine may be purified by ion-exchange chromatography on a column of Dowex 50W-X8 (100–200 mesh, H$^+$ form) as described (Klein and Morris, 1978; Ruegg and Russell, 1980).)

[^{14}C]Arginine substrate; prepare by adding L-[guanidino-^{14}C]arginine to 250 mM L-arginine hydrochloride to give final concentration of 3 μCi/ml

0.2 M KPi, pH 6.6; prepare by mixing 0.2 M dibasic and 0.2 M monobasic potassium phosphate to yield a final pH of 6.6

2 N HCl

300 units/ml jack bean urease ($\geq$50 units/mg) in 1 mM EDTA, pH 7.0–7.5; prepare fresh each time

200 mM glycine, pH 9.7; store aliquots at $-20°$

5.4 mM MnCl$_2$

Reaction buffer: prepare by mixing equal volumes of 200 mM glycine, pH 9.7, and 5.4 mM MnCl$_2$ immediately before use

8% (w/v) NaOH

15-ml disposable screw-cap polypropylene tubes

0.5-ml Eppendorf tubes

2.3-cm discs of Whatman #3 filter paper. Roll or fold and insert into 0.5-ml Eppendorf tubes

Extra-long disposable 1000-μl pipette tips

Scintillation fluid capable of providing reliable results with samples containing NaOH [e.g., Hionic-Fluor scintillation fluid (Packard Bioscience)]

Commercial preparation of beef liver arginase (optional) dissolved in 0.1% Triton X-100 containing protease inhibitors; store aliquots at $-76°$

5. Assay Protocol I

On the day that step 4 of the assay will be performed, add 125 µl 8% NaOH to 0.5-ml Eppendorf tubes containing Whatman filters; do not close tube cap. Keep tubes with wet filters in covered beaker or other container to prevent filters from drying.

1. Add 50 µl cell extract or serum (for nonhepatic tissue samples, we suggest starting with 200–300 µg protein) to 150 µl of reaction buffer in 15-ml tubes; set up a blank reaction by adding 50 µl of 0.1% Triton X-100 to 150 µl of reaction buffer. Cap tubes and incubate at 55° for 10 min to maximally activate arginase.
2. Initiate the reaction by adding 50 µl of [^{14}C]arginine substrate (=0.15 µCi). This yields the following final concentrations for the arginase reaction: 60 mM glycine (pH 9.7)–1.62 mM MnCl$_2$–50 mM L-arginine. Cap tubes and incubate at 37° for 30 to 150 min with gentle shaking. The incubation period can be varied according to the level of arginase activity present in the samples.
3. Incubate samples in boiling water bath for 3 min to stop reaction. Let cool to room temperature and then add 1 ml 0.2 M KPi to each tube. Samples may be frozen at this step to allow final processing at a later time.
4. Add 100 µl urease stock to each sample and immediately insert an open 0.5-ml Eppendorf tube containing the NaOH-soaked filter paper disc so that it is suspended near the top of the 15-ml tube. The open hinged top of the Eppendorf tube provides sufficient tension to keep it from falling to the bottom of the 15-ml tube. Seal top of 15-ml tube with Parafilm. Incubate at 37° for 45 min with *gentle* shaking. *Note:* Take care in this step and the subsequent step to avoid agitation that could splash reaction mix on the outside of the Eppendorf tube.
5. Using an extra-long disposable pipette tip, add 200 µl 2 N HCl through the Parafilm and into the bottom of the 15-ml tube, taking care to avoid the Eppendorf tube. Immediately screw a 15-ml tube cap over the Parafilm and incubate at 37° for 30 min with *gentle* shaking. (*Note:* This step and the subsequent one can be used in radiochemical assays of enzymes that catalyze decarboxylation reactions; e.g., ornithine decarboxylase.)
6. Remove the Eppendorf tube with filter (cut off cap if necessary), place into a 20-ml scintillation vial, and add 15 to 20 ml scintillation fluid.

Arginase activity, as µmoles urea formed/min/mg protein, is calculated as µmoles L-arginine substrate in reaction $\times (A - B) \div (C \times D \times T)$, where A is dpm ^{14}CO$_2$ produced in sample reaction, B is dpm in blank reaction, C is total dpm L-[guanidino-^{14}C]arginine in reaction, D is milligrams of protein added to reaction, and T is time of incubation in minutes. Using this assay, arginase activities as low as 60 pmol/min can be quantified readily.

Based on experience in our laboratory, we suggest using an incubation period of 2 h when assaying 50-μl samples of cell or tissue extracts prepared as described earlier or of human plasma or serum. In the case of liver extracts, use 1:20–1:50 dilutions of the original extract and incubate for only 30 min. When assaying extracts of cells or tissues for which no information on arginase activity is available, we suggest performing an initial assay with an amount of extract sufficient to provide 100 to 300 μg of protein and an incubation period of 2 h. We recommend that aliquots of a stock solution of liver extract (a commercial beef liver arginase preparation or an extract of rat liver prepared as described earlier) be used as a reference to check day-to-day consistency of the assay.

6. Buffers, Reagents, and Other Materials for Assay Protocol II

0.1 M L-arginine hydrochloride
L-[Guanidino-^{14}C]arginine from American Radiolabeled Chemicals, Inc., or Moravec Biochemicals and Radiochemicals. See note under assay protocol I regarding purification of L-[guanidino-^{14}C]arginine should background levels be unacceptably high.
200 mM 2-(N-cyclohexylamino)-ethanesulfonic acid (CHES), pH 9.0
1 mM MnCl$_2$
Reaction mix per assay: 25 μl CHES buffer, 5 μl MnCl$_2$, 5 μl L-arginine, 0.5 μl L-[guanidino-^{14}C]arginine (= 0.05 μCi), 9.5 μl H$_2$O. Total volume 45 μl.
Stop solution: 7 M urea, 10 mM L-arginine, 0.25 M acetic acid; adjust to pH 4.5 with NaOH
Dowex 50W-X8–400: 300 g washed with 2 liters 6 N HCl and then 4 liters H$_2$O. Decant fines and filter to remove excess H$_2$O.
Stop/Dowex solution: 60 g of Dowex 50W-X8, 60 ml H$_2$O, 120 ml stop solution. May be stored at room temperature.
1.5-ml Eppendorf tubes
Water-miscible scintillation fluid such as Ecoscint (National Diagnostics)

7. Assay Protocol II

1. On the day of the assay, reaction mixture is prepared and 45-μl aliquots are added to 1.5-ml Eppendorf tubes. A total counts sample, which will

not be treated with Dowex, is prepared identically, as well as a blank, which will receive no cell extract.
2. The reaction is initiated by the addition of 5 μl of cell extract. Extracts are prepared as described earlier. Depending on the tissue type, reactions are incubated for 30 to 150 min at 37°.
3. Reactions are stopped by the addition of 400 μl of the stop/Dowex solution. Solutions are mixed vigorously by vortexing and then rotated end over end gently for 20 min to ensure adequate mixing.
4. Samples are then centrifuged for 10 min at 6000 rpm in a microcentrifuge to pack the Dowex resin.
5. A 200-μl aliquot of the supernatant is carefully removed and mixed with scintillation fluid for quantification of [^{14}C]urea by scintillation counting. (*Note:* It is essential not to disturb the pellet, as spurious high activity values will be obtained if any particles of Dowex resin are pipetted along with the supernatant.)

Arginase activity, as μmoles urea formed/min/mg protein, is calculated as μmoles L-arginine substrate in the reaction $\times\ 2.25(A - B) \div (C \times D \times T)$, where A is dpm [^{14}C]urea produced in the sample reaction, B is dpm in the blank reaction, C is total dpm L-[guanidino-^{14}C]arginine in reaction, D is milligrams protein added to reaction, and T is time of incubation in minutes. The assay is generally linear up to 10% conversion of the substrate. Unusually high backgrounds require additional purification of the L-[guanidino-^{14}C] arginine as described in assay protocol I. Using this method, activities as low as 2 nmol/min/mg of protein have been determined in tissue extracts (Baggio *et al.*, 1999) or 100 pmol urea formed/min (Ruegg and Russell, 1980).

8. Limitations of Assay

As standard practice, it is essential to determine activities within the linear range of the assay. It is important to note, however, that the linear range may vary for extracts from different tissues or cell types (Fig. 13.2). At least some inhibition at high percentages of arginine conversion is due to the accumulation of ornithine, which is a competitive inhibitor (Pace and Landers, 1981; Reczkowski and Ash, 1994). It is possible that some inhibition may be due also to other compounds present in crude extracts. In the case of human plasma or serum samples, hemolysis that occurs during sample preparation or handling will result in artifactually high activity values because of the presence of arginase I in erythrocytes.

As indicated earlier, the sensitivity of both assay protocols is affected adversely by the level of trace contamination of [^{14}C]urea in the L-[guanidino-^{14}C]arginine substrate. Assay protocol II is not suitable for assaying arginase activity in extracts of microorganisms that also express urease. If

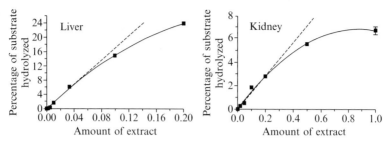

Figure 13.2 Test of arginase assay linearity as a function of amount of tissue extract. Mouse liver and kidney were homogenized in 20 and 15 volumes, respectively, of 0.1% Triton X-100 containing 2 mM Pefabloc SC-10 μg/ml leupeptin–2 μg/ml pepstatin. Aliquots of extracts (a 1:5 dilution of the liver extract or undiluted kidney extract) in reaction buffer were activated at 55° and then diluted serially in reaction buffer. Activities were determined by assay protocol I. The amount of extract is expressed relative to the amount of original extract in the initial reaction mixture. Results are expressed as means and range for assays of duplicate samples.

L-[guanidino-^{14}C]arginine is not available, L-[U-^{14}C]arginine may be used in assay protocol II but not in assay protocol I, particularly when evaluating arginase activity in intact cells, because of potentially confounding effects of ornithine or arginine decarboxylase activities. Commercially available arginase inhibitors [e.g., N^{ω}-hydroxy-L-arginine (NOHA), S-(2-boronoethyl)-L-cysteine, or 2(S)-amino-6-boronohexanoic acid] may be added to the reactions to confirm that any activity is because of arginase. Although N^{ω}-hydroxy-nor-L-arginine (nor-NOHA) is a more potent inhibitor than NOHA, it is unstable at the basic pH of the assays and thus is best used under conditions of neutral pH (e.g., when assaying arginase activity in intact cells).

9. Determination of Arginase Activity in Cultured Cells

Both assay protocols can be adapted to evaluate the arginase activity of intact cells. After adding an aliquot of L-[guanidino-^{14}C]arginine to standard tissue culture medium, cells are incubated at 37° for the desired period of time. Using assay protocol I, the culture medium containing [^{14}C]urea is aspirated and treated with urease to release ^{14}CO$_2$ as described earlier in assay step 4. This method has been used to measure arginase activity in the RAW 264.7 macrophage cell line (Kepka-Lenhart et al., 2000). In this particular study, cells are incubated for 24 h in a 96-well culture plate containing 0.05 μCi of L-[guanidino-^{14}C]arginine per 150 μl of culture medium. Similarly, assay protocol II also has been used to evaluate arginase activity in cultured peritoneal macrophages using 10^4 cpm L-[guanidino-^{14}C]arginine added per 0.5 ml medium (Russell and Ruegg, 1980). Regardless of the protocol used, it is essential to run parallel incubations of complete labeled medium without cells as a control because fetal calf serum contains arginase (Holtta and Pohjanpelto, 1982; Kepka-Lenhart et al., 2000; Ruegg and Russell, 1980) whose activity can be equal to or greater than that of the cells themselves, depending on the particular cell type.

ACKNOWLEDGMENT

This work was supported in part by NIH Grants GM RO1 57384 (SMM) and GM RO1 O67788 (DEA).

REFERENCES

Ash, D. E. (2004). Structure and function of arginases. *J. Nutr.* **134,** 2760S–2764S.
Baggio, R., Emig, F. A., Christianson, D. W., Ash, D. E., Chakder, S., and Rattan, S. (1999). Biochemical and functional profile of a newly developed potent and isozyme-selective arginase inhibitor. *J. Pharmacol. Exp. Ther.* **290,** 1409–1416.
Bansal, V., and Ochoa, J. B. (2003). Arginine availability, arginase, and the immune response. *Curr. Opin. Clin. Nutr. Metab. Care* **6,** 223–228.
Carulli, N., Kaihara, S., and Wagner, H. N., Jr. (1968). Radioisotopic assay of arginase activity. *Anal. Biochem.* **24,** 515–522.
Cederbaum, S. D., Yu, H., Grody, W. W., Kern, R. M., Yoo, P., and Iyer, R. K. (2004). Arginases I and II: Do their functions overlap? *Mol. Genet. Metab.* **81**(Suppl. 1), S38–S44.
Crombez, E. A., and Cederbaum, S. D. (2005). Hyperargininemia due to liver arginase deficiency. *Mol. Genet. Metab.* **84,** 243–251.
Gotoh, T., Sonoki, T., Nagasaki, A., Terada, K., Takiguchi, M., and Mori, M. (1996). Molecular cloning of cDNA for nonhepatic mitochondrial arginase (arginase II) and comparison of its induction with nitric oxide synthase in a murine macrophage-like cell line. *FEBS Lett.* **395,** 119–122.
Holtta, E., and Pohjanpelto, P. (1982). Polyamine dependence of Chinese hamster ovary cells in serum-free culture is due to deficient arginase activity. *Biochim. Biophys. Acta* **721,** 321–327.
Ikemoto, M., Tsunekawa, S., Awane, M., Fukuda, Y., Murayama, H., Igarashi, M., Nagata, A., Kasai, Y., and Totani, M. (2001). A useful ELISA system for human liver-type arginase, and its utility in diagnosis of liver diseases. *Clin. Biochem.* **34,** 455–461.
Kepka-Lenhart, D., Mistry, S. K., Wu, G., and Morris, S. M., Jr. (2000). Arginase I: A limiting factor for nitric oxide and polyamine synthesis by activated macrophages? *Am. J. Physiol.* **279,** R2237–R2242.
Klein, D., and Morris, D. R. (1978). Increased arginase activity during lymphocyte mitogenesis. *Biochem. Biophys. Res. Commun.* **81,** 199–204.
Langle, F., Steininger, R., Roth, R., Winkler, S., Andel, H., Acimovic, S., Fugger, R., and Muhlbacher, F. (1995). L-arginine deficiency and hemodynamic changes as a result of arginase efflux following orthotopic liver transplantation. *Transplant Proc.* **27,** 2872–2873.
Mori, M. (2007). Regulation of nitric oxide synthesis and apoptosis by arginase and arginine recycling. *J. Nutr.* **137,** 1616S–1620S.
Morris, C. R., Kato, G. J., Poljakovic, M., Wang, X., Blackwelder, W. C., Sanchdev, V., Hazen, S. L., Vichinsky, E. P., Morris, S. M., Jr., and Gladwin, M. T. (2005).

Dysregulated arginine metabolism, hemolysis-associated pulmonary hypertension and mortality in sickle cell disease. *JAMA* **294,** 81–90.

Morris, S. M., Jr. (2002). Regulation of enzymes of the urea cycle and arginine metabolism. *Annu. Rev. Nutr.* **22,** 87–105.

Morris, S. M., Jr. (2004). Recent advances in arginine metabolism. *Curr. Opin. Clin. Nutr. Metab. Care* **7,** 45–51.

Morris, S. M., Jr. (2007). Arginine metabolism: Boundaries of our knowledge. *J. Nutr.* **137,** 1602S–1609S.

Morris, S. M., Jr., Bhamidipati, D., and Kepka-Lenhart, D. (1997). Human type II arginase: Sequence analysis and tissue-specific expression. *Gene* **193,** 157–161.

Morris, S. M., Jr., Kepka-Lenhart, D., and Chen, L. C. (1998). Differential regulation of arginases and inducible nitric oxide synthase in murine macrophage cells. *Am. J. Physiol.* **275,** E740–E747.

Pace, C. N., and Landers, R. A. (1981). Arginase inhibition. *Biochim. Biophys. Acta* **658,** 410–412.

Reczkowski, R., and Ash, D. E. (1994). Rat liver arginase: Kinetic mechanism, alternate substrates, and inhibitors. *Arch. Biochem. Biophys.* **312,** 31–37.

Ruegg, U. T., and Russell, A. S. (1980). A rapid and sensitive assay for arginase. *Anal. Biochem.* **102,** 206–212.

Russell, A. S., and Ruegg, U. T. (1980). Arginase production by peritoneal macrophages: A new assay. *J. Immunol. Methods* **32,** 375–382.

Schimke, R. T. (1964). Enzymes of arginine metabolism in mammalian cell culture. I. Repression of argininosuccinate synthetase and argininosuccinase. *J. Biol. Chem.* **239,** 136–145.

Spector, E. B., Kiernan, M., Bernard, B., and Cederbaum, S. D. (1980). Properties of fetal and adult red blood cell arginase: A possible prenatal diagnostic test for arginase deficiency. *Am. J. Hum. Genet.* **32,** 79–87.

Spector, E. B., Rice, S. C., Kern, R. M., Hendrickson, R., and Cederbaum, S. D. (1985). Comparison of arginase activity in red blood cells of lower mammals, primates, and man: Evolution to high activity in primates. *Am. J. Hum. Genet.* **37,** 1138–1145.

Vockley, J. G., Jenkinson, C. P., Shukula, H., Kern, R. M., Grody, W. W., and Cederbaum, S. C. (1996). Cloning and characterization of the human type II arginase gene. *Genomics* **38,** 118–123.

CHAPTER FOURTEEN

MEASUREMENT OF PROTEIN S-NITROSYLATION DURING CELL SIGNALING

Joan B. Mannick* and Christopher M. Schonhoff[†]

Contents

1. Introduction	232
2. Biotin Switch Technique	233
3. Protocol for Analyzing Protein S-Nitrosylation in Tissue Samples Using the Biotin Switch Assay	235
4. Chemical Reduction/Chemiluminescence	236
5. Protocol for Chemical Reduction/Chemiluminescence Measurements of S-Nitrosylation of Immunoprecipitated Proteins	238
6. Conclusion	240
References	240

Abstract

S-Nitrosylation, the modification of a cysteine thiol by a nitric oxide (NO) group, has emerged as an important posttranslational modification of signaling proteins. An impediment to studying the regulation of cell signaling by S-nitrosylation has been the technical challenge of detecting endogenously S-nitrosylated proteins. Detection of S-nitrosylated proteins is difficult because the S–NO bond is labile and therefore can be lost or gained artifactually during sample preparation. Nevertheless, several methods have been developed to measure endogenous protein S-nitrosylation, including the biotin switch assay and the chemical reduction/chemiluminescence assay. This chapter describes these two methods and provides examples of how they have been used successfully to elucidate the role of protein S-nitrosylation in cell physiology and pathophysiology.

---

* Departments of Medicine and Cell Biology, University of Massachusetts Medical School, Worcester, Massachusetts
[†] Department of Biomedical Sciences, Tufts University Cummings School of Veterinary Medicine, North Grafton, Massachusetts

1. INTRODUCTION

It has become increasingly apparent that S-nitrosylation, the modification of a cysteine thiol by a nitric oxide (NO) group, is an important posttranslational modification that has profound effects on protein function. To date, over 200 different proteins have been found to be regulated by S-nitrosylation (Derakhshan et al., 2007; Hess et al., 2005). They include enzymes such as caspases-3 and -9, protein tyrosine phosphatase 1B (PTB1B), thioredoxin, Jun N-terminal kinase (JNK), human homolog of mouse double minute-2 (Hdm2), N-ethylmaleimide-sensitive factor (NSF) ATPase, and the monomeric GTPase p21^{Ras} (Ras) (Haendeler et al., 2002; Lander et al., 1997; Li and Whorton, 2003; Mannick et al., 1999, 2001; Matsushita et al., 2003; Park et al., 2000; Schonhoff et al., 2002). Membrane receptors and ion channels, such as the N-methyl-D-aspartate receptor (NMDAR), the ryanodine receptor (RyR1), and the olfactory cyclic nucleotide-gated Ca^{2+} channel-α (CNGα), have also been shown to be regulated by S-nitrosylation (Broillet, 2000; Choi et al., 2000; Eu et al., 2000). Finally, several transcription factors, such as OxyR, hypoxia-inducible factor α, and nuclear factor κB (NF-κB) are regulated by S-nitrosylation (Kim and Tannenbaum, 2004; Marshall and Stamler, 2001; Palmer et al., 2000). More recently, aberrant S-nitrosylation has been shown to contribute to several disease states, including Parkinson's disease, cerebral ischemia, asthma, sickle cell anemia, and amyotrophic lateral sclerosis (ALS) (Chung et al., 2004; Gu et al., 2002; Pawloski et al., 2005; Que et al., 2005; Schonhoff et al., 2006). The diverse array of proteins regulated by S-nitrosylation along with the emerging role of S-nitrosylation in disease pathogenesis underscores its importance as a regulatory mechanism during signal transduction.

Elucidation of the role of protein S-nitrosylation in cell signaling has been hampered by the technical difficulty of detecting endogenously S-nitrosylated proteins. The S–NO bond is highly labile and redox sensitive. Therefore, S–NO bonds are lost easily or gained artifactually during sample preparation. In addition, the levels of endogenously S-nitrosylated proteins in cells are at the limits of detection of currently available assays. Nevertheless, several techniques have been developed that accurately detect endogenous protein S-nitrosylation in cells. In particular, the biotin switch assay and the chemical reduction/chemiluminescence assays have proven to be sensitive and specific methods to measure low levels of intracellular protein S-nitrosylation.

2. BIOTIN SWITCH TECHNIQUE

The biotin switch technique was first described by Jaffrey et al. (2001) and has rapidly emerged as an important method to detect S-nitrosylated proteins. The assay consists of three steps. First, free thiols on proteins are blocked by incubating samples with the thiol-reacting reagent methylmethanethiosulfonate (MMTS) in the presence of sodium dodecyl sulfate (SDS) at 50°C. The SDS partially denatures the protein(s) so that buried cysteines become accessible. Second, S-nitrosothiols are reduced by ascorbic acid to yield free thiols. Third, the newly reduced thiols are biotinylated with N-[6-(biotinamido)hexyl]-3'-(2'-pyridyldithio)propionamide (biotin-HPDP), a thiol-specific biotinylating reagent. The assay results in ascorbic acid-dependent biotinylation of S-nitrosylated proteins. Biotinylated proteins are then purified on streptavidin-agarose and separated on nonreducing gels. Total levels of protein S-nitrosylation can be measured by analyzing the samples on anti-biotin immunoblots. Alternatively, S-nitrosylation of a specific protein of interest can be assessed by determining if the streptavidin-purified sample contains the protein of interest on immunoblot analysis. We have utilized the biotin switch technique to study how aberrant S-nitrosylation may be involved in the pathogenesis of ALS.

Amyotrophic lateral sclerosis is one of the most common adult-onset neurodegenerative diseases characterized by the degeneration of motor neurons in the spinal cord, brain stem, and cortex. A subset of patients with familial ALS has mutations in the superoxide dismutase 1 (SOD1) gene. We have shown previously that ALS-associated SOD1 mutants have increased denitrosylase activity (Johnson et al., 2001). We hypothesized that the increased denitrosylase activity of SOD1 mutants aberrantly decreases intracellular protein S-nitrosylation and thereby disrupts cell signaling and contributes to motor neuron degeneration in ALS. To test this hypothesis, we analyzed protein S-nitrosylation in the spinal cord of mutant SOD1 transgenic mice using the biotin switch assay. Mutant SOD1 transgenic mice develop motor neuron degeneration very similar to patients with ALS and represent the best animal model of the disease. Data indicate that protein S-nitrosylation was decreased aberrantly, particularly within mitochondria, in the spinal cords of G93A mutant SOD1 transgenic mice (Fig. 14.1) (Schonhoff et al., 2006). Moreover, results suggest that S-nitrosothiol (SNO) depletion disrupts the function and/or subcellular localization of proteins regulated by S-nitrosylation, such as glyceraldehyde-3-phosphate dehydrogenase (GAPDH) (Fig. 14.2) (Hara et al., 2005), and thereby contributes to ALS pathogenesis (Schonhoff et al., 2006). Repletion of intracellular SNO levels with SNO donor compounds rescues cells from mutant SOD1-induced death. These findings suggest that increased denitrosylase activity is a toxic gain of function of ALS-associated

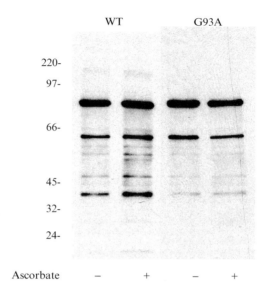

Figure 14.1 SNO levels are decreased in the spinal cords of mutant SOD1 transgenic mice. Protein S-nitrosylation in spinal cord lysates obtained from G93A mutant SOD1 transgenic mice at the time of paralysis (G93A) and from age-matched wild-type SOD1 transgenic mice (WT) was assessed by the biotin switch method. Increased biotin labeling after ascorbate treatment of samples (+ Ascorbate) is indicative of protein S-nitrosylation. S-Nitrosylation of proteins is detected in the WT but not the G93A mutant SOD1 transgenic spinal cord lysates.

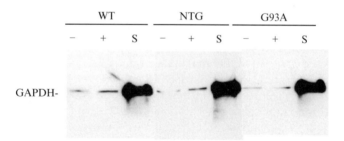

Figure 14.2 GAPDH S-nitrosylation is decreased in the spinal cords of G93A mutant SOD1 transgenic mice. GAPDH S-nitrosylation was assessed by the biotin switch assay in spinal cords obtained from mutant (G93A) SOD1 transgenic mice at the onset of muscle weakness or from age-matched WT SOD1 transgenic (WT) or nontransgenic (NTG) control mice. Levels of biotin-labeled GAPDH in the absence (−) and presence (+) of ascorbate and total levels of GAPDH in the starting lysates (S) of each sample are shown. Increased levels of biotin-labeled GAPDH in samples treated with ascorbic acid are indicative of GAPDH S-nitrosylation. S-Nitrosylated GAPDH is detected in WT transgenic and nontransgenic spinal cords but not in spinal cords obtained from G93A mutant SOD1 transgenic mice.

SOD1 mutants that depletes intracellular SNOs, leading to a disruption in cell signaling pathways regulated by S-nitrosylation and subsequent motor neuron degeneration in ALS.

We found that the biotin switch technique was very useful for analyzing S-nitrosylation levels in mutant SOD1 transgenic mice. A protocol for using the assay to measure total levels of protein S-nitrosylation and/or S-nitrosylation of specific proteins in tissues is provided.

3. Protocol for Analyzing Protein S-Nitrosylation in Tissue Samples Using the Biotin Switch Assay

1. Harvest a tissue of interest as quickly as possible, immediately freeze the tissue in liquid nitrogen, and store in the dark at $-80°C$ to minimize the loss of endogenously S-nitrosylated proteins. The samples should be protected from light during steps 1–4 of the assay.
2. Add 0.5 ml of HEN buffer (250 mM HEPES, pH 7.7, 1 mM EDTA, 0.1 mM neocuproine) per 30 to 50 mg of tissue and homogenize the samples using a Polytron homogenizer.
3. Following homogenization, centrifuge the samples for 10 min at 2000 g at 4°C. Determine the protein concentration of the sample using the MicroBCA kit from Pierce. Dilute the extracts to 0.8 μg/ml with HEN buffer plus 0.4% CHAPS, 1% SDS, and 20 mM MMTS and incubate at 50°C for 20 min with frequent vortexing to block free thiols.
4. Remove excess MMTS by acetone precipitation, and resuspend samples in HEN buffer containing 1% SDS at a protein concentration of 1 to 2 mg/ml. Incubate the samples with 5 mM ascorbate and 0.4 mM biotin-HPDP for 1 h at room temperature. Omit ascorbate or biotin-HPDP from some samples as a control.
5. Perform another acetone precipitation to remove excess biotin-HPDP. Resuspend samples in 300 μl of HEN buffer containing 1% SDS. To this solution, add an additional 600 μl of neutralization buffer (20 mM HEPES, pH 7.7, 0.1 M NaCl, 1 mM EDTA, 0.5% Triton X-100). Add 50 μl of streptavidin-agarose beads to the samples and incubate for 1 h at room temperature while rocking.
6. Wash the samples five times in wash buffer (neutralization buffer plus 600 mM NaCl) followed by one phosphate-buffered saline (PBS) wash.
7. For experiments in which total endogenous S-nitrosylated proteins will be measured, add 50 μl of PBS and 50 μl of 2× SDS sample buffer without 2-mercaptoethanol to the beads. After boiling, perform SDS-PAGE and detect biotinylated proteins by immunoblot analysis using the anti-Biotin-M antibody (Vector Laboratories), Vectastain Elite ABC kit (Vector Laboratories), and ECL reagent (Amersham Pharmacia Biotech). Biotin labeling of proteins that increase in the presence of ascorbic acid is indicative of protein S-nitrosylation.

8. To assess S-nitrosylation of a specific protein, add 50 μl of 2× SDS sample buffer containing 2-mercaptoethanol and 50 μl of elution buffer (20 mM HEPES, pH 7.7, 0.1 M NaCl, 1 mM EDTA, 0.1 M 2-mercaptoethanol) to the beads washed in step 6. After boiling, perform SDS-PAGE and perform an immunoblot of the protein of interest. If treatment of samples with ascorbic acid increases the detection of the protein of interest on immunoblot analysis, this indicates that the protein is S-nitrosylated.

The validity of the biotin switch technique has come under fire from some groups (Gladwin *et al.*, 2006; Huang and Chen, 2006). These investigators present data that false positives can arise from incomplete blocking of free thiols by MMTS or by the ascorbate-dependent reduction of disulfides. These concerns have been addressed by Forrester *et al.* (2007). They present convincing data that ascorbic-dependent false-positive signals are generated when performing the experiment in the presence of indirect sunlight but not when the experimental samples were shielded from sunlight (Forrester *et al.*, 2007). They speculate that ascorbic acid in the presence of indirect sunlight may reduce biotin-HPDP to biotin-SH that, in turn, could integrate into protein disulfides to generate a false-positive SNO-independent S-biotinylation (Forrester *et al.*, 2007). This problem can be avoided by performing the assay in the dark as specified in the original protocol (Jaffrey *et al.*, 2001). It has also been noted that the reduction of disulfide bonds by ascorbic acid is thermodynamically unfavorable (Forrester *et al.*, 2007; Huang and Chen, 2006). A concern has also been raised about whether the biotin switch technique is able to distinguish between S-nitrosylated proteins and S-oxidized or S-glutathionylated proteins (Gladwin *et al.*, 2006). However, several groups have confirmed that the biotin switch assay preferentially detects S-nitrosylated proteins as opposed to S-oxidized or S-glutathionylated proteins (Forrester *et al.*, 2007; Gladwin *et al.*, 2006; Hara *et al.*, 2005).

False negatives may also arise using the biotin switch technique. Ascorbic acid may not fully reduce all S–NO bonds, resulting in a false-negative result. The developers of the assay used 1 mM ascorbic acid, which was sufficient to detect many S-nitrosylated proteins (Jaffrey *et al.*, 2001). A recent report, however, demonstrates that higher ascorbic acid concentrations and/or longer incubation times may be necessary improve the sensitivity of the assay (Zhang *et al.*, 2005).

4. Chemical Reduction/Chemiluminescence

All currently available methods for measuring protein S-nitrosylation, including the biotin switch assay, have potential disadvantages and none is considered the gold standard. Therefore, it is useful to confirm

S-nitrosylation measurements with two separate assays. Another sensitive and specific method for detecting intracellular protein nitrosylation is to immunoprecipitate a protein of interest from cell lysates and then measure the NO content of the immunoprecipitated protein by chemical reduction/chemiluminescence. This is a very useful method to determine if a specific protein is S-nitrosylated or denitrosylated during cell signaling. Several different methods of chemical reduction can be used in this assay (Fang et al., 1998; Mannick et al., 1999; Ruiz et al., 1998; Samouilov and Zweier, 1998). A disadvantage of the chemical reduction/chemiluminescence method is that it requires a nitric oxide analyzer (NOA), which is not available in many laboratories.

For chemical reduction/chemiluminescence experiments, immunoprecipitations should be performed with antibodies directed against the protein of interest or with equal concentrations of isotype-matched control antibodies. Reducing agents must be eliminated from all buffers because reducing agents break S–NO bonds. In addition, metal chelators such as EDTA (0.1 mM) must be added to the buffers to chelate free metals that catalyze the loss and/or gain of S–NO bonds. Immunoprecipitation should be performed in the dark at 4°C, and pH shifts should be avoided, as ultraviolet light and pH shifts will disrupt S–NO bonds. Finally, immunoprecipitated proteins need to be eluted from Sepharose beads prior to chemical reduction/chemiluminescence analysis because the Sepharose beads interfere with the assay. The eluted immunoprecipitated proteins should be analyzed on silver-stained gels to evaluate the concentration of protein in the immunoprecipitates and the purity of the immunoprecipitates. It is important for subsequent S-nitrosylation measurements to confirm that the protein of interest is the only protein immunoprecipitated by its specific antibody and not by a control antibody. If this is not the case, it will be very difficult to determine if the protein of interest or an irrelevant contaminating protein is S-nitrosylated in subsequent chemical reduction/chemiluminescence measurements.

In the chemical reduction/chemiluminescence assay, NO is displaced from S-nitrosylated proteins in the immunoprecipitates by mercuric ion ($HgCl_2$). The released NO reacts rapidly with dissolved oxygen in samples to form the stable end product nitrite (NO_2^-). In the presence of potassium iodide (KI) in acetic acid, NO_2^- is reduced back to NO.

$$I^- + NO_2^- + 2H^+ \rightarrow NO + 1/2 I_2 + H_2O$$

Free NO is detected by chemiluminescence in a NOA (Sievers). In the NOA, NO reacts with ozone, forming excited-state NO2 that emits light (Fang et al., 1998).

5. Protocol for Chemical Reduction/ Chemiluminescence Measurements of S-Nitrosylation of Immunoprecipitated Proteins

Prepare the NOA for S-nitrosylation measurements by adding 5 ml of glacial acetic acid into the purge vessel and purging the vessel with helium or argon gas for several minutes. Then add a freshly made solution of approximately 50 mg of KI in 1 ml of deionized water to the purge vessel. The KI is in excess, so accurate measurement of the quantity is not necessary. Finally, add 100 µl of antifoaming agent to the purge vessel.

When the baseline reading of the NOA has stabilized, generate standard curves using serial twofold dilutions of $NaNO_2$ in water. The standard curves will be used to calculate the concentration of NO in subsequent samples.

Divide the eluted immunoprecipitates of the protein of interest and control immunoprecipitates into two equal aliquots. Add 0.1 M $HgCl_2$ to one aliquot for a final concentration of 2.2 mM, and then incubate for 10 min at room temperature in the dark. The $HgCl_2$ displaces NO from S–NO bonds. Fifty microliters of each sample is then injected into the NOA reaction chamber. The NOA will determine the concentration of NO generated by each sample. The NO signal in the samples that have not been treated with $HgCl_2$ is generated from background nitrite in the samples. The increase in NO signal above background in the $HgCl_2$-treated samples is the SNO-specific signal. The increase in SNO-specific signal in the immunoprecipitate of the protein of interest over the SNO signal generated by the control immunoprecipitate is the SNO-specific signal that can be attributed to the protein of interest.

A variety of other chemical reduction methods can be used to detect protein S-nitrosylation, including the use of a saturated solution of CuCl in 0.1 M cysteine (Fang et al., 1998). The optimal chemical reduction method for detecting protein S-nitrosylation will depend on the protein of interest, so if S-nitrosylation is not detected with one method, an alternate method should be employed.

We have used the chemical reduction/chemiluminescence method to detect S-nitrosylation/denitrosylation of caspase-3 during Fas-induced apoptosis (Mannick et al., 1999). Caspase-3 is a member of the caspase family of cysteine proteases that plays a critical role in the execution of apoptotic cell death. To determine if caspase-3 activity is regulated by endogenous S-nitrosylation, we immunoprecipitated caspase-3 from cell lines before and at various time points after Fas-induced apoptosis (Fig. 14.3A). Our findings indicated that caspase-3 is S-nitrosylated on its catalytic site cysteine in

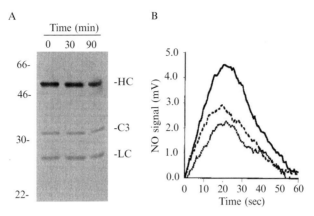

Figure 14.3 Denitrosylation of caspase-3 after Fas activation. (A) Silver stain of caspase-3 immunoprecipitates obtained from 10C9 cells at 0, 30, and 90 min after Fas stimulation. Bands corresponding to caspase-3 (C3), immunoglobulin heavy chain (HC), and light chain (LC) are indicated. Molecular weight markers are shown on the left. (B) Chemical reduction/chemiluminescence measurements of the NO signal generated by caspase-3 immunoprecipitates obtained before (solid line) or 90 min after (dashed line) Fas stimulation of 10C9 cells. The area under the curve is proportional to the NO generated by each sample. Fas stimulation lowered the NO content of the caspase-3 immunoprecipitates to near baseline levels generated by buffer alone (dotted line).

resting cells. S-Nitrosylation of the catalytic site cysteine prevents caspase-3 from being activated in nonapoptotic cells. Upon Fas stimulation, caspase-3 is denitrosylated, allowing the catalytic site to function and apoptosis to proceed (Fig. 14.3B).

Although the chemical reduction/chemiluminescence method is a sensitive and specific method for measuring protein S-nitrosylation, it also has several important limitations. One limitation is that it detects NO derived from nitrite as well as from SNOs. Detection of a SNO-specific NO signal is difficult if the concentration of nitrite in a sample is far higher than the concentration of SNOs. Some investigators add 5% sulfanilamide in 1 N HCl to samples prior to S-nitrosylation measurements in order to reduce nitrite background (Feelisch et al., 2002). However, the addition of sulfanilamide also eliminates signals derived from some S—NO bonds (presumably due to HCl-induced protein denaturation and/or reaction of sulfanilamide with NO^+ equivalents released from SNO bonds). Therefore, we do not pretreat samples with sulfanilamide prior to SNO measurements.

Another limitation of the chemical reduction/chemiluminescence method is that it does not distinguish between a NO signal derived from a protein of interest and a NO signal derived from an irrelevant protein contaminating a sample. Therefore, the protein of interest must be the only protein immunoprecipitated by its specific antibody and not by the control antibody. If a clean immunoprecipitate is not obtainable, then an alternative approach is to determine if site-directed mutagenesis of a cysteine(s) on the protein of interest decreases the NO signal.

A final limitation is that the addition of mercuric chloride may acidify samples, thereby generating a SNO-independent NO signal. Thus, the pH of samples must be monitored after the addition of $HgCl_2$ and buffers adjusted if necessary to avoid significant decreases in pH.

6. Conclusion

Like phosphorylation, S-nitrosylation is a rapidly reversible and precisely targeted posttranslational modification that serves as an on/off switch for protein function during cell signaling. However, unlike phosphorylation, S-nitrosylation is a redox-sensitive and labile protein modification. Therefore, measurement of endogenous protein S-nitrosylation is technically more challenging than measuring stable protein modifications such as phosphorylation. Currently, two of the most reliable methods for detecting protein S-nitrosylation are the biotin switch method and the chemical reduction/chemiluminescence method. In the future, technical advances that facilitate the detection and quantification of protein S-nitrosylation will help us more fully understand the role of S-nitrosylation in signal transduction.

REFERENCES

Broillet, M. C. (2000). A single intracellular cysteine residue is responsible for the activation of the olfactory cyclic nucleotide-gated channel by NO. *J. Biol. Chem.* **275,** 15135–15141.

Choi, Y. B., Tenneti, L., Le, D. A., Ortiz, J., Bai, G., Chen, H. S., and Lipton, S. A. (2000). Molecular basis of NMDA receptor-coupled ion channel modulation by S-nitrosylation. *Nat. Neurosci.* **3,** 15–21.

Chung, K. K., Thomas, B., Li, X., Pletnikova, O., Troncoso, J. C., Marsh, L., Dawson, V. L., and Dawson, T. M. (2004). S-nitrosylation of parkin regulates ubiquitination and compromises parkin's protective function. *Science* **304,** 1328–1331.

Derakhshan, B., Hao, G., and Gross, S. S. (2007). Balancing reactivity against selectivity: The evolution of protein S-nitrosylation as an effector of cell signaling by nitric oxide. *Cardiovasc. Res.* **75,** 210–219.

Eu, J. P., Sun, J., Xu, L., Stamler, J. S., and Meissner, G. (2000). The skeletal muscle calcium release channel: Coupled O_2 sensor and NO signaling functions. *Cell* **102,** 499–509.

Fang, K., Ragsdale, N. V., Carey, R. M., MacDonald, T., and Gaston, B. (1998). Reductive assays for S-nitrosothiols: Implications for measurements in biological systems. *Biochem. Biophys. Res. Commun.* **252,** 535–540.

Feelisch, M., Rassaf, T., Mnaimneh, S., Singh, N., Bryan, N. S., Jourd'Heuil, D., and Kelm, M. (2002). Concomitant S-, N-, and heme-nitros(yl)ation in biological tissues and fluids: Implications for the fate of NO *in vivo. FASEB J* **16,** 1775–1785.

Forrester, M. T., Foster, M. W., and Stamler, J. S. (2007). Assessment and application of the biotin switch technique for examining protein S-nitrosylation under conditions of pharmacologically induced oxidative stress. *J. Biol. Chem.* **282,** 13977–13983.

Gladwin, M. T., Wang, X., and Hogg, N. (2006). Methodological vexation about thiol oxidation versus S-nitrosation: A commentary on "An ascorbate-dependent artifact that interferes with the interpretation of the biotin-switch assay". *Free Radic. Biol. Med.* **41,** 557–561.

Gu, Z., Kaul, M., Yan, B., Kridel, S. J., Cui, J., Strongin, A., Smith, J. W., Liddington, R. C., and Lipton, S. A. (2002). S-nitrosylation of matrix metalloproteinases: Signaling pathway to neuronal cell death. *Science* **297,** 1186–1190.

Haendeler, J., Hoffmann, J., Tischler, V., Berk, B. C., Zeiher, A. M., and Dimmeler, S. (2002). Redox regulatory and anti-apoptotic functions of thioredoxin depend on S-nitrosylation at cysteine 69. *Nat. Cell Biol.* **4,** 743–749.

Hara, M. R., Agrawal, N., Kim, S. F., Cascio, M. B., Fujimuro, M., Ozeki, Y., Takahashi, M., Cheah, J. H., Tankou, S. K., Hester, L. D., Ferris, C. D., Hayward, S. D., *et al.* (2005). S-nitrosylated GAPDH initiates apoptotic cell death by nuclear translocation following Siah1 binding. *Nat. Cell Biol.* **7,** 665–674.

Hess, D. T., Matsumoto, A., Kim, S. O., Marshall, H. E., and Stamler, J. S. (2005). Protein S-nitrosylation: Purview and parameters. *Nat. Rev. Mol. Cell Biol.* **6,** 150–166.

Huang, B., and Chen, C. (2006). An ascorbate-dependent artifact that interferes with the interpretation of the biotin switch assay. *Free Radic. Biol. Med.* **41,** 562–567.

Jaffrey, S. R., Erdjument-Bromage, H., Ferris, C. D., Tempst, P., and Snyder, S. H. (2001). Protein S-nitrosylation: A physiological signal for neuronal nitric oxide. *Nat. Cell Biol.* **3,** 193–197.

Johnson, M. A., Macdonald, T. L., Mannick, J. B., Conaway, M. R., and Gaston, B. (2001). Accelerated S-nitrosothiol breakdown by amyotrophic lateral sclerosis mutant copper, zinc-superoxide dismutase. *J. Biol. Chem.* **276,** 39872–39878.

Kim, J. E., and Tannenbaum, S. R. (2004). S-Nitrosation regulates the activation of endogenous procaspase-9 in HT-29 human colon carcinoma cells. *J. Biol. Chem.* **279,** 9758–9764.

Lander, H. M., Hajjar, D. P., Hempstead, B. L., Mirza, U. A., Chait, B. T., Campbell, S., and Quilliam, L. A. (1997). A molecular redox switch on p21(ras): Structural basis for the nitric oxide-p21(ras) interaction. *J. Biol. Chem.* **272,** 4323–4326.

Li, S., and Whorton, A. R. (2003). Regulation of protein tyrosine phosphatase 1B in intact cells by S-nitrosothiols. *Arch. Biochem. Biophys.* **410,** 269–279.

Mannick, J. B., Hausladen, A., Liu, L., Hess, D. T., Zeng, M., Miao, Q. X., Kane, L. S., Gow, A. J., and Stamler, J. S. (1999). Fas-induced caspase denitrosylation. *Science* **284,** 651–654.

Mannick, J. B., Schonhoff, C., Papeta, N., Ghafourifar, P., Szibor, M., Fang, K., and Gaston, B. (2001). S-Nitrosylation of mitochondrial caspases. *J. Cell Biol.* **154,** 1111–1116.

Marshall, H. E., and Stamler, J. S. (2001). Inhibition of NF-kappa B by S-nitrosylation. *Biochemistry* **40,** 1688–1693.

Matsushita, K., Morrell, C. N., Cambien, B., Yang, S. X., Yamakuchi, M., Bao, C., Hara, M. R., Quick, R. A., Cao, W., O'Rourke, B., Lowenstein, J. M., Pevsner, J., *et al.* (2003). Nitric oxide regulates exocytosis by S-nitrosylation of N-ethylmaleimide-sensitive factor. *Cell* **115,** 139–150.

Palmer, L. A., Gaston, B., and Johns, R. A. (2000). Normoxic stabilization of hypoxia-inducible factor-1 expression and activity: Redox-dependent effect of nitrogen oxides. *Mol. Pharmacol.* **58,** 1197–1203.

Park, H. S., Huh, S. H., Kim, M. S., Lee, S. H., and Choi, E. J. (2000). Nitric oxide negatively regulates c-Jun N-terminal kinase/stress-activated protein kinase by means of S-nitrosylation. *Proc. Natl. Acad. Sci. USA* **97,** 14382–14387.

Pawloski, J. R., Hess, D. T., and Stamler, J. S. (2005). Impaired vasodilation by red blood cells in sickle cell disease. *Proc. Natl. Acad. Sci. USA* **102,** 2531–2536.

Que, L. G., Liu, L., Yan, Y., Whitehead, G. S., Gavett, S. H., Schwartz, D. A., and Stamler, J. S. (2005). Protection from experimental asthma by an endogenous bronchodilator. *Science* **308,** 1618–1621.

Ruiz, F., Corrales, F. J., Miqueo, C., and Mato, J. M. (1998). Nitric oxide inactivates rat hepatic methionine adenosyltransferase *in vivo* by S-nitrosylation. *Hepatology* **28,** 1051–1057.

Samouilov, A., and Zweier, J. L. (1998). Development of chemiluminescence-based methods for specific quantitation of nitrosylated thiols. *Anal. Biochem.* **258,** 322–330.

Schonhoff, C. M., Daou, M. C., Jones, S. N., Schiffer, C. A., and Ross, A. H. (2002). Nitric oxide-mediated inhibition of Hdm2-p53 binding. *Biochemistry* **41,** 13570–13574.

Schonhoff, C. M., Matsuoka, M., Tummala, H., Johnson, M. A., Estevez, A. G., Wu, R., Kamaid, A., Ricart, K. C., Hashimoto, Y., Gaston, B., Macdonald, T. L., Xu, Z., *et al.* (2006). S-nitrosothiol depletion in amyotrophic lateral sclerosis. *Proc. Natl. Acad. Sci. USA* **103,** 2404–2409.

Zhang, Y., Keszler, A., Broniowska, K. A., and Hogg, N. (2005). Characterization and application of the biotin-switch assay for the identification of S-nitrosated proteins. *Free Radic. Biol. Med.* **38,** 874–881.

CHAPTER FIFTEEN

PIVOTAL ROLE OF ARACHIDONIC ACID IN THE REGULATION OF NEURONAL NITRIC OXIDE SYNTHASE ACTIVITY AND INDUCIBLE NITRIC OXIDE SYNTHASE EXPRESSION IN ACTIVATED ASTROCYTES

Orazio Cantoni,[*] Letizia Palomba,[*] Tiziana Persichini,[†] Sofia Mariotto,[‡] Hisanori Suzuki,[‡] *and* Marco Colasanti[†]

Contents

1. Introduction	244
2. Materials and Methods	245
2.1. Cell culture and treatment conditions	245
2.2. Nitric oxide detection system	245
2.3. Statistical analysis	246
3. Results and Discussion	246
Acknowledgments	250
References	250

Abstract

Astrocytes respond to agents leading to progressively greater increases in the intracellular concentration of Ca^{2+} ($[Ca^{2+}]_i$) with a linear release of arachidonic acid (ARA), due to activation of cytosolic phospholipase A_2, and with a bell-shaped curve of nitric oxide (NO) release, due to Ca^{2+}-dependent activation/inhibition of neuronal NO synthase (nNOS). Inhibition of nNOS is mediated by a signaling driven by ARA, either extensively released at high $[Ca^{2+}]_i$ or supplemented to the cultures at nanomolar levels. Proinflammatory factors, as bacterial lipopolysaccharide/interferon-γ, cause rapid ARA-dependent nNOS inhibition, critical for the delayed expression of nuclear factor-κB (NF-κB)-dependent genes as inducible NOS. We therefore propose that the onset of

---

[*] Istituto di Farmacologia e Farmacognosia, Università di Urbino "Carlo Bo," Urbino, Italy
[†] Dipartimento di Biologia, Università di Roma Tre, Rome, Italy
[‡] Dipartimento di Scienze Neurologiche e della Visione, Sezione di Chimica Biologica, Università degli Studi di Verona, Verona, Italy

Methods in Enzymology, Volume 440
ISSN 0076-6879, DOI: 10.1016/S0076-6879(07)00815-4

the neuroinflammatory response is strictly regulated by the relative amounts of NO and ARA produced by their constitutive enzymes. In particular, the inflammatory product ARA initiates the inflammatory response via inhibition of nNOS, thereby allowing NF-κB activation. Astrocytes contribute to the regulation of this process by producing both constitutive NO and ARA, as well as by expressing NF-κB-dependent genes.

1. INTRODUCTION

In the nervous system, activation of cell surface receptors triggers the Ca^{2+}-dependent stimulation of neuronal nitric oxide synthase (nNOS), resulting in the release of low levels of nitric oxide (NO). This basal NO pool, further implemented by endothelial NO synthase (eNOS)-generated NO, regulates an array of physiological functions (Bredt and Snyder, 1992; Moncada et al., 1989; Yun et al., 1997) and is also critically involved in the prevention of neurovascular inflammation (Endres et al., 1998; Zhang and Iadecola, 1994). The latter effect is of pivotal importance, as inflammation associated with oxidative/nitrosative stress is both a common feature of stroke (Senes et al., 2007) and numerous neurodegenerative diseases (Moncada and Bolanos, 2006) and is heavily involved in the pathogenesis/severity of these conditions. Under this perspective, nNOS- and, possibly, eNOS-derived NO would be neuroprotective since able to prevent neuroinflammation, most likely via the inhibition of nuclear factor-κB (NF-κB), a transcription factor regulating the expression of an array of inflammatory genes (Colasanti et al.,1995; Togashi et al., 1997), including inducible NO synthase (iNOS or type II NOS).

The notion that nanomolar levels of NO inhibit NF-κB is documented by numerous studies (Colasanti and Persichini, 2000; Kroncke, 2003), thereby suggesting that the onset of the neuroinflammatory response requires prior depletion of the NO pool, critically regulated by the activity of n/eNOS. Hence proinflammatory factors, such as bacterial lipopolysaccharide (LPS) and/or different cytokines [e.g., interferon (IFN)-γ], should cause rapid inhibition of constitutive NOS in order to evoke a delayed expression of NF-κB-dependent genes. This hypothesis has been clearly validated in cultured cells (Colasanti and Suzuki, 2000) and by in vivo studies documenting the disappearance of NO at the early phases of the inflammatory response (Pahl, 1999; Yamamoto and Gaynor, 2001).

In the brain, NO from constitutive NOS derives primarily from astrocytes and neurons, although it can be released by these cells and eventually reach other cell types, such as the microglia. NO may therefore promote NF-κB inhibition within the same cell in which it has been generated (e.g., astrocytes), as well as in cells devoid of constitutive NOS activity (e.g., microglia). Thus, basal levels of NO (i.e., nNOS-derived NO) are critical for preventing the activation of NF-κB in glial cells, and proinflammatory stimuli suppress nNOS activity in order to elevate the activity of this transcription factor.

This chapter briefly summarizes some of the work from our laboratory that led to identification of the mechanism(s) involved in the suppression of nNOS activity in activated astrocytes.

2. Materials and Methods

2.1. Cell culture and treatment conditions

Primary cultures of cortical astrocytes, derived from 1- or 2-day-old Sprague–Dawley rats (Charles River, Calco, Italy), are prepared as described previously (Rose et al., 1992). Dissected cerebral cortices are incubated for 30 min (37°) with media stock (MS) (Eagle's minimal essential medium with 2.66 g/liter sodium bicarbonate, 4.22 g/liter glucose) containing 0.25% trypsin (Sigma, Milan, Italy). After centrifugation (1000g for 8 min), digested cortices are rinsed with MS supplemented with 10% horse serum, 10% fetal bovine serum (HyClone Laboratories, Logan, UT), penicillin (100 U/ml), and streptomycin (100 μg/ml) (HyClone) and dissociated mechanically with a flame-narrowed Pasteur pipette (the bore size is reduced to one-third of normal). Cells are next seeded in 35-mm Falcon Primaria (VWR, Milan, Italy) dishes with MS supplemented as indicated earlier and allowed to grow for 2 weeks. Cultures are then shaken in an orbital shaker for approximately 8 h (120g) to remove other cell types, such as microglia and oligodentrocytes, thereby leading to a pure astrocyte preparation, as progenitor cells at this age have lost the ability to differentiate into neuronal cells (Dugan et al., 1995). The purity of cultures is confirmed by immunocytochemistry using antibodies against the glial fibrillary acidic protein. At this treatment stage, the total cell number is between 1.0 and 1.5 $\times$ 10^5 cells/dish.

2.2. Nitric oxide detection system

The production of NO is assayed using 4,5-diaminofluorescein diacetate (DAF-2DA), as described by Lopez-Figueroa et al. (2000). DAF-2DA is a nonfluorescent membrane-permeant probe converted within the cell to DAF-2 by cytosolic esterases. DAF-2, also devoid of intrinsic fluorescence, reacts with NO in the presence of O$_2$ to produce DAF-2T, a triazole fluorescent derivative. Thus, the onset of a fluorescence response is indicative of NO formation.

In our experiments, cells are first loaded for 20 min (at 37°) with 10 μM DAF-2DA in saline A (8.182 g/liter NaCl, 0.372 g/liter KCl, 0.336 g/liter NaHCO$_3$, and 0.9 g/liter glucose), rinsed with the same saline, treated as indicated in the text, and finally analyzed as detailed later. In some experiments, the cells are first treated and DAF-2DA is added to the cultures in the last 10 min of incubation. Note that an effective and reproducible DAF-2DA loading requires accurate washing procedures [twice with either saline A or phosphate-buffered saline (PBS)] to remove bovine serum albumin and phenol red, which may affect fluorescence and loading of the dye. In addition, because of the light sensitivity of the probe, it is advisable to work in a low-light environment when handling DAF-2DA or stained samples.

After treatments, cells are washed twice with PBS and glass coverslips are put onto dishes to prevent dehydration of the cells. Cellular fluorescence is then imaged with a BX-51 fluorescence microscope (Olympus Italia, Milan, Italy) equipped with a SPOT-RT camera unit (Diagnostic Instruments, Delta Sistemi, Rome, Italy). The excitation and emission wavelengths are 495 and 515 nm, respectively. To prevent photobleaching, the excitation light provided by a 100-W mercury lamp is attenuated by a neutral density filter (U-25ND6, Olympus). An Olympus LCAch 40× objective is used in all experiments. Images are collected with exposure times of 100 to 400 ms, digitally acquired, and processed for fluorescence determination at the single cell level on a personal computer using Scion Image software (Scion Corp., Frederick, MD). Mean fluorescence values are determined by averaging the fluorescence values of at least 50 cells/treatment condition/experiment.

2.3. Statistical analysis

Statistical analysis of data for multiple comparison is performed by ANOVA followed by a Dunnett's test.

3. Results and Discussion

Agents affecting the intracellular concentration of Ca^{2+} ($[Ca^{2+}]_i$), as the ionophore A2387 or the sarcoplasmic/endoplasmic reticulum Ca^{2+} ATPase blocker thapsigargin, produce biphasic effects in nNOS expressing cells preloaded with the NO-sensitive probe DAF-2DA (Palomba et al., 2004a). At low concentrations, these agents promote a significant fluorescence response that declines, and eventually disappears, by progressively increasing the concentrations of the Ca^{2+}-mobilizing agents. We first described these responses in PC12 cells, a rat pheochromocytoma cell line expressing very low levels of nNOS, undetectable by either immunocytochemical or

Western blot assays, in NIE-115 neuronal cells and in C6 glioma cells (Palomba *et al.*, 2004a).

Figure 15.1A shows the bell-shaped fluorescence response mediated by increasing concentrations of A23187 in rat astrocytes. Figure 15.1B provides evidence indicating that fluorescence elicited by 2.5 μM A23187 is abolished by the NOS inhibitor N^{ω}-nitro-L-arginine methyl ester

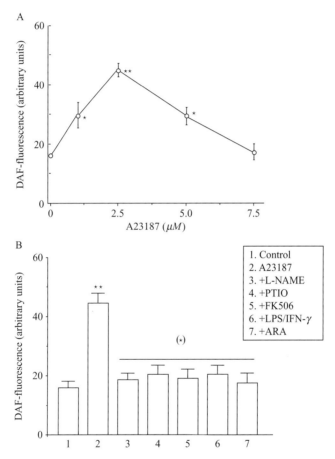

Figure 15.1 Ca^{2+} dependence of nNOS-derived NO release in astrocytes. DAF-2DA-preloaded astrocytes were treated for 15 min with increasing concentrations of A23187 (A) or with 2.5 μM A23187 alone or associated with L-NAME (1 mM), PTIO (50 μM), FK506 (1 μM), LPS (1 μg/ml)/IFN-γ (1000 U/ml), or ARA (0.1 μM) (B). Agents affecting NO release were added to the cultures 5 min prior to A23187 and left during ionophore exposure. After treatments, cells were immediately analyzed for the assessment of the DAF-2 fluorescence response, as detailed in the text. Data points are the means ± SEM from three to five separate experiments, each performed in duplicate. ★$P < 0.05$ and ★★$P < 0.01$ vs control; (★)$P < 0.01$ vs A23187-treated cells (one-way ANOVA followed by Dunnett's test).

(L-NAME), the NO scavenger 2-phenyl-4,4,5,5-tetramethylimidazoline-1-oxyl 3-oxide (PTIO), and FK506, a phosphatase (calcineurin) inhibitor promoting NOS hyperphosphorylation associated with inhibition of activity (Dawson et al., 1993; Sabatini et al., 1997). Hence, the fluorescence response is attributable to the formation of NO.

Collectively, these results, in addition to establishing the DAF-2DA assay as a sensitive and reliable technique for the assessment of constitutive NOS activity, imply that the $[Ca^{2+}]_i$, critical for activation of the Ca^{2+}/calmodulin-dependent enzyme nNOS (Alderton et al., 2001), may in fact also promote its inhibition. In particular, high levels of Ca^{2+} would appear to trigger some events leading to the suppression of nNOS activity. This notion is further confirmed using astrocytes exposed to proinflammatory stimuli, as LPS/IFN-γ, promoting an early increase in the $[Ca^{2+}]_i$ (Hoffmann et al., 2003) associated with nNOS inhibitory signaling (Palomba et al., 2004b). LPS/IFN-γ indeed suppresses NO formation induced by A231287 (Fig. 15.1B).

It therefore appears that the nNOS inhibitory signaling is mediated by a Ca^{2+}-dependent mechanism driven by high $[Ca^{2+}]_i$. Low levels of A23187 (i.e., promoting low $[Ca^{2+}]_i$) stimulate nNOS activity, however, inhibited by high levels of A23187 (i.e., promoting high $[Ca^{2+}]_i$). Under this perspective, LPS/IFN-γ, would fail to promote activation and in fact causes nNOS inhibition because of the high $[Ca^{2+}]_i$ evoked by the proinflammatory stimulus.

Additional studies from our laboratory revealed that contrary to nNOS, exhibiting a bell-shaped curve of activation at increasing $[Ca^{2+}]_i$, another Ca^{2+}-dependent enzyme, cytosolic phospholipase A_2 (cPLA$_2$), releases arachidonic acid (ARA) as a direct function of the $[Ca^{2+}]_i$ (Palomba et al., 2004a). The lipid messenger turned out to be the trigger of the nNOS inhibitory signaling. The notion that critical levels of ARA are necessary to elicit this response is consistent with the observed suppression of NO release mediated by nanomolar levels of ARA in A23187-stimulated cells (Fig. 15.1B). Also consistent with the pivotal role of ARA in nNOS inhibition are findings indicating that pharmacological inhibition of cPLA$_2$, or knock down of the enzyme, was invariably associated with maximal NO formation after exposure to levels of A23187 otherwise causing nNOS inactivation (Palomba et al., 2004a). Fig. 15.2B shows representative micrographs of DAF-2DA-preloaded astrocytes incubated for 15 min with a high concentration of A23187 and AACOCF$_3$ (untreated cells in Fig. 15.A). A23187, or AACOCF$_3$, failed to promote effects when supplemented separately to the cultures (not shown), and the response mediated by their combination was abolished by ARA (Fig. 15.2C). AACOCF$_3$ is an inhibitor of both cPLA$_2$ and Ca^{2+}-independent PLA$_2$ with hardly any effect against the low molecular weight, Ca^{2+}-dependent isoforms (Ackermann et al., 1995). Under the aforementioned conditions, as detailed extensively in previous studies (Palomba et al., 2004a,b), AACOCF$_3$, however, promotes

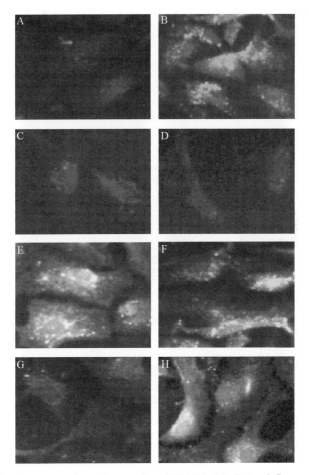

Figure 15.2 Representative micrographs of DAF 2-DA-derived fluorescence from nNOS/iNOS-released NO. DAF-2-DA-preloaded astrocytes were exposed for 15 min to a concentration (7.5 μM) of A23187 failing to promote fluorescence (not shown). The effect of the same treatment associated with $AACOCF_3$ (75 μM) is shown in B. Images from untreated cells are shown in A. The fluorescence response obtained in B was inhibited by 0.1 μM ARA (C). $AACOCF_3$ and ARA were added to the cultures 5 min prior to A23187 and left during ionophore exposure. Also shown is the effect of a 15-min exposure to LPS/IFN-γ alone (D) or associated with $AACOCF_3$ (E). Images shown in the remaining panels were from cells exposed for 8 h to LPS/IFN-γ alone (F), to associated $AACOCF_3$ (G), or to their combination plus ARA (H). In all of these experiments, $AACOCF_3$ and ARA were added to the cultures 5 min prior to LPS/IFN-γ and left during the entire incubation time. After treatments, cells were analyzed for DAF-2DA-derived fluorescence as detailed in the text.

its effects via selective $cPLA_2$ inhibition. It should also be noted that use of a rather high concentration of $AACOCF_3$ is justified by its poor uptake in astrocytes (Palomba *et al.*, 2004b).

Figure 15.2D shows that a 15-min exposure to LPS/IFN-γ does not promote DAF-2DA-derived fluorescence however promptly detected after AACOCF$_3$ supplementation (Fig. 15.2E). Obviously this is only a potential effect due to the increase in the [Ca^{2+}]$_i$ but is nevertheless never observed because it is prevented by parallel ARA-dependent NOS inhibitory signaling. Thus, proinflammatory stimuli in astrocytes promote an early increase in [Ca^{2+}]$_i$ associated with a rapid and robust activation of cPLA$_2$, causing an immediate ARA-dependent inhibition of nNOS. The drop in the NO concentration is critical for the ensuing activation of NF-κB and expression of inflammatory genes such as iNOS. Figure 15.2F shows that indeed LPS/IFN-γ caused delayed NO formation via a mechanism sensitive to AACOCF$_3$ (Fig. 15.2G) and promptly reestablished by the addition of nanomolar levels of exogenous ARA (Fig. 15.2H). Note that astrocytes were exposed to LPS/IFN-γ for 8 h and that DAF-2DA was added to the cultures in the last 10 min. Hence, the DAF-2DA assay can also be conveniently employed to measure iNOS-derived NO over a limited time window.

In conclusion, the aforementioned results provide evidence for cross talk between ARA, released from cPLA$_2$, and NO, released from nNOS, in the regulation of neuroinflammation. We speculate that, under normal conditions, transient increases in the [Ca^{2+}]$_i$ in astrocytes and neurons (or endothelial cells) promote the release of NO, thereby causing inhibition of NF-κB in glial cells, e.g., astrocytes and microglia. The Ca^{2+} signal also evokes the release of ARA from cPLA$_2$ in different brain cell types, and the lipid messenger, as NO, can diffuse freely from one cell to another. Proinflammatory stimuli remarkably affect this equilibrium based on the low Ca^{2+} release by remarkably elevating the [Ca^{2+}]$_i$, thereby overstimulating cPLA$_2$ activity in cells types expressing the appropriate receptors, e.g., astrocytes or microglia. Under these conditions, ARA suppresses nNOS activity in astrocytes/neurons, and the ensuing drop of the NO pool creates favorable conditions for the activation of NF-κB in glial cells.

ACKNOWLEDGMENTS

This work was supported by grants from the Associazione Italiana per la Ricerca sul Cancro (OC) and from Ministero dell'Università e della Ricerca Scientifica e Tecnologica, Progetti di Interesse Nazionale (OC, MC, HS).

REFERENCES

Ackermann, E. J., Conde-Freiboes, K., and Dennis, E. A. (1995). Inhibition of macrophage Ca$^{(2+)}$-independent phospholipase A$_2$ by bromoenol lactone and trifluoromethyl ketones. *J. Biol. Chem.* **270,** 445–450.

Alderton, W. K., Cooper, C. E., and Knowles, R. G. (2001). Nitric oxide synthases: Structure, function and inhibition. *Biochem. J.* **357,** 593–615.

Bredt, D. S. and Snyder, S. H. (1992). Nitric oxide, a novel neuronal messenger. *Neuron* **8**, 3–11.

Colasanti, M. and Persichini, T. (2000). Nitric oxide: An inhibitor of NF-κB/Rel system in glial cells. *Brain Res. Bull.* **52**, 155–161.

Colasanti, M., Persichini, T., Menegazzi, M., Mariotto, S., Giordano, E., Caldarera, C. M., Sogos, V., Lauro, G. M., and Suzuki, H. (1995). Induction of nitric oxide synthase mRNA expression: Suppression by exogenous nitric oxide. *J. Biol. Chem.* **270**, 26731–26733.

Colasanti, M. and Suzuki, H. (2000). The dual personality of NO. *Trends Pharmacol. Sci.* **21**, 249–252.

Dawson, T. M., Steiner, J. P., Dawson, V. L., Dinerman, J. L., Uhl, G. R., and Snyder, S. H. (1993). Immunosuppressant FK506 enhances phosphorylation of nitric oxide synthase and protects against glutamate neurotoxicity. *Proc. Natl. Acad. Sci. USA* **90**, 9808–9812.

Dugan, L. L., Bruno, V. M., Amagasu, S. M., and Giffard, R. G. (1995). Glia modulate the response of murine cortical neurons to excitotoxicity: Glia exacerbate AMPA neurotoxicity. *J. Neurosci.* **15**, 4545–4555.

Endres, M., Laufs, U., Huang, Z., Nakamura, T., Huang, P., Moskowitz, M. A., and Liao, J. K. (1998). Stroke protection by 3-hydroxy-3-methylglutaryl (HMG)-CoA reductase inhibitors mediated by endothelial nitric oxide synthase. *Proc. Natl. Acad. Sci. USA* **95**, 8880–8885.

Hoffmann, A., Kann, O., Ohlemeyer, C., Hanisch, U. K., and Kettenmann, H. (2003). Elevation of basal intracellular calcium as a central element in the activation of brain macrophages (microglia): Suppression of receptor-evoked calcium signaling and control of release function. *J. Neurosci.* **23**, 4410–4419.

Kroncke, K. D. (2003). Nitrosative stress and transcription. *Biol. Chem.* **384**, 1365–1377.

Lopez-Figueroa, M. O., Day, H. E., Lee, S., Rivier, C., Akil, H., and Watson, S. J. (2000). Temporal and anatomical distribution of nitric oxide synthase mRNA expression and nitric oxide production during central nervous system inflammation. *Brain Res.* **852**, 239–246.

Moncada, S. and Bolanos, J. P. (2006). Nitric oxide, cell bioenergetics and neurodegeneration. *J. Neurochem.* **97**, 1676–1689.

Moncada, S., Palmer, R. M., and Higgs, E. A. (1989). Biosynthesis of nitric oxide from L-arginine: A pathway for the regulation of cell function and communication. *Biochem. Pharmacol.* **38**, 1709–1715.

Pahl, H. L. (1999). Activators and target genes of Rel/NF-κB transcription factors. *Oncogene* **18**, 6853–6866.

Palomba, L., Bianchi, M., Persichini, T., Magnani, M., Colasanti, M., and Cantoni, O. (2004a). Downregulation of nitric oxide formation by cytosolic phospholipase A$_2$-released arachidonic acid. *Free Radic. Biol. Med.* **36**, 319–329.

Palomba, L., Persichini, T., Mazzone, V., Colasanti, M., and Cantoni, O. (2004b). Inhibition of nitric-oxide synthase-I (NOS-I)-dependent nitric oxide production by lipopolysaccharide plus interferon-γ is mediated by arachidonic acid: Effects on NF-κB activation and late inducible NOS expression. *J. Biol. Chem.* **279**, 29895–29901.

Rose, K., Goldberg, M. P., and Choi, D. V. (1992). Cytotoxicity in murine neocortical cell culture. *In* "Methods in Toxicology" (C. A. Tyson and J. M. Frazier, eds.), Vol. 1, pp. 46–60. Academic Press, New York.

Sabatini, D. M., Lai, M. M., and Snyder, S. H. (1997). Neuronal roles of immunophilins and their ligands. *Mol. Neurobiol.* **15**, 223–239.

Senes, M., Kazan, N., Coskun, O., Zengi, O., Inan, L., and Yucel, D. (2007). Oxidative and nitrosative stress in acute ischaemic stroke. *Ann. Clin. Biochem.* **44**, 43–47.

Togashi, H., Sasaki, M., Frohman, E., Taira, E., Ratan, R. R., Dawson, T. M., and Dawson, V. L. (1997). Neuronal (type I) nitric oxide synthase regulates nuclear factor κB activity and immunologic (type II) nitric oxide synthase expression. *Proc. Natl. Acad. Sci. USA* **94,** 2676–2680.

Yamamoto, Y. and Gaynor, R. B. (2001). Therapeutic potential of inhibition of the NF-κB pathway in the treatment of inflammation and cancer. *J. Clin. Invest.* **107,** 135–142.

Yun, H. Y., Dawson, V. L., and Dawson, T. M. (1997). Nitric oxide in health and disease of the nervous system. *Mol. Psychiatry* **2,** 300–310.

Zhang, F. and Iadecola, C. (1994). Reduction of focal cerebral ischemic damage by delayed treatment with nitric oxide donors. *J. Cereb. Blood Flow Metab.* **14,** 574–580.

CHAPTER SIXTEEN

RED BLOOD CELLS AS A MODEL TO DIFFERENTIATE BETWEEN DIRECT AND INDIRECT OXIDATION PATHWAYS OF PEROXYNITRITE

Maurizio Minetti,* Donatella Pietraforte,* Elisabetta Straface,[†] Alessio Metere,* Paola Matarrese,[†] *and* Walter Malorni[†]

Contents

1. Introduction	254
2. Direct Reactions of Peroxynitrite with Biological Targets	255
3. Peroxynitrite Homolysis: *Indirect* Radical Chemistry	256
4. Red Blood Cells as an Experimental Model to Test the Fate of Peroxynitrite in a Biological Environment	257
5. Red Blood Cell Modifications Induced by Extracellular Peroxynitrite Decay	260
6. Red Blood Cell Modifications Induced by Intracellular Peroxynitrite Decay	260
7. Peroxynitrite-Dependent Phosphorylation Signaling of RBC	261
8. Peroxynitrite-Induced Biomarkers of RBC Senescence	262
9. Peroxynitrite-Induced Biomarkers of RBC Apoptosis	264
10. Methods	265
10.1. Peroxynitrite oxidation of RBCs	265
10.2. Red blood cell structural alterations	265
10.3. Red blood cell morphometric analysis	266
10.4. Detection of glycophorins A, C, and CD47	266
10.5. Detection of band 3	266
10.6. Glutathione content	267
10.7. Apoptosis evaluation by phosphatidylserine detection	267
10.8. Caspase activity	267
11. Data and Statistics	268
11.1. Flow cytometry data	268
11.2. Statistics	269

* Departments of Cell Biology and Neurosciences, Istituto Superiore di Sanità, Rome, Italy
[†] Drug Research and Evaluation, Istituto Superiore di Sanità, Rome, Italy

Methods in Enzymology, Volume 440   © 2008 Elsevier Inc.
ISSN 0076-6879, DOI: 10.1016/S0076-6879(07)00816-6   All rights reserved.

Acknowledgments 269
References 269

Abstract

Red blood cells are the major physiological scavengers of reactive nitrogen species and have been proposed as real-time biomarkers of some vascular-related diseases. This chapter proposes that the erythrocyte is a suitable cell model for studying the modifications induced by peroxynitrite. Peroxynitrite decays both extra- and intracellularly as a function of cell density and CO_2 concentration, inducing the appearance of distinct cellular biomarkers, as well as the modulation of signaling and metabolism. Intracellular oxidations are due mostly to direct reactions of peroxynitrite with hemoglobin but also lead to the appearance of apoptotic biomarkers. Surface/membrane oxidations are due principally to indirect radical reactions generated by CO_2-catalyzed peroxynitrite homolysis.

1. Introduction

The discovery that superoxide anion ($O_2^{\bullet-}$) and nitric oxide ($^{\bullet}NO$) are the main radical species utilized as transducer molecules to regulate differentiation, apoptosis, cell–cell communication, vasodilation, and so on has revolutionized our concept of the toxicity of free radicals. Because these molecules are, in effect, relatively mild oxidants, they can be safely handled by molecules and their levels finely tuned to achieve redox regulation of cell processes. Nevertheless, an excessive or uncontrolled production of these radicals can setup a plethora of unwanted pathways involved in cell transformation, aging, and degeneration. This scenario further changed after the proposition of Beckman and co-workers (1990) that tissue injury was due in part to the extremely rapid reaction of $^{\bullet}NO$ with $O_2^{\bullet-}$ to produce peroxynitrite. Peroxynitrite is a strong oxidant and reacts directly with several cellular targets, including thiols (Quijano et al., 1997; Radi et al., 1991) and hemoproteins (Exner and Herold, 2000; Floris et al., 1993; Minetti et al., 2000; Thomson et al., 1995). These *direct* reactions are biologically relevant, especially with substrates present at high concentrations in tissues, as in the case of CO_2 (1.3 mM at pH 7.4). Indeed, CO_2 is the species that reacts with peroxynitrite to form about 35% of a geminate pair of radicals, $CO_3^{\bullet-}$ and $^{\bullet}NO_2$, and 65% NO_3^- through the homolysis of the O–O bond (Bonini et al., 1999; Goldstein et al., 2005; Lymar and Hurst, 1998). The CO_2-catalyzed homolysis of peroxynitrite outcompetes many other reactions and is a relevant *indirect* peroxynitrite oxidation pathway (Augusto et al., 2002). A second effect of CO_2 is the reduction in the half-life of peroxynitrite from about 1–2 to 0.01–0.02 s (Radi, 1998). On account of the shorter half-life of $CO_3^{\bullet-}$ and $^{\bullet}NO_2$, CO_2 thus drastically limits the diffusion distance and cell-killing activity of peroxynitrite (Alvarez *et al.*, 2002; Lymar and Hurst, 1996; Zhu *et al.*, 1992).

Red blood cells (RBCs) are the major scavengers of peroxynitrite in blood, and the formation of metHb clearly indicates that it can diffuse into the cell (Romero and Radi, 2005; Romero *et al.*, 1999).

After briefly reviewing the *direct* and *indirect* mechanisms of peroxynitrite-mediated oxidations, this chapter outlines the experiments that have been performed to elucidate whether the appearance of some biomarkers in peroxynitrite-treated RBCs is due mainly to extra- or intracellular reactions. The aim is to provide a more general strategy that can also be adapted to different cell types.

2. Direct Reactions of Peroxynitrite with Biological Targets

Peroxynitrite can react with targets in biological systems via one- or two-electron oxidations. Because these reactions are first order in peroxynitrite and in the target, oxidation is faster at higher target concentrations. Thiols and metal centers are the preferential targets of *direct* peroxynitrite reactions. Thiols may be present at high concentrations (up to 5–10 mM in the case of glutathione) with rate constants for peroxynitrite ranging from 10^3 to 10^7 M^{-1} s^{-1} depending on the specific thiol and its own pK_a. Consequently, thiols are the principal route of peroxynitrite oxidation, except for tissues containing a high intracellular concentration of hemoproteins, as in the case of blood. Peroxynitrite-dependent one- and two-electron oxidation pathways of thiols coexist and compete and are able, respectively, to generate disulfides through the formation of sulfur-centered radicals as the intermediate (thiyl, peroxyl radical, and disulfide radicals) or sulfenic acid (Quijano *et al.*, 1997). The peroxynitrite-dependent oxidation of critical cysteine residues has been reported to inactivate some enzymes, including glyceraldehyde-3-phosphate dehydrogenase, tyrosine phosphatases, and creatine kinase but, conversely, to activate matrix metalloproteinases (Denicola and Radi, 2005).

Other potential targets of peroxynitrite are proteins containing transition metal centers (aconitase, Mn-SOD) and ferrous/ferric forms of hemoproteins (hemoglobin, myoglobin, peroxidases, cytochrome *c*). These proteins show fast rate constants for peroxynitrite (second-order rate constants in the order 10^4–10^7 M^{-1} s^{-1}) (Alvarez and Radi, 2003; Radi, 1998). In general, the peroxynitrite-mediated oxidation of metal centers proceeds through a variety of different mechanisms, which may ultimately diminish or amplify the oxidative outcome. The one-electron oxidations of peroxynitrite with metal centers lead to •NO$_2$ and secondary oxidizing species at the metal centers (Ferrer-Sueta *et al.*, 2002; Quijano *et al.*, 2001; Romero *et al.*, 2003).

Peroxynitrite can also oxidize reduced metal centers of hemoproteins by a two-electron oxidation mechanism with the formation of corresponding ferric forms through oxo-ferryl intermediates (Alayash et al., 1998; Exner and Herold, 2000; Floris et al., 1993; Mehl et al., 1999; Minetti et al., 1999; Romero et al., 2003). Other important targets of *direct* peroxynitrite oxidation in tissues are selenocompounds (e.g., glutathione peroxidase), which have been proposed to have a detoxification role for peroxynitrite (Sies and Arteel, 2000).

One of the most relevant *direct* reactions of peroxynitrite is that with CO_2, which leads to the formation of $CO_3^{\bullet-}$ and $^{\bullet}NO_2$ through the short-lived intermediate nitrosoperoxocarbonate anion ($ONOOCO_2^-$). This reaction redirects the peroxynitrite reactivity toward an *indirect* radical mechanism, the subject of the next paragraph.

3. Peroxynitrite Homolysis: *Indirect* Radical Chemistry

The *indirect* radical chemistry of peroxynitrite has its origin in its *direct* reaction with CO_2 and represents one of the major routes of peroxynitrite activity *in vivo*. For example, in biological fluids, where thiols and metal-containing compounds are present at low concentrations, about 90% of the initial peroxynitrite reactivity is directed toward CO_2, while intracellularly the contribution of CO_2 to peroxynitrite reactivity has been estimated to be around 30 to 40% (Radi, 1998). The reaction of peroxynitrite with CO_2 is fast ($5.8 \times 10^4\ M^{-1}\ s^{-1}$), and the short-lived $ONOOCO_2^-$, through the O–O bond homolysis, generates the strong one-electron oxidant $CO_3^{\bullet-}$ and the nitrating/oxidizing agent $^{\bullet}NO_2$ (Santos et al., 2000). Nitration reactions are accelerated (by a factor approximately ≥ 2) in the presence of 1 to 2 mM CO_2 likely due to the concerted $CO_3^{\bullet-}$ selective one-electron oxidation of phenolic rings to phenoxyl radicals (e.g., tyrosyl radicals) and the termination reaction mediated by $^{\bullet}NO_2$ (Radi, 1998). Oxidants other than $CO_3^{\bullet-}$, such as oxo-metal complexes and, to a lesser extent, $^{\bullet}OH$, are able to induce tyrosyl radical formation, the first step in the pathway leading to 3-nitrotyrosine. Importantly, $^{\bullet}NO_2$ per se is an inefficient nitration-promoting agent because its reaction with tyrosine to form tyrosyl radical is slow compared with other $^{\bullet}NO_2$-involving reactions. Addition of a -NO_2 group to a tyrosine can alter protein function (Minetti et al., 2002) and conformation, induce steric alteration, and inhibit enzyme activity (Radi, 2004).

4. Red Blood Cells as an Experimental Model to Test the Fate of Peroxynitrite in a Biological Environment

Peroxynitrite crosses the RBC membrane either by passive diffusion or through the Band 3 anion channel (Denicola *et al.*, 1998) and rapidly induces Hb oxidation to metHb [$k \approx 2 \times 10^4\ M^{-1}s^{-1}$ (Romero *et al.*, 1999)] (Fig. 16.1). Both of these properties of peroxynitrite, together with the high intracellular Hb concentration (about 20 mM as heme) and high RBC density ($\approx 5 \times 10^9$ cell/ml), strongly support the hypothesis that this cell is the major "sink" of peroxynitrite generated in the vasculature (Minetti *et al.*, 2000; Romero and Radi, 2005; Romero *et al.*, 2006).

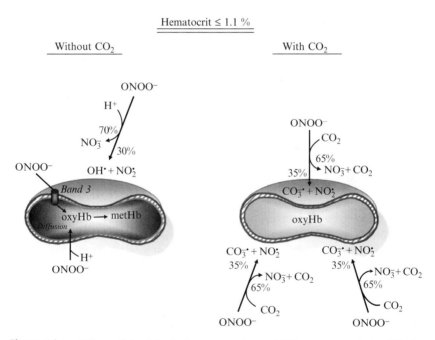

Figure 16.1 Effects of physiological concentrations of CO_2 on peroxynitrite diffusion across erythrocytes. In the absence of CO_2, peroxynitrite crosses the RBC membrane either by passive diffusion or through the Band 3 anion channel. Peroxynitrite intracellular oxidations are due mostly to direct oxidation of hemoglobin to metHb, while only a fraction of peroxynitrite will decay through proton-catalyzed homolysis before entering RBCs. The presence of CO_2 drastically limits the diffusion distance of peroxynitrite due to the shorter half-life of $CO_3^{\bullet-}$ and $^\bullet NO_2$. In this case, peroxynitrite will decay for the most part in the extracellular space and will cross only the few nearest cells. The surface/membrane oxidations are due principally to the indirect reactions, mediated by the $^\bullet NO_2/CO_3^{\bullet-}$ radical couple, with some targets and to the resulting triggering of downstream processes.

This makes Hb a convenient tool for the study of peroxynitrite biochemistry (Romero and Radi, 2005). However, other peroxynitrite-dependent modifications of RBCs have been described in addition to the oxidation to Hb (Celedon et al., 2007; Matarrese et al., 2005): in general, these modifications may either be due to *direct* or *indirect* peroxynitrite-mediated oxidations or be downstream cellular processes triggered by the oxidation of cellular targets. Downstream cellular processes may be even more relevant for studies with other cell types that, unlike RBCs, lack a concentrated intracellular target. Without an intracellular scavenger working as a "sink," it is conceivable that more cellular targets may be modified by peroxynitrite. Although buffers containing CO_2 complicate the chemistry of peroxynitrite, relatively simple considerations may help predict the fate of extracellular peroxynitrite in a cell suspension. This chapter uses the RBC as a cell model, but these considerations can also be applied to other cell types.

Romero and colleagues (Romero and Radi, 2005; Romero et al., 1999) reported the equations describing the percentage of peroxynitrite that would reach a target located at a certain distance from the site of peroxynitrite production. These equations assume a peroxynitrite diffusion coefficient of 1500 μm^2 s^{-1} and a peroxynitrite half-life at pH 7.4 and 25° of 24 ms and 2.7 s with/without 1.3 mM CO_2, respectively. Therefore, in the presence of a physiological concentration of CO_2 [i.e., 1.3 mM derived from 25 mM (bi)carbonate and pH 7.4], it can be calculated easily that 99% of peroxynitrite decays within a radius of approximately 20 μm from the site of addition, whereas in the absence of CO_2, the 99% decay is predicted to occur within a radius of approximately 250 μm. For example, at ≤1.1% RBC (≤110 × 10^6 cell/ml), the mean distance, Δd, between two cells is ≤250 μm: therefore, in the presence of CO_2 most peroxynitrite will decay in the extracellular space (Fig. 16.1) and will cross only a few cells (those fortuitously nearer than 20 μm to the site of addition). Without CO_2, peroxynitrite is expected to cross a distance comparable to Δd, following which part will oxidize Hb and part will decay in the extracellular space through proton-catalyzed decomposition (Fig. 16.1). The metHb formed under these conditions may amount to up to 20–30% of the peroxynitrite added. At 45% hematocrit (≈5 × 10^9 cell/ml), the Δd is about ≈5 μm so that peroxynitrite can cross up to ≈4- and ≈50-fold the Δd in the presence and in the absence of CO_2, respectively. Under these conditions, it has been calculated (Romero and Radi, 2005; Romero et al., 1999) that about half of the peroxynitrite will cross the RBC membrane.

This model predicts that the extracellular decay of peroxynitrite depends exponentially on cell density. An example is given in Fig. 16.2. Without CO_2, the metHb induced by 50 μM peroxynitrite increases by augmenting the cell density, but above 2 to 3% RBC the metHb concentration does not depend on the RBC hematocrit. The observation that no more than ≈65% of peroxynitrite is converted to metHb depends on (i) the competition with

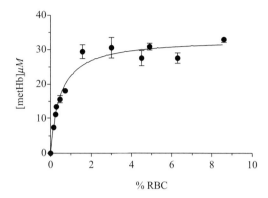

Figure 16.2 Effects of RBC density on peroxynitrite-dependent metHb formation. Peroxynitrite (50 μM) was added to RBCs suspended in PBS, pH 7.2, containing 50 mM phosphate buffer and 0.1 mM diethylenetriaminepentaacetic acid. MetHb content was determined spectrophotometrically in the cell lysate.

other cellular components, (ii) the metabolism of RBCs that can reduce metHb, and (iii) the fraction of peroxynitrite that does not cross the membrane and decays extracellularly.

From these simple considerations, it can be expected that in physiological conditions (i.e., ≈1.3 mM CO_2) and for cell densities ≲10 to 100 × 10^6 cell/ml (depending on cell size) the cellular modifications induced by extracellular peroxynitrite depend for the most part on the $^{\bullet}NO_2/^{\bullet}CO_3$ radical couple. Clearly, these radicals, by reacting with the cell surface/membrane, can modify some targets and trigger downstream cellular processes. This may be the most probable interaction that occurs, for example, in blood, between circulating nucleated white cells (≈7 × 10^6 cell/ml) and peroxynitrite formed extracellularly. However, this conclusion excludes specific conditions, such as those taking place, for instance, at sites of inflammation, that trigger the recruitment of white cells and cell activation to produce intracellular reactive species. Platelets in blood have a relatively high cell density (250 × 10^6 cell/ml), but to envisage a direct reaction between peroxynitrite and intracellular platelet targets it is necessary to know whether peroxynitrite crosses the platelet membrane as easily as it does that of RBCs. In other words, to hypothesize a direct reaction the permeability of the cell to peroxynitrite should be considered.

This topic has been investigated in some detail in the case of RBCs (Pietraforte *et al.*, 2007), and it has been demonstrated experimentally that peroxynitrite modifies different targets according to whether it acts intra- or extracellularly. The possibility that peroxynitrite-modified extracellular molecules of biological fluids (e.g., oxidized plasma components) can also activate downstream cellular processes cannot be excluded, but this topic has not yet been investigated.

5. Red Blood Cell Modifications Induced by Extracellular Peroxynitrite Decay

As outlined previously, in general the presence of CO_2 reduces the toxicity of peroxynitrite and this effect is particularly evident at low cell density (Alvarez et al., 2002; Hurst and Lymar, 1997). In the case of RBCs, it has been reported (Pietraforte et al., 2007) that at low cell density and physiological CO_2, Hb is not the major target of peroxynitrite and that other biomarkers can be revealed. The authors reported that peroxynitrite induced 3-nitrotyrosine, the formation of adducts with the spin trap 5,5-dimethyl-1-pyrroline-N-oxide (DMPO) and the oxidation of surface thiols. The appearance of these biomarkers can easily be reconciled with the *indirect* oxidation pathway of peroxynitrite, as 3-nitrotyrosine and DMPO adducts are well-known radical-dependent processes (Augusto et al., 2002; Lymar and Hurst, 1998), and the oxidation of surface thiols can be due, at least in part, to a radical mechanism (Bonini and Augusto, 2001). Nevertheless, because extracellular and intracellular thiols are in a dynamic equilibrium (Ciriolo et al., 1993), cross talk between different thiol pools cannot be completely ruled out.

6. Red Blood Cell Modifications Induced by Intracellular Peroxynitrite Decay

Bearing in mind the high intracellular concentration of Hb, this protein is certainly the principal target of peroxynitrite. The reaction with peroxynitrite generates metHb and NO_3^- as final products with the intermediate production of several reactive species, including $O_2^{\bullet-}$, $^{\bullet}NO_2$, and ferrylHb (Exner and Herold, 2000; Minetti et al., 2000; Romero et al., 2003). However, it should be considered that the formation of metHb is also a signaling process for the RBC, since this cell counteracts the oxidative insult to Hb by activating cell glycolysis. This process is aimed at the regeneration of NADH and NADPH, which are cofactors of NADH-dependent cytochrome $b5$-metHb reductase and NADPH-dependent metHb reductase, respectively. Two key enzymes are involved in the regeneration of these cofactors, glyceraldehyde-3-phosphate dehydrogenase, which reduces NAD, and glucose-6-phosphate dehydrogenase, which reduces NADP. The latter enzymatic process, which is a crucial step in the pentose phosphate pathway (PPP), is of minor relevance in the reduction of metHb under normal physiological conditions, but takes part in metHb reduction in the presence of oxidative stress. NADH and NADPH have relatively high second-order rate constants for peroxynitrite

($k \approx 4 \times 10^3$ M^{-1} s^{-1}), but their low intracellular concentrations (<100 μM) (Kinoshita et al., 2007) make these cofactors unlikely as direct targets of peroxynitrite.

An additional consideration is that the PPP is responsible for the maintenance of reduced glutathione (GSH), given that glutathione reductase is a NADPH-dependent enzyme, thus linking the GSH-dependent antioxidant machinery to glycolysis. Without describing the multifaceted metabolic pathways of RBCs in detail, it is interesting to note that GSH may also be a direct target of peroxynitrite, albeit a less relevant one than Hb. Indeed, because both the rate constant ($k \approx 1 \times 10^3$ M^{-1} s^{-1}) and the intracellular concentration ($\approx$2–4 mM) of glutathione are lower than those of Hb, it is possible to calculate that only one molecule of GSH for every 100 to 200 molecules of oxyHb can be oxidized directly by peroxynitrite. However, as the major antioxidant machinery of RBCs is GSH dependent and GSH is the principal redox buffer, its involvement in peroxynitrite-dependent downstream cellular processes is a hypothesis to be considered.

Another likely direct intracellular target of peroxynitrite is peroxyredoxin 2. This two-cysteine-containing enzyme is present at a relatively high concentration ($\approx$240 μM) in RBCs, and its role as a scavenger of peroxides, including hydrogen peroxide (Low et al., 2007) and peroxynitrite (Ogusucu et al., 2007), is beginning to emerge. The second-order rate constant of peroxyredoxin 2 for peroxynitrite has been reported to be in the range of 10^5 to 10^6 M^{-1} s^{-1} (Ogusucu et al., 2007; Romero et al., 2006), and it is therefore probable that RBC peroxyredoxin 2 competes with Hb for peroxynitrite. Additionally, peroxyredoxin 2 is linked to the GSH-dependent antioxidant machinery through thioredoxin and thioredoxin reductase and is also linked to the PPP, given that NADPH is the cofactor of thioredoxin reductase [for a review, see Burke-Gaffney et al. (2005)]. An additional feature of this complex picture is represented by peroxynitrite-dependent transmembrane signaling, which, as illustrated in the next paragraph, links protein phosphorylation, thiol oxidation, and RBC glycolysis.

7. Peroxynitrite-Dependent Phosphorylation Signaling of RBC

There are many ways to convert an extracellular signal into significant changes in the cytoplasmic milieu. Some of these signals originate at cell surface transmembrane proteins, which trigger a "start signal" that sets off numerous intracellular signaling pathways. More recently, redox modifications have gained importance as major chemical processes that modulate early events in the process of signal transduction.

The activation of phosphotyrosine kinases (PTK) in RBCs by peroxynitrite has been the subject of a recent review (Serafini et al., 2005), but in this context it is interesting to emphasize that PTK activation, unlike phosphotyrosine phosphatase (PTP) inhibition, has been reported not to be inhibited in the presence of CO_2 and, at least in part, not reversed on the reactivation of glycolysis (Mallozzi et al., 1997, 1999). This is probably a consequence of the complex regulation of src tyrosine kinase activity (see Serafini et al., 2005), which involves both SH-dependent and -independent regulation. Whatever the activation mechanism, it is interesting to stress that direct and indirect effects of peroxynitrite on PTK and PTP are directed toward the upregulation of tyrosine-dependent signaling. This signaling is aimed at bursting RBC glycolysis in conditions of intense oxidative stress. More work is necessary, however, to elucidate the mechanical features of the coordination and dynamics among Band 3 tyrosine phosphorylation signaling, PTP inactivation, PTK activation, GSH homeostasis, and RBC metabolism. These mechanisms are currently being investigated in our laboratory. Band 3 phosphorylation and PTK activation can be analyzed by Western blot analysis and immunoprecipitation, as reported by Serafini et al. (2005).

8. Peroxynitrite-Induced Biomarkers of RBC Senescence

Oxidatively modified RBCs are normally present in the circulation, and these modifications, attributed to differences in the "age" of a cell, can be separated by density gradient centrifugation. The process of cell senescence seems to be a continuous process, possibly ending in a nonfunctional cell that is removed by the reticuloendothelial system. In any case, the meaning of cell modifications in senescent RBCs, as well as the critical event(s) responsible for cell removal at the end of their 120 ± 4-day life span, is still a matter of debate (Berg et al., 2001; Daugas et al., 2001; Lang et al., 2006). Morphological alterations, glycophorin A loss, and Band 3 clustering have been considered events leading to the senescence of RBCs and, interestingly, these changes are also among the key modifications observed after treatment with peroxynitrite (Pietraforte et al., 2007). Cell shape maintenance is strictly associated with RBC function and, in particular, with its deformability and the maintenance of redox homeostasis. Scanning electron microscopy (SEM) allows us to visualize cell surface alterations occurring in "senescent" RBCs. To quantify these altered forms, morphometric analysis can be used. In response to peroxynitrite treatment, RBCs undergo normal discoid shape loss and appear elongated (leptocytes) or characterized by surface blebbing (acanthocytes) (Matarrese et al., 2005).

Interestingly, the CO_2-catalyzed extracellular decay of peroxynitrite also causes a downregulation of glycophorins A (shown in Fig. 16.3) and C, likely through the stimulation of RBC membrane vesiculation (Pietraforte

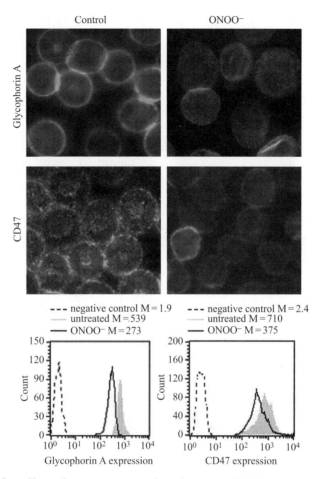

Figure 16.3 Effects of peroxynitrite on glycophorin A and CD47 expression and organization as detected by flow cytometry and static cytometry analysis obtained by IVM, respectively. Immunofluorescence analysis shows that both glycophorin A (first row) and CD47 (second row) decrease their surface expression in 2.5% RBCs treated with 250 μM peroxynitrite (right) compared with control (left). Semiquantitative flow cytometry analysis (third row) indicates a significant ($P < 0.01$) peroxynitrite-dependent downregulation of glycophorin A (−49%, left) and CD47 (−47%, right). Numbers represent median fluorescence intensity obtained in a representative experiment.

et al., 2007). Also of interest is the fact that vesiculation occurs normally during the lifetime of RBCs and that the loss of glycophorins is considered a marker of senescence (Bosman et al., 2005). For this reason, the hypothesis that free radicals derived from the extracellular decay of peroxynitrite may induce vesiculation and cell aging is fascinating. This hypothesis is, however, speculative, as the mechanism(s) leading to RBC membrane remodeling is still largely unknown.

Peroxynitrite treatment of RBCs induces not only the loss of glycophorins, but also the clustering of Band 3 (Pietraforte *et al.*, 2007). Interestingly, peroxynitrite treatment of RBCs also induces a downregulation of CD47 (Fig. 16.3). This is a surface glycoprotein that in RBCs has been found to be an important link in the interaction between Band 3 and Rh in the maintenance of cell membrane integrity (Oldenborg, 2004). CD47 is also considered a self-antigen that inhibits the recognition and phagocytosis of RBCs by macrophages. Thus, a downregulation of this antigen may favor the removal of RBCs by the reticuloendothelial system. The analysis of glycophorins, Band 3, and CD47 can be carried out by both static and flow-cytometry analysis.

9. Peroxynitrite-Induced Biomarkers of RBC Apoptosis

It has been demonstrated that, similarly to nucleated cells, RBCs may undergo cell death (Lang *et al.*, 2005). Cell shrinkage, membrane blebbing, and the breakdown of cell membrane phosphatidylserine (PS) asymmetry have in fact been detected in RBCs. The externalization of PS at the cell surface is considered a typical feature of programmed cell death or apoptosis (Bratosin *et al.*, 2001; Daugas *et al.*, 2001; Kuypers, 1998) and contributes to the shortened life span of defective erythrocytes (Connor *et al.*, 1994; De Jong *et al.*, 2001), as well as to important changes occurring in RBC function, such as hypercoagulability (Schlegel and Williamson, 2001). Importantly, the loss of PS asymmetry has been observed in RBCs in human diseases (De Jong *et al.*, 2001; Kuypers, 1998). In this anucleated cell the appearance of apoptotic biomarkers is probably a slightly different phenomenon than that in nucleated cells and may be considered a particular form of apoptosis. The phenomenon of RBC apoptosis has been suggested to be mediated by an ionic imbalance, for example, from alterations in Ca^{2+}-sensitive channels (Schneider *et al.*, 2007). The increase in intracellular Ca^{2+} can induce the activation of calpains, major enzymes regulated by Ca^{2+}. In particular, μ-calpain, a calcium-dependent cysteine protease normally localized as an inactive pro-enzyme in the cytosol of RBCs, translocates to the membrane, where it undergoes autoproteolytic activation (Berg *et al.*, 2001). A key role for μ-calpain in the apoptosis of RBCs induced by Ca^{2+} entry has thus been proposed (Daugas *et al.*, 2001). However, cathepsin aspartyl-protease (cathepsin E) activation and, strikingly, the activation of apoptosis-specific proteases, that is, caspases, have also been associated with RBC apoptosis induced by peroxynitrite (Matarrese *et al.*, 2005). This is in accordance with results obtained in nucleated cells, in which two distinct apoptotic pathways, one caspase and one noncaspase dependent, were found (Johnson, 2000). Surprisingly, both of these distinct apoptotic pathways seem to be activated in peroxynitrite-treated RBCs. These data depict a complex scenario in which, in addition to μ-calpain, which could conceivably be activated by increasing intracellular Ca^{2+} ions or, alternatively, by other activated proteases, other actors can actively participate in the death of RBCs: cathepsins and caspases. The hallmark of all these pathological conditions is believed to be represented by an irreversible alteration in the redox state of the cell.

10. METHODS

10.1. Peroxynitrite oxidation of RBCs

Procedure

1. Remove plasma and buffy coat from heparinized fresh human blood by centrifugation (10 min, 1000g). Wash RBCs three times with isotonic phosphate-buffered saline, pH 7.4 (PBS), and suspend at the desired hematocrit. The choice of RBC density should be made according to the equations reported by Romero et al. (1999).
2. Add peroxynitrite (prepared as reported by Radi et al., 1991) as a bolus and under rapid stirring to cells suspended in PBS, pH 7.2, containing 50 mM phosphate buffer and 0.1 mM diethylenetriaminepentaacetic acid in the absence or in the presence of 25 mM sodium bicarbonate. Incubate at 37° for 5 min. The phosphate buffer should be pretreated extensively with Chelex 100 to remove trace amounts of contaminating transition metals. Decomposed peroxynitrite is obtained by adding peroxynitrite to the phosphate buffer and leaving for 5 min at room temperature before the addition of RBCs (reversed order of addition). Incubate at 37° for 5 min. Wash RBCs twice in PBS.
3. Evaluate the metHb content spectrophotometrically in RBC lysate as reported by Winterbourn's equations.

10.2. Red blood cell structural alterations

Procedure

1. Wash erythrocytes three times with PBS, pH 7.4, and fix with 2.5% glutaraldehyde in 0.1 M cacodylate buffer, pH 7.4, at room temperature for 20 min.
2. Collect and plate RBCs on poly-L-lysine-coated slides for 15 min. Fix with 1% OsO_4 for 30 min at room temperature. Wash three times with 0.1 M cacodylate buffer (pH 7.4) at room temperature.
3. Dehydrate through graded ethanol and critical point dried in CO_2.

4. Gold coat by sputtering with the Balzers Union SCD 040 apparatus and examine samples with a scanning electron microscope.

10.3. Red blood cell morphometric analysis

Procedure

1. Count ≥500 cells (50 RBCs for each different SEM field at a magnification of ×3000) from control and peroxynitrite-treated RBCs in triplicate.
2. Use Student's t test for statistical analysis.

10.4. Detection of glycophorins A, C, and CD47

Antibodies: monoclonal antibody to glycophorins conjugated directly to FITC (Dako, Glostrup, Denmark), monoclonal antibody to CD47 (Santa Cruz Biotechnology, CA), and Alexa Fluor 488-conjugated antimouse IgG (Molecular Probes/Invitrogen, Eugene, OR)

Procedure

1. Stain unfixed RBCs with FITC-labeled antiglycophorins for 30 min at 4°. Wash three times with PBS at 4°.
2. Mount with glycerol–PBS (2:1) and observe by intensified charge-coupled video microscopy (IVM) or analyze by flow cytometry.

10.5. Detection of band 3

Antibodies: monoclonal antibody to Band 3 and FITC-labeled antimouse antibody (Sigma, St. Louis, MO).

Procedure

1. Wash RBCs three times with PBS, pH 7.4, and fix with 3.7% formaldehyde in PBS for 10 min at room temperature. Wash three times with PBS at room temperature.
2. Permeabilize with 0.5% Triton X-100 in PBS for 5 min at room temperature. Wash three times with PBS at room temperature.
3. Stain with anti-Band 3 antibodies for 30 min at 37°. Wash three times with PBS at room temperature.
4. Incubate with FITC-labeled antimouse antibody for 30 min at 37°. Wash three times with PBS at room temperature.
5. Mount with glycerol–PBS (2:1) and observe by intensified charge-coupled video microscopy (IVM) or analyze by flow cytometry.

10.6. Glutathione content

Reagent: Monochlorobimane (MCB, Molecular Probes)

Procedure
Monochlorobimane staining is performed for GSH as described previously (Sahaf *et al.*, 2003). Median values of fluorescence intensity histograms are used to provide semiquantitative assessment of GSH content in comparison with untreated control cells.

1. Resuspend cells at 5×10^5 in 500 μl PBS, pH 7.4, containing 25 μg/ml MCB for 20 min at 37°. Wash samples twice in PBS.
2. Resuspend cells in 200 μl ice-cold PBS and analyze samples immediately with a LRS II cytometer (Becton & Dickinson, San Jose, CA) equipped with a UVB laser.

10.7. Apoptosis evaluation by phosphatidylserine detection

Reagent: annexin-V-FITC apoptosis detection kit (Eppendorf s.r.l., Milan, Italy)

Procedure
Quantitative evaluation of apoptosis is performed by flow cytometry after staining with the annexin-V-FITC apoptosis detection kit.

1. Collect fresh control and treated RBCs (5×10^5 cells) by centrifugation.
2. Resuspend cells in 490 μl of 1× binding buffer containing 5 μl annexin V-FITC iodide and incubate for 5 min in the dark at room temperature.
3. Analyze immediately on a FACScan flow cytometer (Becton Dickinson) equipped with a 488-argon laser using FL1 signal detector.

10.8. Caspase activity

Reagent: Colorimetric protease assay kit (MBL, Woburn, MA)

Procedure
The activity of caspases is measured in cells extracted using a colorimetric assay. The assay is based on spectrophotometric detection of the chromophore *p*-nitroanilide (*p*-NA) after cleavage from labeled substrates. The *p*-NA light

emission can be quantified using a microtiter plate reader at 405 nm. Comparison of the absorbance of *p*-NA from treated samples with an untreated control allows determination of the fold increase in caspase activity.

1. Collect control and treated cells (2×10^6) by centrifugation.
2. Resuspend cells in 100 µl extraction buffer, incubate on ice for 20 min, and centrifuge samples for 1 min at 10,000*g*.
3. Analyze the supernatant for protein concentration.
4. Resuspend cell lysate (200 µg) in 85 µl of extraction buffer. For a positive control, 2 µl of active caspases is resuspended in extraction buffer, whereas for a negative control, a treated cell lysate incubated with caspase inhibitor zVAD is used.
5. Add 10 µl of reaction buffer and 5 µl of specific caspase substrate, incubate for 1 h in the dark, and read samples on a spectrophotometer.

Alternatively, the activation state of caspases can be evaluated in intact living cells by flow cytometry using the CaspGLOW fluorescein active caspase staining kit (MBL). This kit provides a sensitive means for detecting activated caspases in living cells. The assay utilizes specific FMK–caspase inhibitors conjugated to FITC as the fluorescent marker. These inhibitors are cell permeant, nontoxic, and bind irreversibly to the caspase-active form. The FITC label allows detection of activated caspases in apoptotic cells directly by flow cytometry.

1. Collect control and treated cells by centrifugation.
2. Resuspend control and treated cells (3×10^5) in 300 µl of wash buffer containing 1 µl of the specific FITC-conjugated caspase substrate for 1 h at 37°.
3. Wash three times in the same buffer and resuspend cells in 300 µl of wash buffer.
4. Analyze immediately on a cytometer equipped with a 488-nm argon laser using a FL1 signal detector.

11. Data and Statistics

11.1. Flow cytometry data

For FACS analysis, at least 20,000 events are acquired. Data are recorded and analyzed statistically by a Macintosh computer using CellQuest software or a PC computer using DIVA software (for GSH analysis) (both by Becton & Dickinson). Data regarding intracellular redox balance are reported as mean values of the median fluorescence among at least four separate experiments ± standard deviation; data regarding apoptosis and caspase activation are reported as percentages of cells.

11.2. Statistics

Statistical analysis of apoptosis data is performed by the nonparametric ANOVA test. Statistical significance of flow cytometry studies of redox balance is calculated using the nonparametric Kolmogorov–Smirnov test. As a general rule, only P values less than 0.01 are considered significant.

ACKNOWLEDGMENTS

This work was supported in part by ISS-NIH collaborative project "Peripheral blood determinants of redox changes in human respiratory diseases: Biochemical and pathophysiological evaluations" Rif. 0F14 and a grant from Italian Ministry of Health Rif. 6AEF.

REFERENCES

Alayash, A. I., Brockner Ryan, B. A., and Cashon, R. E. (1998). Peroxynitrite-mediated heme oxidation and protein modification of native and chemically modified hemoglobins. *Arch. Biochem. Biophys.* **349,** 65–73.

Alvarez, B. and Radi, R. (2003). Peroxynitrite reactivity with amino acids and proteins. *Amino Acids* **25,** 295–311.

Alvarez, M. N., Piacenza, L., Irigoin, F., Peluffo, G., and Radi, R. (2002). Macrophage-derived peroxynitrite diffusion and toxicity to *Trypanosoma cruzi*. *Arch. Biochem. Biophys.* **432,** 222–232.

Augusto, O., Lopez de Menezes, S., Linares, E., Romero, N., and Radi, R. (2002). EPR detection of glutathiyl and hemoglobin-cysteinyl radicals during the interaction of peroxynitrite with human erythrocytes. *Biochemistry* **41,** 14323–14328.

Beckman, J. S., Beckman, T. W., Chen, J., Marshall, P. A., and Freeman, B. A. (1990). Apparent hydroxyl radical production by peroxynitrite: Implications for endothelial injury from nitric oxide and superoxide. *Proc. Natl. Acad. Sci. USA* **87,** 1620–1624.

Berg, C. P., Engels, I. H., Rothbart, A., Lauber, K., Renz, A., Schlosser, S. F., Schulze-Osthoff, K., and Wesselborg, S. (2001). Human mature red blood cells express caspase-3 and capase-8, but are devoid of mitochondrial regulators of apoptosis. *Cell Death Differ.* **8,** 1197–1206.

Bonini, M. G. and Augusto, O. (2001). Carbon dioxide stimulates the production of thiyl, sulfinyl, and disulfide radical anion from thiol oxidation by peroxynitrite. *J. Biol. Chem.* **276,** 9749–9754.

Bonini, M. G., Radi, R., Ferrer-Sueta, G., Ferreira, A. M., and Augusto, O. (1999). Direct EPR detection of the carbonate radical anion produced from peroxynitrite and carbon dioxide. *J. Biol. Chem.* **274,** 10802–10806.

Bosman, G. J., Willekens, F. L., and Weere, J. M. (2005). Erythrocyte aging: A more than superficial resemblance to apoptosis? *Cell Physiol. Biochem.* **16,** 1–8.

Bratosin, D., Estaquier, J., Petit, F., Arnoult, D., Quatannens, B., Tissier, J. P., Slomianny, C., Sartiaux, C., Alonso, C., Huart, J. J., Montreuil, J., and Ameisen, J. C. (2001).

Programmed cell death in mature erythrocytes: A model for investigating death effector pathways operating in the absence of mitochondria. *Cell Death Differ.* **8,** 1143–1156.

Burke-Gaffney, A., Callister, M. E. J., and Nakamura, H. (2005). Thioredoxin: Friend or foe in human disease? *Trends Pharmacol. Sci.* **26,** 398–404.

Celedon, G., Gonzalez, G., Pino, J., and Lissi, E. A. (2007). Peroxynitrite oxidizes erythrocyte membrane and 3 protein and diminishes its anion transport capacity. *Free Radic. Res.* **41,** 316–323.

Ciriolo, M. R., Paci, M., Sette, M., De Martino, A., Bozzi, A., and Rotilio, G. (1993). Transduction of reducing power across the plasma membrane by reduced glutathione: A 1H-NMR spin-echo study of intact human erythrocytes. *Eur. J. Biochem.* **215,** 711–718.

Connor, J., Pak, C. C., and Schroit, A. J. (1994). Exposure of phosphatidylserine in the outer leaflet of human red blood cells: Relationship to cell density, cell age, and clearance by mononuclear cells. *J. Biol. Chem.* **269,** 2399–2404.

Daugas, E., Cande, C., and Kroemer, G. (2001). Erythrocytes: Death of a mummy. *Cell Death Differ.* **8,** 1131–1133.

De Jong, K., Emerson, R. K., Butler, J., Bastacky, J., Mohandas, N., and Kuypers, F. A. (2001). Short survival of phosphatidylserine-exposing red blood cells in murine sickle cell anemia. *Blood* **98,** 1577–1584.

Denicola, A. and Radi, R. (2005). Peroxynitrite and drug-dependent toxicity. *Toxicology* **208,** 273–288.

Denicola, A., Souza, J. M., and Radi, R. (1998). Diffusion of peroxynitrite across erythrocyte membranes. *Proc. Natl. Acad. Sci. USA* **95,** 3566–3571.

Exner, M. and Herold, S. (2000). Kinetic and mechanistic studies of the peroxynitrite-mediated oxidation of oxymyoglobin and oxyhemoglobin. *Chem. Res. Toxicol.* **13,** 287–293.

Ferrer-Sueta, G., Quijano, C., Alvarez, B., and Radi, R. (2002). Reactions of manganese porphyrins and manganese-superoxide dismutase with peroxynitrite. *Methods Enzymol.* **349,** 23–37.

Floris, R., Piersma, S. R., Yang, G., Jones, P., and Wever, R. (1993). Interaction of myeloperoxidase with peroxynitrite: A comparison with lactoperoxidase, horseradish peroxidase and catalase. *Eur. J. Biochem.* **215,** 767–777.

Goldstein, S., Lind, J., and Merenyi, G. (2005). Chemistry of peroxynitrites as compared to peroxynitrates. *Chem. Rev.* **105,** 2457–2470.

Hurst, J. K. and Lymar, S. V. (1997). Toxicity of peroxynitrite and related reactive nitrogen species toward *Escherichia coli*. *Chem. Res. Toxicol.* **10,** 802–810.

Johnson, D. E. (2000). Noncaspase proteases in apoptosis. *Leukemia* **14,** 1695–1703.

Kinoshita, A., Tsukada, K., Soga, T., Hishiki, T., Ueno, Y., Nakayama, Y., Tomita, M., and Suematsu, M. (2007). Roles of hemoglobin allostery in hypoxia-induced metabolic alterations in erythrocytes: Simulation and its verification by metabolome analysis. *J. Biol. Chem.* **282,** 10731–10741.

Kuypers, F. A. (1998). Phospholipid asymmetry in healt and disease. *Curr. Opin. Hematol.* **5,** 122–131.

Lang, F., Lang, K. S., Lang, P. A., Huber, S. M., and Wieder, T. (2006). Mechanisms and significance of eryptosis. *Antioxid. Redox Signal.* **8,** 1183–1192.

Lang, K. S., Lang, P. A., Bauer, C., Duranton, C., Wieder, T., Huber, S. M., and Lang, F. (2005). Mechanisms of suicidal erythrocyte death. *Cell Physiol. Biochem.* **15,** 195–202.

Low, F. M., Hampton, M. B., Alexander, V., Peskin, A. V., and Winterbourn, C. C. (2007). Peroxiredoxin 2 functions as a non-catalytic scavenger of low level hydrogen peroxide in the erythrocyte. *Blood* **109,** 2611–2617.

Lymar, S. V. and Hurst, J. K. (1996). Carbon dioxide: Physiological catalyst for peroxynitrite-mediated cellular damage or cellular protectant? *Chem. Res. Toxicol.* **9,** 845–850.

Lymar, S. V. and Hurst, J. K. (1998). Radical nature of peroxynitrite reactivity. *Chem. Res. Toxicol.* **11,** 714–715.

Mallozzi, C., Di Stasi, A. M. M., and Minetti, M. (1997). Peroxynitrite modulates tyrosine-dependent signal transduction pathway of human erythrocyte band 3. *FASEB J.* **11,** 1281–1290.

Mallozzi, C., Di Stasi, A. M. M., and Minetti, M. (1999). Activation of src tyrosine kinases by peroxynitrite. *FEBS Lett.* **456,** 201–206.

Matarrese, P., Straface, E., Pietraforte, D., Gambardella, L., Vona, R., Maccaglia, A., Minetti, M., and Malorni, W. (2005). Peroxynitrite induces senescence and apoptosis of red blood cells through the activation of aspartyl and cysteinyl proteases. *FASEB J.* **19,** 416–418.

Mehl, M., Daiber, A., Herold, S., Shoun, H., and Ullrich, V. (1999). Peroxynitrite reaction with heme proteins. *Nitric Oxide* **3,** 142–152.

Minetti, M., Mallozzi, C., and Di Stasi, A. M. M. (2002). Peroxynitrite activates kinases of the src family and upregulates tyrosine phosphorylation signaling. *Free Radic. Biol. Med.* **33,** 744–754.

Minetti, M., Pietraforte, D., Carbone, V., Salzano, A. M., Scorza, G., and Marino, G. (2000). Scavenging of peroxynitrite by oxyhemoglobin and identification of modified globin residues. *Biochemistry* **39,** 6689–6697.

Minetti, M., Scorza, G., and Pietraforte, D. (1999). Peroxynitrite induces long-lived tyrosyl radical(s) in oxyhemoglobin of red blood cells through a reaction involving CO_2 and a ferryl species. *Biochemistry* **38,** 2078–2087.

Ogusucu, R., Rettori, D., Munhoz, D. C., Netto, L. E. S., and O., A. (2007). Reactions of yeast thioredoxin peroxidases I and II with hydrogen peroxide and peroxynitrite: Rate constants by competitive kinetics. *Free Radic. Biol. Med.* **42,** 326–334.

Oldenborg, P.-A. (2004). Role of CD47 in erythroid cells and in autoimmunity. *Leukemia Lymphoma* **45,** 1319–1327.

Pietraforte, D., Matarrese, P., Straface, E., Gambardella, L., Metere, A., Scorza, G., Leto, T. L., Malorni, W., and Minetti, M. (2007). Two different pathways are involved in peroxynitrite-induced senescence and apoptosis of human erythrocytes. *Free Radic. Biol. Med.* **42,** 202–214.

Quijano, C., Alvarez, B., Gatti, R. M., Augusto, O., and Radi, R. (1997). Pathways of peroxynitrite oxidation of thiol groups. *Biochem. J.* **322,** 167–173.

Quijano, C., Hernandez-Saavedra, D., Castro, L., McCord, J. M., Freeman, B. A., and Radi, R. (2001). Reaction of peroxynitrite with Mn-superoxide dismutase: Role of the metal center in decomposition kinetics and nitration. *J. Biol. Chem.* **276,** 11631–11638.

Radi, R. (1998). Peroxynitrite reactions and diffusion in biology. *Chem. Res. Toxicol.* **11,** 720–721.

Radi, R. (2004). Nitric oxide, oxidants, and protein tyrosine nitration. *Proc. Natl. Acad. Sci. USA* **101,** 4003–4008.

Radi, R., Beckman, J. S., Bush, K. M., and Freeman, B. A. (1991). Peroxynitrite oxidation of sulfhydryls: The cytotoxic potential of superoxide and nitric oxide. *J. Biol. Chem.* **266,** 4244–4250.

Romero, N., Denicola, A., and Radi, R. (2006). Red blood cells in the metabolism of nitric oxide-derived peroxynitrite. *IUBMB Life* **58,** 572–580.

Romero, N., Denicola, A., Souza, J. M., and Radi, R. (1999). Diffusion of peroxynitrite in the presence of carbon dioxide. *Arch. Biochem. Biophys.* **368,** 23–30.

Romero, N. and Radi, R. (2005). Hemoglobin and red blood cells as tools for studying peroxynitrite biochemistry. *Methods Enzymol.* **396,** 229–245.

Romero, N., Radi, R., Linares, E., Augusto, O., Detweiler, C. D., Mason, R. P., and Denicola, A. (2003). Reaction of human hemoglobin with peroxynitrite: Isomerization to nitrate and secondary formation of protein radicals. *J. Biol. Chem.* **278,** 44049–44057.

Sahaf, B., Heydari, K., Herzenberg, L. A., and Herzenberg, L. A. (2003). Lymphocyte surface thiol levels. *Proc. Natl. Acad. Sci. USA* **100,** 4001–4005.

Santos, C. X. C., Bonini, M. G., and O., A. (2000). Role of the carbonate radical anion in tyrosine nitration and hydroxylation by peroxynitrite. *Arch. Biochem. Biophys.* **377,** 146–152.

Schlegel, R. A. and Williamson, P. (2001). Phosphatidylserine, a death knell. *Cell Death Differ.* **8,** 551–563.

Schneider, J., Nicolay, J. P., Föller, M., Wieder, T., and Lang, F. (2007). Suicidal erythrocyte death following cellular K+ loss. *Cell Physiol. Biochem.* **20,** 35–44.

Serafini, M., Mallozzi, C., Di Stasi, A. M. M., and Minetti, M. (2005). Peroxynitrite-dependent upregulation of Src kinases in red blood cells: Strategies to study the activation mechanisms. *Methods Enzymol.* **396,** 215–229.

Sies, H. and Arteel, G. E. (2000). Interaction of peroxynitrite with selenoproteins and glutathione peroxidase mimics. *Free Radic. Biol. Med.* **28,** 1451–1455.

Thomson, L., Trujillo, M., Telleri, R., and Radi, R. (1995). Kinetics of cytochrome c2+ oxidation by peroxynitrite: Implications for superoxide measurements in nitric oxide–producing biological systems. *Arch. Biochem. Biophys.* **319,** 491–497.

Zhu, L., Gunn, C., and Beckman, J. S. (1992). Bactericidal activity of peroxynitrite. *Arch. Biochem. Biophys.* **298,** 452–457.

CHAPTER SEVENTEEN

DETECTION AND PROTEOMIC IDENTIFICATION OF S-NITROSATED PROTEINS IN HUMAN HEPATOCYTES

Laura M. López-Sánchez,* Fernando J. Corrales,[†] Manuel De La Mata,* Jordi Muntané,* *and* Antonio Rodríguez-Ariza*

Contents

1. Introduction	274
2. Preparation of CSNO	276
3. Preparation of Primary Human Hepatocytes and Cell Culture	276
4. Treatment of Hepatocytes and Sample Preparation	277
5. Biotin Switch Assay	277
6. Detection and Purification of Biotinylated Proteins	278
7. Final Considerations	279
Acknowledgments	280
References	280

Abstract

The S-nitrosation of protein thiols is a redox-based posttranslational modification that modulates protein function and cell phenotype. Although the detection of S-nitrosated proteins is problematical because of the lability of S-nitrosothiols, an increasing range of proteins has been shown to undergo S-nitrosation with the improvement of molecular tools. This chapter describes the methodology used to identify potential targets of S-nitrosation in cultured primary human hepatocytes using proteomic approaches. This methodology is based on the biotin switch method, which labels S-nitrosated proteins with an affinity tag, allowing their selective detection and proteomic identification.

* Liver Research Unit, Hospital Universitario Reina Sofía, Córdoba, Spain
[†] Hepatology and Gene Therapy Unit, Universidad de Navarra, Pamplona, Spain

1. Introduction

Under physiological aerobic conditions, nitric oxide (NO) reacts with O_2 to yield NO_2 and N_2O_3, powerful electrophiles that S-nitrosate cysteines to form S-nitrosothiols. The S-nitrosation of protein thiols, often referred to as "S-nitrosylation," is a form of posttranslational modification that has been regarded as a mechanism by which NO can transmit signals both within and between cells and tissues (Stamler et al., 2001). S-Nitrosation is being advanced as a central biological signaling mechanism, and first attempts are underway to identify potential targets of S-nitrosation by using proteomic approaches. Jaffrey et al. (2001) developed an original approach to assess S-nitrosation in which S-nitrosoproteins are selectively tagged with biotin and hence the method is termed the "biotin switch" method (Fig. 17.1). The assay consists of three steps. In the first step, free thiols are blocked with the thiol-specific methylating agent methyl methanethiosulfonate (MMTS). The second step involves the selective reduction

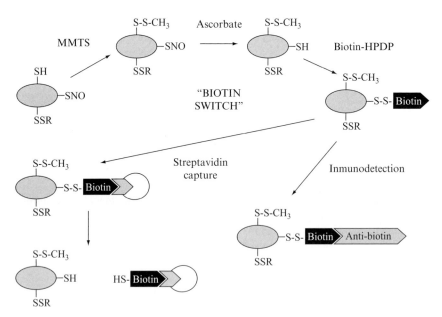

Figure 17.1 During the biotin switch assay, S-nitrosoproteins are selectively tagged with biotin. In the first step, free thiols (-SH) are blocked with the thiol-specific methylating reagent MMTS. During the second step, and under conditions where disulfides (-SSR) are not reduced, S-nitrosothiols (-SNO) are selectively reduced by ascorbate. In the final step, these newly formed thiols are reacted with the thiol-specific biotinylating reagent biotin-HPDP. Biotinylated proteins can be inmunodetected or, alternatively, can be captured on streptavidin resins for their purification and proteomic identification.

of protein nitrosothiols to thiols with ascorbate. In the final step, these newly formed thiols are reacted with the sulfhydryl-specific biotinylating reagent N-[6-(biotinamido)hexyl]-3′-(2′-pyridyldithio)propionamide (biotin-HPDP). This method offers significant advantages in that biotinylated proteins can be detected on Western blots following incubation with antibiotin antibodies or recognition via streptavidin. Moreover, biotinylated proteins can be captured on streptavidin matrices for identification by mass spectrometry (Fig. 17.1).

Detection of S-nitrosated proteins is problematical due to the lability of S-nitrosothiols, which are unstable under certain conditions. It has been established that they are sensitive to both photolytic and transition metal ion-dependent breakdown and that samples must be kept in the dark and in the presence of transition metal chelators.

Several issues have been raised regarding the sensitivity and specificity of the biotin switch method. The complete blocking of free thiols is imperative to minimize background biotinylation, and therefore to improve signal-to-noise ratio and sensitivity. Another worrisome issue is the possibility that the ascorbate treatment may reduce other cysteine oxidation-derived modifications such as S-glutathionylation or S-oxidations, including sulfenic, sulfinic, and sulfonic acids (Kettenhofen et al., 2007). Importantly, Forrester et al. (2007) showed that ascorbate does not directly reduce the S–NO bond, but rather undergoes transnitrosation by SNO to generate O-nitrosoascorbate as the intermediate, which homolyzes rapidly to yield the semidehydroascorbate radical and NO. This reaction with ascorbate is unique among cysteine oxidation products, thus conferring specificity to the biotin switch assay.

As in other assays, the biotin switch method poses potential problems and limitations. However, once they are recognized, this method represents a useful tool for the study of S-nitrosothiol biology and may facilitate the isolation and analysis of the S-nitrosoproteome. We have used this technique to detect and identify S-nitrosated proteins during the alteration of SNO homeostasis in human hepatocytes (Lopez-Sanchez et al., 2007; Ranchal et al., 2006). We use S-nitroso-L-cysteine (CSNO), a nitrosothiol that is effective at relatively low concentrations and has very rapid effects increasing the intracellular pool of S-nitrosothiols. This effectiveness has been related to the direct transfer of this intact molecule across cellular membranes (Mallis and Thomas, 2000). CSNO can be taken up into cells via amino acid transport system L, whereas nitrosoglutathione (GSNO) is not transported directly, but requires the presence of cysteine and/or cystine (Zhang and Hogg, 2005) or cleavage by γ-glutamyl transpeptidase to the membrane-permeable S-nitroso-cysteinyl glycine dipeptide (Gaston et al., 2006). Once inside the cell, transnitrosation to both protein thiols and glutathione occurs, but GSNO once formed is broken down rapidly by GSNO reductase (Liu et al., 2001).

2. Preparation of CSNO

We synthesize CSNO following a previously described method (Shah et al., 2003; Mallis and Thomas, 2000). The CSNO solution is freshly prepared immediately before use by mixing 220 μl of 220 mM L-cysteine and 220 μl of 220 mM NaNO$_2$ with 25 μl of 4.0 N HCl. This solution is incubated in the dark at room temperature for 10 min and is then neutralized by adding 25 μl of 4.0 N NaOH and maintained on ice. The final concentration is calculated from absorbance at 334 nm using a molar absorption coefficient of 740 M^{-1} cm^{-1}. Usually, the yield is between 60 and 70% with CSNO solutions in the range of 60 to 70 mM.

3. Preparation of Primary Human Hepatocytes and Cell Culture

Liver resections are obtained after written consent from patients submitted to surgical intervention for a primary or secondary liver tumor. Cell isolation is carried through *ex vivo* collagenase perfusion. The following steps are carried out maintaining the piece of liver and solutions at 37° under strict sterile conditions. We first perfuse the piece of liver with nonrecirculating washing solution I (20 mM HEPES, 120 mM NaCl, 5 mM KCl, 0.5% glucose, 100 μM sorbitol, 100 μM manitol, 100 μM GSH, 100 U/ml penicillin, 100 μg/ml streptomycin, and 0.25 μg/ml amphotericin B), pH 7.4, at a flow of 75 ml/min in order to remove blood cells. Afterward, the liver is perfused with nonrecirculating chelating solution II (0.5 mM EGTA, 58.4 mM NaCl, 5.4 mM KCl, 0.44 mM KH$_2$PO$_4$, 0.34 mM NaHPO$_4$, 25 mM N-tris[hydroxymethyl]methylglycine, 100 μM sorbitol, 100 μM manitol, 100 μM GSH, 100 U/ml penicillin, 100 μg/ml streptomycin, and 0.25 μg/ml amphotericin B), pH 7.4, at a flow of 75 ml/min. The liver is finally perfused with recirculating isolation solution III (0.050% collagenase, 20 mM HEPES, 120 mM NaCl, 5 mM KCl, 0.7 mM CaCl$_2$, 0.5% glucose, 100 μM sorbitol, 100 μM manitol, 100 μM GSH, 100 U/ml penicillin, 100 μg/ml streptomycin, and 0.25 μg/ml amphotericin B), pH 7.4, at a flow of 75 ml/min. The obtained cell suspension is filtered through nylon mesh (250 μm) and washed three times at 50g 5 min 4° in supplemented culture medium. DEM:Ham-F12 and William's E mediums (1:1) are supplemented with 26 mM NaHCO$_3$, 15 mM HEPES, 0.292 g/liter glutamine, 50 mg/liter vitamin C, 0.04 mg/liter dexamethasone, 2 mg/liter insulin, 200 μg/liter glucagon, 50 mg/liter transferrin, and 4 ng/liter ethanolamine). We usually obtain a consistent cell viability >85%, as determined by trypan blue exclusion. Hepatocytes

(150,000 cells/cm^2) are seeded in type I collagen-coated 100-mm diameter dishes and cultured in culture medium containing 5% fetal calf serum for 12 h. Afterward, the medium is removed and replaced by fresh culture medium without fetal bovine serum. Treatment of cells is initiated 48 h after seeding of the cells to allow stabilization of the culture.

4. Treatment of Hepatocytes and Sample Preparation

After culture stabilization, the medium is refreshed, CSNO is added at different final concentrations (0.5, 2, and 5 mM), and cells are incubated for 2 to 24 h at 37°. After washing hepatocytes with 0.1 M phosphate-buffered saline, 1 ml of nondenaturing lysis solution (50 mM Tris-HCl, pH 7.4, 300 mM NaCl, 5 mM EDTA, 0.1 mM neocuproine, 1% Triton X-100, 1 mM phenylmethylsulfonyl fluoride, 5 μg/ml aprotinin, and 10 μg/ml leupeptin) is added to a 100-mm diameter dish, and cells are scraped, harvested, and incubated on ice for 15 min. From now on, it is imperative to protect samples from light to avoid artifacts and decomposition of S-nitrosothiols. After centrifugation at 10,000g, 4° for 15 min, supernatants are collected and protein is quantified with the Bradford reagent.

N-Ethylmaleimide is included at a final concentration of 5 mM in the lysis solution for blockade of thiols during homogenization to prevent *ex vivo* transnitrosation of protein thiols. However, any thiol-blocking agent must be excluded from the lysis solution if *in vitro* transnitrosation experiments are planned. For example, as a positive control, we suggest incubating cellular lysates with 1 mM GSNO for 1 h at room temperature before performing the biotin switch method.

5. Biotin Switch Assay

Hepatocyte lysates are adjusted to 300–400 μg of protein per milliliter, and equal volumes are mixed with 3 volumes of blocking buffer [9 volumes of HEN buffer (250 mM HEPES, pH 7.7, 1 mM EDTA, and 0.1 mM neocuproine), 1 volume of SDS [25% (v/v) in H$_2$O], and 0.1 volumes of 2 M MMTS (stock solution in N,N-dimethylformamide)]. Free thiols are then blocked at 50° for 1 h with agitation. The SDS and temperature of 50° are designed to ensure that buried thiols are alkylated. The protein must be washed to remove MMTS so that any remaining blocking agent is not able to then block the newly exposed thiol group following the ascorbate reduction of S-nitrosothiols. To remove MMTS, proteins are precipitated twice with 2 volumes of prechilled (-20°) acetone, incubated at -20° for

15 min, and centrifuged at 2000g for 10 min followed by gentle rinsing of the pellet with 4 × 1 ml 70% acetone/H$_2$O.

In the following step, nitrosothiols are reduced and labeled at once. To this end, precipitates obtained as described earlier are resuspended in 60 μl HENS buffer (250 mM HEPES, pH 7.7, 1 mM EDTA, 0.1 mM neocuproine, and 1% SDS); following the addition of 2 μl of 50 mM ascorbic acid (from a freshly prepared 50 mM stock solution in deionized water) and 20 μl of 4 mM biotin-HPDP (prepared fresh by diluting a 50 mM stock solution in N,N-dimethylformamide), they are incubated for 1 h at room temperature. Proteins are acetone precipitated as described earlier and resuspended in 40 μl of HENS buffer. After this step, S-nitrosated proteins are biotin tagged and there is no need to protect samples from light.

6. Detection and Purification of Biotinylated Proteins

The detection of biotinylated proteins can be carried out by Western blot with antibiotin antibodies or by using streptavidin conjugated to alkaline phosphatase or horseradish peroxidase. Because biotin-HPDP is cleavable under reducing conditions, prepared samples are loaded onto SDS-PAGE gels without reducing agents and, to prevent nonspecific reactions of biotin-HPDP, are not boiled before electrophoresis. We usually separate samples from the biotin switch assay on 10% SDS-PAGE gels; after transfer of proteins to nitrocellulose membranes, they are incubated with a primary monoclonal antibiotin antibody (Sigma-Aldrich, St. Louis, MO; 1/2500 dilution) for 1 h and a secondary antibody coupled to horseradish peroxidase (Santa Cruz Biotechnology Inc., Santa Cruz, CA). Inmunoreactive proteins are then visualized using the ECL Advance detection system (Amersham Biosciences, Uppsala, Sweden). Appropriate controls in which hepatocyte lysates are processed in the absence of HPDP-biotin should be included to distinguish endogenously biotinylated proteins.

Biotinylated proteins can be alternatively purified for proteomic identification. In this case, the starting material for the biotin switch assay should be in the order of 3 to 4 mg protein. Samples are diluted in 2 volumes of neutralization buffer (20 mM HEPES, pH 7.7, 0.1 M NaCl, 1 mM EDTA, and 0.5% Triton X-100). Add 18 μl of streptavidin agarose/mg of protein used in the initial protein sample. We use EZview Red streptavidin-agarose (Sigma-Aldrich) washed previously in neutralization buffer. Use of a colored resin helps minimize accidental aspiration of the resin during the washing and discarding steps that follow. Biotinylated proteins are incubated with the resin for 3 h at room temperature with continuous gentle inversion agitation. Samples are then centrifuged at 8200g, 4° for 30 s, and supernatants containing unbound proteins are discarded. The resin is washed five times with washing buffer (20 mM HEPES, pH 7.7, 600 mM NaCl, 1 mM EDTA, and 0.5% Triton X-100) and centrifuging as described previously. To elute bound proteins, the resin is incubated with 1 volume of elution buffer (20 mM HEPES, pH 7.7, 0.1 M NaCl, 1 mM EDTA, and 0.1 M 2-mercaptoethanol) for 20 min at 37° with gentle inversion agitation. As the biotin is incorporated via a disulfide bond and elution is carried out in reducing conditions, endogenously biotinylated proteins are not purified. Supernatants are collected and purified proteins are concentrated with a Microcon YM-3 centrifugal filter unit (3 kDa NMWL, Millipore, Billerica, MA) and separated on 10% SDS-PAGE. In our experience, two-dimensional electrophoresis results in a low yield for protein identification. Gels are stained with Sypro Ruby protein stain (Biorad) and protein bands excised, in-gel digested with trypsin and proteins identified by mass spectrometry (MS) using matrix-assisted laser desorption/ionization–time of flight (MALDI-TOF) and peptide mass fingerprinting. Gel bands are excised using a robotic workstation (Investigator Propic, Genomics Solutions, Ann Harbor, MI) and trypsin digested using a robotic digestion system (ProGest, Genomic Solutions). Tryptic digests are then analyzed on a MALDI-TOF/TOF 4700 proteomics analyzer (Applied Biosystems, Foster City, CA). Alternatively, eluted proteins can be precipitated using trichloroacetic acid/acetone, and trypsin digested for liquid chromatography-MS/MS sequencing of the resulting peptides. This alternative has the advantage over gel studies in that a high yield results in a higher number of identified proteins. An ESI-MS/MS analysis is performed using a microcapillary reversed-phase LC system (CapLC, Waters, Milford, MA) coupled online to a Q-TOF Micro (Waters) using a PicoTip nanospray ionization source (Waters). MS/MS data are collected in an automated data-dependent mode, and the three most intense ions in each survey scan are fragmented sequentially by collision-induced dissociation. Data processing is performed with MassLynx 4 and ProteinLynx Global Server 2 (Waters).

7. Final Considerations

The levels of protein S-nitrosothiols generated *in vivo* are too low for a proteomic analysis with the current available methods. Therefore, most studies have relied on exogenous treatments to increase the intracellular pool of S-nitrosothiols. It is important to assess if there is enough S-nitrosothiol in the sample under study to make the biotin switch assay results meaningful. The total S-nitrosothiol content of our protein samples is measured by a displacement chemiluminescence technique using tri-iodide as the reductant (Lopez-Sanchez et al., 2007). S-Nitrosothiols decompose rapidly in the presence of mercuric ion Hg^{2+}, and $HgCl_2$ has been used as a diagnostic marker for the presence of S-nitrosothiols in chemiluminescence techniques. However, it has been demonstrated that Hg^{2+} interferes with thiol labeling by biotin-HPDP (Zhang et al., 2005), and we do not recommend the pretreatment of samples with $HgCl_2$ as a diagnostic indicator of S-nitrosothiols in the biotin switch assay.

Variations on the methodology pioneered by Jaffrey et al. (2001) demonstrate the versatility of this method in the study of S-nitrosothiol biology. For example, biotinylated proteins can be trypsin digested before streptavidin purification of biotinylated peptides. Then, after MS analysis, cysteine residues in identified peptides will be putative cysteine S-nitrosation sites (Hao et al., 2006). Alternatively, if peptides bound to streptavidin are eluted with formic acid instead of with mercaptoethanol, biotin remains associated to them and MS analysis can identify biotinylation sites directly (Greco et al., 2006). However, radioactive probes can be used instead of biotin-HPDP with the advantage that the stoichiometry of S-nitrosation can be quantified and experiments of peptide mapping can be performed using this approach (Jaffrey et al., 2002). A detailed description of some of these methodological modifications can be found in other chapters in this volume.

ACKNOWLEDGMENTS

This work was supported by grants from the Programa de Promoción de la Investigación en Salud del Ministerio de Sanidad y Consumo (PI04/1470) and Consejería de Salud de la Junta de Andalucía (099/06).

REFERENCES

Forrester, M. T., Foster, M. W., and Stamler, J. S. (2007). Assessment and application of the biotin switch technique for examining protein S-nitrosylation under conditions of pharmacologically induced oxidative stress. *J. Biol. Chem.* **282**, 13977–13983.

Gaston, B., Singel, D., Doctor, A., and Stamler, J. S. (2006). S-nitrosothiol signaling in respiratory biology. *Am. J. Respir. Crit. Care Med.* **173**, 1186–1193.

Greco, T. M., Hodara, R., Parastatidis, I., Heijnen, H. F., Dennehy, M. K., Liebler, D. C., and Ischiropoulos, H. (2006). Identification of S-nitrosylation motifs by site-specific mapping of the S-nitrosocysteine proteome in human vascular smooth muscle cells. *Proc. Natl. Acad. Sci. USA* **103**, 7420–7425.

Hao, G., Derakhshan, B., Shi, L., Campagne, F., and Gross, S. S. (2006). SNOSID, a proteomic method for identification of cysteine S-nitrosylation sites in complex protein mixtures. *Proc. Natl. Acad. Sci. USA* **103**, 1012–1017.

Jaffrey, S. R., Erdjument-Bromage, H., Ferris, C. D., Tempst, P., and Snyder, S. H. (2001). Protein S-nitrosylation: A physiological signal for neuronal nitric oxide. *Nat. Cell Biol.* **3**, 193–197.

Jaffrey, S. R., Fang, M., and Snyder, S. H. (2002). Nitrosopeptide mapping: A novel methodology reveals S-nitrosylation of Dexras1 on a single cysteine residue. *Chem. Biol.* **9,** 1329–1335.

Kettenhofen, N. J., Broniowska, K. A., Keszler, A., Zhang, Y., and Hogg, N. (2007). Proteomic methods for analysis of S-nitrosation. *J. Chromatogr. B Analyt. Technol. Biomed. Life Sci.* **851,** 152–159.

Liu, L., Hausladen, A., Zeng, M., Que, L., Heitman, J., and Stamler, J. S. (2001). A metabolic enzyme for S-nitrosothiol conserved from bacteria to humans. *Nature* **410,** 490–494.

Lopez-Sanchez, L. M., Collado, J. A., Corrales, F. J., Lopez-Cillero, P., Montero, J. L., Fraga, E., Serrano, J., De la Mata, M., Muntane, J., and Rodriguez-Ariza, A. (2007). S-Nitrosation of proteins during D-galactosamine-induced cell death in human hepatocytes. *Free Radic. Res.* **41,** 50–61.

Mallis, R. J., and Thomas, J. A. (2000). Effect of S-nitrosothiols on cellular glutathione and reactive protein sulfhydryls. *Arch. Biochem. Biophys.* **383,** 60–69.

Ranchal, I., Gonzalez, R., Lopez-Sanchez, L. M., Barrera, P., Lopez-Cillero, P., Serrano, J., Bernardos, A., De la Mata, M., Rodriguez-Ariza, A., and Muntane, J. (2006). The differential effect of PGE(1) on D-galactosamine-induced nitrosative stress and cell death in primary culture of human hepatocytes. *Prostaglandins Other Lipid Mediat.* **79,** 245–259.

Shah, C. M., Locke, I. C., Chowdrey, H. S., and Gordge, M. P. (2003). Rapid S-nitrosothiol metabolism by platelets and megakaryocytes. *Biochem. Soc. Trans.* **31,** 1450–1452.

Stamler, J. S., Lamas, S., and Fang, F. C. (2001). Nitrosylation: The prototypic redox-based signaling mechanism. *Cell* **106,** 675–683.

Zhang, Y., and Hogg, N. (2005). S-Nitrosothiols: Cellular formation and transport. *Free Radic. Biol. Med.* **38,** 831–838.

Zhang, Y., Keszler, A., Broniowska, K. A., and Hogg, N. (2005). Characterization and application of the biotin-switch assay for the identification of S-nitrosated proteins. *Free Radic. Biol. Med.* **38,** 874–881.

CHAPTER EIGHTEEN

IDENTIFICATION OF S-NITROSYLATED PROTEINS IN PLANTS

Simone Sell, Christian Lindermayr, *and* Jörg Durner

Contents

1. Introduction 283
2. Generation of Protein Nitrosothiols 285
3. Blocking Reaction of Free Thiols 287
4. Reduction of Nitrosothiols and S-Biotinylation 288
5. Affinity Purification of Biotinylated Proteins by NeutrAvidin 289
6. Modified Techniques Related to the Biotin Switch Assay 290
References 291

Abstract

Posttranslational protein modifications affect the function or the activity of proteins and exhibit important mechanisms in regulating cellular events. A broad spectrum of modifications is known, including redox-dependent alterations. During the last decade, covalent binding of nitric oxide (NO) to protein cysteines, termed S-nitrosylation, seems especially an evident process for redox-related signaling. To reveal potential target proteins for S-nitrosylation, the biotin switch method gains more and more in importance. This technique is a tool used for analyzing the nitrosylome as well as the examination of single candidates. It is based on substitution of the NO group by a biotin linker that simplifies the detection and the purification of recently S-nitrosylated proteins in a three-step procedure.

1. INTRODUCTION

Nitric oxide (NO) is a highly reactive molecule, which has a large-scale influence on redox-regulated signal transduction in mammalians, plants, and microorganisms. It affects processes such as defense, growth and development, neurotransmission, vasodilation, and inflammation (Bethke

Institute of Biochemical Plant Pathology, Helmholtz Zentrum München, German Research Center for Environmental Health, Munich-Neuherberg, Germany *et al.*, 2007; Bove and van der Vliet, 2006; Cohen and Adachi, 2006; Garcia-Brugger *et al.*, 2006; Moncada and Bolanos, 2006; Tuteja *et al.*, 2004; Wendehenne *et al.*, 2004). Under physiological conditions NO is generated enzymatically, e.g., by NO synthases or nitrate reductase, or nonenzymatically from nitrate/nitrite. An important receptor of NO is the soluble guanylate cyclase. Binding of NO to the prosthetic heme Fe^{2+} of guanylate cyclase yields in the production of cyclic guanosine monophosphate, which acts as a second messenger in diverse biological processes (Russwurm and Koesling, 2004). However, within the cyclic guanosine monophosphate-independent signaling pathway, NO reacts with redox-sensitive cysteine residues of proteins or low molecular weight thiols such as glutathione (GSH) or cysteine. In detail, NO is attached covalently on the sulfur atom of a cysteine residue, which results in the formation of nitrosothiols (SNO). Since this reaction is reversible, *S*-nitrosylation exhibits an ubiquitous redox-related posttranslational protein modification (Hess *et al.*, 2005, Stamler *et al.*, 1992, 2001). In contrast to, for example, phosphorylation, the mechanism of *S*-nitrosylation does not follow a general reaction scheme. On the molecular level, its mode of action notably depends on the microenvironment of the protein being modified, the concentration of reactants, and the redox state of the biological system (Foster *et al.*, 2003). The generation of nitrosothiols requires an electron acceptor such as oxygen, NAD^+, Fe^{2+}, Zn^{2+}, or Cu^{2+} (Gergel and Cederbaum, 1996; Gow *et al.*, 1997; Vanin *et al.*, 1997). Because of its relative high instability, this modification represents a very important cellular redox-based signaling mechanism and several stress-related, metabolic, and signaling proteins, as well as transcription factors, were already identified as targets for this type of modification (Belenghi *et al.*, 2007, Hashemy *et al.*, 2007, Lindermayr *et al.*, 2006). Generally, S-nitrosylation of proteins can result in altering their function, activity, or stability (Azad *et al.,* 2006, Hara et al., 2005, Kim *et al.,* 2004, Li *et al.,* 2007).

Factors that determine the specificity for *S*-nitrosylation targets are diverse. Nucleophilicity of thiols, their hydrophobic neighborhood, and allosteric regulators affect the accessibility and reactivity of cysteine residues (Eu *et al.*, 2000; Hess *et al.*, 2001, 2005). Additionally, flanking amino acids, which fit the proposed "acid–base" motif, were also shown to be an important modulator for *S*-nitrosylation (Greco *et al.*, 2006; Stamler *et al.*, 1997).

Because *S*-nitrosothiols are quite instable, Jaffrey and colleagues developed a three-step procedure, named the biotin switch method, where nitroso residues are stably labeled with a biotin linker via a disulfide bridge (Fig. 18.1) (Jaffrey and Snyder, 2001; Jaffrey *et al.*, 2001). Afterward, previously *S*-nitrosylated proteins can be proved simply by immunoblot analysis using commercially available antibiotin antibodies. Furthermore, they can be purified by affinity chromatography and identified using either protein-specific antibodies or by mass spectrometry. The biotin switch assay

Identification of S-Nitrosylated Proteins in Plants

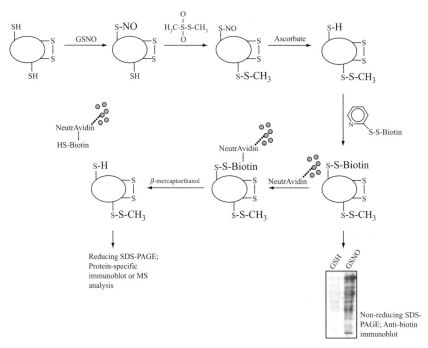

Figure 18.1 Schematic demonstration of the biotin switch technique. A target cysteine residue of a hypothetical protein is transnitrosylated by GSNO, followed by the blocking of free thiols using MMTS. Afterward, nitrosylated cysteine residues are reduced selectively with ascorbate, and the newly generated free thiols are finally S-biotinylated with biotin-HPDP. The biotinylated proteins can be detected directly by immunoblotting with an antibiotin antibody or can be purified by affinity chromatography using the biotin-binding NeutrAvidin matrix.

was applied successfully in diverse organisms and tissues (Jaffrey et al., 2001; Kuncewicz et al., 2003; Lindermayr et al., 2005; Martinez-Ruiz and Lamas, 2007; Rhee et al., 2005).

2. Generation of Protein Nitrosothiols

Crude extracts from plant tissues or cell compartments, as well as recombinant purified proteins, can be used as starting material to analyze protein S-nitrosylation. To avoid nonspecific S-nitrosylation, proteins should be prepared carefully to guarantee their native conformation. Furthermore, if reducing agents such as dithiothreitol, β-mercaptoethanol, or GSH are used for protein extraction, they have to be removed to allow effective S-nitrosylation of proteins. To protect nitrosothiols from metal-catalyzed denitrosylation, working buffers should contain EDTA and neocuproine to chelate divalent metals and Cu(I). As S-nitrosylating agents, S-nitrosoglutathione (GSNO), an endogenous NO donor, or other NO-releasing compounds, such as sodium nitroprusside, S-nitroso-N-acetyl-DL-penicillamine, or 2,2-(hydroxynitrosohydrazono)bis-ethanamine, can be used. Some laboratories might have equipment to use gaseous NO (Fig. 18.2) (Lindermayr et al., 2005). To reduce light-induced decomposition of nitrosothiols, the samples should be kept in the dark. Apart from in vitro S-nitrosylation, the endogenous NO-producing systems can be induced to analyze and identify in vivo S-nitrosylated proteins. This is a challenging approach, as it gives hints to the physiological importance of protein S-nitrosylation. The constitutively endogenous S-nitrosylation of an Arabidopsis metacaspse, for example, was demonstrated by the biotin switch method (Belenghi et al., 2007).

1. Extract proteins from the tissue of interest directly with HEN buffer [25 mM HEPES-NaOH, pH 7.7, 1 mM EDTA, 0.1 mM neocuproine (Sigma)]. If isolated proteins of subcellular compartments or purified recombinant proteins are used, a buffer exchange with HEN buffer by gel filtration [e.g., PD-10 columns (GE Healthcare) or Micro Biospin columns (Bio-Rad)] is strongly recommended.
2. Determine the protein concentration (e.g., according to Bradford). To ensure an optimal signal-to-noise ratio, adjust the protein concentration to a maximum of 0.8 μg/μl HEN, as otherwise S-nitrosylation is ineffective and subsequently the blocking of unmodified free thiols may be incomplete.

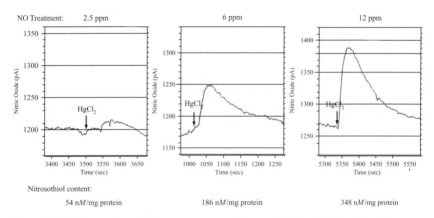

Figure 18.2 Electrochemical measurement of nitrosothiols in *Arabidopsis* leaf extracts. *Arabidopsis* plants were treated with 2.5, 6, or 12 ppm gaseous nitric oxide. After 20 min, leaves were harvested and proteins were extracted with HEN buffer. Nitric oxide was released from the nitrosothiols by the addition of mercury(II) chloride at a final concentration of 0.06% and determined by a nitric oxide-sensitive electrode (Apollo 4000, World Precision Instruments). Based on sensor calibration, the nitrosothiol content was calculated.

3. For immunoblot detection, the use of 100 μg crude extract or 5 to 10 μg purified recombinant protein is sufficient, whereas proteomic identification by mass spectrometry requires 10 to 50 mg starting material.
4. Treat the protein sample with 10 to 500 μM GSNO [stock solution: 10 mM GSNO (Alexxis) in water] at 25° in the dark for 20 to 30 min. As a negative control, incubate a sample aliquot with GSH (Sigma). Keep the samples in the dark to minimize light-induced decomposition of nitrosothiols.
5. To demonstrate the specificity of the biotin switch method, several control treatments should be included. Add 10 mM dithiothreitol (DTT) to a GSNO-treated sample and continue with the incubation for 10 min. Another independent control that verifies protein S-nitrosylation is the SNO photolysis by ultraviolet (UV) light. After the nitrosylation reaction, one sample aliquot is incubated in a polypropylene tube on a UV transilluminator for 5 min. Both treatments lead to NO release from protein SNOs and, as consequence, should not show any biotinylation signal after subjecting them to the biotin switch assay.
6. Optional: Perform an additional gel filtration step to remove NO donors and/or reducing agents. This serves to prevent artificial S-nitrosylation or ineffective trapping of free thiols during the following blocking reaction. However, because of the high molar excess of the blocking reagent, we could not determine any differences between samples that were or were not desalted at this stage.

3. Blocking Reaction of Free Thiols

During the biotin switch assay, S-nitrosylated cysteine residues are reduced and labeled with biotin. To avoid unspecific biotinylation the free cysteine sulfur groups, these residues have to be blocked. S-Methyl methanethiosulfonate (MMTS) traps thiol groups highly efficiently and specifically and does not react with other amino acid residues. The addition of sodium dodecyl sulfate (SDS) guarantees the access of MMTS to sterically concealed free thiols. The formed S-methylthiolates, however, are potential targets for nucleophilic attack by further free protein thiols, which could result in the formation of intra- and intermolecular disulfide bridges (Karala and Ruddock, 2007). However, these reactions do not generate false positives during the biotin switch method.

1. Denature the proteins and trap nonmodified free thiols by adding final concentrations of 2% SDS [stock solution: 25% (w/v) in ddH$_2$O] and 20 mM MMTS [stock solution: 2 M MMTS (Sigma) in dimethylformamide (DMF), freshly prepared] to the sample. Incubate the reaction at 50° for 20 min and vortex frequently to guarantee complete blocking of

free thiol groups. If the blocking is inefficient, increase the MMTS concentration instead of prolonging the incubation time to prevent SNO degradation.
2. Remove residual MMTS from the protein sample by adding 2 volumes of ice-cold acetone and precipitate the proteins at $-20°$ for at least 20 min. To increase the yield of protein recovery, the precipitation time can be extended overnight.
3. Recover the proteins by centrifugation at $10,000\,g$ and $4°$ for 10 min. Rinse the pellet and the walls of the tubes with acetone to remove residual MMTS and dry the pellet for about 5 min.

4. Reduction of Nitrosothiols and S-Biotinylation

Ascorbate is a physiological antioxidant that serves as a SNO-specific reducing agent in the biotin switch assay. After ascorbate treatment the newly generated SH groups are S-biotinylated using biotin-HPDP. Other redox modifications on cysteines should remain unaffected by ascorbate. However, because the biotin switch technique becomes more and more important for the detection of S-nitrosothiols, the number of critical publications arises. Most of them doubt the specificity of reactivity of ascorbate and argue that ascorbate leads to false positives. For example, prereduced bovine serum albumin, which was subjected to the biotin switch procedure, was biotinylated and even disulfide bridges were reduced by ascorbate (Dahm *et al.*, 2006; Huang and Chen, 2006; Kuncewicz *et al.*, 2003; Landino *et al.*, 2006). Based on these reports, Forrester *et al.* (2007) performed comprehensive analyses that examined the reducing behavior of ascorbate. They recognized that indirect sunlight during the biotinylation reaction might be responsible for SNO-independent and ascorbate-dependent protein biotinylation. Thus, in indirect sunlight, ascorbate reduces biotin-HPDP to biotin-SH, which is responsible for unspecific protein biotinylation via thiol/disulfide exchange at blocked cysteines.

1. Resuspend the precipitated proteins in 10 μl HENS buffer [25 mM HEPES-NaOH, pH 7.7, 1 mM EDTA, 0.1 mM neocuproine (Sigma), 1% SDS] per 100 μg starting material of protein.
2. Add 1 mM ascorbate and 1 mM biotin-HPDP (stock solution: 4 mM N-[6-(biotin amido)hexyl]-3'-(2'-pyridyldithio)propionamide, Perbio) in water-free DMF) and incubate the reaction mixture for 1 h at 25°. For high abundant proteins, 1 mM ascorbate should be sufficient for the detection of biotinylation; however, higher ascorbate concentrations might be necessary to increase sensitivity. In contrast to some earlier publications, it is strongly recommended to perform the reduction/biotinylation reaction

in the darkness to prevent unspecific biotinylation. Regarding the doubtful specificity of SNO reduction by ascorbate, further controls might be useful.

  a. Perform the S-biotinylation reaction in the absence of ascorbate to verify the ascorbate dependency of the biotin modification.
  b. Moreover, treat one sample without biotin-HPDP or with DTT after biotinylation to check endogenous biotinylation.

3. Separate the biotinylated proteins by SDS-PAGE under nonreducing conditions (without boiling the sample prior to SDS-PAGE loading) and transfer them onto PVDF or a nitrocellulose membrane. Biotinylated proteins can be detected using commercially available antibiotin antibodies (Fig. 18.3). For purification of biotinylated proteins, remove excessive biotin-HPDP by precipitating the proteins with 2 volumes of acetone at $-20°$. Recover the proteins by centrifugation, wash the pellet with acetone, and resuspend them in HENS buffer (≈ 10 mg/ml).

5. AFFINITY PURIFICATION OF BIOTINYLATED PROTEINS BY NEUTRAVIDIN

NeutrAvidin is a deglycosylated version of avidin with a tetrameric structure and a strong affinity to biotin. Deglycosylation does not affect the biotin-binding activity of NeutrAvidin, but shifts its isoelectric point to 6.3, resulting in minimized nonspecific interactions. The elution of bound

Treatment	ddH$_2$O	GSH	GSNO					
Conc. [mM]	+	0.5	0.05	0.1	0.25	0.5	0.5	0.5
Additives	-	-	-	-	-	-	10 mM DTT a.n.	w/o Biotin	10 mM DTT a.b.
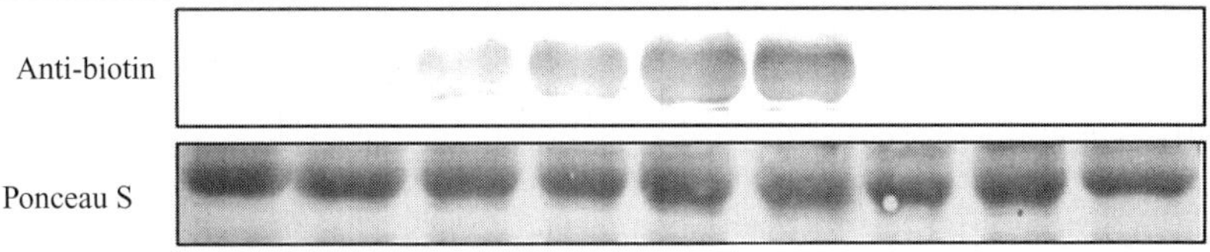 Anti-biotin									
Ponceau S									

Figure 18.3 Phosphoglycerate dehydrogenase (PGDH) is specifically S-nitrosylated *in vitro* by the biotin switch assay. Nine micrograms of recombinant, His-tagged, and purified PGDH was treated with increasing GSNO concentrations (0.05–0.5 mM; lanes 3–6), with 0.5 mM GSH (lane 2), or left untreated (lane 1). DTT (10 mM) was added to S-nitrosylated PGDH before being subjected to the biotin switch method (10 mM DTT a.n., lane 7). As controls, the S-biotinylation reaction was performed without biotin-HPDP (lane 8). Moreover, biotinylated PGDH was incubated with 10 mM DTT for 10 min before being applied on nonreducing SDS-PAGE (10 mM DTT a.b., lane 11). Electrophoretically separated proteins were transferred onto a nitrocellulose membrane. Ponceau S staining of blotted proteins demonstrates equal sample loading (bottom). Biotinylated PGDH was detected by antibiotin immunoblotting (top).

biotinylated proteins is achieved by adding reducing agents, such as β-mercaptoethanol or DTT, which reduce the disulfide bridge between the protein and the biotin linker.

1. Equilibrate 30 μl of the 50% NeutrAvidin agarose slurry (Perbio) per milligram protein with neutralization buffer [20 mM HEPES-NaOH, pH 7.7, 0.1 M NaCl, 1 mM EDTA, 0.5% (v/v) Triton X-100].
2. Add at least 2 volumes of neutralization buffer to the biotinylated proteins in HENS buffer, as well as the equilibrated NeutrAvidin–agarose. Because SDS can decrease biotin-NeutrAvidin-binding efficiency, up to 20 volumes of neutralization buffer can be added to the biotinylated proteins to dilute the SDS concentration of the HENS buffer. Incubate at 25° for 1 h or overnight at 4° with gentle shaking.
3. Centrifuge samples for 1 min at 200 g, discard the supernatant, wash the matrix two times with 10 matrix volumes of washing buffer [20 mM HEPES-NaOH, pH 7.7, 600 mM NaCl, 1 mM EDTA, 0.5% (v/v) Triton X-100]. Centrifuge for 1 min at 200 g between each washing step. Transfer the matrix into an empty column and perform a third washing step on the column.
4. Close the column, add at least 2 matrix volumes of elution buffer (20 mM HEPES-NaOH, pH 7.7, 0.1 M NaCl, 1 mM EDTA, 0.1 M β-mercaptoethanol), and eluate the bound proteins after incubation for 5 min.
5. Precipitate the purified proteins with 2 volumes of acetone at $-20°$ preferably overnight, recover the protein by centrifugation (30 min at 4° and 25,000 g), and follow up the reducing SDS-PAGE (Fig. 18.4).

6. Modified Techniques Related to the Biotin Switch Assay

Modifications of the biotin switch method have emerged. SNO site identification was developed as a proteomic approach that enables simultaneous identification of SNO cysteine sites and their cognate proteins in complex biological mixtures (Hao et al., 2006). The biotinylated proteins were tryptic digested, and the resulting peptides containing target cysteines for S-nitrosylation were purified by pull down using NeutrAvidin and were identified by mass spectrometry.

Another related technique is the His tag switch method, which comprises (1) the blocking of free thiols by N-ethylmaleimide alkylation, (2) the SNO-specific reducing step with ascorbate, and (3) the addition of a His tag by alkylation for detection (Camerini et al., 2007). This strategy allows purification and unambiguous identification of the modified cysteines by mass spectrometry.

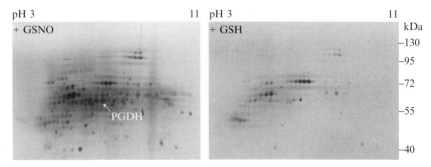

Figure 18.4 S-Nitrosylated proteins of *Arabidopsis thaliana* cell cultures. Thirty-five milligrams of protein extracted from an *A. thaliana* suspension culture was treated with 250 μM GSNO or GSH as negative control and subjected to the biotin switch method. Biotinylated proteins were purified by NeutrAvidin pull down. Eluates containing approximately 200 μg protein were separated by two-dimensional gel electrophoresis. In the first dimension, proteins were separated according to their isoelectric point in a nonlinear gradient (pH 3–11). To separate proteins in the second dimension regarding their molecular mass, a 12% SDS-PAGE was applied. Proteins were visualized by silver stain. The white arrow indicates the S-biotinylated and hence S-nitrosylated phosphoglycerate dehydrogenase (PGDH) that was identified by peptide mass fingerprint analysis with matrix-assisted laser desorption ionization mass spectrometry.

REFERENCES

Azad, N., Vallyathan, V., Wang, L., Tantishaiyakul, V., Stehlik, C., Leonard, S. S., and Rojanasakul, Y. (2006). S-nitrosylation of Bcl-2 inhibits its ubiquitin-proteasomal degradation. A novel antiapoptotic mechanism that suppresses apoptosis. *J. Biol. Chem.* **281,** 34124–34134.

Belenghi, B., Romero-Puertas, M. C., Vercammen, D., Brackenier, A., Inze, D., Delledonne, M., and Van Breusegem, F. (2007). Metacaspase activity of *Arabidopsis thaliana* is regulated by S-nitrosylation of a critical cysteine residue. *J. Biol. Chem.* **282,** 1352–1358.

Bethke, P. C., Libourel, I. G., Aoyama, N., Chung, Y. Y., Still, D. W., and Jones, R. L. (2007). The Arabidopsis aleurone layer responds to nitric oxide, gibberellin, and abscisic acid and is sufficient and necessary for seed dormancy. *Plant Physiol.* **143,** 1173–1188.

Bove, P. F. and van der Vliet, A. (2006). Nitric oxide and reactive nitrogen species in airway epithelial signaling and inflammation. *Free Radic. Biol. Med.* **41,** 515–527.

Camerini, S., Polci, M. L., Restuccia, U., Usuelli, V., Malgaroli, A., and Bachi, A. (2007). A novel approach to identify proteins modified by nitric oxide: The HIS-TAG switch method. *J Proteome Res.* **6,** 3224–3231.

Cohen, R. A. and Adachi, T. (2006). Nitric-oxide-induced vasodilatation: Regulation by physiologic s-glutathiolation and pathologic oxidation of the sarcoplasmic endoplasmic reticulum calcium ATPase. *Trends Cardiovasc. Med.* **16,** 109–114.

Dahm, C. C., Moore, K., and Murphy, M. P. (2006). Persistent S-nitrosation of complex I and other mitochondrial membrane proteins by S-nitrosothiols but not nitric oxide or peroxynitrite: Implications for the interaction of nitric oxide with mitochondria. *J. Biol. Chem.* **281,** 10056–10065.

Eu, J. P., Sun, J., Xu, L., Stamler, J. S., and Meissner, G. (2000). The skeletal muscle calcium release channel: Coupled O2 sensor and NO signaling functions. *Cell* **102,** 499–509.

Forrester, M. T., Foster, M. W., and Stamler, J. S. (2007). Assessment and application of the biotin switch technique for examining protein S-nitrosylation under conditions of pharmacologically induced oxidative stress. *J. Biol. Chem.* **282,** 13977–13983.

Foster, M. W., McMahon, T. J., and Stamler, J. S. (2003). S-nitrosylation in health and disease. *Trends Mol. Med.* **9,** 160–168.

Garcia-Brugger, A., Lamotte, O., Vandelle, E., Bourque, S., Lecourieux, D., Poinssot, B., Wendehenne, D., and Pugin, A. (2006). Early signaling events induced by elicitors of plant defenses. *Mol. Plant Microbe Interact.* **19,** 711–724.

Gergel, D. and Cederbaum, A. I. (1996). Inhibition of the catalytic activity of alcohol dehydrogenase by nitric oxide is associated with S nitrosylation and the release of zinc. *Biochemistry* **35,** 16186–16194.

Gow, A. J., Buerk, D. G., and Ischiropoulos, H. (1997). A novel reaction mechanism for the formation of S-nitrosothiol *in vivo*. *J. Biol. Chem.* **272,** 2841–2845.

Greco, T. M., Hodara, R., Parastatidis, I., Heijnen, H. F., Dennehy, M. K., Liebler, D. C., and Ischiropoulos, H. (2006). Identification of S-nitrosylation motifs by site-specific mapping of the S-nitrosocysteine proteome in human vascular smooth muscle cells. *Proc. Natl. Acad. Sci. USA* **103,** 7420–7425.

Hao, G., Derakhshan, B., Shi, L., Campagne, F., and Gross, S. S. (2006). SNOSID, a proteomic method for identification of cysteine S-nitrosylation sites in complex protein mixtures. *Proc. Natl. Acad. Sci. USA* **103,** 1012–1017.

Hara, M. R., Agrawal, N., Kim, S. F., Cascio, M. B., Fujimuro, M., Ozeki, Y., Takahashi, M., Cheah, J. H., Tankou, S. K., Hester, L. D., Ferris, C. D., Hayward, S. D., *et al.* (2005). S-nitrosylated GAPDH initiates apoptotic cell death by nuclear translocation following Siah1 binding. *Nat. Cell Biol.* **7,** 665–674.

Hashemy, S. I., Johansson, C., Berndt, C., Lillig, C. H., and Holmgren, A. (2007). Oxidation and S-nitrosylation of cysteines in human cytosolic and mitochondrial glutaredoxins: Effects on structure and activity. *J. Biol. Chem.* **282,** 14428–14436.

Hess, D. T., Matsumoto, A., Kim, S. O., Marshall, H. E., and Stamler, J. S. (2005). Protein S-nitrosylation: Purview and parameters. *Nat. Rev. Mol. Cell Biol.* **6,** 150–166.

Hess, D. T., Matsumoto, A., Nudelman, R., and Stamler, J. S. (2001). S-nitrosylation: Spectrum and specificity. *Nat. Cell Biol.* **3,** E46–E49.

Huang, B. and Chen, C. (2006). An ascorbate-dependent artifact that interferes with the interpretation of the biotin switch assay. *Free Radic. Biol. Med.* **41,** 562–567.

Jaffrey, S. R., Erdjument-Bromage, H., Ferris, C. D., Tempst, P., and Snyder, S. H. (2001). Protein S-nitrosylation: A physiological signal for neuronal nitric oxide. *Nat. Cell Biol.* **3,** 193–197.

Jaffrey, S. R. and Snyder, S. H. (2001). The biotin switch method for the detection of S-nitrosylated proteins. *Sci. STKE* **2001,** PL1.

Karala, A. R. and Ruddock, L. W. (2007). Does s-methyl methanethiosulfonate trap the thiol-disulfide state of proteins? *Antioxid. Redox Signal* **9,** 527–531.

Kim, S., Wing, S. S., and Ponka, P. (2004). S-nitrosylation of IRP2 regulates its stability via the ubiquitin-proteasome pathway. *Mol. Cell. Biol.* **24,** 330–337.

Kuncewicz, T., Sheta, E. A., Goldknopf, I. L., and Kone, B. C. (2003). Proteomic analysis of s-nitrosylated proteins in mesangial cells. *Mol. Cell. Proteomics* **2,** 156–163.

Landino, L. M., Koumas, M. T., Mason, C. E., and Alston, J. A. (2006). Ascorbic acid reduction of microtubule protein disulfides and its relevance to protein S-nitrosylation assays. *Biochem. Biophys. Res. Commun.* **340,** 347–352.

Li, F., Sonveaux, P., Rabbani, Z. N., Liu, S., Yan, B., Huang, Q., Vujaskovic, Z., Dewhirst, M. W., and Li, C. Y. (2007). Regulation of HIF-1alpha stability through S-nitrosylation. *Mol. Cell* **26,** 63–74.

Lindermayr, C., Saalbach, G., Bahnweg, G., and Durner, J. (2006). Differential inhibition of Arabidopsis methionine adenosyltransferases by protein s-nitrosylation. *J. Biol. Chem.* **281**, 4285–4291.

Lindermayr, C., Saalbach, G., and Durner, J. (2005). Proteomic identification of s-nitrosylated proteins in Arabidopsis. *Plant Physiol.* **137**, 921–930.

Martinez-Ruiz, A. and Lamas, S. (2007). Proteomic identification of S-nitrosylated proteins in endothelial cells. *Methods Mol. Biol.* **357**, 215–223.

Moncada, S. and Bolanos, J. P. (2006). Nitric oxide, cell bioenergetics and neurodegeneration. *J. Neurochem.* **97**, 1676–1689.

Rhee, K. Y., Erdjument-Bromage, H., Tempst, P., and Nathan, C. F. (2005). S-nitroso proteome of Mycobacterium tuberculosis: Enzymes of intermediary metabolism and antioxidant defense. *Proc. Natl. Acad. Sci. USA* **102**, 467–472.

Russwurm, M., and Koesling, D. (2004). NO activation of guanylyl cyclase. *EMBO J.* **23**, 4443–4450.

Stamler, J. S., Lamas, S., and Fang, F. C. (2001). Nitrosylation, the prototypic redox-based signaling mechanism. *Cell* **106**, 675–683.

Stamler, J. S., Simon, D. I., Osborne, J. A., Mullins, M. E., Jaraki, O., Michel, T., Singel, D. J., and Loscalzo, J. (1992). S-nitrosylation of proteins with nitric oxide: Synthesis and characterization of biologically active compounds. *Proc. Natl. Acad. Sci. USA* **89**, 444–448.

Stamler, J. S., Toone, E. J., Lipton, S. A., and Sucher, N. J. (1997). (S)NO signals: Translocation, regulation, and a consensus motif. *Neuron* **18**, 691–696.

Tuteja, N., Chandra, M., Tuteja, R., and Misra, M. K. (2004). Nitric oxide as a unique bioactive signaling messenger in physiology and pathophysiology. *J. Biomed. Biotechnol.* **2004**, 227–237.

Vanin, A. F., Malenkova, I. V., and Serezhenkov, V. A. (1997). Iron catalyzes both decomposition and synthesis of S-nitrosothiols: Optical and electron paramagnetic resonance studies. *Nitric Oxide* **1**, 191–203.

Wendehenne, D., Durner, J., and Klessig, D. F. (2004). Nitric oxide: A new player in plant signaling and defence responses. *Curr. Opin. Plant Biol.* **7**, 449–455.

CHAPTER NINETEEN

IDENTIFICATION OF 3-NITROTYOSINE-MODIFIED BRAIN PROTEINS BY REDOX PROTEOMICS

D. Allan Butterfield[*,†,‡] *and* Rukhsana Sultana[*,†]

Contents

1. Introduction	296
2. Materials	297
3. Method	298
3.1. Gel electrophoresis apparatus and immobilized pH gradient (IPG) strips	298
3.2. Stock solutions for sample homogenization	299
3.3. Composition of rehydration buffer (isoelectric focusing buffer)	299
3.4. Two-dimensional gel electrophoresis	299
3.5. Immunochemical detection blot	300
3.6. Isoelectric focusing sample preparation	300
3.7. Sample preparation for 3-NT-bound proteins	301
3.8. Isoelectric focusing gel electrophoresis (first dimension)	301
3.9. Sodium dodecyl sulfate-polyacrylamide gel electrophoresis (second dimension)	302
3.10. Protein staining	302
3.11. Western blotting	302
3.12. Image analysis	303
3.13. Identification of proteins with excessive 3-NT bound compared to control	304
3.14. Confirmation of protein identification	305
4. Comments	305
Acknowledgment	305
References	305

[*] Department of Chemistry, University of Kentucky, Lexington, Kentucky
[†] Sanders-Brown Center on Aging, University of Kentucky, Lexington, Kentucky
[‡] Center of Membrane Sciences, University of Kentucky, Lexington, Kentucky

Methods in Enzymology, Volume 440   © 2008 Elsevier Inc.
ISSN 0076-6879, DOI: 10.1016/S0076-6879(07)00819-1   All rights reserved.

Abstract

Two-dimensional (2D) gel electrophoresis allows separation of complex mixtures of proteins based on isoelectric points and relative mobility. This method has not changed much fundamentally since their original description in the late 1970s. Despite several limitations, such as solubilization of membrane proteins and separation of highly basic proteins, this method has been used successfully in many laboratories as part of proteomics protocols. Our laboratory coupled 2D-PAGE with 2D Western blot analysis to identify brain proteins modified oxidatively with excess carbonylation, bound 4-hydroxy-2-nonenal, or 3-nitrotyrosine (3-NT) in various diseases and animal models of these disorders. This chapter describes in detail the protocol used for the identification of 3-NT-modified proteins in biological samples that may help in delineating the role of protein nitration in the progression or pathogenesis of various diseases.

1. Introduction

Oxidative stress, an imbalance in reactive oxygen (ROS) or reactive nitrogen species (RNS) and the antioxidant system, has been shown to play an important role in the pathophysiology of many diseases, such as cardiovascular diseases, neurodegenerative diseases, ischemia, and cancer (Anantharaman *et al.*, 2006; Aulak *et al.*, 2004a; Butterfield and Stadtman, 1997; Dalle-Donne *et al.*, 2005, 2006; Good *et al.*, 1998; Horiguchi *et al.*, 2003; Ischiropoulos and Beckman, 2003; Poladia and Bauer, 2003). ROS and RNS can react with virtually all biological molecules, including proteins, lipids, carbohydrates, DNA, RNA, and antioxidants, leading to their oxidative modification. Proteins constitute one of the major targets of ROS and RNS that may lead to structural changes and loss of protein function, in addition to making a protein more susceptible to protein degradation by the proteasome (Buchczyk *et al.*, 2003; Butterfield *et al.*, 2007a; Grune *et al.*, 1997; Sultana *et al.*, 2006, 2007).

Reactive nitrogen species, such as nitric oxide, react with superoxide free radicals, resulting in the production of peroxynitrite, a highly reactive compound with a half-life of less than a second (Beckman, 2002; Koppenol *et al.*, 1992; Marechal *et al.*, 2007; Smith *et al.*, 1997). One of the pathways of peroxynitrite decomposition includes the production of hydroxyl anion and nitronium cation, a reaction catalyzed by transition metals (Beckman *et al.*, 1992). The electrophilic nitronium species has the ability to react with phenolic compounds, such as tyrosine, leading to the site-specific nitration of protein-bound tyrosine, that is, 3-nitrotyrosine (3-NT) (Ischiropoulos *et al.*, 1992; Souza *et al.*, 2000; van der Vliet *et al.*, 1995). Other pathways for peroxynitrite-mediated 3-NT formation require the presence of CO_2 (Squadrito and Pryor, 2002). The formation of 3-NT in proteins can lead to altered protein conformation, solubility, susceptibility to aggregation, and increased protein degradation (Hyun *et al.*, 2003). Tyrosine nitration could also interfere with phosphorylation signaling, as the relatively bulky nitro group is located near and can sterically interfere with the phosphorylation site of tyrosine (Butterfield and Stadtman, 1997; Di Stasi *et al.*, 1999; Gow *et al.*, 1996; Kong *et al.*, 1996; Mallozzi *et al.*, 1997).

3-Nitrotyrosine has been used a diagnostic marker for the detection of reactive nitrogen species (Dalle-Donne *et al.*, 2005, 2006; Halliwell *et al.*, 1999; Kuhn *et al.*, 2004). Therefore, much effort has been devoted to the development of methods to determine 3-NT in biological samples using immunohistochemistry, high-performance liquid chromatography, gas chromatography, and immunochemical detection (Aulak *et al.*, 2004b; Butterfield *et al.*, 2007b; Castegna *et al.*, 2003; Hensley *et al.*, 1998; Pennathur *et al.*, 1999; Sodum *et al.*, 2000; Sultana *et al.*, 2006, 2007). Analytical techniques, such as gas chromatography–mass spectrometry (GC-MS) and liquid chromatography–mass spectrometry (LC-MS), are used as sensitive techniques for the detection of 3-NT (Amoresano *et al.*, 2007; Jiang and Balazy, 1998; Pennathur *et al.*, 1999; Sodum *et al.*, 2000), but one of the drawbacks of these techniques is the susceptibility of certain derivatization reagents to induce the nitration of tyrosine in the presence of the endogenous nitrate in the samples (Yi *et al.*, 2000).

This chapter describes a method used in our laboratory for the identification of specific protein targets of protein nitration. In order to identify specific targets of nitration, we coupled proteomics with immunochemical detection methods (Fig. 19.1) that allow one to identify a number of protein targets of nitration in brains of subjects with Alzheimer's disease (AD) and mild cognitive impaired (MCI) brain compared to their respective controls (Butterfield *et al.*, 2007a; Castegna *et al.*, 2003; Sultana *et al.*, 2006, 2007). These excessively 3-NT-modified proteins normally regulate various cellular functions. The identification of 3-NT-modified proteins and their dysfunction correlated well with known pathological changes in AD and MCI (Castegna *et al.*, 2003; Sultana *et al.*, 2006, 2007). Despite a number of limitations of this redox proteomics approach (Dalle-Donne *et al.*, 2006), such as efficient solubilization of membrane proteins, analysis of highly basic proteins, and difficulty in detecting low abundance proteins, this technique has been a powerful tool in identifying important protein targets of oxidation and nitration in brain in AD and MCI (Butterfield *et al.*, 2003, 2006a,b, 2007a; Castegna *et al.*, 2003; Sultana *et al.*, 2006, 2007).

2. Materials

Bicinchoninic acid (BCA) assay kit (Pierce)
Immobilized pH gradient gel strip, 3–10 (Bio-Rad, Hercules, CA)

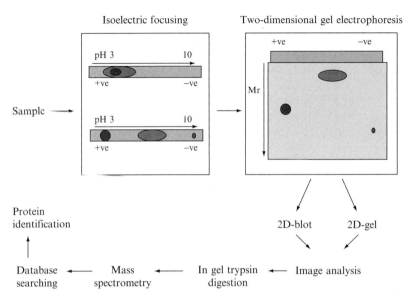

Figure 19.1 Protocol used for the identification of 3-NT-modified proteins in a biological sample. (See color insert.)

Criterion gels, 8–16% (Bio-Rad)
Agarose solution (Bio-Rad)
Dithiothreitol (DTT; Bio-Rad)
Iodoacetamide (Bio-Rad)
Biolytes, isoelectric point (pI) 3–10 (Bio-Rad)
Urea (Bio-Rad)
Thiourea (Bio-Rad)
CHAPS (Sigma Aldrich, St. Louis, MO)
Anti-3-NT antibody (Sigma Aldrich)
Bovine serum albumin (BSA; Sigma Aldrich)
Antirabbit conjugated to alkaline phosphatase antibody (Sigma Aldrich)
Sigma Fast tablet [5-bromo-4-chloro-3-indolyl phosphate/nitro blue tetrazolium (BCIP/NBT)] (Sigma Aldrich)
Nitrocellulose membrane (Bio-Rad)
Whatman filter papers (Bio-Rad)

3. Method

3.1. Gel electrophoresis apparatus and immobilized pH gradient (IPG) strips

We use IPG strips and isoelectric focusing (IEF) equipment from Bio-Rad (Hercules, CA) for first-dimension separation of proteins based on pI. The use of IPG strips improves the reproducibility of results. IPG strips come with different pI ranges that can be selected based on the experimental need. For second-dimension separation of proteins based on relative mobility (that is often related to molecular weight or shape), precasted 8 to 16% linear gradients gels are used from Bio-Rad. The IEF system allows simultaneous IEF focusing of 12 samples, and in our experience they work much better than the tube-gel system. In the tube-gel system the selection of the tube size needs to be done depending on sample size, and also one needs to make sure that the tubes fit perfectly in the apparatus, as the tube gels may slide during isoelectric focusing. Also, if there are any air bubbles on either side of the tube, these will interfere with isoelectric focusing and so on. However, using IPG strips, one can avoid all the aforementioned problems.

3.2. Stock solutions for sample homogenization

1. Homogenization buffer (pH 7.4): 10 mM HEPES, 137 mM NaCl, 4.6 mM KCl, 1.1 mM KH$_2$PO$_4$, 0.6 mM MgSO$_4$, 0.5 μg/ml leupeptin, 0.7 μg/ml pepstatin, 0.5 μg/ml type II S soybean trypsin inhibitor, and 40 μg/ml phenylmethylsulfonyl fluoride. To make this solution, add all the ingredients, dissolve in deionized water, and store at 4°. This solution should be prepared fresh for each use.
2. Wash buffer I: ethanol and ethyl acetate. Add 50:50 (v/v) ethanol:ethyl acetate. This solution should be prepared fresh for each use.

3.3. Composition of rehydration buffer (isoelectric focusing buffer)

Rehydration buffer: 8 M urea, 2 M thiourea, 2% CHAPS, 0.2% biolytes (3/10) (Bio-Rad), 50 mM DTT, and bromophenol blue dissolved in deionized water

To make this solution, weigh an appropriate amount of urea in a small graduated cylinder (10–20 ml). Add all the ingredients in a measuring cylinder or Biologix graduated 15-ml centrifugation tubes. Adjust the final volume. This solution needs to be prepared fresh for each use.

3.4. Two-dimensional gel electrophoresis

1. DTT equilibrium buffer (pH 6.8): 50 mM Tris-HCl, 6 M urea, 1% (m/v) SDS, 30% (v/v) glycerol, and 0.5% DTT. Dissolve the ingredients in deionized water at room temperature. Make fresh before use.
2. IA equilibrium buffer (pH 6.8): 50 mM Tris-HCl, 6 M urea, 1% (w/v), SDS, 30% (v/v), glycerol, and 4.5% iodoacetamide. Dissolve the ingredients in deionized water at room temperature (50-ml Biologix graded centrifugation tubes can be used to prepare this solution). This solution is

light sensitive and should be covered with black or brown paper. Make this solution fresh just before use.
3. Running buffer or electrode buffer (10×): 20% (w/v) glycine, 5% (w/v), tris(hydroxymethyl)aminomethane, and 1% (w/v) SDS. Dissolve the ingredients in deionized water and store at room temperature or purchase 10× running buffer from Bio-Rad.
4. Gel fixing solution: 10% (v/v) methanol and 7% (v/v) acetic acid. Mix the ingredients and make up the volume with deionized water. This solution can be stored at room temperature for a year or until the color of the solution is not yellow.
5. SYPRO Ruby Stain is purchased from Bio-Rad and stored at room temperature.

3.5. Immunochemical detection blot

1. Transfer buffer: 20% (w/v) glycine, 5% (w/v) tris(hydroxymethyl)aminomethane, and 10% (v/v) methanol. Mix all the ingredients and make up the volume with deionized water. Store at 4°. This solution can be reused until the color of the solution changes to yellow.
2. Wash buffer II: 0.01% (w/v) sodium azide and 0.2% (v/v) Tween 20. Dissolve the contents in phosphate-buffered saline, pH 6.8. This solution can be stored at room temperature.
3. Blocking buffer: 2% BSA. Dissolve BSA in wash buffer II. This solution should be prepared fresh before use.
4. Primary antibody solution: Dilute anti-3-NT antibody (1:1000) (Sigma Aldrich, St. Louis, MO/San Antonio, TX) in blocking buffer. Prepare fresh just before use.
5. Secondary antibody solution: Dilute secondary antibody (antirabbit IgG antibody conjugated to alkaline phosphatase, Sigma Aldrich) in wash buffer II (1:3000). Prepare fresh.
6. Developing solution: Dissolve Sigma Fast tablet (BCIP/NBT)(Sigma Aldrich) in 10 ml deionized water. This solution is light sensitive and should be prepared fresh before use.

3.6. Isoelectric focusing sample preparation

Homogenize samples in sample homogenization buffer (20%, w/v). After homogenization, sonicate samples for 10 s on ice to ensure proper breaking of the cells and also to break down chromosomal DNA. Centrifuge the samples at 15,000 g for 5 min at 25° to remove unbroken cells or aggregated DNA. Use the recovered supernatant for determination of protein concentration in the samples by the BCA assay (Pierce).

3.7. Sample preparation for 3-NT-bound proteins

Usually, 150 to 250 μg of the samples is used for the detection of 3-NT-bound proteins. Briefly, add 30% ice-cold acetone to the known concentration of the sample to precipitate the proteins followed by incubation at −80° for 1 h. Centrifuge the samples at 10,000 g for 5 min at 4°. Decant the supernatant and wash the pellet four times with ice-cold wash buffer I at 10,000 g for 5 min at 4°. Dry the final pellet and resuspend it in 200 μl of rehydration buffer. Incubate the samples at room temperature from 1 to 2 h on an orbital shaker to ensure complete solubilization of the proteins. Sonicate the samples at 2 rpm for 10 s.

3.8. Isoelectric focusing gel electrophoresis (first dimension)

For IEF, IPG strips of a required pH are obtained from Bio-Rad depending on the type of experiment desired. Our laboratory normally uses IPG strips of 3 to 10 pH, but sometimes IPG strips of a more narrow pH range are used, i.e., pI 4 to 7, to improve the resolution of specific spots. Using a micropipette, load all 200 μl of the samples into the bottom of the well in an IEF tray carefully. It is important to ensure that no air bubbles are introduced in the tray as they will interfere with current flow and thereby may lead to poor IEF of the sample. Place the IPG strips with gel side facing down on top of the sample and cover the IEF tray and place it in the IEF machine. Carry out active rehydration of the IPG strips at 50 V 20° for 1 h. After 1 h, add 2 ml of mineral oil in each lane and carry out the active rehydration step again for about 16 h. Active rehydration steps aid in swelling of the dried gel on the IPG strip, as well as the absorption of the protein sample into the strip. The addition of mineral oil to the IPG strip prevents sample evaporation due to heat formation. After 16 h of rehydration, wet paper wicks with 8 μl of nanopure water (Bio-Rad), and place the wicks between the electrode and the IPG strip, which will prevent burning of the IPG strips.

Perform isoelectric focusing first at 300 V for a 2-h linear gradient, and then at 500 V for a 2-h linear gradient, at 1000 V for a 2-h linear gradient, at 8000 V for a 8-h linear gradient, and finally at 8000 V for 10-h rapid gradient. This IEF program can be used with any type of samples. After completion of IEF, transfer the IPG strips into an equilibration tray with the gel side facing up. The strips can be equilibrated in equilibration buffer as mentioned in the next section before proceeding to SDS-PAGE. If necessary, freeze the IPG strips at −80° until use. When the strips are frozen, they turn milky white in color due to crystallization of urea, but when the IPG strips are thawed at room temperature, the urea becomes resolubilized and the milky color disappears.

3.9. Sodium dodecyl sulfate-polyacrylamide gel electrophoresis (second dimension)

While waiting for the IPG strips to thaw, warm the agarose solution (Bio-Rad) and wash the precasted criterion gels (linear gradient 8–16%) with water. Remove the water layer from the gel using Kimwipes. Prepare 1× running buffer (Bio-Rad) by diluting 100 ml of running buffer with 900 ml of deionized water in a measuring cylinder. Once the IPG strips come to room temperature, add 4 ml of DTT equilibrium buffer to the IPG strips and incubate at room temperature in the dark for 10 min. After 10 min transfer the strips into the next well and further incubate the IPG strips in 4 ml of IA equilibration buffer in the dark for 10 min. Incubation of the IPG strips with DTT breaks the disulfide bonds and the IA incubation will prevent the reformation of disulfide bond by acetylating the free –SH groups. Wash the IPG strips in 1× running buffer to remove excess equilibration buffer. Place the IPG strip after equilibration on top of the criterion gels with the gel side facing up. Load 2 μl of unstained molecular weight marker into the standard well adjacent to the IPG strip for the gels and precision stained molecular weight on the gel that will be used for blot. Seal the space between the IPG strip and the gel with the agarose solution. Allow 10 min for agarose to solidify, place the gels in a tank filled with running buffer, and then fill the upper tank with running buffer. Run the gels at 200 V for 65 min at room temperature or until the dye front (bromophenol blue) runs off the gel into the lower tank. Disconnect the power supply and disassemble the gel unit. Break open the gel plate and cut one corner of the gel to allow its orientation to be tracked. Of the two gels employed in one electrophoresis unit, the one that will be used for the detection of 3-NT-bound proteins is transferred on the nitrocellulose membrane and the other gel is stained with SYPRO stain for protein level analysis per spot.

3.10. Protein staining

Fix the gels in 50 ml of fixative solution at room temperature for 60 min. Then remove the fixative solution and incubate the gels in 50 ml of SYPRO Ruby gel stain (Bio-Rad/Molecular Probes, Eugene, OR) for 4 h to overnight at room temperature. Wash the gels in deionized water for 1 h and then scan at ex/em = 300, 490/640 nm using an ultraviolet (UV) transilluminator (Molecular Dynamics, Sunnyvale, CA).

3.11. Western blotting

For Western blotting, soak both the precut nitrocellulose membrane and the Whatman filter paper in transfer buffer for 10 min. In our laboratory we use the semidry transfer method, which takes less time and gives good

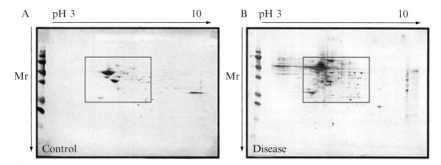

Figure 19.2 A representative two-dimensional blot obtained for 3-NT-modified proteins from control (A) and disease (B) samples. A box is drawn around the area that shows increased 3-NT-modified proteins on the blot.

transfer of the proteins onto the nitrocellulose membrane. For transfer, first make a transfer sandwich in the following order: first place one soaked filter paper on the semidry transfer unit platform, followed by the nitrocellulose membrane, gel, and one more filter paper, and then carry out the transfer of proteins from the gel to the membrane at 15 V for 2 h at room temperature. It is important to ensure that no air bubbles are trapped in the transfer sandwich, which may interfere with the transfer of the proteins. After the completion of transfer, remove the membrane and stain with Ponceau S (a reversible protein-binding stain) to ensure proper transfer of the protein. Add 20 ml of blocking buffer to the membrane and incubate the blots for 1 h at room temperature on a rocking platform or overnight at 4°. Blocking of the membrane will prevent nonspecific binding of the primary or secondary antibody to the membrane. Incubate the membrane with anti-3NT antibody (1: 2000) (Sigma) for 2 h at room temperature on a rocking platform (primary antibody incubation can be done from 1 h at room temperature to overnight at 4°). Remove the primary antibody and wash the membrane three times for 5 min each with 50 ml wash buffer II. Incubate the membrane in secondary antibody (1:3000) (antirabbit ALP conjugated) for 1 h on a rocking platform. Wash the membrane three times for 5 min each with wash buffer II. After the final wash, develop the membrane using a Sigma Fast tablet for 10 to 30 min (Sigma). Once the desirable color intensity is obtained, wash the membrane with deionized water to stop the color development and dry the membrane between Kimwipes. An example of a 2D blot obtained for 3-NT-modified proteins is shown in Fig. 19.2.

3.12. Image analysis

Scan gels and blots using a UV transilluminator ($\lambda_{ex} = 470$ nm, $\lambda_{em} = 618$ nm, Molecular Dynamics) and Adobe Photoshop on a Microtek Scanmaker 4900, respectively.

3.13. Identification of proteins with excessive 3-NT bound compared to control

Carry out identification of the 3-NT-bound proteins using PD Quest image analysis software (Bio-Rad). Briefly, match the gels and blots separately, as can be seen in Fig. 19.3. This procedure is then followed by making a high match set that will allow the investigator to match the corresponding gels and blots. Data obtained from this high match set can be compared statistically to detect significantly nitrated protein spots between subjects and controls. Excise protein spots in subject blots showing a significant increase in 3-NT-bound protein levels from the gel and digest in-gel with trypsin. Submit the resulting peptides for mass spectrometry for mass analysis. Interrogation of appropriate data bases (Butterfield *et al.*, 2006b) leads to identification of the nitrated proteins. The goodness of fit of the identification is imbedded in at least one database (Swissprot) by a MOWSE score, a value of which must be exceeded for significance. This parameter is related to the probability that the identified protein is a randomly identified protein. Typically, *p* values based on MOWSE scores are on the order of 10^{-8}, indicating that the identification is correct and giving confidence to the investigator.

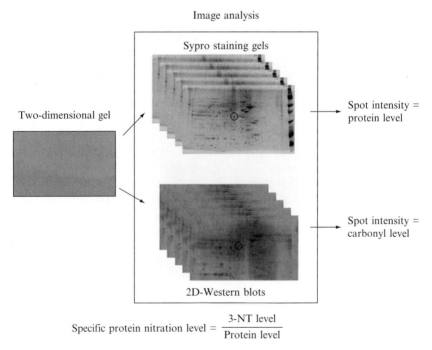

Figure 19.3 Analysis of blots and gels for the identification of specifically nitrated proteins in a biological sample.

3.14. Confirmation of protein identification

Despite the confidence elicited by the MOWSE score, it is prudent to validate the identity of proteomic-identified proteins, particularly when new sample matrices or types are employed for the first time. The identity of the 3-NT-modified protein can be confirmed by immunoprecipitation of the protein, followed by a one-dimensional Western blot or by immunoprecipitation of the 3-NT-modified proteins followed by probing with the antibody of the protein of interest.

4. COMMENTS

It is apparent that proteomics coupled to immunochemical methods illustrated in this chapter, as applied to the identification of 3-NT-modified proteins, is a powerful approach to guide researchers interested in understanding the role of protein nitration at the protein level. Once a protein spot on 2D Western blots is identified as excessively nitrated compared to controls using PD Quest software, that protein spot can be excised from the gel, digested, and identified by mass spectrometry. Immunoprecipitation of the nitrated proteins can be performed to validate mass spectrometric protein identification data. However, as noted earlier, although 2D gel electrophoresis is a powerful technique, it does pose some challenges, such as difficulty observing low-abundance proteins, solubilization of membrane proteins, and analysis of proteins with a high pI. Despite these difficulties, 2D gel-based proteomics still can be used successfully to determine the posttranslational modification of proteins, such as protein nitration, that will help in the elucidation of the biochemical pathways involved in the mechanism of pathogenesis (Butterfield, 2004; Butterfield *et al.*, 2003, 2006a,b, 2007a; Castegna *et al.*, 2003; Sultana *et al.*, 2006, 2007).

ACKNOWLEDGMENT

This work was supported in part by NIH Grants AG-05119 and AG-10836.

REFERENCES

Amoresano, A., Chiappetta, G., Pucci, P., D'Ischia, M., and Marino, G. (2007). Bidimensional tandem mass spectrometry for selective identification of nitration sites in proteins. *Anal. Chem.* **79**, 2109–2117.

Anantharaman, M., Tangpong, J., Keller, J. N., Murphy, M. P., Markesbery, W. R., Kiningham, K. K., and St. Clair, D. K. (2006). Beta-amyloid mediated nitration of manganese superoxide dismutase: Implication for oxidative stress in a APPNLH/NLH

X PS-1P264L/P264L double knock-in mouse model of Alzheimer's disease. *Am. J. Pathol.* **168,** 1608–1618.

Aulak, K. S., Koeck, T., Crabb, J. W., and Stuehr, D. J. (2004a). Dynamics of protein nitration in cells and mitochondria. *Am. J. Physiol. Heart Circ. Physiol.* **286,** H30–H38.

Aulak, K. S., Koeck, T., Crabb, J. W., and Stuehr, D. J. (2004b). Proteomic method for identification of tyrosine-nitrated proteins. *Methods Mol. Biol.* **279,** 151–165.

Beckman, J. S. (2002). Protein tyrosine nitration and peroxynitrite. *FASEB J.* **16,** 1144.

Beckman, J. S., Ischiropoulos, H., Zhu, L., van der Woerd, M., Smith, C., Chen, J., Harrison, J., Martin, J. C., and Tsai, M. (1992). Kinetics of superoxide dismutase- and iron-catalyzed nitration of phenolics by peroxynitrite. *Arch. Biochem. Biophys.* **298,** 438–445.

Buchczyk, D. P., Grune, T., Sies, H., and Klotz, L. O. (2003). Modifications of glyceraldehyde-3-phosphate dehydrogenase induced by increasing concentrations of peroxynitrite: Early recognition by 20S proteasome. *Biol. Chem.* **384,** 237–241.

Butterfield, D. A. (2004). Proteomics: A new approach to investigate oxidative stress in Alzheimer's disease brain. *Brain Res.* **1000,** 1–7.

Butterfield, D. A., Boyd-Kimball, D., and Castegna, A. (2003). Proteomics in Alzheimer's disease: Insights into mechanisms of neurodegeneration. *J. Neurochem.* **86,** 1313–1327.

Butterfield, D. A., Perluigi, M., and Sultana, R. (2006a). Oxidative stress in Alzheimer's disease brain: New insights from redox proteomics. *Eur. J. Pharmacol.* **545,** 39–50.

Butterfield, D. A., Poon, H. F., St. Clair, D., Keller, J. N., Pierce, W. M., Klein, J. B., and Markesbery, W. R. (2006b). Redox proteomics identification of oxidatively modified hippocampal proteins in mild cognitive impairment: Insights into the development of Alzheimer's disease. *Neurobiol. Dis.* **22,** 223–232.

Butterfield, D. A., Reed, T., Newman, S. F., and Sultana, R. (2007a). Roles of amyloid beta-peptide-associated oxidative stress and brain protein modifications in the pathogenesis of Alzheimer's disease and mild cognitive impairment. *Free Radic. Biol. Med.* **43,** 658–677.

Butterfield, D. A., Reed, T. T., Perluigi, M., De Marco, C., Coccia, R., Keller, J. N., Markesbery, W. R., and Sultana, R. (2007b). Elevated levels of 3-nitrotyrosine in brain from subjects with amnestic mild cognitive impairment: Implications for the role of nitration in the progression of Alzheimer's disease. *Brain Res.* **1148,** 243–248.

Butterfield, D. A., and Stadtman, E. R. (1997). Protein oxidation processes in aging brain. *Adv. Cell Aging Gerontol.* **2,** 161–191.

Castegna, A., Thongboonkerd, V., Klein, J. B., Lynn, B., Markesbery, W. R., and Butterfield, D. A. (2003). Proteomic identification of nitrated proteins in Alzheimer's disease brain. *J. Neurochem.* **85,** 1394–1401.

Dalle-Donne, I., Scaloni, A., and Butterfield, D. A. (2006). "Redox Proteomics: From Protein Modifications to Cellular Dysfunction and Diseases." Wiley, Hoboken, NJ.

Dalle-Donne, I., Scaloni, A., Giustarini, D., Cavarra, E., Tell, G., Lungarella, G., Colombo, R., Rossi, R., and Milzani, A. (2005). Proteins as biomarkers of oxidative/nitrosative stress in diseases: The contribution of redox proteomics. *Mass Spectrom. Rev.* **24,** 55–99.

Di Stasi, A. M., Mallozzi, C., Macchia, G., Petrucci, T. C., and Minetti, M. (1999). Peroxynitrite induces tryosine nitration and modulates tyrosine phosphorylation of synaptic proteins. *J. Neurochem.* **73,** 727–735.

Good, P. F., Hsu, A., Werner, P., Perl, D. P., and Olanow, C. W. (1998). Protein nitration in Parkinson's disease. *J. Neuropathol. Exp. Neurol.* **57,** 338–342.

Gow, A. J., Duran, D., Malcolm, S., and Ischiropoulos, H. (1996). Effects of peroxynitrite-induced protein modifications on tyrosine phosphorylation and degradation. *FEBS Lett.* **385,** 63–66.

Grune, T., Reinheckel, T., and Davies, K. J. (1997). Degradation of oxidized proteins in mammalian cells. *FASEB J.* **11,** 526–534.

Halliwell, B., Zhao, K., and Whiteman, M. (1999). Nitric oxide and peroxynitrite. The ugly, the uglier and the not so good: A personal view of recent controversies. *Free Radic. Res.* **31,** 651–669.

Hensley, K., Maidt, M. L., Yu, Z., Sang, H., Markesbery, W. R., and Floyd, R. A. (1998). Electrochemical analysis of protein nitrotyrosine and dityrosine in the Alzheimer brain indicates region-specific accumulation. *J. Neurosci.* **18,** 8126–8132.

Horiguchi, T., Uryu, K., Giasson, B. I., Ischiropoulos, H., LightFoot, R., Bellmann, C., Richter-Landsberg, C., Lee, V. M., and Trojanowski, J. Q. (2003). Nitration of tau protein is linked to neurodegeneration in tauopathies. *Am. J. Pathol.* **163,** 1021–1031.

Hyun, D. H., Lee, M., Halliwell, B., and Jenner, P. (2003). Proteasomal inhibition causes the formation of protein aggregates containing a wide range of proteins, including nitrated proteins. *J. Neurochem.* **86,** 363–373.

Ischiropoulos, H., and Beckman, J. S. (2003). Oxidative stress and nitration in neurodegeneration: Cause, effect, or association? *J. Clin. Invest.* **111,** 163–169.

Ischiropoulos, H., Zhu, L., Chen, J., Tsai, M., Martin, J. C., Smith, C. D., and Beckman, J. S. (1992). Peroxynitrite-mediated tyrosine nitration catalyzed by superoxide dismutase. *Arch. Biochem. Biophys.* **298,** 431–437.

Jiang, H., and Balazy, M. (1998). Detection of 3-nitrotyrosine in human platelets exposed to peroxynitrite by a new gas chromatography/mass spectrometry assay. *Nitric Oxide* **2,** 350–359.

Kong, S. K., Yim, M. B., Stadtman, E. R., and Chock, P. B. (1996). Peroxynitrite disables the tyrosine phosphorylation regulatory mechanism: Lymphocyte-specific tyrosine kinase fails to phosphorylate nitrated cdc2(6–20)NH2 peptide. *Proc. Natl. Acad. Sci. USA* **93,** 3377–3382.

Koppenol, W. H., Moreno, J. J., Pryor, W. A., Ischiropoulos, H., and Beckman, J. S. (1992). Peroxynitrite, a cloaked oxidant formed by nitric oxide and superoxide. *Chem. Res. Toxicol.* **5,** 834–842.

Kuhn, D. M., Sakowski, S. A., Sadidi, M., and Geddes, T. J. (2004). Nitrotyrosine as a marker for peroxynitrite-induced neurotoxicity: The beginning or the end of the end of dopamine neurons? *J. Neurochem.* **89,** 529–536.

Mallozzi, C., Di Stasi, A. M., and Minetti, M. (1997). Peroxynitrite modulates tyrosine-dependent signal transduction pathway of human erythrocyte band 3. *FASEB J.* **11,** 1281–1290.

Marechal, A., Mattioli, T. A., Stuehr, D. J., and Santolini, J. (2007). Activation of peroxynitrite by inducible nitric-oxide synthase: A direct source of nitrative stress. *J. Biol. Chem.* **282,** 14101–14112.

Pennathur, S., Jackson-Lewis, V., Przedborski, S., and Heinecke, J. W. (1999). Mass spectrometric quantification of 3-nitrotyrosine, ortho-tyrosine, and o,o'-dityrosine in brain tissue of 1-methyl-4-phenyl-1,2,3, 6-tetrahydropyridine-treated mice, a model of oxidative stress in Parkinson's disease. *J. Biol. Chem.* **274,** 34621–34628.

Poladia, D. P., and Bauer, J. A. (2003). Early cell-specific changes in nitric oxide synthases, reactive nitrogen species formation, and ubiquitinylation during diabetes-related bladder remodeling. *Diabetes Metab. Res. Rev.* **19,** 313–319.

Smith, M. A., Richey Harris, P. L., Sayre, L. M., Beckman, J. S., and Perry, G. (1997). Widespread peroxynitrite-mediated damage in Alzheimer's disease. *J. Neurosci.* **17,** 2653–2657.

Sodum, R. S., Akerkar, S. A., and Fiala, E. S. (2000). Determination of 3-nitrotyrosine by high-pressure liquid chromatography with a dual-mode electrochemical detector. *Anal. Biochem.* **280,** 278–285.

Souza, J. M., Choi, I., Chen, Q., Weisse, M., Daikhin, E., Yudkoff, M., Obin, M., Ara, J., Horwitz, J., and Ischiropoulos, H. (2000). Proteolytic degradation of tyrosine nitrated proteins. *Arch. Biochem. Biophys.* **380,** 360–366.

Squadrito, G. L., and Pryor, W. A. (2002). Mapping the reaction of peroxynitrite with CO2: Energetics, reactive species, and biological implications. *Chem. Res. Toxicol.* **15,** 885–895.

Sultana, R., Poon, H. F., Cai, J., Pierce, W. M., Merchant, M., Klein, J. B., Markesbery, W. R., and Butterfield, D. A. (2006). Identification of nitrated proteins in Alzheimer's disease brain using a redox proteomics approach. *Neurobiol. Dis.* **22,** 76–87.

Sultana, R., Reed, T. T., Perluigi, M., Coccia, R., Pierce, W. M., and Butterfield, D. A. (2007). Proteomic identification of nitrated brain proteins in amnestic mild cognitive impairment: A regional study. *J. Cell. Mol. Med.* **11,** 839–851.

van der Vliet, A., Eiserich, J. P., O'Neill, C. A., Halliwell, B., and Cross, C. E. (1995). Tyrosine modification by reactive nitrogen species: A closer look. *Arch. Biochem. Biophys.* **319,** 341–349.

Yi, D., Ingelse, B. A., Duncan, M. W., and Smythe, G. A. (2000). Quantification of 3-nitrotyrosine in biological tissues and fluids: Generating valid results by eliminating artifactual formation. *J. Am. Soc. Mass Spectrom.* **11,** 578–586.

CHAPTER TWENTY

SLOT-BLOT ANALYSIS OF 3-NITROTYROSINE-MODIFIED BRAIN PROTEINS

Rukhsana Sultana[*,†] and D. Allan Butterfield[*,†,‡]

Contents

1. Introduction	309
2. Materials	311
3. Solutions	312
4. Sample Preparation for 3-NT Determination	312
5. Comments	313
Acknowledgment	314
References	314

Abstract

3-Nitrotyrosine (3-NT) is used as a biomarker of nitrosative stress. The formation of 3-NT has been reported in a number of diseases, including Alzheimer's disease, Huntington's disease, Parkinson's disease, amyotrophic lateral sclerosis, and cancer. The nitration of proteins is a reversible process, but it can induce a conformational change and thereby functional alterations of the affected protein. 3-NT measurements in biological samples are usually carried out by methodologies such as immunohistochemistry, high-performance liquid chromatography, gas chromatography, and immunochemical detection. This chapter describes the immunochemical method for the determination of protein-bound 3-NT using slot-blot analysis.

1. INTRODUCTION

Oxidative damage to proteins is known to play a role in a number of diseases associated with aging, neurodegeneration, cancer, ischemia, and so on (Butterfield and Stadtman, 1997). Oxidative modification of proteins is

[*] Department of Chemistry, University of Kentucky, Lexington, Kentucky
[†] Sanders-Brown Center on Aging, University of Kentucky, Lexington, Kentucky
[‡] Center of Membrane Sciences, University of Kentucky, Lexington, Kentucky

Methods in Enzymology, Volume 440 © 2008 Elsevier Inc.
ISSN 0076-6879, DOI: 10.1016/S0076-6879(07)00820-8 All rights reserved.

known to contribute to oxidative stress, an imbalance of reactive oxygen species (ROS), reactive nitrogen species (RNS), and antioxidant species. Oxidative damage is caused by ROS such as superoxide, singlet oxygen, hydrogen peroxide, and others. In addition, excess production of nitric oxide (NO) can cause protein nitration; more specifically, tyrosine nitration occurs with the addition of a NO_2 group at the three position of either free or protein-bound tyrosine, leading to the formation of 3-nitrotyrosine (3-NT).

The synthesis of NO occurs via the catalytic conversion of arginine to citrulline by nitric oxide synthases (NOS) (Michel and Feron, 1997; Mori, 2007). There are three canonical isoforms of NOS: neuronal (NOS I or nNOS), inducible (NOS II), and endothelial (NOS III)] and a significant number of spliced and posttranslationally modified variants (Carreras and Poderoso, 2007). In addition, new isoforms or mitochondrial variants of NOS (mtNOS) have been described in rat liver (Carreras and Poderoso, 2007; Ghafourifar and Richter, 1997; Giulivi et al., 1998), and brain (Carreras and Poderoso, 2007; Elfering et al., 2002; Riobo et al., 2002), but this notion of mitochondrial NOS remains controversial. RNS generated by NOS within a physiologically relevant concentration are not toxic and are relatively specific in their cellular targets (Bishop and Anderson, 2005; Lafon-Cazal et al., 1993). For example, under normal physiological conditions, NO mediates vasodilation and neurotransmission, prevents platelet adherence and aggregation, and so on (Cabrales et al., 2005; Mariotto et al., 2004; Rubbo and Freeman, 1996; Zdzisinska and Kandefer-Szerszen, 1998). Protein nitration is a reversible and selective process and may be a cellular signaling mechanism, similar to protein phosphorylation (Aulak et al., 2004; Broillet, 1999; Gow et al., 1997).

Reactive nitrogen species can also be produced in brain via the overexpression of inducible and neuronal-specific nitric oxide synthase, leading to increased levels of NO. Nitric oxide reacts with superoxide anion ($O_2^{\bullet-}$) at a diffusion-controlled rate in the presence of CO_2 to produce peroxynitrite ($ONOO^-$), a product that is highly reactive with biomolecules such as proteins, lipids, and carbohydrates. Peroxynitrite reaction with proteins can result in S-nitrosylation or 3-nitrotyrosine formations (Gow et al., 1997; Koppenol et al., 1992).

3-Nitrotyrosine is a covalent protein modification and has been used as a marker of nitrosative stress under a variety of disease conditions. The involvement of protein nitration has been reported in a number of diseases, including neurodegenerative disorders, aging, cancer, cardiovascular, and ischemia (Beckman, 2002; Butterfield et al., 2007a; Dalle-Donne et al., 2005, 2006; Ischiropoulos and Beckman, 2003; Smith et al., 1997; Sultana et al., 2007; Xu et al., 2001). Since 3-NT was suggested as a biomarker of nitrosative stress, a number of analytical methods, such as immunohistochemistry, high-performance liquid chromatography, gas chromatography, and immunochemical detection, were developed to quantify this marker in biological samples.

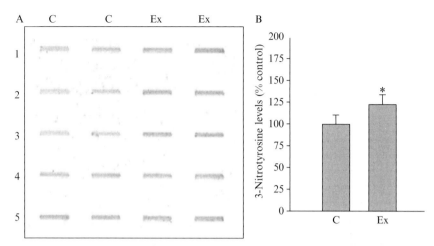

Figure 20.1 (A) A representative slot-blot image and (B) histogram constructed from data obtained from control (C) and experimental (Ex) samples. In this experiment, five samples from controls and five experimental samples were loaded in duplicate. One through five denote sample identification.

This chapter describes the immunochemical approach used to detect and quantify protein-bound 3NT in brain. Briefly, the samples are transferred onto a nitrocellulose membrane after denaturation. This procedure is then followed by incubation of the membrane with an anti-3NT antibody, which is then recognized by using a chromogenic detection method. The amount of nitrated tyrosine is the limiting factor in the detection; hence by employing slot-blot analysis, ≈250 ng of protein is loaded, overcoming this limiting factor. This method is sensitive and reliable and is useful in detecting 3-NT levels in biological samples (Fig. 20.1). Artifactual results can be prevented by avoiding acidification and preventing an increase of temperature in the samples (Yi et al., 2000).

2. Materials

Bicinchoninic acid (BCA) assay kit (Pierce)
Anti-3-NT antibody (Sigma Aldrich, St. Louis, MO)
Bovine serum albumin (Sigma Aldrich)
Antirabbit conjugated to alkaline phosphatase antibody (Sigma Aldrich)
Sigma Fast tablet [5-bromo-4-chloro-3-indolyl phosphate/nitro blue tetrazolium (BCIP/NBT)](Sigma Aldrich)
Nitrocellulose membrane (Bio-Rad, Hercules, CA)
Whatman filter papers (Bio-Rad)

3. Solutions

1. Buffer A: 10 mM HEPES, 137 mM NaCl, 4.6 mM KCl, 1.1 mM KH_2PO_4, 0.6 mM $MgSO_4$, 0.5 µg/ml leupeptin (stored as an aliquot at −20°), 0.7 µg/ml pepstatin (stored as an aliquot at −20°), 0.5 µg/ml type II S soybean trypsin inhibitor, and 40 µg/ml phenylmethylsulfonyl fluoride dissolved in deionized water stored at 4°.
2. Buffer B (Laemmli buffer stock): 0.125 M Trizma (pH 6.8), 4% SDS, and 20% glycerol. (Store at room temperature.)
3. Buffer C (wash buffer 1): 0.01% (w/v) sodium azide and 0.2% (v/v) Tween 20 dissolved in phosphate-buffered saline (PBS) stored at room temperature
4. Buffer D (blocking buffer): Buffer C containing 2% bovine serum albumin. Make fresh before use.
5. Buffer E (primary antibody solution): 20 ml of buffer C containing anti-3-NT antibody (1:5000)
6. Buffer F: 20 ml of buffer C containing antirabbit antibody conjugated to alkaline phosphatase (1:3000). (Make fresh.)
7. Developing solution: Dissolve Sigma Fast tablet (1 tablet/10 ml of deionized water). (Solution is light sensitive.)

4. Sample Preparation for 3-NT Determination

1. Sonicate brain tissue (100 mg) from the control and experimental samples in 200 µl of ice-cold buffer A. It is important to take into account the postmortem interval (PMI). In our hands, brain from animals is harvested immediately after sacrifice; for human studies, brain is obtained with a PMI of <4 h. This rapid isolation of brain mitigates against artifactual oxidative or nitrosative indices.
2. Centrifuge the sonicated samples at 2500 g for 10 min to remove the nuclear fraction and unbroken cells.
3. Determine the protein concentration in the supernatant by the BCA assay and use the supernatant for slot-blot analysis.
4. Sample derivatization: Mix the sample (5 µl) with 10 µl of buffer B and incubate at room temperature for 20 min or boil for 2 min.
5. Dilute the derivatized sample to 1 µg/ ml in PBS.
6. Soak the precut nitrocellulose membrane and three Whatman filters in PBS for 10 min.
7. Assemble a sandwich on the slot-blot apparatus from Bio-Rad in the following order: two filter papers followed by the nitrocellulose membrane.

8. Load 250 ng of the samples per well in duplicate or triplicate and turn on a vacuum to transfer the proteins onto the nitrocellulose membrane.
9. When finished, immerse membrane in buffer D and incubate for 1 h at room temperature or overnight at 4°.
10. Wash 3 × 5 min with buffer C to remove any unbound primary antibody.
11. Incubate for 1 h in buffer E at room temperature.
12. Wash 3 × 5 min with buffer C to remove any unbound primary antibody.
13. Incubate for 1 h in buffer F at room temperature.
14. Wash 3 × 5 min with buffer C to remove unbound secondary antibody.
15. Develop with Sigma Fast (1 tablet per 10 ml of deionized water) at room temperature for 10 to 30 min.
16. Wash the membrane with deionized water to stop color development and dry the membrane between Kimwipes.
17. Scan the blot using Adobe Photoshop on a Microtek Scanmaker 4900 instrument. Total levels of 3-NT are determined using Scion image software.
18. Controls to assure the investigator that quantitation of 3-NT has no artifacts are necessary. These controls include use of the secondary antibody in the absence of the primary antibody to test for specificity of secondary antibody binding and reduction of 3-NT to 3-aminotyrosine using dithionite. Subsequent repetition of steps 1 to 17 should prove negative in the absence of artifacts.

5. Comments

Analysis of 3-NT has proved useful to gain insights into molecular processes in neurodegenerative disorders. For example, in Parkinson's disease (PD), nitration of the protein-bound Tyr may alter neural functions through inactivation of enzymes such as tyrosine hydroxylase, an enzyme involved in dopamine synthesis that could lead to decreased levels of dopamine and neuronal loss, eventually leading to PD (Ara et al., 1998; McFarlane et al., 2005). Increased immmunoreactivity of 3-NT was reported in degenerating neurons in PD showing Lewy bodies (Good et al., 1998).

In amyotrophic lateral sclerosis (ALS), superoxide dismutase is reported to be nitrated, which could enhance the nitration of other critical cellular targets (Ischiropoulos et al., 1992), such as neurofilaments, which will disrupt their assembly and may play a significant role in the development of ALS (Crow et al., 1997; Strong et al., 1998). Free 3-NT levels, but not protein-bound 3-NT levels, were elevated two- to threefold in ALS transgenic mice throughout the progression of the disease (Bruijn et al., 1997). Increased protein nitration also occurs in central nervous system tissue from ALS patients (Abe et al., 1995; Beal et al., 1997; Strong et al., 1998).

Increased protein nitration was reported in neurons containing neurofibrillary tangles in AD brains, suggesting the role of 3-NT formation as an intermediate and involvement of protein nitration in the pathology or progression of AD (Anantharaman et al., 2006; Castegna et al., 2003; Horiguchi et al., 2003; Reynolds et al., 2006; Smith et al., 1997; Sultana et al., 2006). Our laboratory showed an increase in the levels of total protein-bound 3-NT in the inferior parietal lobule and hippocampus in brain from subjects with mild cognitive impairment (MCI). Because MCI is considered a transitory stage between control and AD, it is tempting to speculate that the nitration of brain proteins may play a role in the progression of this disease (Butterfield et al., 2007b).

The slot-blot approach measures total protein-bound 3-NT levels and can provide insight into the mechanisms of protein nitration in the pathogenesis and progression of disease.

ACKNOWLEDGMENT

This work was supported in part by NIH Grants AG-10836 and AG-05119.

REFERENCES

Abe, K., Pan, L. H., Watanabe, M., Kato, T., and Itoyama, Y. (1995). Induction of nitrotyrosine-like immunoreactivity in the lower motor neuron of amyotrophic lateral sclerosis. *Neurosci. Lett.* **199,** 152–154.

Anantharaman, M., Tangpong, J., Keller, J. N., Murphy, M. P., Markesbery, W. R., Kiningham, K. K., and St Clair, D. K. (2006). Beta-amyloid mediated nitration of manganese superoxide dismutase: Implication for oxidative stress in a APPNLH/NLH X PS-1P264L/P264L double knock-in mouse model of Alzheimer's disease. *Am. J. Pathol.* **168,** 1608–1618.

Ara, J., Przedborski, S., Naini, A. B., Jackson-Lewis, V., Trifiletti, R. R., Horwitz, J., and Ischiropoulos, H. (1998). Inactivation of tyrosine hydroxylase by nitration following exposure to peroxynitrite and 1-methyl-4-phenyl-1,2,3,6-tetrahydropyridine (MPTP). *Proc. Natl. Acad. Sci. USA* **95,** 7659–7663.

Aulak, K. S., Koeck, T., Crabb, J. W., and Stuehr, D. J. (2004). Dynamics of protein nitration in cells and mitochondria. *Am. J. Physiol. Heart Circ. Physiol.* **286,** H30–H38.

Beal, M. F., Ferrante, R. J., Browne, S. E., Matthews, R. T., Kowall, N. W., and Brown, R. H., Jr. (1997). Increased 3-nitrotyrosine in both sporadic and familial amyotrophic lateral sclerosis. *Ann. Neurol.* **42,** 644–654.

Beckman, J. S. (2002). Protein tyrosine nitration and peroxynitrite. *FASEB J.* **16,** 1144.

Bishop, A., and Anderson, J. E. (2005). NO signaling in the CNS: From the physiological to the pathological. *Toxicology* **208,** 193–205.

Broillet, M. C. (1999). S-nitrosylation of proteins. *Cell. Mol. Life Sci.* **55,** 1036–1042.

Bruijn, L. I., Beal, M. F., Becher, M. W., Schulz, J. B., Wong, P. C., Price, D. L., and Cleveland, D. W. (1997). Elevated free nitrotyrosine levels, but not protein-bound nitrotyrosine or hydroxyl radicals, throughout amyotrophic lateral sclerosis (ALS)-like disease implicate tyrosine nitration as an aberrant *in vivo* property of one familial ALS-linked superoxide dismutase 1 mutant. *Proc. Natl. Acad. Sci. USA* **94,** 7606–7611.

Butterfield, D. A., Reed, T., Newman, S. F., and Sultana, R. (2007a). Roles of amyloid beta-peptide-associated oxidative stress and brain protein modifications in the pathogenesis of Alzheimer's disease and mild cognitive impairment. *Free Radic. Biol. Med.* **43,** 658–677.

Butterfield, D. A., Reed, T. T., Perluigi, M., De Marco, C., Coccia, R., Keller, J. N., Markesbery, W. R., and Sultana, R. (2007b). Elevated levels of 3-nitrotyrosine in brain from subjects with amnestic mild cognitive impairment: Implications for the role of nitration in the progression of Alzheimer's disease. *Brain Res.* **1148,** 243–248.

Butterfield, D. A., and Stadtman, E. R. (1997). Protein oxidation processes in aging brain. *Adv. Cell Aging Gerontol.* **2,** 161–191.

Cabrales, P., Tsai, A. G., Frangos, J. A., and Intaglietta, M. (2005). Role of endothelial nitric oxide in microvascular oxygen delivery and consumption. *Free Radic. Biol. Med.* **39,** 1229–1237.

Carreras, M. C., and Poderoso, J. J. (2007). Mitochondrial nitric oxide in the signaling of cell integrated responses. *Am. J. Physiol. Cell Physiol.* **292,** C1569–C1580.

Castegna, A., Thongboonkerd, V., Klein, J. B., Lynn, B., Markesbery, W. R., and Butterfield, D. A. (2003). Proteomic identification of nitrated proteins in Alzheimer's disease brain. *J. Neurochem.* **85,** 1394–1401.

Crow, J. P., Ye, Y. Z., Strong, M., Kirk, M., Barnes, S., and Beckman, J. S. (1997). Superoxide dismutase catalyzes nitration of tyrosines by peroxynitrite in the rod and head domains of neurofilament-L. *J. Neurochem.* **69,** 1945–1953.

Dalle-Donne, I., Scaloni, A., and Butterfield, D. A. (2006). "Redox Proteomics: From Protein Modifications to Cellular Dysfunction and Diseases." Wiley, Hoboken, NJ.

Dalle-Donne, I., Scaloni, A., Giustarini, D., Cavarra, E., Tell, G., Lungarella, G., Colombo, R., Rossi, R., and Milzani, A. (2005). Proteins as biomarkers of oxidative/nitrosative stress in diseases: The contribution of redox proteomics. *Mass Spectrom. Rev.* **24,** 55–99.

Elfering, S. L., Sarkela, T. M., and Giulivi, C. (2002). Biochemistry of mitochondrial nitric-oxide synthase. *J. Biol. Chem.* **277,** 38079–38086.

Ghafourifar, P., and Richter, C. (1997). Nitric oxide synthase activity in mitochondria. *FEBS Lett.* **418,** 291–296.

Giulivi, C., Poderoso, J. J., and Boveris, A. (1998). Production of nitric oxide by mitochondria. *J. Biol. Chem.* **273,** 11038–11043.

Good, P. F., Hsu, A., Werner, P., Perl, D. P., and Olanow, C. W. (1998). Protein nitration in Parkinson's disease. *J. Neuropathol. Exp. Neurol.* **57,** 338–342.

Gow, A. J., Buerk, D. G., and Ischiropoulos, H. (1997). A novel reaction mechanism for the formation of S-nitrosothiol *in vivo*. *J. Biol. Chem.* **272,** 2841–2845.

Horiguchi, T., Uryu, K., Giasson, B. I., Ischiropoulos, H., LightFoot, R., Bellmann, C., Richter-Landsberg, C., Lee, V. M., and Trojanowski, J. Q. (2003). Nitration of tau protein is linked to neurodegeneration in tauopathies. *Am. J. Pathol.* **163,** 1021–1031.

Ischiropoulos, H., and Beckman, J. S. (2003). Oxidative stress and nitration in neurodegeneration: Cause, effect, or association? *J. Clin. Invest.* **111,** 163–169.

Ischiropoulos, H., Zhu, L., Chen, J., Tsai, M., Martin, J. C., Smith, C. D., and Beckman, J. S. (1992). Peroxynitrite-mediated tyrosine nitration catalyzed by superoxide dismutase. *Arch. Biochem. Biophys.* **298,** 431–437.

Koppenol, W. H., Moreno, J. J., Pryor, W. A., Ischiropoulos, H., and Beckman, J. S. (1992). Peroxynitrite, a cloaked oxidant formed by nitric oxide and superoxide. *Chem. Res. Toxicol.* **5,** 834–842.

Lafon-Cazal, M., Culcasi, M., Gaven, F., Pietri, S., and Bockaert, J. (1993). Nitric oxide, superoxide and peroxynitrite: Putative mediators of NMDA-induced cell death in cerebellar granule cells. *Neuropharmacology* **32,** 1259–1266.

Mariotto, S., Menegazzi, M., and Suzuki, H. (2004). Biochemical aspects of nitric oxide. *Curr. Pharm. Des.* **10,** 1627–1645.

McFarlane, D., Dybdal, N., Donaldson, M. T., Miller, L., and Cribb, A. E. (2005). Nitration and increased alpha-synuclein expression associated with dopaminergic neurodegeneration in equine pituitary pars intermedia dysfunction. *J. Neuroendocrinol.* **17,** 73–80.

Michel, T., and Feron, O. (1997). Nitric oxide synthases: Which, where, how, and why? *J. Clin. Invest.* **100,** 2146–2152.

Mori, M. (2007). Regulation of nitric oxide synthesis and apoptosis by arginase and arginine recycling. *J. Nutr.* **137,** 1616S–1620S.

Reynolds, M. R., Reyes, J. F., Fu, Y., Bigio, E. H., Guillozet-Bongaarts, A. L., Berry, R. W., and Binder, L. I. (2006). Tau nitration occurs at tyrosine 29 in the fibrillar lesions of Alzheimer's disease and other tauopathies. *J. Neurosci.* **26,** 10636–10645.

Riobo, N. A., Melani, M., Sanjuan, N., Fiszman, M. L., Gravielle, M. C., Carreras, M. C., Cadenas, E., and Poderoso, J. J. (2002). The modulation of mitochondrial nitric-oxide synthase activity in rat brain development. *J. Biol. Chem.* **277,** 42447–42455.

Rubbo, H., and Freeman, B. A. (1996). Nitric oxide regulation of lipid oxidation reactions: Formation and analysis of nitrogen-containing oxidized lipid derivatives. *Methods Enzymol.* **269,** 385–394.

Smith, M. A., Richey Harris, P. L., Sayre, L. M., Beckman, J. S., and Perry, G. (1997). Widespread peroxynitrite-mediated damage in Alzheimer's disease. *J. Neurosci.* **17,** 2653–2657.

Strong, M. J., Sopper, M. M., Crow, J. P., Strong, W. L., and Beckman, J. S. (1998). Nitration of the low molecular weight neurofilament is equivalent in sporadic amyotrophic lateral sclerosis and control cervical spinal cord. *Biochem. Biophys. Res. Commun.* **248,** 157–164.

Sultana, R., Poon, H. F., Cai, J., Pierce, W. M., Merchant, M., Klein, J. B., Markesbery, W. R., and Butterfield, D. A. (2006). Identification of nitrated proteins in Alzheimer's disease brain using a redox proteomics approach. *Neurobiol. Dis.* **22,** 76–87.

Sultana, R., Reed, T. T., Perluigi, M., Coccia, R., Pierce, W. M., and Butterfield, D. A. (2007). Proteomic identification of nitrated brain proteins in amnestic mild cognitive impairment: A regional study journal of cellular and molecular medicine. *J. Cell. Mol. Med.* **11,** 839–851.

Xu, J., Kim, G. M., Chen, S., Yan, P., Ahmed, S. H., Ku, G., Beckman, J. S., Xu, X. M., and Hsu, C. Y. (2001). iNOS and nitrotyrosine expression after spinal cord injury. *J. Neurotrauma* **18,** 523–532.

Yi, D., Ingelse, B. A., Duncan, M. W., and Smythe, G. A. (2000). Quantification of 3-nitrotyrosine in biological tissues and fluids: Generating valid results by eliminating artifactual formation. *J. Am. Soc. Mass Spectrom.* **11,** 578–586.

Zdzisinska, B., and Kandefer-Szerszen, M. (1998). The role of nitric oxide in normal and pathologic immunologic reactions. *Postepy Hig. Med. Dosw.* **52,** 621–636.

CHAPTER TWENTY-ONE

Detection Assays for Determination of Mitochondrial Nitric Oxide Synthase Activity; Advantages and Limitations

Pedram Ghafourifar, Mordhwaj S. Parihar, Rafal Nazarewicz, Woineshet J. Zenebe, *and* Arti Parihar

Contents

1. Introduction	318
2. Colorimetric Nitric Oxide Synthase Assay	319
2.1. Determination of nitrite using the Griess reaction	319
2.2. Advantages and pitfalls	320
3. Determination of Mitochondrial Nitric Oxide Synthase Activity Using Radioassay	320
3.1. Advantages and pitfalls	321
4. Spectrophotometric Determination of Mitochondrial Nitric Oxide Synthase Activity	322
4.1. Advantages and pitfalls	322
5. Polarographic Nitric Oxide Synthase Assays	322
5.1. Advantages and pitfalls	323
6. Chemiluminescence Assay	323
6.1. Advantages and pitfalls	325
7. Fluorescent-Based Nitric Oxide Detection Assays	326
7.1. Diaminofluorescein diacetate	327
7.2. Copper-fluorescein	331
8. Conclusion	331
References	332

Abstract

Nitric oxide (NO) is a reactive radical synthesized by members of the NO synthase (NOS) family, including mitochondrial-specific NOS (mtNOS). Some of the assays used for the determination of cytoplasmic NOS activity have been utilized to detect mtNOS activity. However, it seems that many of those assays

Department of Surgery, The Ohio State University College of Medicine, Columbus, Ohio

Methods in Enzymology, Volume 440
ISSN 0076-6879, DOI: 10.1016/S0076-6879(07)00821-X

© 2008 Elsevier Inc.
All rights reserved.

need to be adjusted and optimized to detect NO in the unique environment of mitochondria. Additionally, most mtNOS detection assays are designed and optimized for isolated mitochondria and may exert inherent pitfalls and limitations once used in living cells. This chapter describes several assays used commonly for mtNOS detection in isolated mitochondria and in mitochondria of live cells. Those include colorimetric and spectrophotometric methods, Griess reaction, radioassay, and polarographic and chemiluminescence assays. It also describes fluorescent-based assays for the detection of mitochondrial NO in live cells. Advantages and limitations of each assay are discussed.

1. Introduction

Twenty years ago, pioneering works of Moncada's group led to the discovery that the endothelium-derived relaxing factor is, indeed, nitric oxide (NO; Palmer et al., 1987). Such a fundamental discovery has opened abundant new fields of studies in a wide range of biological systems. In mammalian cells, the enzymatic synthesis of NO is catalyzed by NO synthase (NOS) isozymes. NOS isozymes are generally referred to as inducible NOS (iNOS), neuronal NOS (nNOS), endothelial NOS (eNOS), and mitochondrial NOS (mtNOS). While iNOS activity is regulated primarily by its expression, the activity of nNOS, eNOS, and mtNOS is regulated by Ca^{2+}. mtNOS activity has been reported for mitochondria isolated from several organs, including liver (Ghafourifar and Richter, 1997), heart (Kanai et al., 2001), brain (Lores-Arnaiz et al., 2004; Riobo et al., 2002), and kidney (Boveris et al., 2003). Numerous functions have been reported for mtNOS-derived NO. Those include regulation of mitochondrial respiration (Boveris et al., 2003; Ghafourifar and Richter, 1997; Ghafourifar et al., 1999), transmembrane potential ($\Delta\psi$; Ghafourifar and Richter, 1997), Ca^{2+} retention (Dedkova and Blatter, 2005), matrix pH (Ghafourifar and Richter, 1999), ATP synthesis (Giulivi, 1998), and the mitochondrial pathway of apoptosis (Nazarewicz et al., 2007a; Zenebe et al., 2007).

The detection of NO produced by mtNOS can be challenging. Most commonly used conventional NO detection assays, including spectrophotometry, colorimetery, polarography, chemiluminescence, fluorometry, and radioassay, do not differentiate between cytoplasmic and mitochondrial NO. Therefore, most studies on mtNOS activity and its functions have been conducted on isolated mitochondria. An ideal mtNOS assay that could be carried out in most laboratory settings would detect low concentrations of NO and distinguish between cytoplasmic and mitochondrial NO. This chapter reviews commonly used NOS detection assays, discusses their advantages and pitfalls, and reviews possible selectivity of those assays for detecting mtNOS activity in cells.

2. Colorimetric Nitric Oxide Synthase Assay

The half-life of NO in biological systems ranges from less than 1 s to about 30 s (Eich *et al.*, 1996; Feldman *et al.*, 1993; Knowles and Moncada, 1992). Therefore, the detection of NO in most biological systems requires appropriate considerations. The wide range of reactions of NO in biology produces several nitrogen oxide congeners (NOx), most of which ultimately convert to two final metabolites, nitrite and nitrate. Using proper control experiments, total NO production can be estimated reliably by determining the concentrations of nitrite and nitrate as indicators of NOS activity (Moshage *et al.*, 1995). Nitrate and/or nitrite can be quantified directly by ultraviolet spectroscopy, mass spectroscopy, chromatography, ion selective electrodes, or capillary electrophoresis. Other commonly used NOS assays with varying advantages and limitations include colorimetric Griess (Green *et al.*, 1982; Moshage *et al.*, 1995), fluorescent (Casey and Hilderman, 2000; Kojima *et al.*, 1998), chemiluminescence (Yang *et al.*, 1997), and electrochemical (Zhang and Broderick, 2003) methods. A colorimetric simple and accurate method for measuring total NOx is known as the Griess assay (Griess, 1879). This assay is based on reducing NOx to nitrite and then detecting the nitrite. The Griess assay can be used for most biological systems, including organs, tissues, cells, and subcellular compartments. This method is discussed briefly.

2.1. Determination of nitrite using the Griess reaction

The Griess assay detects the red–pink color produced by the reaction of Griess reagents (discussed later) with nitrites. Therefore, all nitrates in the sample should be reduced to nitrite to be detected in this assay. Different laboratories have used various methods for the conversion of all nitrates to nitrite. Those include formate–nitrate reductase (Teniguchi *et al.*, 2003) and NADPH-dependent nitrate reductase (Verdon *et al.*, 1995) and cadmium-based reduction systems (Davison and Woof, 1978). After reduction to nitrite, samples are reacted with the Griess reagent consisting of equal volumes of sulfanilamide solution and N-(1-napthyl)ethylenediamine (NED) solution (Fig. 21.1). The Griess reagent should be made fresh before use. The sulfanilamide solution is prepared by dissolving 1 g of sulfanilamide in 100 ml of 5% orthophosphoric acid. This reagent can be stored at 4° and should be protected from light. The NED solution is prepared by dissolving 0.1 g NED hydrochloride in 100 ml H_2O. This reagent can be stored at 4° and should be protected from light. Nitrite standard solutions can be prepared from a stock solution of sodium nitrite of about 10 mM. To perform the assay, the sample is reacted with an equal volume of Griess reagent for 15 to 30 min, the absorbance is measured at 546 nm (Wink *et al.*, 1993), and the concentration of nitrite is calculated using a standard curve.

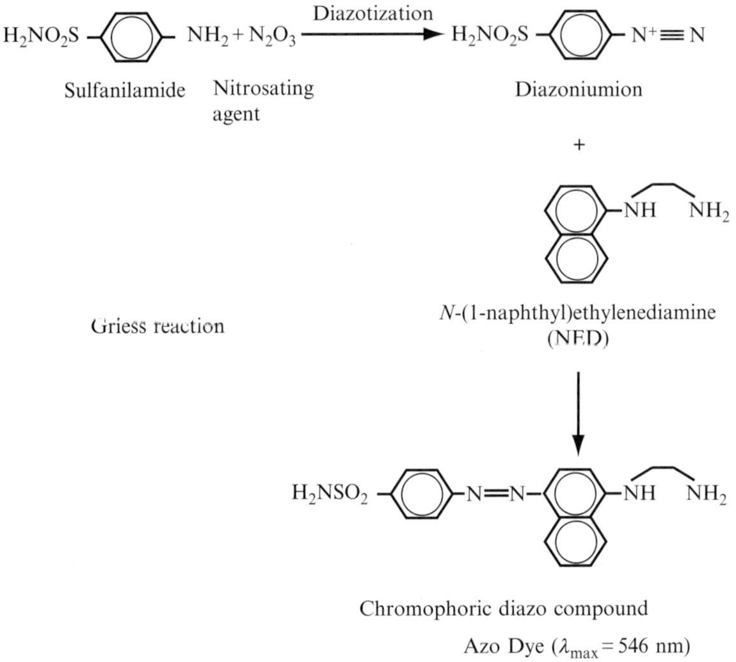

Figure 21.1 Griess reaction. Formation of the chromophoric diazo compound by the Griess reaction is shown.

2.2. Advantages and pitfalls

The Griess reaction utilizes few reagents, is performed easily under most laboratory environments, does not require sophisticated instruments, and is reliable and reproducible. However, important pitfalls restrict use of this assay as a primary NOS activity determination assay. First and foremost, the Griess reaction detects nitrite that can be formed by various non-NOS pathways. Therefore, only the NOS inhibitor-sensitive portion of the signal can be considered an indicator of NOS activity. Second, although the Griess reaction can be used to detect mtNOS activity in isolated mitochondria, so far there are no reports showing successful use of this assay to measure mtNOS activity in a cell system.

3. Determination of Mitochondrial Nitric Oxide Synthase Activity Using Radioassay

Nitric oxide synthase isozymes utilize L-arginine to produce NO and L-citrulline as the final coproduct. This reaction exerts a 1:1 stoichiometry, that is, each mole of L-arginine produces 1 mol NO and 1 mol L-citrulline

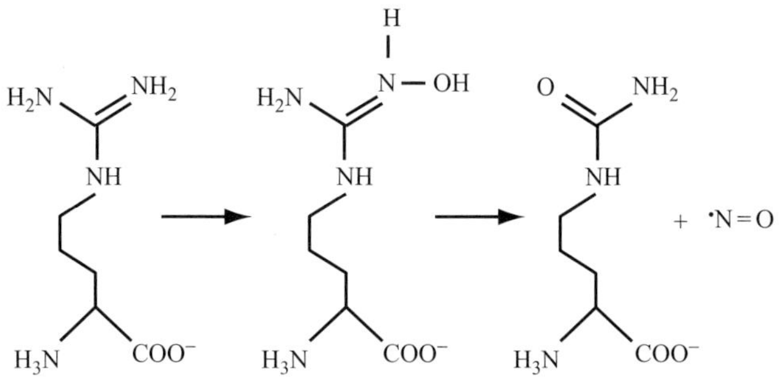

Figure 21.2 NOS-catalyzed NO synthesis. Conversion of L-arginine to L-citrulline and NO takes place in two steps with N^G-hydroxy-L-arginine as an intermediate product.

(Fig. 21.2). Therefore, measurement of L-citrulline has been widely used as a reliable NOS activity assay. Use of the radiolabeled L-arginine to L-citrulline assay for NOS activity determination was first described by Bredt and Snyder (1990). Typically, this assay is based on the formation of ^{3}H- or ^{14}C-labeled L-citrulline from labeled L-arginine. L-Citrullin is separated from L-arginine by passing the sample through a Na$^+$ form of an acidic cation exchanger such as Dowex 50, as described by Mayer *et al.* (1994). This assay can also be used for the detection of mtNOS activity in isolated mitochondria (Ghafourifar and Richter, 1997).

3.1. Advantages and pitfalls

The radioassay is sensitive and detects basal, stimulated, or inhibited mtNOS activity (Ghafourifar and Richter, 1997). The drawback is that this method requires utilizing radioactive materials that sometimes require regulations and are not available in all laboratory setups. More importantly, the radioassay in a cell system is not selective, as it detects total citrulline produced by cytoplasmic NOS, mtNOS, and non-NOS pathways. Therefore, using the radioassay for the detection of mtNOS while mitochondria are within the cells does not seem practical. Moreover, some extracts, particularly those from the liver, produce citrulline through arginase. Because the K_m of NOS for L-arginine is 2 to 10 μM, which is 1000 to 4000 times lower than 8 to 10 mM, which is the K_m of arginase for L-arginine, it is important to use low amounts of radiolabeled L-arginine in mtNOS assays (Ghafourifar *et al.*, 2005).

4. Spectrophotometric Determination of Mitochondrial Nitric Oxide Synthase Activity

The oxyhemoglobin assay is a spectrophotometric technique that allows rapid and sensitive quantification of NO. The first reports on selective reaction of NO with hemoglobin were published in the 1920s (Anson and Mirsky, 1925). The principles and technical aspects for using the oxyhemoglobin assay for NOS activity measurements have been described by Feelisch and Noack (1987). The assay is based on conversion by NO of oxyhemoglobin to methemoglobin as follows:

$$\text{OxyHb} + \text{NO} \rightarrow \text{metHb} + \text{NO}_3^-$$

Methemoglobin and oxyhemoglobin exert different optical spectra and can be distinguished from each other using spectrophotometric assays (Knowles et al., 1990; van Assendelft, 1970). This method has been widely used to measure mtNOS activity quantitatively using the molar extinction coefficient of 0.1 $M^{-1}\text{cm}^{-1}$ for methemoglobin (Ghafourifar et al., 2005).

4.1. Advantages and pitfalls

This method is very sensitive, rapid, relatively inexpensive, and detects NO real time. The major drawback for using oxyhemoglobin to detect mtNOS activity is that in intact mitochondria, mtNOS-derived NO reacts with abundant mitochondrial reactive oxygen species and the total amount of NO reaching the extra mitochondria and detected by oxyhemoglobin is smaller than NO produced in the mitochondria. However, broken mitochondria or submitochondrial particles can be utilized for the detection of mtNOS activity using this assay (Ghafourifar et al., 2005). The use of relatively large amounts of superoxide dismutase (≥ 1 kU/ml) is recommended when broken mitochondria or mitoplasts are used.

5. Polarographic Nitric Oxide Synthase Assays

Polarographic methods utilize sensors generally referred to Clark-type electrodes. NO sensors are designed based on the electrochemical potential for the oxidation of NO to NO_3 at the anode of the electrode. The kinetics of this reaction is noted as an alteration of electric potential that is recorded by a sensitive amperometer connected to the electrode. The electrode placed in the test sample is covered with a membrane that is highly selective and readily permeable to NO. This allows selective measurement of NO at low nanomolar concentrations.

5.1. Advantages and pitfalls

The advantages of this method are that NO synthesis is detected in real time and that the NO electrode does not respond to many other reactive species, such as nitrates, hydrogen peroxide, or superoxide generated in mitochondria at high levels. Another advantage of this method is the quantification of NO that can be achieved using NO donors or NO solution to generate the standard solution. NO electrodes respond quickly to NO in liquid samples by generating an electrochemical signal directly proportional to the concentration of NO. To measure mitochondrial NO, this method is limited due to the reaction of NO in mitochondria with different compounds such as superoxide or glutathione. Thus, the amount of NO diffusing outside the intact mitochondrial membrane is generally well below the detection limits of most NO electrodes. To date, NO electrodes are not designed to enter the mitochondria and NO produced within the mitochondria hardly reaches extramitochondrial compartments. One laboratory utilized a homemade NO-sensitive microsensor and detected NO produced by an isolated intact mouse heart mitochondrion (Kanai et al., 2001). Amperometric methods can be performed using broken mitochondria or submitochondrial particles (Nazarewicz et al., 2007a). In those preparations, the addition of superoxide dismutase is required to neutralize excess superoxide that potently interacts with NO to form peroxynitrite. Thus, isolated mitochondria results obtained using NO electrodes may relate only moderately to the cell environment. Therefore, amperometric methods do not seem to be the preferred method to measure NO synthesis within mitochondria of intact cells. Polarographic methods for NO detection require relatively expensive instruments, and most electrodes have short lifetimes.

6. Chemiluminescence Assay

Chemiluminescence is a phenomenon where a chemical reaction generates and emits light. Most chemiluminescence events follow the following equation:

$$S_1 + S_2 \rightarrow P_1^* + P_2$$

$$P_1^* \rightarrow P_1 + h\nu$$

where S_1 and S_2 are substrates for the reaction, P_1 and P_2 are the products of the reaction, and $h\nu$ is the quantum of light (photon). An asterisk denotes the excited state of the product.

A gas-phase chemiluminescence reaction between NO and ozone (O_3) can be used to detect low amounts of NO. The following equation shows the principle of the chemiluminescence NO detection:

$$NO + O_3 \rightarrow NO_2^* + O_2$$

$$NO_2^* \rightarrow NO_2 + h\nu$$

Long before physiological functions of NO were discovered, the chemiluminescence measurement of NO reacted with O_3 was utilized to determine the amounts of NO as an air pollutant (Fontijn et al., 1970). In the 1980s, the chemiluminescence NO detection method was advanced and used for biological samples (Braman and Hendrix, 1989; Cox and Frank, 1982) and finally applied for detection of NOS activity (Ignarro et al., 1987; Palmer et al., 1987). Biological samples such as mitochondria contain various nitrogen species, including NO congeners. NO reacts with superoxide to form peroxynitrite, with oxygen to form NO_2, with heme to form metal-nitrosyl complexes, and with SH groups to form nitrosothiols. Using specific approaches to the powerful chemiluminescence assay, most of the compounds at low picomolar concentrations can be detected and differentiated. Chemiluminescence can be observed only when O_3 reacts with NO. Therefore, in liquid samples, other nitrogen species that are metabolites of NO should be converted to NO. Total NO is then released from the sample to react with O_3 to generate chemiluminescence. Alternatively, certain conditions can be used to determine a specific NO congener. The following conditions are used most commonly:

- Nitrites, nitrates: vanadium chloride at 95°
- Nitrites: vanadium chloride at 30°
- Nitrosothiols: KI, $CuSO_4$ in acetic acid at 50°
- Metal-nitrosyl complexes: KCN, $K_3Fe(CN)_6$, KI, $CuSO_4$ in acetic acid at 50°.

To determine NOS activity or total NO in a sample, the sample is injected into a vessel containing reducing agents that reduce all nitrogen oxides to NO. NO is then released and reacted with O_3 supplied by an O_3 generator to generate the chemiluminescent signal. In order to increase the quality of the measurement and to minimize the background signal, light generated in the reaction chamber is filtered by a red band-pass optical filter of >600 nm, and the intensity of light is amplified by a photomultiplier. The chemiluminescent signal from liquid samples is correlated linearly with the NO concentration in the sample. A schematic presentation of the chemiluminescence instrument is presented in Fig. 21.3.

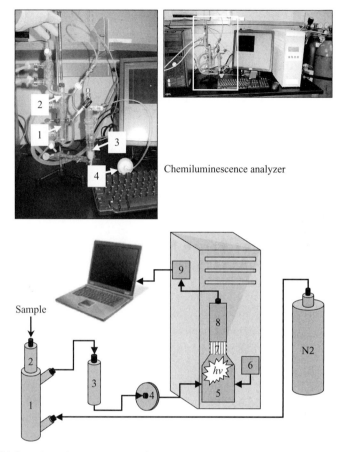

Figure 21.3 Chemiluminescent analyzer. Reducing agent in purge vessel (1) is purged with nitrogen provided from a tank (N_2). Samples are injected into the purge vessel through the sample inlet (2). Nitric oxide (NO) gas released from the sample is passed through a NaOH filter (3) neutralizing the remnant of reducing agents that could accidentally come with the sample and through a paper filter (4) removing moisture. The sample is then transferred to the reaction chamber (5) where NO reacts with O_3 provide by the O_3 generator (6). Chemiluminescence (hv) emitted during this reaction passes through a light filter (7), amplified by the photomultiplier (8), and detected by the sensor (9). The signal is processed and sent to a computer. NO content in the sample calculated by software is expressed as area under the curve.

6.1. Advantages and pitfalls

In mitochondria, mtNOS-derived NO reacts with various mitochondrial components to produce various nitrogen oxide species (NOx). Stimulation of mtNOS by loading mitochondria with Ca^{2+} causes a generation of significantly higher amounts of NOx that can be detected easily by chemiluminescence. In order to detect NOx, mitochondrial samples should be incubated in the dark. Conventional NOS inhibitors containing nitro groups or calcium chelators such as EGTA interfere with chemiluminescence and should be avoided in this assay. Therefore, use of the chemiluminescence method as the primary mtNOS detection assay is limited and chemiluminescence should be used along with other mtNOS assays where NOS inhibitors could be utilized (Nazarewicz et al., 2007a; Zenebe et al., 2007). Figure 21.4 is a representative trace for mtNOS activity measurement. As shown in Fig. 21.4, stimulation of mtNOS can be achieved by elevating mitochondrial Ca^{2+}.

Although chemiluminescence is very sensitive and selectively detects nitro compounds, in cell systems this assay does not distinguish between NO generated by cytoplasmic NOS isozymes or mtNOS.

7. Fluorescent-Based Nitric Oxide Detection Assays

Among various methods used to measure NO in cells, those based on fluorogenic probes have been gaining popularity. The most frequently used fluorescence probe is the cell-permeable fluorescent probe 4,5-diaminofluorescein diacetate (DAF-2DA) (Kojima et al., 1998; Nazarewicz et al., 2007b;

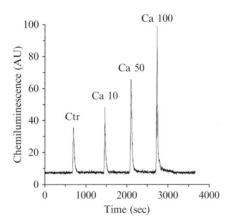

Fig. 21.4 Detection of mtNOS activity using chemiluminescence assay. Isolated rat liver mitochondria (1 mg) were incubated for 20 min in the dark at room temperature in 100 μl of HEPES buffer (0.1 M, pH 7.10). mtNOS was stimulated by loading mitochondria with Ca^{2+} (10–100 μM). Control (Ctr) or mtNOS-stimulated samples (Ca) were injected into a purge vessel containing 0.8% vanadium chloride in 1 M HCl and antifoam thermostated at 95°. The vessel was depleted of oxygen by purging with N_2 for 20 min prior to sample injection and during the entire measurement. Chemiluminescence of NO released was measured using the NO analyzer (Sievers 280i GE, Boulder, CO).

Parihar et al., 2008a). Other NO-sensitive probes have also been used to detect NO. The Cu(II) fluorescein-based compound CuFL(I) (Lim et al., 2006; Parihar et al., 2008b) and 8-(3′,4′-diaminophenyl)-difluoroboradiaza-S-indacence (Zhang et al., 2004) are used often. Several laboratories have used DAF-2DA to detect mitochondrial NO (Dedkova et al., 2004; Manser and Houghton, 2006). Our laboratory has used both DAF-2DA (Nazarewicz et al., 2007; Parihar et al., 2008a) and 2-{2-chloro-6-hydroxy-5-[(2-methyl-quinolin-8-ylamino)-methyl]-3-oxo-3 H-xanthen-9-yl}-benzoic acid (FL; Parihar et al., 2008b) to detect mitochondrial NO in various cells. The following sections describe the procedures and discuss the limitations.

7.1. Diaminofluorescein diacetate

7.1.1. Assay principle

4,5-Diaminofluorescein diacetate, which is membrane permeable, crosses the cell membrane readily and distributes within the cell nonselectively. Once inside the cell, the diacetate group (DA) is hydrolyzed by esterases, and the active fluorophore DAF-2 is released. NO reacts with the nonfluorescent DAF-2 to produce its fluorescent triazole adduct, triazolofluorescein (DAF-2T). DAF-2T can be detected by excitation at 488 nm and by measuring the emission at 516 nm (Fig. 21.5). DAF-2 and DAF-2T have almost identical absorbance and emission; however, the quantum yield from DAF-2T fluorescence is about 200-fold higher than that of DAF-2 (Espey et al., 2001; Rathel et al., 2003). Thus, autofluorescence of DAF-2 is negligible with regard to the high fluorescence of DAF-2T.

7.1.2. Loading of dyes and fluorescent image acquisition

4,5-Diaminofluorescein diacetate can be used to detect NO in live cells. Initially, cells grown on a cover glass in an appropriate medium (regular culture medium for any given cell type) are incubated with DAF-2DA. We typically load cells with 5 μM DAF-2DA and incubate for 30 min as described (Nazarewicz et al., 2007b; Parihar et al., 2008a). Those conditions generally allow compartmentalization of the dye into mitochondria; however, they can be modified to accommodate the loading of different cells. During the last 10 min of loading with DAF-2DA, a mitochondrial marker, mitotracker red (200 nM), is added to the medium. After incubation, the coverslips containing cells are washed twice with low fluorescence medium, and laser scanning confocal microscopy is used to detect the fluorescence. The coverslip is mounted on the slide chamber filled with low fluorescence medium and placed on the stage of an inverted microscope equipped with a water objective for imaging. Both fluorescent probes are excited with a 488-nm line of argon, 543-nm line of HeNe1, and 633-nm line of HeNe2 lasers at room temperature, and emitted fluorescent signals are measured simultaneously at 516 nm for DAF-2 and 579 nm for

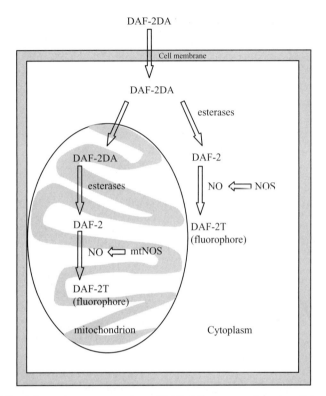

Figure 21.5 Schematic representation of DAF-2 fluorescent imaging. Diaminofluorescein diacetate (DAF-2DA) penetrates through cell and mitochondrial membranes. Intracellular or intramitochondrial esterases convert DAF-2DA to diaminofluorescein (DAF-2), which is membrane impermeable. NO reacts with DAF-2 and converts it to the fluorescent product triazolofluorescein (DAF-2T). DAF-2T is also membrane impermeable.

mitotracker red. Activation of DAF-2 by NO is irreversible, and the fluorescent signal remains steady even if NO levels decrease. Wells without cells but containing imaging medium should be used to measure the autofluorescence of DAF-2, medium, and other components. Autofluorescence signals should be subtracted from fluorescence intensity obtained from cell or mitochondria. By use of analyzing software, mitochondria can be selected on an image and fluorescence intensity can be quantitated.

Representative images of cell labeled with DAF-2 and mitotracker red at resting are shown in Fig. 21.6A. As shown, both cytoplasmic and mitochondrial NO fluorescent are detected. As described earlier, both cytoplasmic NOS isozymes and mtNOS are Ca^{2+} sensitive. Figure 21.6B shows that adding Ca^{2+} (10 μM) to the medium increases both cytoplasmic and mitochondrial NO signals. As shown, the NO signal is not selective for

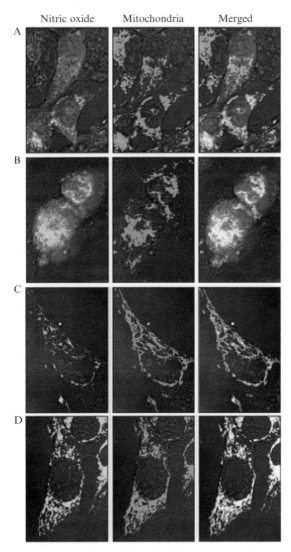

Figure 21.6 Measurements of intramitochondrial NO. Cardiomyocyte HL-1 cells were cultured on glass coverslips in Claycomb medium supplemented with norepinephrine (0.1 mM), L-glutamine (2 mM), and fetal bovine serum (10%) under 5% CO_2 at 37°. Cells were loaded with 4,5-diaminofluorescein diacetate (DAF-2 DA; 5 μM) and mitotracker red (MTR; 200 nM) in Claycomb medium for 30 min at 37°. Fluorescent-labeled cells were washed twice with low fluorescence medium. Fluorescence was measured for nonpermeabilized and permeabilized cells without or with Ca^{2+} (10 μM). (A and B) Nonpermeabilized cells: (A) basal cytoplasmic and mitochondrial NO and (B) NO stimulated by Ca^{2+}. (C and D) Permeabilized cells: (C) basal mitochondrial NO and (D) mitochondrial NO upon stimulation of mtNOS by Ca^{2+}. Permeabilization was performed using digitonin (10 μM, 3 min), followed by twice washing as described in the text. (See color insert.)

mitochondria and the distribution of the DAF-2 signal is detected in both cytoplasm and mitochondria.

To perform selective NO signal localized to mitochondria, permeabilization of the plasma membrane with mild detergents can be used. A wide variety of detergents have been used to selectively permeabilize the plasma membrane. For example, saponin is a relatively mild detergent used to solubilize cholesterol of various plasma membranes. Triton X-100 is another detergent used commonly in immunofluorescent staining that can solubilize cell membranes efficiently without disturbing protein–protein interactions. Sodium dodecyl sulfate, which is a nonionic detergent, is not suitable because it denatures proteins. Digitonin, which is a weak nonionized detergent, is used most frequently to permeabilize plasma membranes of various cells. Using an appropriate concentration and incubation time allows selective permeabilization of the cell membrane while allowing internal membranes to remain intact. Different cells may require different timings for permeabilization with digitonin. In our hands, cardiac myocyte HL-1 cells require 3 min (Nazarewicz et al., 2007b), whereas dopaminergic neuroblastoma SHSY-5Y cells require 2 min of permeabilization (Parihar et al., 2008b). Thus, optimization of permeabilization timings should be individualized for each cell type. A long permeabilization time can adversely affect the integrity of the subcellular membranous structure, including mitochondria. After permeabilization, the permeabilized medium should be replaced quickly with fresh medium.

As shown in Figs. 21.6C and 21.6D, permeabilization of HL-1 cells with digitonin (10 μM for 3 min) allows selective removal of cytoplasmic DAF-2 without affecting the mitochondrial distribution of the fluorescent probe. As shown, the DAF-2 fluorescence signal is punctuated typical for mitochondria. As shown in Fig. 21.6D, permeabilized cells respond to Ca^{2+} by increasing the NO signal in mitochondria. By using NOS-specific inhibitors or NO donors, the specificity of DAF-2 for NO can be tested. Prior to loading with DAF, collapsing the mitochondrial membrane potential with a protonophore such as FCCP or blocking the mitochondrial Ca^{2+} uniporter with ruthenium red, which causes a decrease of mitochondrial Ca^{2+}, can be considered to validate mtNOS activity measurements. Likewise, lowering mitochondrial Ca^{2+} by inhibiting the respiratory chain with antimycin A in combination with oligomycin can be used to validate mtNOS assays.

7.1.3. Technical notes

- Perform all operations with DAF-2DA and mitotracker red in dark. Make sure that all samples are protected from light while being transported to the confocal microscope and until measurement is completed.

- Remove all media used for dye loading and plasma membrane permeabilization. Wash cells two or three times with low fluorescence medium before performing the imaging.

7.1.4. Advantages and pitfalls

As mentioned, an advantage of plasma membrane permeabilization is that it allows selective mitochondrial localization of the fluorescence signal. Fluorescent intensity quantitation of individual mitochondria is possible after permeabilization. A pitfall for this method is that upon permeabilization, cells become semi-intact and changes in cytoplasmic constituents may affect mitochondrial fluorescence.

7.2. Copper-fluorescein

CuFL is another NO detection probe that has been described and can also be used for the detection of mitochondrial NO. FL (0.5 mg) is dissolved in dimethyl sulfoxide and mixed with a copper chloride (1.0 mM) solution (1:1) immediately prior to loading cells. Cells are probed with mitotracker red (200 nM) in Dulbecco's modified Eagle medium/F12 medium and incubated at 37° under 5% CO_2 for 10 min. Cells are further incubated for 15 min with mixed copper–FL (CuFL, 1 μM). Probed cells are permeabilized with digitonin (10 μM) to remove the cytosolic fraction of FL. After washing cells with medium, images are acquired using 543 nm excitation and 579 nm emission for mitotracker red and 488 nm excitation and 516 nm emission for CuFL (Parihar *et al.*, 2008b).

7.2.1. Advantages and pitfalls

The CuFL probe detects NO in a reaction that generates Cu(I) and NO^+ rather than a derivative reactive nitrogen species. NO^+ reacts irreversibly with the fluorescent-based ligand FL, which is nitrosated and released from copper with substantial turn-on fluorescence (Lim *et al.*, 2006).

8. Conclusion

This chapter reviewed currently available NOS detection assays that potentially could be used to detect mtNOS activity. Advantages and pitfalls of each assay need to be considered carefully to yield appropriate and accurate results. Individual experimental conditions may require fine-tuning each assay to accommodate the specific needs of each experimental setup. Inherent complications associated with testing mitochondria require the need for utilizing more than one assay to lower the probability of potential artifacts and to provide more decisive conclusion.

REFERENCES

Anson, M. L. and Mirsky, A. E. (1925). On the combination of nitric oxide with hemoglobin. *J. Physiol.* **60,** 100–102.
Boveris, A., Valdez, L. B., Alvarez, S., Zaobornyj, T., Boveris, A. D., and Navarro, A. (2003). Kidney mitochondrial nitric oxide synthase. *Antioxid. Redox Signal.* **5,** 265–271.
Braman, R. S. and Hendrix, S. A. (1989). Nanogram nitrite and nitrate determination in environmental and biological materials by vanadium (III) reduction with chemiluminescence detection. *Anal. Chem.* **61,** 2715–2718.
Bredt, D. S. and Snyder, S. H. (1990). Isolation of nitric oxide synthetase, a calmodulin-requiring enzyme. *Proc. Natl. Acad. Sci. USA* **87,** 682–685.
Casey, T. E. and Hilderman, R. H. (2000). Modification of the cadmium reduction assay for detection of nitrite production using fluorescence indicator 2,3 diaminonaphthalene. *Nitric Oxide* **4,** 67–74.
Cox, R. D. and Frank, C. W. (1982). Determination of nitrate and nitrite in blood and urine by chemiluminescence. *J. Anal. Toxicol.* **6,** 148–152.
Davison, W. and Woof, C. (1978). Comparison of different forms of cadmium as reducing agent for the batch determination of nitrate. *Analyst* **103,** 403–406.
Dedkova, E. N. and Blatter, L. A. (2005). Modulation of mitochondrial Ca^{2+} by nitric oxide in cultured bovine vascular endothelial cells. *Am. J. Physiol. Cell. Physiol.* **289,** C836–C845.
Dedkova, E. N., Ji, X., Lipsius, S. L., and Blatter, L. A. (2004). Mitochondrial calcium uptake stimulates nitric oxide production in mitochondria of bovine vascular endothelial cells. *Am. J. Physiol. Cell. Physiol.* **286,** C406–C415.
Eich, R. F., Li, T., Lemon, D. D., Doherty, D. H., Curry, S. R., Aitken, J. F., Mathews, A. J., Johnson, K. A., Smith, R. D., Phillips, G. N., Jr., and Olson, J. S. (1996). Mechanism of NO-induced oxidation of myoglobin and hemoglobin. *Biochemistry* **35,** 6976–6983.
Espey, M. G., Miranda, K. M., Thomas, D. D., and Wink, D. A. (2001). Distinction between nitrosating mechanisms within human cells and aqueous solution. *J. Biol. Chem.* **276,** 30085–30091.
Feelisch, M. and Noack, E. A. (1987). Correlation between nitric oxide formation during degradation of organic nitrates and activation of guanylate cyclase. *Eur. J. Pharmacol.* **139,** 19–30.
Feldman, P. L., Griffith, O. W., and Stuehr, D. J. (1993). The surprising life of nitric oxide. *C&EN,* December. 26–38.
Fontijn, A., Sabadell, A. J., and Ronco, R. J. (1970). Homogenous chemiluminescent measurement of nitric oxide with ozone: Implication for continuous selective monitoring of gaseous air pollutants. *Anal. Chem.* **42,** 575–579.
Ghafourifar, P., Asbury, M. L., Joshi, S. S., and Kincaid, E. D. (2005). Determination of mitochondrial nitric oxide synthase activity. *Methods Enzymol.* **396,** 424–444.
Ghafourifar, P. and Richter, C. (1997). Nitric oxide synthase activity in mitochondria. *FEBS Lett.* **418,** 291–296.
Ghafourifar, P. and Richter, C. (1999). Mitochondrial nitric oxide synthase regulates mitochondrial matrix pH. *Biol. Chem.* **380,** 1025–1028.
Ghafourifar, P., Schenk, U., Klein, S. D., and Richter, C. (1999). Mitochondrial nitric-oxide synthase stimulation causes cytochrome c release from isolated mitochondria: Evidence for intramitochondrial peroxynitrite formation. *J. Biol. Chem.* **274,** 31185–31188.
Giulivi, C. (1998). Functional implications of nitric oxide produced by mitochondria in mitochondrial metabolism. *Biochem. J.* **332,** 673–679.

Green, L. C., Wagner, D. A., Glogowski, J., Skipper, P. L., Wishnok, J. S., and Tannenbaum, S. R. (1982). Analysis of nitrate, nitrite, and [15N]nitrate in biological samples. *Anal. Biochem.* **126,** 131–138.

Griess, P. (1879). Benerkungen zu der Abhandlung der HH. Wesley and Benedikt ueber einige Azoverbindungen. *Ber. Dtsch. Chem. Ges.* **12,** 426.

Ignarro, L. J., Buga, G. M., Wood, K. S., Byrns, R. E., and Chaudhuri, G. (1987). Endothelium-derived relaxing factor produced and released from artery and vein is nitric oxide. *Proc. Natl. Acad. Sci. USA* **84,** 9265–9269.

Kanai, A. J., Pearce, L. L., Clemens, P. R., Birder, L. A., VanBibber, M. M., Choi, S. Y., de Groat, W. C., and Peterson, J. (2001). Identification of a neuronal nitric oxide synthase in isolated cardiac mitochondria using electrochemical detection. *Proc. Natl. Acad. Sci. USA* **98,** 14126–14131.

Knowles, R. G., Merrett, M., Salter, M., and Moncada, S. (1990). Differential induction of brain, lung and liver nitric oxide synthase by endotoxin in the rat. *Biochem. J.* **270,** 833–836.

Knowles, R. G. and Moncada, S. (1992). Nitric oxide as a signal in blood vessels. *TIBS* **17,** 399–402.

Kojima, H., Nakatsubo, N., Kikuchi, K., Kawahara, S., Kirino, Y., Nagoshi, H., Hirata, Y., and Nagano, T. (1998). Detection and imaging of nitric oxide with novel fluorescent indicators: Diaminofluoresceins. *Anal. Chem.* **70,** 2446–2453.

Lim, M. H., Xu, D., and Lippard, S. J. (2006). Visualization of nitric oxide in living cells by a copper-based fluorescent probe. *Nat. Chem. Biol.* **2,** 375–380.

Lores-Arnaiz, S., Amico, G. D., Czerniczyniec, A., Bustamante, J., and Boveris, A. (2004). Brain mitochondrial nitric oxide synthase: *In vitro* and *in vivo* inhibition by chlorpromazine. *Arch. Biochem. Biophys.* **430,** 170–177.

Manser, R. C. and Houghton, F. D. (2006). Ca^{2+}-linked upregulation and mitochondrial production of nitric oxide in the mouse preimplantation embryo. *J. Cell. Sci.* **119,** 2048–2055.

Mayer, B., Klatt, P., Werner, E. R., and Schmidt, K. (1994). Molecular mechanisms of inhibition of porcine brain nitric oxide synthase by the antinociceptive drug 7-nitroindazole. *Neuropharmacology* **33,** 1253–1259.

Moshage, H., Kok, B., Huizenga, J. R., and Jansen, P. L. (1995). Nitrite and nitrate determination in plasma: A critical evaluation. *Clin. Chem.* **41,** 892–896.

Nazarewicz, R. R., Zenebe, W. J., Parihar, A., Larson, S. K., Alidema, E., Choi, J., and Ghafourifar, P. (2007a). Tamoxifen induces oxidative stress and mitochondrial apoptosis via stimulating mitochondrial nitric oxide synthase. *Cancer Res.* **67,** 1282–1290.

Nazarewicz, R. R., Zenebe, W. J., Parihar, A., Parihar, M. S., Vaccaro, M., Rink, C., Sen, C. K., and Ghafourifar, P. (2007b). 12(S)-Hydroperoxyeicosatetraenoic acid (12-HETE) increases mitochondrial nitric oxide by increasing intramitochondrial calcium. *Arch. Biochem. Biophys.* **468,** 114–120.

Palmer, R. M., Ferrige, A. G., and Moncada, S. (1987). Nitric oxide release accounts for the biological activity of endothelium-derived relaxing factor. *Nature* **327,** 524–526.

Parihar, A., Parihar, M. S., and Ghafourifar, P. (2008a). Significance of mitochondrial calcium and nitric oxide for apoptosis of human breast cancer cells induced by tamoxifen and etoposide. *Int. J. Mol. Med.* **21,** 317–324.

Parihar, M. S., Parihar, A., Villamena, F. A., Vaccaro, P. S., and Ghafourifar, P. (2008b). Inactivation of mitochondrial respiratory chain complex I leads mitochondrial nitric oxide synthase to become pro-oxidative. *Biochem. Biophys. Res. Commun.* **367,** 761–767.

Rathel, T. R., Leikert, J. J., Vollmar, A. M., and Dirsch, V. M. (2003). Application of 4,5-diaminofluorescein to reliably measure nitric oxide released from endothelial cells *in vitro*. *Biol. Proc.* **5,** 136–142.

Riobo, N. A., Melani, M., Sanjuan, N., Fiszman, M. L., Gravielle, M. C., Carreras, M. C., Cadenas, E., and Poderoso, J. J. (2002). The modulation of mitochondrial nitric-oxide synthase activity in rat brain development. *J. Biol. Chem.* **277,** 42447–42455.

Teniguchi, S., Takahashi, K., and Noji, S. (2003). Nitrate. *In* "Methods of Enzymatic Analysis" (J. Bergmeyer and M. Grassl, eds.), Vol. 7, p. 5. VCH Verlagsgesellschaft mbH, Weiheim, Germany.

Van Assendelft, O. (1970) (ed.)."Spectrophotometry of Hemoglobin Derivatives." Charles C. Thomas, Springfield, IL.

Verdon, C. P., Burton, B. A., and Prior, R. L. (1995). Sample pretreatment with nitrate reductase and glucose-6-phosphphate dehydrogenase quantitatively reduces nitrate while avoiding interference by $NADP^+$ when the Griess reaction is used to assay for nitrite. *Anal. Biochem.* **224,** 502–508.

Wink, D. A., Darbyshire, J F., Nims, R. W., Saavedra, J. E., and Ford, P. C. (1993). Reaction of the bioregulatory agent nitric oxide in oxygenated aqueous media: Determination of the kinetics for oxidation and nitrosation by intermediates generated in the NO/O_2 reaction. *Chem. Res. Toxicol.* **6,** 23–27.

Yang, F., Troncy, E., Francoeur, M., Vinet, B., Vinay, P., Czaika, G., and Blaise, G. (1997). Effect of reducing agents and temperatures on conversion of nitrite and nitrate to nitric oxide and detection of NO by chemiluminescence. *Clin. Chem.* **43,** 657–662.

Zenebe, W. J., Nazarewicz, R. R., Parihar, M. S., and Ghafourifar, P. (2007). Hypoxia/reoxygenation of isolated rat heart mitochondria causes cytochrome c release and oxidative stress; evidence for involvement of mitochondrial nitric oxide synthase. *J. Mol. Cell. Cardiol.* **43,** 411–419.

Zhang, X. and Broderick, M. (2003). Electrochemical NO sensors and their applications in biomedical research. *Biomed. Significance Nitric Oxide International Scientific Literature, Inc.*

Zhang, X., Chi, R., Zou, J., and Zhang, H. S. (2004). Development of a novel fluorescent probe for nitric oxide detection: 8-(3′,4′-diaminophenyl)-difluoroboradiaza-S-indacence. *Spectrochim. Acta A Mol. Biomol. Spectrosc.* **60,** 3129–3134.

SECTION THREE

ORGANISM METHODS

CHAPTER TWENTY-TWO

ASSAY OF 3-NITROTYROSINE IN TISSUES AND BODY FLUIDS BY LIQUID CHROMATOGRAPHY WITH TANDEM MASS SPECTROMETRIC DETECTION

Naila Rabbani *and* Paul J. Thornalley

Contents

1. 3-Nitrotyrosine (3-NT) in Physiological Systems	338
2. Measurement of 3-NT	339
3. Liquid Chromatography with Tandem Mass Spectrometric Detection (LC-MS/MS) Assay of 3-NT Residues and 3-NT-Free Adducts: Thornalley Group Method	340
3.1. Materials	340
3.2. Preparation of ultrafiltrate of physiological fluid for assay of free 3-NT	341
3.3. Preparation of exhaustive enzymatic hydrolysates for assay 3-NT residues	341
3.4. LC-MS/MS assay sample and standard preparation	341
3.5. Sample analysis by LC-MS/MS	342
4. Estimates of 3-NT Residues and Free 3-NT in Plasma and Red Blood Cells under Basal Conditions and Effect of Disease	342
5. 3-Nitrotyrosine Residues in Lipoproteins	351
6. 3-Nitrotyrosine Residues and Free Adduct in Cerebrospinal Fluid	353
7. 3-Nitrotyrosine Residue Content of Tissues	353
8. Concluding Remarks	354
Acknowledgment	355
References	355

Abstract

3-Nitrotyrosine (3-NT) is a marker of protein nitration in physiological systems. It is present as 3-nitrotyrosine residues in proteins of tissue, extracellular matrix, plasma, and other body fluids and food. It is also present in body fluids

Protein Damage and Systems Biology Research Group, Clinical Sciences Research Institute, Warwick Medical School, University of Warwick, University Hospital, Coventry, United Kingdom

Methods in Enzymology, Volume 440
ISSN 0076-6879, DOI: 10.1016/S0076-6879(07)00822-1

© 2008 Elsevier Inc.
All rights reserved.

and some beverages as free nitrotyrosine and is excreted in urine with the major urinary metabolite 3-nitro-4-hydroxyphenylacetic acid. Quantitation of 3-nitrotyrosine requires tandem mass spectrometry for specific detection. The method developed to determine 3-nitrotyrosine (along with protein glycation and oxidation adducts in a quantitative screening assay) by liquid chromatography with tandem mass spectrometric detection is described. The 3-NT residue contents of plasma protein, hemoglobin, lipoproteins, and cerebrospinal fluid protein and the concentrations of free 3-nitrotyrosine in plasma, urine, and cerebrospinal fluid are given. Changes of 3-nitrotyrosine residue and free 3-nitrotyrosine in diabetes, cirrhosis, acute and chronic renal failure, and neurological disorders, including Alzheimer's disease, are presented and compared with independent estimates.

1. 3-Nitrotyrosine (3-NT) in Physiological Systems

3-Nitrotyrosine is present in proteins of tissues, extracellular matrix, plasma, and other body fluids formed by the nitration of tyrosine residues. This has been referred to as "protein-bound" 3-NT but is appropriately called a 3-NT residue by analogy with conventional nomenclature of amino acids and to distinguish this from free 3-NT (nitrated tyrosine or 3-NT free adduct) that may bind reversibly to proteins. 3-NT residues are also present in proteins of ingested food and, similar to oxidation and glycation adducts of ingested proteins, are absorbed as free 3-NT or 3-NT residue-containing peptides, which are hydrolyzed rapidly to free 3-NT in the venous circulation. Free 3-NT is also present in plasma, urine, cerebrospinal fluid (CSF), synovial fluid, and presumably interstitial fluid and other body fluids. Free 3-NT is formed from 3-NT residues of nitrated proteins by cellular proteolysis (Greenacre *et al.*, 1999; Grune *et al.*, 2001). Free 3-NT is also present in some beverages—milk and cola drinks, for example. The cellular compartmentalization of 3-NT residue-containing proteins does not usually change compared to the unmodified protein, although albumin-containing 3-NT residues showed preferentially transport across the capillary endothelium (Predescu *et al.*, 2002). Free 3-NT is transported across cell membranes by the L-aromatic amino acid transporter and dopamine transporters (Blanchard-Fillion *et al.*, 2006). Free 3-NT is metabolized in rats and human subjects, leading to formation of the major urinary metabolite 3-nitro-4-hydroxyphenylacetic acid (3-NHA)—with minor formation of 3-nitro-4-hydroxyphenyllactic acid (Ohshima *et al.*, 1990; Tabrizi-Fard *et al.*, 1999). 3-NHA is not a good marker of flux of formation of 3-NT, however, as it may be formed by 3-NT-independent pathways (Pannala *et al.*, 2006). At pharmacological doses (10 mg/kg) in rats, the half-life of 3-NT in plasma was 69 min and the volume of distribution was 0.2 liter/kg

(Tabrizi-Fard et al., 1999). 3-NT residues in plasma protein nitrated *ex vivo* had a much longer half-life (63 h), which relates to the half-life for turnover of the plasma proteins (Hitomi et al., 2007) but is slightly longer than the half-life of rat albumin (46 h) (Gaizutis et al., 1975). Although in pharmacokinetic studies it has been claimed that minimal free 3-NT is excreted intact, free 3-NT is indeed present in urine (Thornalley et al., 2003; Tsikas et al., 2005).

2. Measurement of 3-NT

3-Nitrotyrosine has been quantified in tissues and body fluids by ELISA, high-performance liquid chromatography (HPLC) with absorbance, fluorescence or electrochemical detection, and mass spectrometric techniques [gas chromatography with mass spectrometric detection (GC–MS), gas chromatography with tandem mass spectrometric detection (GC–MS/MS), and liquid chromatography with tandem mass spectrometric detection (LC–MS/MS)](reviewed in Ryberg and Caidahl, 2007). Problems with the generation of 3-NT during sample processing, particularly during acid hydrolysis of proteins and derivatization of amino acids prior to chromatographic separation, have been identified (Yi et al., 2000). Thornalley et al. (2003) recommend enzymatic hydrolysis of proteins with a cocktail of enzymes under anaerobic conditions, although there is a slightly decreased recovery compared to acid hydrolysis (Delatour et al., 2007); enzymatic hydrolysis enables the concurrent assay of glycation and oxidation adduct residues in proteins—applicable to quantitative screening of protein damage in many disease and abnormal physiological states (Thornalley, 2006). Chromatography is best performed by HPLC without prior derivatization, avoiding formation of 3-NT during the derivatization procedure (Thornalley, 2006; Thornalley et al., 2003), and detection and quantitation are best performed by LC–MS/MS with a stable isotope-substituted standard ([^{2}H$_3$]3-NT or similar) (Delatour, 2004; Thornalley et al., 2003). The highly specific detection and internal standardization that this LC–MS/MS (and GC–MS/MS) approach gives are critical in avoiding overestimation of 3-NT (Ryberg et al., 2007).

Estimates of 3-NT residues in protein of tissues, plasma, and other body fluids by ELISA have overestimated 3-NT residue content by at least 10- to 100-fold, in comparison to estimates by GC–MS/MS and LC–MS/MS with validated good analytical recovery (Ceriello et al., 2001; Khan et al., 1998) and remain in use (Rossner et al., 2007). Similarly, HPLC with absorbance or electrochemical detection of 3-NT has produced erroneous results—overestimating 3-NT residues and free 3-NT by 5- to 100-fold (Hensley et al., 1998; Hoeldtke et al., 2002; Kaur et al., 1998; Tohgi et al., 1999).

There is significant doubt over the interpretation of 3-NT residue estimates using ELISA—and also proteomics studies using antibodies to pull down putative 3-NT-modified proteins where peptide fragment modification with a mass increase of 45 mass units ($+NO_2$, $-H$) has not been confirmed by mass spectrometry. Recent improvements in specificity and sensitivity have brought estimates of plasma protein 3-NT residue content by ELISA close to levels that are expected to corroborate with "gold standard" estimation by GC-MS/MS and LC-MS/MS methods (Bo *et al.*, 2005; Sun *et al.*, 2007). A definitive study optimizing conditions using ELISA and LC-MS/MS detection of 3-NT residues in proteins is required. Such corroboration is likely to be matrix specific. The failure of immunoassays of 3-NT residues is likely attributable to poor specificity of the antibodies used—detecting a different adduct in 3-NT-modified protein with high affinity. This chapter compares studies where 3-NT residue and free 3-NT estimates have been performed by mass spectrometric techniques with stable isotopic internal standardization and multiple reaction monitoring (detector response monitoring selective for parent and fragment mass; GC-MS/MS and LC-MS/MS)—a view accepted by experienced investigators in 3-NT estimations (Gaut *et al.*, 2002; Ryberg *et al.*, 2007; Tsikas *et al.*, 2002).

3. Liquid Chromatography with Tandem Mass Spectrometric Detection (LC-MS/MS) Assay of 3-NT Residues and 3-NT-Free Adducts: Thornalley Group Method

3.1. Materials

Tyrosine and the 3-NT standard are from Sigma Chem Co. (Poole, Dorset, UK). L-[$^2H_4$]Tyrosine is from Cambridge Isotopes, Inc. (Andover, MA). [$^2H_3$]-3-Nitrotyrosine ([$^2H_3$]3-NT) is prepared from L-[$^2H_4$]tyrosine and purified as described (Delatour *et al.*, 2002a). Solutions of 3-NT are prepared in 0.1% trifluoroacetic acid and calibrated by spectrophotometry; $\epsilon_{381\ nm} = 2200\ M^{-1}\ cm^{-1}$ (Sokolovsky *et al.*, 1966). Proteases, peptidases, and other enzymes are from Sigma. Pronase E (EC 3.4.24.31)–type XIV from bacterial *Streptomyces griseus* with a specific activity of 4.4 units/mg protein (1 unit of activity hydrolyzed casein forming 1.0 μmol of tyrosine per min at pH 7.5 and 37°), pepsin (EC 3.4.23.1)–porcine stomach mucosa with a specific activity of 3460 units/mg protein (1 unit hydrolyzed hemoglobin with an increase in A_{280} of 0.001 AU per minute of trichloroacetic acid-soluble products, at pH 2 and 37°), leucine aminopeptidase (EC 3.4.11.2)–type VI from porcine kidney microsomes with a specific activity of 22 units/mg protein (1 unit of activity hydrolyzed 1.0 μmol of L-leucine-*p*-nitroanilide to L-leucine and *p*-nitroaniline per minute at pH 7.2 and 37°)

and prolidase (EC 3.4.13.9)–from porcine kidney with a specific activity of 145 units/mg protein (1 unit of activity hydrolyzes 1.0 μmol of Gly-Pro per minute, at pH 8 at 40°) are used. Penicillin-streptomycin solution, containing 5,000 units/ml penicillin and 5 mg/ml streptomycin, was from Sigma.

3.2. Preparation of ultrafiltrate of physiological fluid for assay of free 3-NT

Plasma, urine, CSF, or other physiological fluid (100 μl) is placed in a microspin filter (12-kDa filter cutoff, Waters or Whatman) and centrifuged at 4°, and the ultrafiltrate (50 μl) is collected.

3.3. Preparation of exhaustive enzymatic hydrolysates for assay 3-NT residues

Plasma and lipoprotein are delipidified. The protein sample (250 μg, 2 mg/ml) is extracted with 2 × 1 volume of water-saturated diethyl ether to remove lipids, and the residual ether is removed by centrifugal evaporation. An aliquot of protein sample (ca. 200 μg) is then diluted to 500 μl with water and concentrated to about 50 μl by ultrafiltration (12-kDa cutoff membrane). The sample is diluted to 500 μl and concentrated to 50 μl a further two times to complete the washing process. The protein concentration is then determined by the Bradford method.

An aliquot of washed protein sample containing 100 μg of protein is diluted to 20 μl with water. Aliquots of 40 mM HCl (25 μl), pepsin solution (2 mg/ml in 20 mM HCl; 5 μl), and thymol solution (2 mg/ml in 20 mM HCl; 5 μl) are added, and the sample is incubated at 37° for 24 h. The sample is then neutralized and buffered at pH 7.4 by the addition of 25 μl of 0.5 M potassium phosphate buffer, pH 7.4, and 5 μl of 260 mM KOH. Subsequent steps are performed under nitrogen (or carbon monoxide for hemoglobin) to inhibit the oxidative degradation of protein substrate (and inactivate heme moieties). Pronase E solution (2 mg/ml in 10 mM KH_2PO_4, pH 7.4; 5 μl) and penicillin-streptomycin solution (1000 units per ml penicillin and 1 mg/ml streptomycin; 5 μl) are added, and the samples is incubated at 37 °C for 24 h. Aminopeptidase solution (2 mg/ml in 10 mM KH_2PO_4, pH 7.4; 5 μ) and prolidase solution (2 mg/ml in 10 mM KH_2PO_4, pH 7.4; 5 μl) are added, and the sample is incubated at 37° for 48 h. This gives the final enzymatic hydrolysate (105 μl) for the LC-MS/MS assay.

3.4. LC-MS/MS assay sample and standard preparation

Aliquots of plasma, urine, and CSF ultrafiltrate (50 μl) and exhaustive enzymatic hydrolysates are mixed with [$^2H_4$]tyr (10 nmol) and [$^2H_3$]3-NT (10 pmol) in 50 μl water. A calibration curve is constructed from seven standard samples containing [²H₄]tyr (10 nmol), [²H₃]3-NT (10 pmol), and 0 to 20 nmol tyr and 0 to 10 pmol 3-NT. The total volume of test samples and standards is 100 μl.

3.5. Sample analysis by LC-MS/MS

Samples are assayed by LC-MS/MS for tyrosine and 3-NT by the method described previously (Thornalley et al., 2003) with modifications. The column is a 5-μm particle size Hypercarb column (Thermo Hypersil, Runcorn, Cheshire, UK), 2.1 × 50 mm. The mobile phase is 0.1% trifluoroacetic acid with isocratic 10% acetonitrile from 0 to 10 min and then a linear gradient of 10 to 50% acetonitrile from 10 to 20 min; the flow rate is 0.2 ml/min. The interbatch coefficient of variation is 10.5% and the recovery is 88%. The parent ion–fragment ion mass transitions are given in Table 22.1. LC-MS/MS analysis is performed on suitable high-sensitivity LC-MS/MS systems. We have used a Waters 2690 Separation module with a Quattro Ultima triple quadrupole mass spectrometric detector. The ionization source temperature is 120°, and the desolvation gas temperature is 350°. The cone gas and desolvation gas–flow rates are 150 and 550 liters/h, respectively. The capillary voltage is 3.55 kV, and the cone voltage is 80 V. Argon gas (2.7×10^{-3} mbar) is in the collision cell. Programmed molecular ion and fragment ion masses and collision energies are optimized to ±0.1 Da and ±1 eV for multiple reaction monitoring detection of tyr and 3-NT analytes and [²H₄]tyr (10 nmol) and [²H₃]3-NT isotopomer standards (Fig. 22.1). The limits of detection for tyrosine and 3-NT are 490 and 11 fmol, respectively.

4. Estimates of 3-NT Residues and Free 3-NT in Plasma and Red Blood Cells under Basal Conditions and Effect of Disease

3-Nitrotyrosine residues are present in plasma protein—mostly on serum albumin (Hitomi et al., 2007; Mitrogianni et al., 2004). Peptide mapping studies of human serum albumin indicated nitration with peroxynitrite produced nitration preferentially on tyr-138 and tyr-413 and in subdomains IB and IIIA—corroborated in an independent study (Nikov et al., 2003), respectively (Jiao et al., 2001). Whereas nitration of human serum albumin with myeloperoxidase-catalyzed oxidation of nitrite produced nitration preferentially on tyr-162 (Willard et al., 2003). The concentration of 3-NT residues in plasma protein is in the range of 0.6 to 3.1 μmol/mol tyr (Table 22.1). This equates to less than 1 in 10,000 molecules of human serum albumin in plasma containing a single 3-NT residue. Where the 3-NT residue content of extracted human serum

Table 22.1 Estimates of 3-NT residue content of plasma protein, hemoglobin, and lipoproteins by the LC-MS/MS method and comparison with other reports of tandem mass spectrometric analysis

Protein	Species	3-NT residue content (μmol/mol)	Comment and related 3-NT residue estimate (μmol/mol tyr)	Authors	Reference
Plasma protein	Human	0.6 ± 0.3 ($n = 12$)	Type 1 diabetic patients: 1.2 ± 0.9 ($n = 21$; $P < 0.05$)	Thornalley group	Ahmed et al. (2005)
		3.1 ± 0.9 ($n = 8$)	Increased 3-fold in PD patients ($n = 8$; $P < 0.001$) and HD patients ($n = 8$; $P < 0.01$)	Thornalley group	Agalou et al. (2005)
		9 ± 6 ($n = 10$; portal)	Cirrhotic patients: 14 ± 4 ($n = 10$, hepatic; $P < 0.05$)	Thornalley group	Ahmed et al. (2004b)
		1.55 ± 0.45 ($n = 18$)[a]	Range: 0.5–3.5	Comparison report	Tsikas et al. (2003)
		2.1 ± 1.4 ($n = 12$)		Comparison report	Soderling et al. (2003)
		1.21 ± 0.57 ($n = 20$)	No effect of organic nitrates	Comparison report	Keimer et al. (2003)
		0.7–3.2 ($n = 22$)	Premature infant population—no effect after inhaled nitric oxide gas	Comparison report	Lorch et al. (2003)
Plasma protein	Rat	13 ± 2 ($n = 5$)	No effect of BNX and BUL	Thornalley group	Rabbani et al. (2007)
		0.67 ± 0.29 ($n = 13$)	Not increased in streptozotocin-induced diabetes	Thornalley group	P. J. Thornalley, unpublished results

(*continued*)

Table 22.1 (continued)

Protein	Species	3-NT residue content (μmol/mol)	Comment and related 3-NT residue estimate (μmol/mol tyr)	Authors	Reference
		4–5 ($n = 2$)		Comparison report	Delatour et al. (2002b)
		5–18 ($n = 6$)		Comparison report	Delatour et al. (2002c)
Hemoglobin	Human	24 ± 2 ($n = 12$)	Type 1 diabetic patients: 10 ± 3 ($n = 21$; $P < 0.05$)	Thornalley group	Ahmed et al. (2005)
LDL (apoB-100)	Human	12 ± 4 ($n = 6$)	Diabetic patients: 21 ± 10 ($n = 8$; $P < 0.05$)	Thornalley group	Rabbani and Thornalley, unpublished results
HDL (apoA-1 and others)	Human	4.8 ± 1.9 ($n = 4$)	Diabetic patients: 5.1 ± 1.6 ($n = 8$; $P > 0.05$)	Thornalley group	Rabbani and Thornalley, unpublished results
CSF	Human	8.9 ± 4.7 ($n = 18$)	Subjects with Alzheimer's disease: 14.2 ± 9.4 ($n = 32$; $P < 0.05$)	Thornalley group	Ahmed et al. (2004a)

[a] Determined in affinity chromatography-extracted serum albumin. For estimation of 3-NT residues and 3-NT in plasma, peripheral venous plasma was used unless otherwise stated.

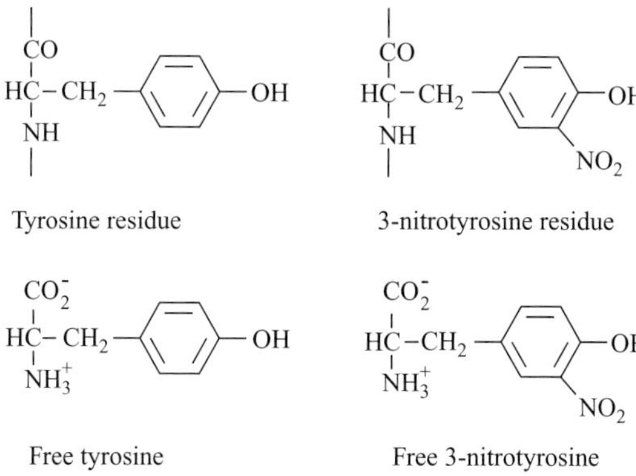

Figure 22.1 Mass spectrometric multiple reaction monitoring 3-nitrotyrosine and tyrosine analytes. From Thornalley et al. (2003).

albumin was determined, the concentration of 3-NT residue-containing albumin was 24 nM or approximately 0.003% (Tsikas et al., 2003) (Table 22.1). The concentration of free 3-NT in human plasma is in the range of 0.6 to 9.4 nM (Table 22.2). Estimates of the urinary excretion of free 3-NT in normal healthy human subjects varies widely (0.5–21 nmol/mmol creatinine; Table 22.2). This variation may reflect differences in rates of metabolism of 3-NT and also varied exposure through endogenous protein nitration and ingestion of varied amounts of 3-NT residues and free 3-NT in food and beverages. The renal clearance of free 3-NT is 8 to 13 ml/min (Agalou et al., 2005; Thornalley et al., 2003), declining in chronic renal failure (Agalou et al., 2005).

3-Nitrotyrosine residues are also detected in hemoglobin of red blood cell lysates. The 3-NT residue content of hemoglobin of normal healthy human subjects is 24 ± 2 μmol/mol tyr or 0.029 ± 0.003 %Hb. This is about 10-fold higher than the percentage 3-NT content of albumin in plasma (Ahmed et al., 2005)(Table 22.1).

The concentration of 3-NT residues and free 3-NT in plasma of normal healthy rats is 4 to 18 μmol/mol tyr and 1.4 to 2.1 nM, respectively (Tables 22.1 and 22.2). Urinary excretion of free 3-NT of rats is low relative to excretion of human subjects (0.071 ± 0.047 nmol/mmol creatinine; Table 22.3).

Diabetes is viewed as a disease state with increased oxidative and nitrosative stress implicated in the development of vascular complications, microvascular complications (nephropathy, retinopathy, and neuropathy), and macrovascular complications (heart disease and stroke). We determined the 3-NT residue content of the plasma protein of streptozotocin-induced diabetic rats (Babaei-Jadidi et al., 2003) by the LC-MS/MS method (Thornalley et al., 2003) and there was no significant increase from normal

Table 22.2 Estimates of free 3-NT content of plasma by the LC-MS/MS method and comparison with other reports of tandem mass spectrometric analysis

Species	3-NT (nM)	Comment and related 3-NT estimate (nM)	Authors	Reference
Human	6.5 ± 2.5 ($n = 6$)	Renal clearance 8.3 ± 2.8 ml/min	Thornalley group	Thornalley et al. (2003)
	0.74 ± 0.66 ($n = 8$)	Subjects with cirrhosis: 0.71 ± 0.49 ($n = 10$)	Thornalley group	Ahmed et al. (2004b)
	2.66 ± 0.32 ($n = 8$)	Increased in PD patients (4.02 ± 0.42; $n = 8$, $P < 0.05$) but not HD of CRF patients	Thornalley group	Agalou et al. (2005)
	9.4 ± 0.4 ($n = 6$)	Not increased in triosephosphate isomerize deficiency	Thornalley group	Ahmed et al. (2003)
	0.73 ± 0.53 ($n = 18$)	Range: 0.3–4.2	Comparison report	Tsikas et al. (2003)
	0.74 ± 0.30 ($n = 12$)		Comparison report	Soderling et al. (2003)
	0.71 ± 0.04 ($n = 10$)	Subjects with sleep apnea: 0.66 ± 0.04 ($n = 10$)	Comparison report	Svatikova et al. (2004)
	2.8 ± 0.84 ($n = 11$)		Comparison report	Schwedhelm et al. (1999)
	0.64 ± 0.15 ($n = 20$)	No effect of organic nitrates	Comparison report	Keimer et al. (2003)
	2.05 ± 0.37 ($n = 5$)	Increased 17-fold in BNX and 11-fold in BUL	Thornalley group	Rabbani et al. (2007)
Rat	$1.4 – 1.5$ ($n = 6$)		Comparison report	Delatour et al. (2002a)

Table 22.3 Estimates of free 3-NT content of urine by the LC-MS/MS method and comparison with other reports of tandem mass spectrometric analysis

Species	3-NT content (nmol/mmol creatinine)	Other 3-NT estimate (nmol/mmol creatinine)	Authors	Reference
Human	21 ± 10 ($n = 12$)	Type 1 diabetic patients: 30 ± 7 ($n = 21$, $P < 0.01$)	Thornalley group	Ahmed et al. (2005)
	5.16 ± 0.67	Increased in TPI-deficient propositus to 35.6 ± 2.5	Thornalley group	Ahmed et al. (2003)
	0.46 ± 0.49 ($n = 10$)	Range 0.05–1.30	Comparison report	Tsikas et al. (2005)
Rat	0.071 ± 0.047 ($n = 13$)	Not increased significantly in streptozotocin-induced diabetes	Thornalley group	Thornalley, unpublished results

control levels (Table 22.1). In clinical type 1 diabetic patients, however, plasma protein 3-NT residues were increased twofold in patients with respect to normal controls (1.2 ± 0.9 versus 0.6 ± 0.3 μmol/mol tyr; Table 22.1). We studied the effect of decreased postprandial hyperglycemia with insulin lispro (LP) in a crossover study (Beisswenger et al., 2001), but this was without significant effect. Remarkably, the 3-NT residue content of hemoglobin was decreased in diabetic patients (−59%). This decrease was partially reversed by LP therapy (−42%) (Ahmed et al., 2005) (Fig. 22.2). The interactions of nitric oxide (NO) with human serum albumin and hemoglobin and the effect of hemoglobin compartmentalization in blood on the availability of NO are controversial (Gladwin et al., 2003; Reiter et al., 2002). The explanation for the decreased 3-NT residue content of hemoglobin in diabetes, as for the availability of NO, probably lies in the compartmentalization of hemoglobin in red blood cells (RBCs).

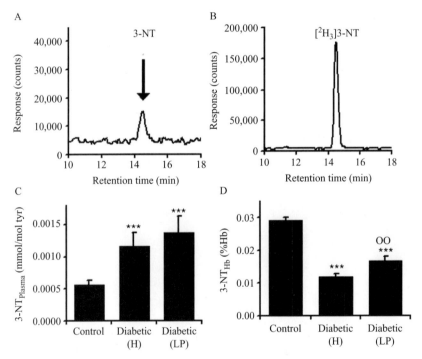

Figure 22.2 Specimen analytical chromatogram for the determination of 3-NT in hemoglobin enzymatic digest: (A) 3-NT (152 fmol, 50 μg protein equivalent) in normal control hemoglobin; (B) [^{2}H$_3$]3-NT standard (10 pmol); and (C and D) concentration of 3-NT residues in plasma protein and hemoglobin, respectively. Data are mean ± SEM. Significance: ***$P < 0.001$ with respect to control subjects (t test); oo, $P < 0.01$ with respect to human insulin therapy (paired t test). From Ahmed et al. (2005).

The increased formation of 3-NT residues in plasma in diabetes is thought to be a consequence of increased formation of peroxynitrite arising from increased production of superoxide in vascular cells suffering hyperglycemia-induced mitochondrial dysfunction (Nishikawa et al., 2000). Increased vascular formation of peroxynitrite was implicated in the increased extracellular matrix content of 3-NT and vascular complications of diabetes (Soriano et al., 2001). Peroxynitrite may be formed from superoxide and NO inside RBCs—NO originating from outside RBCs and superoxide originating from intra- and/or extra-RBCs sources. Peroxynitrite is also formed in the plasma and may diffuse into RBCs, including in the presence of carbon dioxide, physiologically (Romero et al., 1999). NO is formed by constitutive nitric oxide synthase in endothelial cells (Kuboki et al., 2000). As NO crosses from the endothelium through the plasma, it is intercepted by superoxide with increased efficiency in diabetes, increasing the formation of peroxynitrite and 3-NT in plasma proteins (Fig. 22.3). Less NO then enters RBCs to form 3-NT in hemoglobin. This explains the decreased 3-NT residue content of hemoglobin in diabetes concomitant with increased plasma protein 3-NT residues. The addition of LP probably increased hemoglobin 3-NT residues because it decreased postprandial hyperglycemia, related superoxide formation and hence more NO then enters RBCs, forming peroxynitrite therein. LP may also increase NO formation by increased expression of endothelial nitric oxide synthase (Kuboki et al., 2000), but this was not sufficient to increase the levels of 3-NT residues in plasma protein significantly. The effects of peroxynitrite, as well as NO, as evidenced by the formation of 3-NT residues in plasma protein and hemoglobin, and the influence of vascular cell superoxide formation in diabetes are compartmentalized in blood physiologically.

In liver cirrhosis, hepatic oxidative stress is linked to severe inflammation (Kharbanda et al., 2001; Willis et al., 2002). The 3-NT residue content of plasma protein of cirrhotic patients was increased with respect to controls (Table 22.1). The 3-NT residue content of plasma protein also increased with the increasing severity of cirrhosis (Child-Pugh index of C and B versus A): 14 ± 4 versus 9 ± 6 μmol/mol tyr. The mean plasma free 3-NT concentration was not increased from normal control values (Ahmed et al., 2004b).

Changes in 3-NT residues and free 3-NT in plasma in acute and chronic renal failure have been studied as markers of nitrosative stress. The kidney has an important role in the handling of 3-NT. In studies of experimental models of acute renal failure—bilateral nephrectomy (BNX) and bilateral ureteral ligation (BUL)—we found that 3-NT residues of plasma protein were not changed markedly with respect to sham-operated controls, whereas the plasma concentration of 3-NT increased 17-fold in BNX and 11-fold in BUL (Fig. 22.4). This suggests that the kidney has an important role in clearing 3-NT from the plasma and retains partial activity for doing

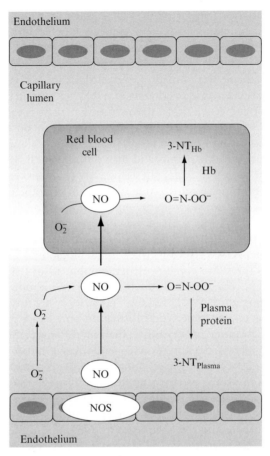

Figure 22.3 Scheme showing putative compartmentalization of the formation of 3-NT in blood and effect of diabetes. The schematic diagram shows a cross section through a blood capillary lumen illustrating the flows of nitric oxide from the endothelium into the red blood cells and the formation of peroxynitrite and 3-NT residues in plasma protein and hemoglobin. From Ahmed et al. (2005).

this in ureteral obstruction (Rabbani et al., 2007). In chronic renal failure of patients with renal replacement therapy—peritoneal dialysis (PD) or hemodialysis (HD)—we found that the plasma protein content of 3-NT residues was increased 3-fold in PD patients and HD patients before hemodialysis. The plasma protein content of 3-NT residues was decreased during hemodialysis (Agalou et al., 2005). The decrease of 3-NT residue content of plasma protein after a HD session may be due to preferential endothelial transcytosis of 3-NT-modified albumin (Predescu et al., 2002) during the increased transcytosis of plasma proteins as part of the inflammatory

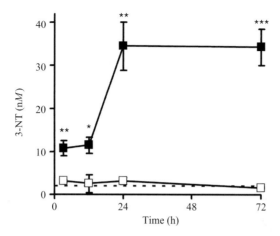

Figure 22.4 Free 3-nitrotyrosine in plasma of rats after sham operation, bilateral nephrectomy, and unoperated controls. Dotted line, unoperated controls; hollow squares, sham-operated controls; and solid squares, BNX. Data are mean ± SEM ($n = 4$). Free 3-NT concentration in controls: 2.05 ± 0.37 nM ($n = 5$). Significance: an asterisk indicates significance of difference with respect to unoperated and sham-operated controls where one, two, and three symbols reflect $P < 0.05$, $P < 0.01$, and $P < 00.1$, respectively. Adapted from Rabbani et al. (2007).

response associated with dialysis. Free 3-NT of plasma was increased 51% in PD patients. Free 3-NT was excreted by transfer into the peritoneal cavity and elimination in the peritoneal dialysate and by residual diuresis. The 24-h excretion rate of free 3-NT was increased 4-fold with respect to controls, which may be because of increased free 3-NT formation and/or decreased free 3-NT metabolism in PD. The proportion of free 3-NT excreted in the peritoneal dialysate was 65%. The plasma concentration of free 3-NT was not increased in HD patients (Agalou et al., 2005).

5. 3-Nitrotyrosine Residues in Lipoproteins

We applied the LC-MS/MS assay to the quantitation of 3-NT residues in low-density lipoprotein (LDL) and high-density lipoprotein (HDL) isolated from human plasma of normal healthy control subjects and diabetic patients. We found 3-NT residue contents of 12 ± 4 μmol/mol tyr in LDL, increased 75% in diabetic patients, and 4.8 μmol/mol tyr in HDL (Table 22.1). The estimate of 3-NT residues in HDL is much lower than reported previously by Pennathur et al. (1999) (68 μmol/mol tyr) using GC-MS and may indicate further the requirement for tandem mass spectrometry for specific 3-NT detection.

Table 22.4 Estimates of free 3-NT content of CSF by the LC-MS/MS method and comparison with other reports of tandem mass spectrometric analysis

Neurological disorder	Free 3-NT content (nM)	Other free 3-NT estimate (nM)	Authors	Reference
Alzheimer's disease	Control group: 0.40 ± 0.28 ($n = 18$)	AD subjects: 1.03 ± 0.46 ($n = 32$, $P < 0.001$)	Thornalley group	Ahmed et al. (2004a)
	Control group: 0.35 ± 0.02 ($n = 19$)	AD subjects: 0.44 ± 0.031 ($n = 17$)	Comparison report	Ryberg et al. (2004)
Amyotrophic lateral sclerosis	Control group: 0.35 ± 0.02 ($n = 19$)	ALS subjects: 0.38 ± 0.03 ($n = 14$)	Comparison report	Ryberg et al. (2004)

6. 3-Nitrotyrosine Residues and Free Adduct in Cerebrospinal Fluid

We studied the 3-NT residue content and free 3-NT concentration in CSF of 32 human subjects diagnosed with Alzheimer's disease (AD) and 18 age-matched control subjects. In the AD group, the 3-NT residue content of CSF protein was increased 59% and the concentration of free 3-NT in CSF was increased 93% with respect to normal controls. Multiple linear regression gave a regression model of Mini-Mental State Examination score on 3-NT residues and glycation-related variables, suggesting that protein nitration and glycation adduct residue estimates of CSF may provide an indicator for the diagnosis of Alzheimer's disease (Ahmed et al., 2004a). Several targets of protein nitration in AD have been identified (Castegna et al., 2003). Ryberg et al. (2004) found similar levels of free 3-NT in CSF of control subjects but found no significant increase of free 3-NT in CSF of subjects with AD or amyotrophic lateral sclerosis (ALS) (Table 22.4).

7. 3-Nitrotyrosine Residue Content of Tissues

The 3-NT residue content of rat tissues by tandem mass spectrometric methods has been estimated in several studies. The 3-NT content of rat liver is 51.1 ± 0.6 μmol/mol tyr (Delatour et al., 2002c). The basal level of 3-NT residues of protein extracts of whole rat kidney samples is 50 to 68 μmol/mol tyr (Delatour et al., 2002c). We studied the 3-NT residue content of cytosolic protein extracts of several rats tissues by the LC-MS/MS method ($n = 13$). The 3-NT residue content estimates (μmol/mol tyr) were renal glomeruli 0.69 ± 0.20, retina 2.1 ± 1.1, sciatic nerve 1.2 ± 0.2, and brain 3.8 ± 1.6. In a corresponding study group of streptozotocin-induced diabetic rats, only sciatic nerve showed a significant increase of 3-NT residue content (3.1 ± 0.7 μmol/mol tyr, n = 11; $P < 0.01$).

3-Nitrotyrosine residues of brain protein extracts were found to be enriched sixfold in microvessel-enriched brain tissue versus tissue devoid of microvessels (Althaus et al., 2000). Proteomics analysis indicated that flotillin-1 and α-tubulin are important sites of 3-NT residue formation (Dremina et al., 2005). Proteomic analysis of cerebellar homogenate from Fisher 344/Brown Norway (BN/F1) rats showed an age-dependent increase in 3-NT residue content. Seven proteins were found to be nitrated by corroboration of anti-3-NT antibody Western blotting and 3-NT-modified tryptic peptide detection by mass spectrometry and four were identified: ryanodine receptor 3, LDL-related receptor 2, nebulin-related anchoring protein isoform C, and 2,3 cyclic nucleotide 3-phosphodiesterase

(Gokulrangan et al., 2007). In mice, a comprehensive proteomic study identified 7792 proteins from a whole mouse brain that contained 3-NT residues. Further analysis identified 31 hot-spot sites of 3-NT residue formation in 29 different proteins. Many of these proteins have been linked to Parkinson's disease, Alzheimer's disease, or other neurodegenerative disorders (Unlu et al., 2001).

There was also age-related protein nitration in skeletal muscle of Fisher 344 and Fisher 344/Brown Norway F1 rats. Western blot showed an age-related increase in the nitration of a few specific proteins: β-enolase, α-fructose aldolase, and creatine kinase, succinate dehydrogenase, rab GDP dissociation inhibitor β, triosephosphate isomerase, troponin I, α-crystallin, and glyceraldehyde-3-phosphate dehydrogenase (GAPDH) (Kanski et al., 2003). Other studies showed an age-dependent nitration of glycogen phosphorylase b with 3-NT residue accumulation on tyr-161 and tyr-573 (Sharov et al., 2006).

Proteomic techniques were used to identify age-dependent nitration of cardiac proteins from whole heart homogenates and heart mitochondria of Fisher 344/Brown Norway F1 rats. Proteins identified as nitrated were α-enolase, α-aldolase, desmin, aconitate hydratase, methylmalonate semi-aldehyde dehydrogenase, 3-ketoacyl-CoA thiolase, acetyl-CoA acetyltransferase, GAPDH, malate dehydrogenase, creatine kinase, electron-transfer flavoprotein, manganese-superoxide dismutase, F1-ATPase, and the voltage-dependent anion channel. Further MS/MS analysis located nitration at Y105 of the electron-transfer flavoprotein (Kanski et al., 2005).

In mice, the hepatic basal level of 3-NT residues is 58.2 ± 5.3 μmol/mol tyr. This was increased modestly after acetaminophen intoxication (300 mg/kg, ip). 3-NT residue content increased to 72 to 82 μmol/mol tyr (Ishii et al., 2006).

8. Concluding Remarks

This chapter presented methodological details of the estimation of 3-NT residue contents and 3-NT concentrations in tissues and body fluids developed and employed by a research team using LC-MS/MS. The authors restricted presentation and comparison of estimates of these analytes to methods using tandem mass spectrometric detection. These are the most reliable methods. To provide wider access to at least 3-NT residue measurements, it is important that studies corroborating latest generation, high sensitivity immunoassays for 3-NT residues with LC-MS/MS and GC-MS/MS estimates are performed for defined and controlled sample matrix conditions. It is equally important that the shortcomings of some ELISA kits for 3-NT residue estimations are widely appreciated. We are currently modifying our procedure for ultra performance liquid chromatography–MS/MS with improved sample throughput. From current knowledge, it is clear that nitration of tyrosine residues in proteins of physiological systems is a minor modification, particularly compared alongside protein glycation, which is 100- to 1000-fold higher (Thornalley, 2006). Tyrosine nitration increases in certain disease states, however, and may be important when functionally important sites are favored or are hot spots for tyrosine nitration. Ongoing and future research will characterize the "nitrotyrosine proteome"—proteins with hot-spot functional sites of tyrosine nitration. For 3-NT free adduct analysis and variation, it will be important to quantify 3-NT residue content and 3-NT free adduct concentration in foods and beverages, which is a contributory factor in the variation of 3-NT free adduct concentrations in plasma, urine, and urinary excretion rates.

ACKNOWLEDGMENT

We thank the Wellcome Trust (U.K.) for support for our research into protein damage in mechanism of disease.

REFERENCES

Agalou, S., Ahmed, N., Babaei-Jadidi, R., Dawnay, A., and Thornalley, P. J. (2005). Profound mishandling of protein glycation degradation products in uremia and dialysis. *J. Am. Soc. Nephrol.* **16,** 1471–1485.

Ahmed, N., Ahmed, U., Thornalley, P. J., Hager, K., Fleischer, G. A., and Munch, G. (2004a). Protein glycation, oxidation and nitration marker residues and free adducts of cerebrospinal fluid in Alzheimer's disease and link to cognitive impairment. *J. Neurochem.* **92,** 255–263.

Ahmed, N., Babaei-Jadidi, R., Howell, S. K., Beisswenger, P. J., and Thornalley, P. J. (2005). Degradation products of proteins damaged by glycation, oxidation and nitration in clinical type 1 diabetes. *Diabetologia* **48,** 1590–1603.

Ahmed, N., Battah, S., Karachalias, N., Babaei-Jadidi, R., Horanyi, M., Baroti, K., Hollan, S., and Thornalley, P. J. (2003). Increased formation of methylglyoxal and protein glycation, oxidation and nitrosation in triosephosphate isomerase deficiency. *Biochim. Biophys. Acta* **1639,** 121–132.

Ahmed, N., Thornalley, P. J., Luthen, R., Haussinger, D., Sebekova, K., Schinzel, R., Voelker, W., and Heidland, A. (2004b). Processing of protein glycation, oxidation and nitrosation adducts in the liver and the effect of cirrhosis. *J. Hepatol.* **41,** 913–919.

Althaus, J. S., Schmidt, K. R., Fountain, S. T., Tseng, M. T., Carroll, R. T., Galatsis, P., and Hall, E. D. (2000). LC-MS/MS detection of peroxynitrite-derived 3-nitrotyrosine in rat microvessels. *Free Radic. Biol. Med.* **29,** 1085–1095.

Babaei-Jadidi, R., Karachalias, N., Ahmed, N., Battah, S., and Thornalley, P. J. (2003). Prevention of incipient diabetic nephropathy by high dose thiamine and Benfotiamine. *Diabetes* **52,** 2110–2120.

Beisswenger, P. J., Wood, M. E., Howell, S. K., Touchette, A. D., O'Dell, R. M., and Szwergold, B. S. (2001). α-Oxoaldehydes increase in the postprandial period and reflect the degree of hyperglycaemia. *Diabetes Care* **24,** 726–732.

Blanchard-Fillion, B., Prou, D., Polydoro, M., Spielberg, D., Tsika, E., Wang, Z., Hazen, S. L., Koval, M., Przedborski, S., and Ischiropoulos, H. (2006). Metabolism of 3-nitrotyrosine induces apoptotic death in dopaminergic cells. *J. Neurosci.* **26**, 6124–6130.

Bo, S., Gambino, R., Guidi, S., Silli, B., Gentile, L., Cassader, M., and ad Pagano, G. F. (2005). Plasma nitrotyrosine levels, antioxidant vitamins and hyperglycaemia. *Diabet. Med.* **22**, 1185–1189.

Castegna, A., Thongboonkerd, V., Klein, J. B., Lynn, B., Markesbery, W. R., and Butterfield, D. A. (2003). Proteomic identification of nitrated proteins in Alzheimer's disease brain. *J. Neurochem.* **85**, 1394–1401.

Ceriello, A., Mercuri, F., Quagliaro, L., Assaloni, R., Motz, E., Tonutti, L., and Taboga, C. (2001). Detection of nitrotyrosine in the diabetic plasma: Evidence of oxidative stress. *Diabetologia* **44**, 834–838.

Delatour, T. (2004). Performance of quantitative analyses by liquid chromatography–electrospray ionisation tandem mass spectrometry: From external calibration to isotopomer-based exact matching. *Anal. Bioanal. Chem.* **380**, 515–523.

Delatour, T., Fenaille, F., Parisod, V., Richoz, J., Vuichoud, J., Mottier, P., and Buetler, T. (2007). A comparative study of proteolysis methods for the measurement of 3-nitrotyrosine residues: Enzymatic digestion versus hydrochloric acid-mediated hydrolysis. *J. Chromatogr. B* **851**, 268–276.

Delatour, T., Guy, P. A., Stadler, R. H., and Turesky, R. J. (2002a). 3-Nitrotyrosine butyl ester: A novel derivative to assess tyrosine nitration in rat plasma by liquid chromatography-tandem mass spectrometry detection. *Anal. Biochem.* **302**, 10–18.

Delatour, T., Richoz, J., Vouros, P., and Turesky, R. J. (2002b). Simultaneous determination of 3-nitrotyrosine and tyrosine in plasma proteins of rats and assessment of artifactual tyrosine nitration. *J. Chromatogr. B* **779**, 189–199.

Delatour, T., Richoz, J., Vuichoud, J., and Stadler, R. H. (2002c). Artifactual nitration controlled measurement of protein-bound 3-nitrotyrosine in biological fluids and tissues by isotope dilution liquid chromatography electrospray ionization tandem mass spectrometry. *Chem. Res. Toxicol.* **15**, 1209–1217.

Dremina, E. S., Sharov, V. S., and Schoneich, C. (2005). Protein tyrosine nitration in rat brain is associated with raft proteins, flotillin-1 and alpha-tubulin: Effect of biological aging. *J. Neurochem.* **93**, 1262–1271.

Gaizutis, M., Pesce, A. J., and Pollak, V. E. (1975). Renal clearance of human and rat albumins in the rat. *Proc. Soc. Exp. Biol. Med.* **148**, 947–952.

Gaut, J. P., Byun, J., Tran, H. D., and Heinecke, J. W. (2002). Artifact-free quantitation of free 3-chlorotyrosine, 3-bromotyrosine, and 3-nitrotyrosine in human plasma by electron capture-negative chemical ionization gas chromatography mass spectrometry and liquid chromatography-electrospray ionization tandem mass spectrometry. *Anal. Biochem.* **300**, 252–259.

Gladwin, M. T., Lancaster, J. R., Freeman, B. A., and Schechter, A. N. (2003). Nitric oxide's reactions with hemoglobin: A view through the SNO-storm. *Nat. Med.* **9**, 496–500.

Gokulrangan, G., Zaidi, A., Michaelis, M. L., and Schoneich, C. (2007). Proteomic analysis of protein nitration in rat cerebellum: Effect of biological aging. *J. Neurochem.* **100**, 1494–1504.

Greenacre, S. A. B., Evans, P., Halliwell, B., and Brain, S. D. (1999). Formation and loss of nitrated proteins in peroxynitrite-treated rat skin *in vivo*. *Biochem. Biophys. Res. Commun.* **262**, 781–786.

Grune, T., Klotz, L. O., Gieche, J., Rudeck, M., and Sies, H. (2001). Protein oxidation and proteolysis by the nonradical oxidants singlet oxygen or peroxynitrite. *Free Radic. Biol. Med.* **30**, 1243–1253.

Hensley, K., Maidt, M. L., Yu, Z., Sang, H., Markesbery, W. R., and Floyd, R. A. (1998). Electrochemical analysis of protein nitrotyrosine and dityrosine in the Alzheimer brain indicates region-specific accumulation. *J. Neurosci.* **18,** 8126–8132.

Hitomi, Y. H., Okuda, J., Nishino, H., Kambayashi, Y., Hibino, Y., Takemoto, K., Takigawa, T., Ohno, H., Taniguchi, N., and Ogino, K. (2007). Disposition of protein-bound 3-nitrotyrosine in rat plasma analysed by a novel protocol for HPLC-ECD. *J. Biochem. (Tokyo)* **141,** 495–502.

Hoeldtke, R. D., Bryner, K. D., McNeill, D. R., Hobbs, G. R., Riggs, J. E., Warehime, S. S., Christie, I., Ganser, G., and Van Dyke, K. (2002). Nitrosative stress, uric acid, and peripheral nerve function in early type 1 diabetes. *Diabetes* **51,** 2817–2825.

Ishii, Y., Iijima, M., Umemura, T., Nishikawa, A., Iwasaki, Y., Ito, R., Saito, K., Hirose, M., and Nakazawa, H. (2006). Determination of nitrotyrosine and tyrosine by high-performance liquid chromatography with tandem mass spectrometry and immunohistochemical analysis in livers of mice administered acetaminophen. *J. Pharm. Biomed. Anal.* **41,** 1325–1331.

Jiao, K., Mandapati, S., Skipper, P. L., Tannenbaum, S. R., and Wishnok, J. S. (2001). Site-selective nitration of tyrosine in human serum albumin by peroxynitrite. *Anal. Biochem.* **293,** 43–52.

Kanski, J., Alterman, M. A., and Schoneich, C. (2003). Proteomic identification of age-dependent protein nitration in rat skeletal muscle. *Free Radic. Biol. Med.* **35,** 1229–1239.

Kanski, J., Behring, A., Pelling, J., and Schoneich, C. (2005). Proteomic identification of 3-nitrotyrosine-containing rat cardiac proteins: Effects of biological aging. *Am. J. Physiol. Heart Circ. Physiol.* **288,** H371–H381.

Kaur, H., Lyras, L., Jenner, P., and Halliwell, B. (1998). Artefacts in HPLC detection of 3-nitrotyrosine in human brain tissue. *J. Neurochem.* **70,** 2220–2223.

Keimer, R., Stutzer, F. K., Tsikas, D., Troost, R., Gutzki, F. M., and Frolich, J. C. (2003). Lack of oxidative stress during sustained therapy with isosorbide dinitrate and pentaerythrityl tetranitrate in healthy humans: A randomized, double-blind crossover. *J. Cardiovasc. Pharmacol.* **41,** 284–292.

Khan, J., Brennan, D. M., Bradley, N., Gao, B. R., Bruckdorfer, R., and Jacobs, M. (1998). 3-nitrotyrosine in the proteins of human plasma determined by an ELISA method. *Biochem. J.* **330,** 795–801.

Kharbanda, K. K., Todero, S. L., Shubert, K. A., Sorrell, M. F., and Tuma, D. J. (2001). Malondialdehyde–acetaldehyde–protein adducts increase secretion of chemokines by rat hepatic stellate cells. *Alcohol* **25,** 123–128.

Kuboki, K., Jiang, Z. Y., Takahara, N., Ha, S. W., Igarashi, M., Yamauchi, T., Feener, E. P., Herbert, T. P., Rhodes, C. J., and King, G. L. (2000). Regulation of endothelial constitutive nitric oxide synthase gene expression in endothelial cells and *in vivo*: A specific vascular action of insulin. *Circulation* **101,** 676–681.

Lorch, S. A., Banks, B. A., Chritsie, J., Merrill, J. D., Althaus, J., Schmidt, K., Ballard, P. L., Ischiropoulos, H., and Ballard, R. A. (2003). Plasma 3-nitrotyrosine and outcome in neonates with severe bronchopulmonary dysplasia after inhbaled nitric oxide. *Free Radic. Biol. Med.* **34,** 1146–1152.

Mitrogianni, Z., Barbouti, A., Galaris, D., and Siamopoulos, K. C. (2004). Tyrosine nitration in plasma proteins from patients undergoing hemodialysis. *Am. J. Kidney Dis.* **44,** 286–292.

Nikov, G., Bhat, V., Wishnok, J. S., and Tannenbaum, S. R. (2003). Analysis of nitrated proteins by nitrotyrosine-specific affinity probes and mass spectrometry. *Anal. Biochem.* **320,** 214–222.

Nishikawa, T., Edelstein, D., Liang Du, X., Yamagishi, S., Matsumura, T., Kaneda, Y., Yorek, M. A., Beede, D., Oates, P. J., Hammes, H.-P., Giardino, I., and Brownlee, M.

(2000). Normalizing mitochondrial superoxide production blocks three pathways of hyperglycaemia damage. *Nature* **404,** 787–790.

Ohshima, H., Friesen, M., Brouet, I., and Bartsch, H. (1990). Nitrotyrosine as a new marker for endogenous nitrosation and nitration of proteins. *Food Chem. Toxicol.* **28,** 647–652.

Pannala, A. S., Mani, A. R., Rice-Evans, C. A., and Moore, K. P. (2006). pH-dependent nitration of para-hydroxyphenylacetic acid in the stomach. *Free Radic. Biol. Med.* **41,** 896–901.

Pennathur, S., Jackson-Lewis, V., Przedborski, S., and Heinecke, J. W. (1999). Mass spectrometric quantification of 3-nitrotyrosine, ortho- tyrosine, and o,o'-dityrosine in brain tissue of 1-methyl-4-phenyl-1,2,3,6-tetrahydropyridien-treated mice, a model of oxidative stress in Parkinson's disease. *J. Biol. Chem.* **274,** 34621–34628.

Predescu, D., Predescu, S., and Malik, A. B. (2002). Transport of nitrated albumin across continuous vascular endothelium. *Proc. Natl. Acad. Sci. USA* **99,** 13932–13937.

Rabbani, N., Sebekova, K., Sebekova, K., Jr., Heidland, A., and Thornalley, P. J. (2007). Accumulation of free adduct glycation, oxidation, and nitration products follows acute loss of renal function. *Kidney Int.* **72,** 1113–1121.

Reiter, C. D., Wang, X. D., Tanus-Santos, J. E., Hogg, N., Cannon, R. O., Schechter, A. N., and Gladwin, M. T. (2002). Cell-free hemoglobin limits nitric oxide bioavailability in sickle-cell disease. *Nat. Med.* **8,** 1383–1389.

Romero, N., Denicola, A., Souza, J. M., and Radi, R. (1999). Diffusion of peroxynitrite in the presence of carbon dioxide. *Arch. Biochem. Biophys.* **368,** 23–30.

Rossner, J., Svecova, V., Milcova, A., Lnenickova, Z., Solansky, I., Santella, R. M., and Sram, R. (2007). Oxidative and nitrosative stress markers in bus drivers. *Mutat. Res./Fundam. Mol. Mech. Mutagen.* **617,** 23–32.

Ryberg, H. and Caidahl, K. (2007). Chromatographic and mass spectrometric methods for quantitative determination of 3-nitrotyrosine in biological samples and their application to human samples. *J. Chromatogr. B* **851,** 160–171.

Ryberg, H., Soderling, A. S., Davidsson, P., Blennow, K., Caidahl, K., and Persson, L. I. (2004). Cerebrospinal fluid levels of free 3-nitrotyrosine are not elevated in the majority of patients with amyotrophic lateral sclerosis or Alzheimer's disease. *Neurochem. Int.* **45,** 57–62.

Schwedhelm, E., Tsikas, D., Gutzki, F. M., and Frolich, J. C. (1999). Gas chromatographic-tandem mass spectrometric quantification of free 3-nitrotyrosine in human plasma at the basal state. *Anal. Biochem.* **276,** 195–203.

Sharov, V. S., Galeva, N. A., Kanski, J., Williams, T. D., and Schoneich, C. (2006). Age-associated tyrosine nitration of rat skeletal muscle glycogen phosphorylase b: Characterization by HPLC-nanoelectrospray-tandem mass spectrometry. *Exp. Gerontol.* **41,** 407–416.

Soderling, A. S., Ryberg, H., Gabrielsson, A., Larstad, M., Toren, K., Niari, S., and Caidahl, K. (2003). A derivatization assay using gaschromatography/negative chemical ionization tandem mass spectrometry to quantify 3-nitrotyrosine in human plasma. *J. Mass Spectrom.* **38,** 1187–1196.

Sokolovsky, M., Riordan, J. F., and Vallee, B. L. (1966). Tetranitromethane: A reagent for the nitration of tyrosyl residues in proteins. *Biochemistry* **5,** 3582–3589.

Soriano, F. G., Virag, L., Jagtap, P., Szabo, E., Mabley, J. G., Liaudet, L., Marton, M., Hoyt, D. G., Murthy, K. G. K., Salzman, A. L., Southan, G. J., and Szabo, C. (2001). Diabetic endothelial dysfunction: The role of poly(ADP-ribose) polymerase activation. *Nat. Med.* **7,** 108–113.

Sun, Y. C., Chang, P. Y., Tsao, K. C., Wu, T. L., Sun, C. F., Wu, L. L., and Wu, J. T. (2007). Establishment of a sandwich ELISA using commercial antibody for plasma or serum 3-nitrotyrosine (3NT): Elevation in inflammatory diseases and complementary between 3NT and myeloperoxidase. *Clin. Chim. Acta* **378,** 175–180.

Svatikova, A., Wolk, R., Wang, H. H., Otto, M. E., Bybee, K. A., Singh, R. J., and Somers, V. K. (2004). Circulating free nitrotyrosine in obstructive sleep apnea. *Am. J. Physiol. Regul. Integr. Comp. Physiol.* **287,** R284–R287.

Tabrizi-Fard, M. A., Maurer, T. S., and Fung, H. L. (1999). In vivo disposition of 3-nitro-L-tyrosine in rats: Implications on tracking systemic peroxynitrite exposure. *Drug Metab. Dispos.* **27,** 429–431.

Thornalley, P. J. (2006). Quantitative screening of protein glycation, oxidation, and nitration adducts by LC-MS/MS: Protein damage in diabetes, uremia, cirrhosis, and Alzheimer's disease. In "Redox Proteomics" (I. Dalle-Donne, A. Scaloni, and D. A. Butterfield, eds.), pp. 681–728. Wiley, Hoboken.

Thornalley, P. J., Battah, S., Ahmed, N., Karachalias, N., Agalou, S., Babaei-Jadidi, R., and Dawnay, A. (2003). Quantitative screening of advanced glycation endproducts in cellular and extracellular proteins by tandem mass spectrometry. *Biochem. J.* **375,** 581–592.

Tohgi, H., Abe, T., Yamazaki, K., Murata, T., Ishizaki, E., and Isobe, C. (1999). Alterations of 3-nitrotyrosine concentration in the cerebrospinal fluid during aging and in patients with Alzheimer's disease. *Neurosci. Lett.* **269,** 52–54.

Tsikas, D., Mitschke, A., Suchy, M. T., Gutzki, F. M., and Stichtenoth, D. O. (2005). Determination of 3-nitrotyrosine in human urine at the basal state by gas chromatography-tandem mass spectrometry and evaluation of the excretion after oral intake. *J. Chromatogr. B* **827,** 146–156.

Tsikas, D., Schwedhelm, E., and Frolich, J. C. (2002). Methodological considerations on the detection of 3-nitrotyrosine in the cardiovascular system. *Circ. Res.* **90,** e70.

Tsikas, D., Schwedhelm, E., Stutzer, F. K., Gutzki, F. M., Rode, I., Mehls, C., and Frolich, J. C. (2003). Accurate quantification of basal plasma levels of 3-nitrotyrosine and 3-nitrotyrosinoalbumin by gas chromatography-tandem mass spectrometry. *J. Chromatogr. B Anal. Technol. Biomed. Life Sci.* **784,** 77–90.

Unlu, A., Turkozkan, N., Cimen, B., Karabicak, U., and Yaman, H. (2001). The effect of *Escherichia coli*-derived lipopolysaccharides on plasma levels of malondialdehyde and 3-nitrotyrosine. *Clin. Chem. Lab. Med.* **39,** 491–493.

Willard, B. B., Ruse, C. I., Keightley, J. A., Bond, M., and Kinter, M. (2003). Site-specific quantitation of protein nitration using liquid chromatography/tandem mass spectrometry. *Anal. Chem.* **75,** 2370–2376.

Willis, M. S., Klassen, L. W., Tuma, D. J., and Thiele, G. M. (2002). Malondialdehyde-acetaldehyde-haptenated protein induces cell death by induction of necrosis and apoptosis in immune cells. *Int. Immunopharmacol.* **2,** 519–535.

Yi, D., Ingelse, B. A., Duncan, M. W., and Smythe, G. A. (2000). Quantification of 3-nitrotyrosine in biological tissues and fluids: Generating valid results by eliminating artifactual formation. *J. Am. Soc. Mass Spectrom.* **11,** 578–586.

CHAPTER TWENTY-THREE

Nitrite and Nitrate Measurement by Griess Reagent in Human Plasma: Evaluation of Interferences and Standardization

Daniela Giustarini,* Ranieri Rossi,* Aldo Milzani,[†] *and* Isabella Dalle-Donne[†]

Contents

1. Introduction	362
2. Experimental Procedures	364
2.1. Materials	364
2.2. Blood collection and plasma separation	364
2.3. Evaluation of the effect of pH and NADPH on the measurement of nitrite	364
2.4. Evaluation of the effect of plasma dilution	365
2.5. The effect of ethanol addition	365
2.6. Nitrate reductase activity	365
2.7. Recovery of added nitrite	365
2.8. Effect of incubation time with nitrate reductase	366
2.9. Measurement of NOx in plasma samples	366
3. Results	366
4. Discussion	373
Acknowledgment	378
References	378

Abstract

Nitrite and nitrate represent the final products of nitric oxide (NO) oxidation pathways, and their hematic concentrations are frequently assessed as an index of systemic NO production. However, their intake with food can influence their levels. Nitrite and nitrate could have a role by producing NO, because nitrite can release NO after reaction with deoxyhemoglobin and dietary nitrate can be reduced substantially to nitrite by commensal bacteria in the oral cavity.

* Department of Evolutionary Biology, University of Siena, Siena, Italy
[†] Department of Biology, University of Milan, Milan, Italy

Methods in Enzymology, Volume 440 © 2008 Elsevier Inc.
ISSN 0076-6879, DOI: 10.1016/S0076-6879(07)00823-3 All rights reserved.

Different methods have been applied for nitrite/nitrate detection, with the most commonly used being the spectrophotometric assay based on the Griess reagent. However, a reference methodology for these determinations is still missing and many possible interferences have been reported. This chapter assesses how different experimental conditions can influence the results when detecting nitrite and nitrate in human plasma by the Griess assay and provides a simple method characterized by high reproducibility and minimized interferences by plasma constituents.

1. INTRODUCTION

Nitric oxide (NO) mediates many physiological functions and, consequently, it is commonly investigated under many different pathological conditions. Nevertheless, NO, per se, is difficult to quantify, as a consequence of its short half-life (milliseconds or less, depending on the environment) in the presence of O_2 and other scavenging molecules, for example, hemoglobin (Gally et al., 1990). Therefore, assays that indicate the presence of NO indirectly are commonly carried out. In particular, accumulation of the stable degradation products of NO, nitrite (NO_2^-) and nitrate (NO_3^-), S-nitrosothiols, or the increase in cGMP levels (due to NO-dependent activation of guanylyl cyclase) are measured preferably as an index of NO production (Forstermann and Ishii, 1996; Giustarini et al., 2003; Tsikas, 2005). Nitrite and nitrate represent the final products of the NO oxidation pathways and, consequently, their concentration in human body fluids depends on NO production itself. Other variables can influence the levels of nitrite/nitrate, such as alimentary intake (van Vliet et al., 1997).

Plasma nitrite/nitrate content is assessed frequently as an index of NO production in the whole organism, as plasma is in a dynamic equilibrium with all organs and interstitial fluids. NO_2^- and NO_3^- have always been considered only as NO end products, without any physiological meaning. It has been proposed that NO_2^- may have a role as a signaling molecule by producing NO, either in the stomach at acidic pH or by reaction with deoxyhemoglobin, thus inducing hypoxic vasodilation (Gladwin, 2005). It also seems to act as a modulator of ischemia/reperfusion tissue injury and infarction (Lundberg and Weitzberg, 2005). Finally, because plasma NO_2^- sensitively reflects changes in endothelial NOS activity following shear stress in healthy subjects, but not when endothelial dysfunction occurs, its detection in human plasma may represent a valid diagnostic tool (Rassaf et al., 2006). Analogously, it has been suggested that nitrate of dietary origin can have a physiological role, as it is reduced substantially to nitrite by commensal bacteria in the oral cavity following an enterosalivary cycle (Lundberg et al., 2006). In fact, plasma levels of nitrite increase significantly after an oral load of sodium nitrate, corresponding to 300 g of spinach (or other nitrate-rich green leafy vegetables). This could ensure enough systemic NO levels for being protective for the cardiovascular system (Lundberg and Govoni, 2004). Furthermore, dietary NO_3^- can reduce diastolic blood pressure in healthy subjects (Larsen et al., 2006). As a consequence, interest regarding NO_2^- and NO_3^- is increasing and the needs of simple, time-sparing, and inexpensive methods to detect them would be welcome.

A reliable blood measurement of nitrite is difficult because it is unstable, being oxidized rapidly to nitrate by hemoglobin, with a $t_{1/2}$ of about 180 s (Giustarini et al., 2004). Once plasma is separated from blood, both nitrite and nitrate are stable at $-20°C$ for at least 1 year (without hemolysis) (Moshage et al., 1995). A combined measure of nitrate + nitrite (NOx) can be performed, as the procedure is easier than that for nitrite analysis and some critical steps in sample manipulation needed for nitrite analysis can be avoided. Because NO_3^- is largely more abundant than NO_2^- in body fluids, NOx is almost synonymous with nitrate in most cases (Tsikas, 2007). Different methods have been applied for nitrite/nitrate detection, namely colorimetric and fluorometric assays, chemiluminescence, gas chromatography/mass spectrometry, and capillary electrophoresis (Bryan and Grisham, 2007; Ellis, 1998). The most commonly used method is the spectrophotometric assay, which is based on formation of an azo dye by reaction of NO_2^- with the Griess reagent. Specifically, NO_2^- reacts directly with sulfanilamide under acidic conditions and is then revealed after diazotization with N-(1-naphthyl)ethylenediamine (NED) (Fig. 23.1). Nitrate can also be measured by this method after its reduction to nitrite. This can be performed enzymatically (by nitrate reductase) or by metallic reduction (Cd, Vn) (Ellis, 1998). Notwithstanding their widespread

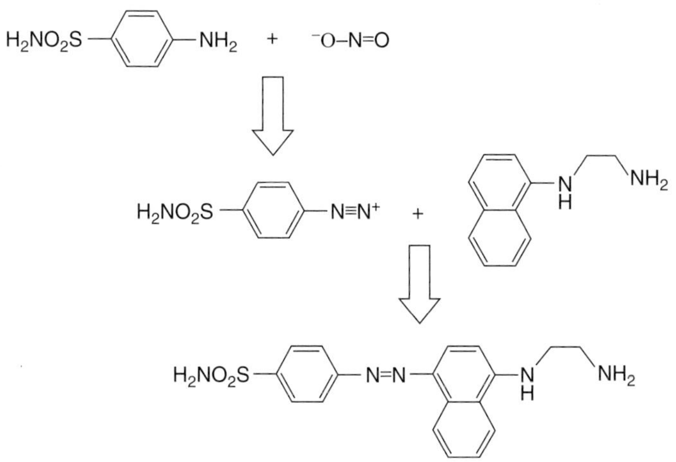

Figure 23.1 Schematic diagram representing the Griess reaction principle. Under acidic conditions, nitrite reacts with sulfanilamide to produce a diazonium ion, which is then coupled to NED to produce a chromophoric azo product, which strongly absorbs at 545 nm.

application, a reference methodology for these determinations is still lacking and many different variants of the Griess reaction are available. Significant differences can be observed for the type of reagents used to prepare the Griess solution and their relative ratio, sample manipulation during the preanalytical phase, protein removal, and reduction of NO_3^- to NO_2^- (Tsikas, 2007).

This chapter assesses how different experimental conditions can influence the results when detecting NOx in human plasma by Griess reagents and provides an improved method, characterized by high reproducibility and minimized (and quantified) interferences by plasma constituents.

2. Experimental Procedures

2.1. Materials

Nitrate reductase from *Aspergillus*, FAD, NADPH, sulfanilamide, *N*-ethylmaleimide (NEM), *N*-(1-naphthyl)ethylenediamine, and all other reagents of highest purity available are from Sigma-Aldrich Chemie GmbH. PD10 gel-filtration columns are from Amersham, and ultrafiltration cartridges (MW 10,000 cutoff) are from Supelco. Before use, cartridges are washed extensively with abundant MilliQ water.

2.2. Blood collection and plasma separation

Human blood is obtained from healthy volunteers after informed consensus by venipuncture, using K_3EDTA as an anticoagulant. Blood is collected and centrifuged rapidly at 15,000g for 10 s to separate plasma. This rapid plasma separation has been shown not to provoke any visible red blood cell hemolysis.

2.3. Evaluation of the effect of pH and NADPH on the measurement of nitrite

Griess solutions at different pHs are prepared by adding 1% (w/v) sulfanilamide and 0.1% (w/v) NED to different acid solutions: (a) pH 3.3: 2.5% (v/v) acetic acid brought to the indicated pH with 5 M NaOH; (b) pH 2.5: 2.5% (v/v) acetic acid; (c) pH 1.46: 2% (v/v) phosphoric acid; (d) pH 1.16: 6% (v/v) phosphoric acid; (e) pH 0.75: 10% (w/v) trichloroacetic acid (TCA); and (f) pH 0.6: 0.25 M HCl.

Sodium nitrite solutions in MilliQ water are freshly prepared and reacted with the different Griess solutions (1:1 ratio) for 30 min in the dark. Samples are then analyzed in the 650- to 480-nm range by a Jasco V550 spectrophotometer. Spectra are recorded against a blank prepared in the same conditions but with a Griess reagent where NED is omitted. The same experiment is also performed by reacting the different Griess solutions under the same conditions (a to f) with sodium nitrite solutions containing 80 μM NADPH [stock solution 10 mM in 0.5% (w/v) sodium bicarbonate].

The dose dependence of NADPH interference on nitrite detection is assessed by preparing 10 μM sodium nitrite solutions (in MilliQ water) in the presence of 0 to 100 μM NADPH. These solutions are incubated with the Griess reagent (1% sulfanilamide, 0.1% NED, 6% phosphoric acid) and analyzed as indicated earlier.

2.4. Evaluation of the effect of plasma dilution

Nitrite-free plasma is obtained after the storage of human blood at 37°C for 1 h by a 10-s centrifugation at 15,000g. Plasma is diluted up to 80% with MilliQ, and 500 μl of each dilution is added with 10 μl of 500 μM sodium nitrite and 5 μl of 300 mM NEM; after 2 min, 500 μl Griess solution (1% sulfanilamide, 0.1% NED, 6% phosphoric acid) is added. After a 30-min incubation in the dark, samples are deproteinized by the addition of 3% (w/v, final concentration) TCA and centrifuged for 2 min at 10,000g, and supernatants are analyzed by a spectrophotometer as described earlier.

2.5. The effect of ethanol addition

Aliquots (100 μl) of nitrite-free plasma are added with 1 μl of 5 mM sodium nitrite and 5 μl of 300 mM NEM. After 2 min, samples are diluted with 400 μl of water/ethanol solutions (0–62.5% ethanol), 1:1 reacted with the Griess solution (1% sulfanilamide, 0.1% NED, 6% phosphoric acid), TCA treated, and measured as described earlier.

2.6. Nitrate reductase activity

Nitrate reductase is dissolved in 0.05 M phosphate buffer, pH 7.5, and assayed by mixing in a cuvette 1 ml of 0.05 M phosphate buffer, pH 7.5, containing 1 mM sodium nitrate, 0.2 mM NADPH, 0.005 mM FAD, and 3 μl of the enzyme-containing solution. NADPH oxidation is evaluated at a 340-nm wavelength. Aliquots of the enzyme-containing solution are then stored at 4, -20, or -80°C, and enzyme activity is measured with time.

2.7. Recovery of added nitrite

Aliquots (100 μl) of nitrite-free plasma are added with 1 μl of 1 to 4 mM sodium nitrite solutions, 1 μl of 12.5 mM NADPH, and 5 μl of 300 mM NEM. After 2 min, samples are diluted with 400 μl of a 62.5% ethanol solution. All samples are then 1:1 reacted with the Griess solution

(1% sulfanilamide, 0.1% NED, 6% phosphoric acid), TCA treated, and measured as described previously.

2.8. Effect of incubation time with nitrate reductase

Plasma aliquots (0.8 ml) are treated with 8 µl of 1 to 4 mM sodium nitrate solutions. Samples are then added with 8 µl of 12.5 mM NADPH, 4 µl of 1 mM FAD, and 32 µl of nitrate reductase (5 U/ml dissolved in 50 mM phosphate buffer, pH 7.5). At specified times, a 0.1-ml sample is added with 5 µl of 300 mM NEM. After 2 min, samples are diluted with 400 µl of a 62.5% ethanol solution, 1:1 reacted with the Griess solution (1% sulfanilamide, 0.1% NED, 6% phosphoric acid), TCA treated, and measured as described earlier.

2.9. Measurement of NOx in plasma samples

One hundred microliters of plasma is added with 1 µl of 12.5 mM NADPH, 1 µl of 0.5 mM FAD, and 4 µl of nitrate reductase (5 U/ml dissolved in 50 mM phosphate buffer, pH 7.5); after a 90-min incubation, samples are added with 5 µl of 300 mM NEM. After 2 min, samples are diluted with 400 µl of a 62.5% ethanol solution, 1:1 reacted with the Griess reagent (1% sulfanilamide, 0.1% NED, 6% phosphoric acid), TCA treated, and measured as described previously. For experiments where nitrate is added, 100 µl of plasma is added with 1 µl of 0.5 to 8 mM sodium nitrate. Nitrite/nitrate-free plasma is obtained by gel filtration on PD10 columns equilibrated with 50 mM phosphate buffer, pH 7.5. After gel filtration, the initial protein concentration is restored by ultrafiltration on Supelco Centrisart (10 kDa cutoff) cartridges. Protein concentration is measured by the Bradford assay.

3. Results

Colorimetric methods based on the Griess reaction fundamentally detect NO_2^- that, under acidic conditions, reacts with sulfanilamide and NED (Fig. 23.1) to produce an azo compound, which strongly absorbs in the visible region with a peak around 545 nm. To obtain the acidic pH necessary for the reaction to proceed, different acids (HCl, acetic acid, or phosphoric acid) at different final concentrations (and, consequently, at different final pH values) have been used indifferently (Granger et al., 1996; Tsikas, 2007). However, little is known about the influence of various acids on the yield of the reaction.

This chapter evaluated the reaction of NO_2^- with sulfanilamide and NED under different pH values. Incubation of sodium nitrite solutions (2–30 μM) with the Griess reagent (1% sulfanilamide, 0.1% NED) prepared with HCl, phosphoric acid, acetic acid, or TCA in the 0.6 to 3.3 pH range gave different results and indicated that lower pH values resulted in a lower final absorbance (Fig. 23.2). These differences were not due to an insufficient incubation time to complete the reactions, as measurements repeated after a further 30 min did not show any significant increase in absorbance (data not shown). Data from Fig. 23.2 thus suggest that higher pH values (2.5–3), resulting in a higher response, are preferred. However, nitrite is not usually measured by the Griess reaction in water or in buffers, but in biological samples, which may contain several molecules possibly interfering with the assay. Among these molecules, NADPH, used to reduce nitrate to nitrite, is one of the most important and studied (Verdon *et al.*, 1995). Therefore, we repeated the experiment shown in Fig. 23.2 in the presence of 80 μM NADPH. Figure 23.3 shows that NADPH interference with NO_2^- detection is higher at higher pH values. Data from Figs. 23.2 and 23.3 suggest that the Griess reagent prepared using 6% phosphoric acid could be the most suitable solution to measure nitrite in complex biological samples and in the presence of significant amounts of NADPH. Because NADPH significantly decreases the slope of the curve even when a 6%

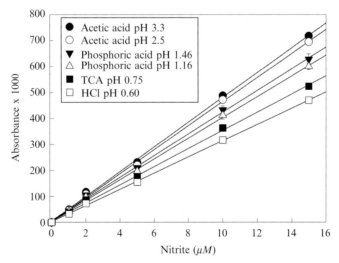

Figure 23.2 Nitrite detection by the Griess reagent prepared at different pH values. Sodium nitrite solutions in MilliQ water were freshly prepared, reacted with the Griess solutions (1:1 ratio) for 30 min, and then analyzed by a spectrophotometer in the 650- to 480-nm range; the height of the peak at 545 nm is reported. The value of nitrite in the abscissa is referred to its final concentration (after 1:1 dilution with the reagent). The number of replicates is three.

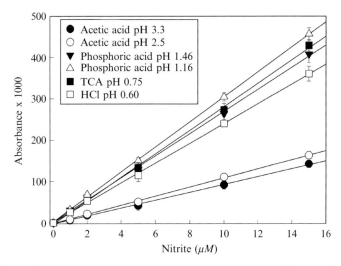

Figure 23.3 Nitrite detection by the Griess reagent prepared at different pH values in the presence of NADPH. Sodium nitrite solutions containing 80 μM NADPH (final concentration) were freshly prepared, reacted with the Griess solutions (1:1 ratio) for 30 min, and then analyzed by a spectrophotometer in the 650- to 480-nm range; the height of the peak at 545 nm is reported. The value of nitrite in the abscissa is referred to its final concentration (after 1:1 dilution with the reagent). The number of replicates is four.

solution of phosphoric acid is used (Figs. 23.2 and 23.3), we assessed the dose dependence of NADPH interference on nitrite detection. Solutions containing the same amount of sodium nitrite and 0 to 100 μM NADPH were analyzed for the NO_2^- content with the Griess reagent prepared with 6% phosphoric acid (Fig. 23.4). The interference of NADPH in these conditions was minimal at concentrations below 25 μM. Furthermore, the possible effect of plasma on nitrite colorimetric measurement was analyzed by adding a NO_2^- standard concentration to human plasma diluted to different final percentages with MilliQ. Samples were then assayed with the Griess reagent. Figure 23.5 indicates that plasma constituents decreased the 545-nm peak; in particular, in the presence of 100% plasma, the height of the peak was less than half of that obtained in the absence of plasma (i.e., 100% MilliQ).

In the experiments carried out with plasma (Fig. 23.5), proteins were removed before spectra analysis by TCA precipitation: TCA was added to samples after a 30-min reaction with the Griess reagent. The use of TCA is necessary, as sample acidification with the Griess reagent alone does not allow its deproteinization; consequently, the sample turbidity is excessively high. After protein precipitation, we noticed that the pellet had an intense purple color. This was likely a consequence of the coprecipitation of the diazo compound with the proteins. This phenomenon could be minimized by adding an alcoholic solution to the final reaction mixture. Plasma samples

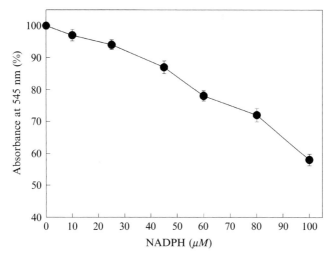

Figure 23.4 Dose dependence of NADPH interference on nitrite detection. Solutions of 10 μM sodium nitrite and NADPH at different final concentrations in MilliQ water were reacted with the Griess reagent (1% sulfanilamide, 0.1% NED, 6% phosphoric acid) for 30 min and then analyzed by a spectrophotometer in the 650- to 480-nm range. In the graph, the percentage of the absorbance of peaks measured at 545 nm in the presence of NADPH with respect to those recorded with nitrite alone is reported. The number of replicates is four.

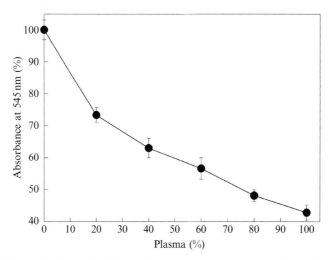

Figure 23.5 Evaluation of the effect of plasma on the measurement of nitrite by the Griess reaction. Nitrite-free plasma was diluted from 0 to 80% with MilliQ. Five hundred microliters of each sample was then added with 10 μl of 500 μM sodium nitrite and 5 μl of 300 mM NEM; after 2 min, 500 μl of the Griess reagent (1% sulfanilamide, 0.1% NED, 6% phosphoric acid) was added. After 30 min, samples were deproteinized by the addition of TCA and analyzed by a spectrophotometer in the 650- to 480-nm range. In the graph, the percentage of the absorbance of peaks measured at 545 nm in the presence of plasma with respect to those recorded with nitrite alone is reported. The number of replicates is four.

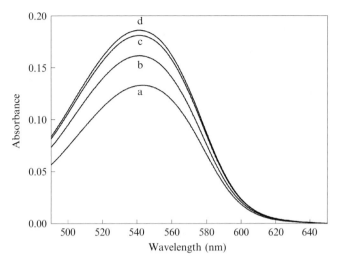

Figure 23.6 Effect of ethanol addition on the measurement of nitrite by the Griess reagent in the presence of plasma. Aliquots (100 μl) of nitrite-free plasma were added with 1 μl of 5 mM sodium nitrite and 5 μl of 300 mM NEM. After 2 min, samples were diluted with 400 μl of water/ethanol solutions: (a) water, (b) 25% ethanol, (c) 50% ethanol, and (d) 62.5% ethanol; 500 μl of the Griess reagent (1% sulfanilamide, 0.1% NED, 6% phosphoric acid) was then added. After 30 min, samples were deproteinized by TCA addition and analyzed by a spectrophotometer in the 650- to 480-nm range; in the graph, a typical experiment of four replicates is shown.

containing a fixed amount of nitrite were added with different ratios of ethyl alcohol/water and then allowed to react with the Griess reagent (Fig. 23.6). The height of the recorded peaks indicates that the addition of ethyl alcohol at a 62.5% final concentration to human plasma helps the efficiency of the reaction, probably diminishing the unspecific binding of the azo dye to albumin.

The experiments shown in Figs. 23.2 to 23.6 as a whole suggest that use of the Griess reagent containing 6% phosphoric acid, NADPH below 30 μM, four- to fivefold diluted plasma, and 62.5% ethyl alcohol could be the best condition to minimize both NADPH and plasma constituent interferences. We tested the efficacy of these conditions by applying them to detect known amounts of nitrite added both to plasma and to buffer only (Fig. 23.7). At the three concentrations of $NaNO_2$ tested (namely, 10, 20, and 40 μM), spectra recorded in the presence of plasma were only about 12% lower than those obtained in buffer. Similar values were measured when plasma from different healthy donors was used, as evidenced by the low standard deviation reported in Fig. 23.7, indicating a reproducible event. In addition, the intersample coefficient of variation and the intraday variation on measurements carried out on the same sample were as low as 2%, suggesting a high reproducibility.

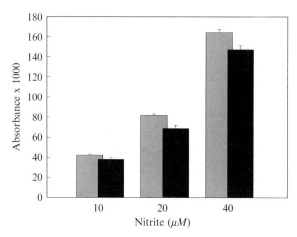

Figure 23.7 Analysis of the recovery of the added nitrite. Aliquots (100 μl) of nitrite-free plasma were added with 1 μl of 1 to 4 mM sodium nitrite solutions, 1 μl of 10 mM NADPH, and 5 μl of 300 mM NEM. After 2 min, samples were diluted with 400 μl of a 62.5% ethanol solution and added with 500 μl of the Griess reagent (1% sulfanilamide, 0.1% NED, 6% phosphoric acid). After 30 min, samples were deproteinized by TCA addition and analyzed by a spectrophotometer in the 650- to 480-nm range. The height of peaks recorded at 545 nm is reported (black bars), together with those obtained in similar experiments in which plasma was omitted (gray bars). The number of replicates is four.

With the experiments shown in Figs. 23.2 through 23.7, we have defined the best conditions to minimize interferences and variability due to plasma constituents during NO_2^- detection. However, nitrate, and not nitrite, is the main component of NOx in plasma, as well as in other biological fluids (Tsikas, 2007). Therefore, it is also crucial to assess the method used to reduce nitrate to nitrite. Either chemical or enzymatic reduction can be performed, with the latter being applied most frequently (Bryan and Grisham, 2007). We have verified the capability of the enzyme nitrate reductase from *Aspergillus niger* to reduce nitrate to nitrite under our conditions. First of all, we have measured the enzyme activity after resolubilization of the lyophilized enzyme and followed it with time after storage under various conditions (Fig. 23.8). It is noteworthy that, once resolubilized, nitrate reductase decreases its activity rapidly, in particular after cycles of freezing and thawing; thus, its storage at $-70°C$, after reconstitution in aliquots, to limit freezing and thawing cycles and checking its activity before use are recommended actions. Known concentrations of nitrate were added to human plasma, and 125 μM NADPH, 5 μM FAD, and 0.2 U/ml of nitrate reductase were used for its reduction. Figure 23.9 indicates that 40 to 50 min was necessary to reduce all nitrate present in the samples. From these experiments, we concluded that 90 min was the time that surely allowed the enzyme to reduce all nitrite under our conditions. This was confirmed by

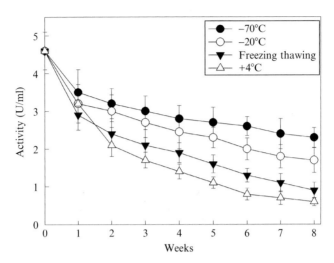

Figure 23.8 Variation of nitrate reductase activity with time and under different storage conditions. Lyophilized enzyme was dissolved in 50 mM phosphate buffer, pH 7.5, and stored at 4, −20, or −70°C divided into aliquots. Enzyme activity was measured weekly; some samples (freezing/ thawing) were maintained at −20°C but frozen/ thawed once every 3 days.

the fact that when nitrite instead of nitrate was added to the samples (Fig. 23.9, stars), the same absorbance values were obtained.

Finally, we tested the capability of our standardized procedure (as detailed earlier) to measure nitrate in human plasma samples spiked with know amounts of NO_3^-. Figure 23.10 shows the values of absorbance obtained following the addition of known amounts of nitrate to four different plasma samples. The perfect linearity obtained for each sample and the similar slope of the curves suggest high reproducibility.

Usually, calibration curves for NOx measurements are carried out in water or buffer, as also recommended by manufacturers of kits for NOx analysis. However, from an analytical point of view, it is more correct to carry out calibration curves in a matrix as similar to that where NOx will be measured. Thus, to perform calibration curves, human plasma was passed through PD10 columns to remove nitrate and nitrite, and samples were then ultrafiltrated to restore the initial protein concentration. Data are reported in Fig. 23.11 (closed circles), where a comparison with mean values obtained from data of Fig. 23.10 after subtraction of the value measured without the addition of nitrate (basal value) for each sample is also shown (Fig. 23.11, open circles). A perfect linearity was obtained, together with high reproducibility of the method, and the slope obtained with plasma passed through PD10 columns was only slightly lower than that obtained with plasma samples spiked with nitrate, thus suggesting little interference by low molecular weight compounds.

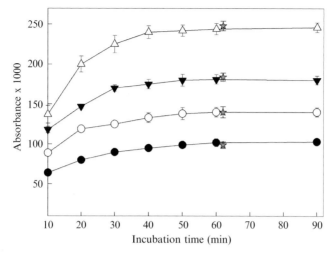

Figure 23.9 Effect of incubation time with nitrate reductase. Plasma aliquots (0.8 ml) were treated with 8 μl of 1 to 4 mM sodium nitrate solutions to obtain the following final concentration of added nitrite: 0 μM (closed circles), 10 μM (open circles), 20 μM (closed triangles), and 40 μM (open triangles). All samples were then added with 8 μl of 12.5 mM NADPH, 4 μl of 1 mM FAD, and 32 μl of nitrate reductase (5 U/ml dissolved in 50 mM phosphate buffer, pH 7.5). At the specified times, a 0.1-ml sample was added with 5 μl of 300 mM NEM. After 2 min, samples were diluted with 400 μl of a 62.5% ethanol solution and then 1:1 reacted with the Griess reagent (1% sulfanilamide, 0.1% NED, 6% phosphoric acid). After 30 min, samples were deproteinized by TCA addition and analyzed by a spectrophotometer in the 650- to 480-nm range. The height of peaks recorded at 545 nm is reported. Stars indicate height of the peaks measured when the same amounts of nitrite instead of nitrate were added to the sample under the same conditions. The number of replicates is four.

Once our experimental conditions were verified as a valid procedure to detect nitrite and nitrate in plasma, we finally measured the NOx concentration in 15 fasting, healthy subjects; the mean plasmatic value was 27.3 ± 8.3 μM (12–45 μM).

4. Discussion

The most commonly applied method to measure nitrite and nitrate is based on the Griess reaction, first described by Johann Peter Griess in 1879. The original Griess reaction has been modified since and, instead of the original reagents, that is, sulfanilic acid and α-naphthylamine (the so-called original Griess reagents), sulfanilamide and NED were introduced, following the observation that NED offers several advances over α-naphthylamine in terms of reproducibility, sensitivity, and pH independence of the color

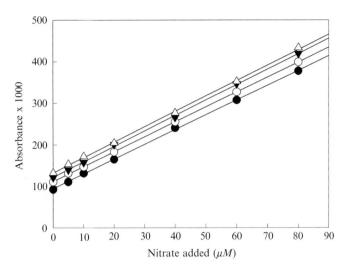

Figure 23.10 Measurement of NOx in plasma samples. One hundred-microliter aliquots of plasma samples were added with 1 μl of 12.5 mM NADPH, 1 μl of 0.5 mM FAD, and 4 μl of nitrate reductase (5 U/ml dissolved in 50 mM phosphate buffer, pH 7.5). After 90 min, samples were treated with 5 μl of 300 mM NEM. After 2 min, samples were diluted with 400 μl of a 62.5% ethanol solution and 1:1 reacted with the Griess reagent (1% sulfanilamide, 0.1% NED, 6% phosphoric acid). After 30 min, samples were deproteinized by TCA addition and analyzed by a spectrophotometer in the 650- to 480-nm range. Aliquots (100 μl) of plasma samples were added with 1 μl of 0.5 to 8 mM sodium nitrate solutions and treated as described in the text. The height of peaks recorded at 545 nm is reported. The number of replicates is four. The value of nitrate in the abscissa is referred to as the concentration of nitrate added to plasma.

(Tsikas, 2007). Several colorimetric methods, essentially based on a batch reaction with Griess reagents with various modifications, for detection of NOx (namely, nitrite and nitrate) in biological fluids, such as plasma and urine, have been introduced and described (Tsikas, 2005, 2007). Specifically, the diazotization reactions are specific for nitrite, but these analyses also allow the nitrate measurement after its reduction to nitrite. Given the high content in protein and the relative lower concentration of NOx in plasma with respect to urine, analytical methods developed to measure NO_2^- and NO_3^- in plasma have to deal with a major number of possible interferences (Tsikas, 2007). Basically, two crucial points in NOx measurements are evident: (i) quantitative nitrite detection by the Griess reaction in different biological samples and (ii) efficiency of reduction of nitrate to nitrite. As a general rule, the more complex nitrite-containing sample, the higher the probability of interferences. Potential, mechanism-based interferences may become significant at any step of the diazotization reaction; in addition, recovery of the azo dye generated by the reaction of nitrite with sulfanilamide and NED depends on many factors, such as pH, temperature,

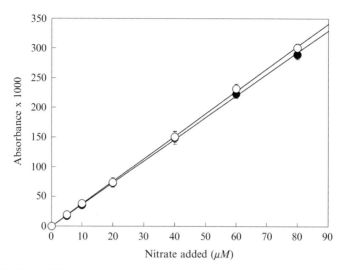

Figure 23.11 Calibration curve. One hundred-microliter aliquots of plasma samples (previously passed through PD10 columns and brought to the initial protein concentration by ultrafiltration) were added with 1 μl of 0.5 to 8 mM sodium nitrate solutions, 1 μl of 12.5 mM NADPH, 1 μl of 0.5 mM FAD, and 4 μl of nitrate reductase (5 U/ml dissolved in 50 mM phosphate buffer, pH 7.5). After 90 min, samples were treated with 5 μl of 300 mM NEM and, after 2 min, were diluted with 400 μl of a 62.5% ethanol solution and 1:1 reacted with the Griess reagent (1% sulfanilamide, 0.1% NED, 6% phosphoric acid). After 30 min, samples were deproteinized by TCA addition and analyzed by a spectrophotometer in the 650- to 480-nm range. The height of peaks recorded at 545 nm is reported. The number of replicates is four. The value of nitrate in the abscissa is referred to as the final concentration of added nitrate to plasma. The calibration curve (closed circles) was compared with the values reported in Fig. 23.10 (open circles) after subtraction of the basal NOx concentration.

and relative concentration of the reagents, only to name a few (Fox, 1979). Interference in the assay by proteins and NADPH is generally recognized, and most of the previously described methods include protein separation by ultrafiltration (Miranda et al., 2001) and NADPH oxidation to NADP$^+$ by incubation with NADPH consuming mixtures (e.g., lactate dehydrogenase + piruvate) (Wang et al., 1997). Alternatively, a NADPH regenerating system is included, that is, glucose-6-phosphate dehydrogenase + glucose-6-phosphate, which allows the use of NADPH at lower concentrations (Verdon et al., 1995).

Nevertheless, there is a general belief that Griess assays are reliable for accurate, quantitative measurement of NOx in biological samples. In particular, the commercial availability of "ready-to-use" kits is extremely tempting, and NO researchers are making increase use thereof. However, it is worth mentioning that commercial availability does not automatically guarantee the accuracy and robustness of the method.

Over the last years, dozens of papers have included the measurement of NOx in biological fluids, particularly in plasma/serum, with values ranging from 4 to 108 μM (Tsikas, 2007). Increased/decreased values for NOx, with respect to healthy controls, were found in various physiopathological conditions, for example, Crohn's disease, stenosis, diabetes, and renal failure (Adachi et al., 1998; Ishibashi et al., 2000; Maejima et al., 2001; Oudkerk Pool et al., 1995).

This chapter evaluated the main interferences to NOx measurements in plasma and provided an inexpensive, easy, and time-sparing standardized procedure for plasma NOx measurement. It is noteworthy that the same methodology could likely be applied to also measure NOx in other biological samples. All the reported measurements were carried out recording the whole 650- to 480-nm spectra and not a single point(s); moreover, spectra were recorded against a blank constituted by the sample to be analyzed and the same reagents, where only NED was omitted.

First, we evaluated the effect of different pH values on the Griess reaction. The pH influence was described extensively by Fox (1979), who reported that many factors may influence the diazotization reaction, among these pH, with a maximal pigment formation in the 2.5 to 3.5 pH range. Commonly used Griess reagents can be prepared by using different acids, namely phosphoric acid, acetic acid, and HCl, and, frequently, the final concentrations are different too (Granger et al., 1996; Tsikas, 2007). We confirmed that pH can influence the yield of nitrite titration significantly, with higher performance at higher pH values (Fig. 23.2). It was also evident that NADPH interference is maximized at 3 to 4 pH values (Fig. 23.3). From our data it is clear that the Griess solution prepared with 6% phosphoric acid (pH 1.16) is the best condition, allowing both a good yield of the reaction and low NADPH interference. We also identified at what concentration both NADPH and plasma interference could be tolerated (Figs. 23.4 and 23.5). In addition, we showed that ethyl alcohol addition to the mixture greatly reduces the coprecipitation of the azo dye with plasmatic proteins (Fig. 23.6), which must be removed before measurements because they can produce turbidity and absorb at 540 nm, particularly if the sample is slightly hemolyzed. Combining all the information obtained from experiments shown in Figs. 23.2 to 23.7, we concluded that sample dilution, ethyl alcohol addition to plasma samples, and use of a Griess solution prepared with a higher phosphoric acid content (6% instead of 2%) minimized all the possible interferences satisfactorily. Nitrite detection under these conditions, both in buffered solutions and in plasma samples (added with the same amounts of nitrite), gave similar results, thus confirming our findings (Fig. 23.7).

Once verified that added nitrite was measured by our method quantitatively, we analyzed the other crucial step in NOx measurement, that is, nitrate reduction by nitrate reductase. Enzymatic reduction is carried out more commonly than chemical reduction, whose major drawback is that some metals lead to incomplete reduction or carry the reduction further than nitrite (Cortas and Wakid, 1990), and various conditions have been reported (Tsikas, 2007). Most commercial kits also include reduction of nitrate by nitrate reductase from *Aspergillus*. However, little is reported about the stability with time of this enzyme. In some kits (e.g., from Cayman Chemical Company), it is reported that the enzyme can be frozen and thawed once after its reconstitution with buffer. Information on nitrate reductase stability was also reported by Ricart-Jané and colleagues (2002), who evidenced how stability of the bacterial nitrate reductase is limited and, after storage at both 4 and $-20\,°C$, is lost rapidly. We pointed out that nitrate reductase is not stable after its reconstitution with buffer (Fig. 23.8). In particular, cycles of freezing and thawing should be avoided and measurement of its activity is highly recommended before use. The reason for such instability, to the authors' knowledge, is unknown. Experiments with nitrate added to plasma (Fig. 23.9) led us to conclude that, under our conditions, the reduction of nitrate is complete within 45 to 60 min and that nitrate added is recovered quantitatively (Figs. 23.10 and 23.11). Moreover, since only a slightly different slope was obtained after nitrate addition both to human plasma and to the same sample after gel filtration, we suggest that calibration curves can be carried out by the addition of nitrate to plasma after G25 gel filtration. The mean nitrate concentration measured ($27.3 \pm 8.3\ \mu M$) in healthy humans is consistent with many other previous findings (Tsikas, 2007).

The method for NOx measurements in plasma/serum described here is based on the following points.

i. Accurate measurement of spectra in the 480- to 650-nm range against the same samples reacted with a modified Griess solution, where NED was omitted. This procedure allowed us to minimize, and eventually evaluate, deviations of recorded spectra due to turbidity or colored protein presence. This procedure allowed us to also detect spectra with peaks at 545 nm as low as 0.02 OD, with the limit of detection in the final mixture being close to $0.5\ \mu M$.
ii. Sample dilution: a fivefold dilution can be carried out, considering the method detection limit and plasma NOx levels. This allowed us to eliminate both ultrafiltration, which is critical, as the material contained in some cartridges could release significant amounts of nitrate and nitrite (Everett *et al.*, 1995), and procedures for NADPH interference minimization (e.g., elimination by lactate dehydrogenase), which are time-consuming and expensive.
iii. Use of a 6% phosphoric acid solution to prepare the Griess reagent minimizes pH variation due to a protein-rich sample addition.

iv. Addition of ethanol to the solution used to dilute samples largely decreases the unspecific binding of the azo compound to plasmatic proteins.

The method was found to be reproducible with intersample coefficient of variance and intraday variation on measurements carried out on the same sample as low as 2%, with high linearity in the range of 5 to 80 μM NOx.

In view of the large number of applications that the Griess reaction has found in clinical and experimental studies, it is crucial to apply easy-to-perform and virtually low artifact-prone methods. Investigators from various disciplines are the users of these analytical methods, and there is no doubt that assays based on the Griess reaction are the most frequently applied analytical methods for the analysis of nitrite and nitrate in the NO area of research. However, as stated by Tsikas (2007), we agree that accurate quantitative determination of NOx in plasma (or serum) is still a challenging analytical task. Many of the common methodological problems may arise from preanalytical factors, particularly from the shortcoming in the assay related to the Griess reaction itself. In addition, it must be emphasized that calibration curves are usually performed with nitrite instead of nitrate (i.e., the main component of NOx) and, moreover, in water or buffer, which are very different from plasma. The most critical point in the measurement of NOx in biological samples by batch Griess assays seems to be the unknown recovery rate (Tsikas, 1997).

This chapter reported on a new Griess assay able to measure NOx in plasma samples with high precision and accuracy and also quantified the recovery rate of both nitrate and nitrite. Because there are no time-consuming and expensive steps (e.g., ultrafiltration), it appears suitable for use in future clinical studies performed to analyze further the role of nitrite/nitrate as potential reservoirs of NO activity and the link between NOx levels and disease conditions.

ACKNOWLEDGMENT

This work was supported by FIRST 2006 (Fondo Interno Ricerca Scientifica e Tecnologica) University of Milan.

REFERENCES

Adachi, J., Morita, S., Yasuda, H., Miwa, A., Ueno, Y., Asano, M., and Tatsuno, Y. (1998). Elevated plasma nitrate in patients with crush syndrome caused by the Kobe earthquake. *Clin. Chim. Acta* **269,** 137–145.

Cortas, N. K., and Wakid, N. W. (1990). Determination of inorganic nitrate in serum and urine by a kinetic cadmium-reduction method. *Clin. Chem.* **36,** 1440–1443.

Ellis, G., Adatia, I., Yazdanpanah, M., and Makela, S. K. (1998). Nitrite and nitrate analyses: A clinical biochemistry perspective. *Clin. Biochem.* **31,** 195–220.

Everett, S. A., Dennis, M. F., Tozer, G. M., Prise, V. E., Wardman, P., and Stratford, M. R. (1995). Nitric oxide in biological fluids: Analysis of nitrite and nitrate by high-performance ion chromatography. *J. Chromatogr. A* **706,** 437–442.

Forstermann, U., and Ishii, K. (1996). Measurement of cyclic GMP as an indicator of nitric oxide production. *In* "Methods in Nitric Oxide Research" (M. Feelisch and J. S. Stamler, eds.), pp. 555–566. Wiley, Chichester.

Fox, J. B., Jr. (1979). Kinetics and mechanisms of the Griess reaction. *Anal. Biochem.* **51,** 1493–1502.

Gally, J. A., Montague, P. R., Reeke, G. N., Jr., and Edelman, G. M. (1990). The NO hypothesis: Possible effects of a short-lived, rapidly diffusible signal in the development and function of the nervous system. *Proc. Natl. Acad. Sci. USA* **87,** 3547–3551.

Giustarini, D., Dalle-Donne, I., Colombo, R., Milzani, A., and Rossi, R. (2004). Adaptation of the Griess reaction for detection of nitrite in human plasma. *Free Radic. Res.* **38,** 1235–1240.

Giustarini, D., Milzani, A., Colombo, R., Dalle-Donne, I., and Rossi, R. (2003). Nitric oxide and S-nitrosothiols in human blood. *Clin. Chim. Acta* **330,** 85–98.

Gladwin, M. T. (2005). Nitrite as an intrinsic signaling molecule. *Nat. Chem. Biol.* **1,** 245–246.

Granger, D. L., Taintor, R. R., Boockvar, K. S., and Hibbs, J. B., Jr. (1996). Measurement of nitrate and nitrite in biological samples using nitrate reductase and Griess reaction. *Methods Enzymol.* **268,** 142–151.

Ishibashi, T., Matsubara, T., Ida, T., Hori, T., Yamazoe, M., Aizawa, Y., Yoshida, J., and Nishio, M. (2000). Negative NO_3^- difference in human coronary circulation with severe atherosclerotic stenosis. *Life Sci.* **66,** 173–184.

Larsen, F. J., Ekblom, B., Sahlin, K., Lundberg, J. O., and Weitzberg, E. (2006). Effects of dietary nitrate on blood pressure in healthy volunteers. *N. Engl. J. Med.* **355,** 2792–2793.

Lundberg, J. O., Feelisch, M., Bjorne, H., Jansson, E. A., and Weitzberg, E. (2006). Cardioprotective effects of vegetables: Is nitrate the answer? *Nitric Oxide* **15,** 359–362.

Lundberg, J. O., and Govoni, M. (2004). Inorganic nitrate is a possible source for systemic generation of nitric oxide. *Free Radic. Biol. Med.* **37,** 395–400.

Lundberg, J. O., and Weitzberg, E. (2005). NO generation from nitrite and its role in vascular control. *Arterioscler. Thromb. Vasc Biol.* **25,** 915–922.

Maejima, K., Nakano, S., Himeno, M., Tsuda, S., Makiishi, H., Ito, T., Nakagawa, A., Kigoshi, T., Ishibashi, T., Nishio, M., and Uchida, K. (2001). Increased basal levels of plasma nitric oxide in type 2 diabetic subjects: Relationship to microvascular complications. *J. Diabetes Complications* **15,** 135–143.

Miranda, K. M., Espey, M. G., and Wink, D. A. (2001). A rapid, simple spectrophotometric method for simultaneous detection of nitrate and nitrite. *Nitric Oxide* **5,** 62–71.

Moshage, H., Kok, B., Huizenga, J. R., and Jansen, P. L. (1995). Nitrite and nitrate determinations in plasma: A critical evaluation. *Clin. Chem.* **41,** 892–896.

Oudkerk Pool, M., Bouma, G., Visser, J. J., Kolkman, J. J., Tran, D. D., Meuwissen, S. G., and Pena, A. S. (1995). Serum nitrate levels in ulcerative colitis and Crohn's disease. *Scand. J. Gastroenterol.* **30,** 784–788.

Rassaf, T., Heiss, C., Hendgen-Cotta, U., Balzer, J., Matern, S., Kleinbongard, P., Lee, A., Lauer, T., and Kelm, M. (2006). Plasma nitrite reserve and endothelial function in the human forearm circulation. *Free Radic. Biol. Med.* **41,** 295–301.

Ricart-Jané, D., Llobera, M., and Lopez-Tejero, M. D. (2002). Anticoagulants and other preanalytical factors interfere in plasma nitrate/nitrite quantification by the Griess method. *Nitric Oxide* **6,** 178–185.

Tsikas, D. (2005). Methods of quantitative analysis of the nitric oxide metabolites nitrite and nitrate in human biological fluids. *Free Radic. Res.* **39,** 797–815.

Tsikas, D. (2007). Analysis of nitrite and nitrate in biological fluids by assays based on the Griess reaction: Appraisal of the Griess reaction in the L-arginine/nitric oxide area of research. *J. Chromatogr. B Analyt. Technol. Biomed. Life Sci.* **851,** 51–70.

Tsikas, D., Gutzki, F. M., Rossa, S., Bauer, H., Neumann, C., Dockendorff, K., Sandmann, J., and Frolich, J. C. (1997). Measurement of nitrite and nitrate in biological fluids by gas chromatography-mass spectrometry and by the Griess assay: Problems with the Griess assay–solutions by gas chromatography-mass spectrometry. *Anal. Biochem.* **244,** 208–220.

van Vliet, J. J., Vaessen, H. A., van den Burg, G., and Schothorst, R. C. (1997). Twenty-four-hour duplicate diet study 1994; nitrate and nitrite: Method development and intake per person per day. *Cancer Lett.* **114,** 305–307.

Verdon, C. P., Burton, B. A., and Prior, R. L. (1995). Sample pretreatment with nitrate reductase and glucose-6-phosphate dehydrogenase quantitatively reduces nitrate while avoiding interference by NADP+ when the Griess reaction is used to assay for nitrite. *Anal. Biochem.* **224,** 502–508.

Wang, J., Brown, M. A., Tam, S. H., Chan, M. C., and Whitworth, J. A. (1997). Effects of diet on measurement of nitric oxide metabolites. *Clin. Exp. Pharmacol. Physiol.* **24,** 418–420.

CHAPTER TWENTY-FOUR

DETECTION OF NITRIC OXIDE AND ITS DERIVATIVES IN HUMAN MIXED SALIVA AND ACIDIFIED SALIVA

Umeo Takahama,* Sachiko Hirota,[†] *and* Oniki Takayuki*

Contents

1. Introduction	382
2. Formation of Reactive Nitrogen Oxide Species (RNOS) in Mixed Whole Saliva and the Bacterial Fraction	382
3. Detection of RNOS in Mixed Whole Saliva and the Bacterial Fraction	384
3.1. Whole saliva filtrate, bacterial fraction, and salivary peroxidase preparation	384
3.2. Preparation of leukocytes	385
3.3. Detection of RNOS by nitration of phenolics	385
3.4. Fluorometric detection of RNOS	386
3.5. Detection of nitric oxide by electron spin resonance (ESR)	389
4. Formation of RNOS in Acidified Saliva	390
5. Detection of RNOS in Acidified Saliva	392
5.1. Detection of nitric oxide by oxygen consumption	392
5.2. Detection of nitric oxide by ESR	392
5.3. Nitration	393
5.4. Detection of NO_2^-/HNO_2, NO_3^-, and SCN^-	393
6. Concluding Remarks	393
References	394

Abstract

Nitrate is secreted into the human oral cavity as a salivary component. The nitrate is transformed to nitrite and nitric oxide (NO) by oral bacteria. NO is oxidized by O_2 producing NO_2 and N_2O_3 and also by O_2^- producing $ONOO^-$. Salivary peroxidase can oxidize nitrite and NO to NO_2 or its equivalent in the oral cavity. Nitrite dissolved in saliva is mixed with gastric juice, generating nitrous acid that is transformed to NO and NO_2 via N_2O_3 by self-decomposition. In addition, nitrous acid can react with ascorbic acid and phenolics producing NO

* Department of Bioscience, Kyushu Dental College, Kitakyushu, Japan
[†] Department of Nutritional Science, Kyushu Women's University, Kitakyushu, Japan

and with H_2O_2 producing ONOOH. This chapter deals with the detection of reactive nitrogen oxide species (RNOS), especially NO, N_2O_3, NO_2, and $ONOO^-$/ONOOH, in mixed whole saliva and acidified saliva using fluorescent probes and spin-trapping reagents. It is also shown that measurements of nitration and oxygen consumption are useful in studying the formation and scavenging of RNOS in the aforementioned systems.

1. INTRODUCTION

Nitrate is normally added as a food preservative to make ham and bacon, and nitrite is added to the processed food to make the color bright pink due to the formation of methomyoglobin (Benjamin, 2000). Nitrate is also found in large quantities in vegetables such as lettuce and spinach but the quantities of nitrite in vegetable are small. The ingestion of food that contains nitrate and nitrite results in their uptake from intestine and is then secreted as a salivary and urinary component. Nitrate is, in general, nonreactive with organic molecules, but nitrite generated by the reduction of nitrate is reactive. For example, nitrite is a nitrating and nitrosating reagent and an antimicrobial agent especially under acidic conditions (Allaker, 2001; Silvia Mendetz, 1999). In the human oral cavity, nitrate secreted as a salivary component is reduced to nitrite and nitric oxide (NO) by certain bacteria, and salivary nitrite may be transformed to NO, NO_2, and N_2O_3 after mixing with gastric juice. It is well known that NO is an antimicrobial compound as well as a physiologically important compound (Halliwell and Gutteridge, 1999). This chapter deals with the detection of reactive nitrogen oxide species (RNOS) derived from nitrite in the human oral cavity and the species formed in acidified saliva using methods that have been employed in the authors' laboratory.

2. FORMATION OF REACTIVE NITROGEN OXIDE SPECIES (RNOS) IN MIXED WHOLE SALIVA AND THE BACTERIAL FRACTION

The concentration of nitrate in saliva (0.2–2.5 mM) is dependent on the amount of nitrate ingested (Pannala *et al.*, 2003; Tanaka *et al.*, 2004). The nitrate in saliva is reduced to nitrite (pK_a of 3.3) and NO successively by bacteria such as *Streptococcus salivarius*, *S. mitis*, and *S. bovis* (Palmerini *et al.*, 2003). The concentration of nitrite in saliva (0.05–1 mM) is dependent on the concentration of nitrate (Pannala *et al.*, 2003). The nitrite and NO formed in the human oral cavity can be oxidized by molecular oxygen (reaction 24.1) (Halliwell and Gutteridge, 1999) and by salivary peroxidase (reactions 24.2 and 24.3) (Takahama et al., 2003a, 2006a,b) producing NO_2:

$$2NO + O_2 \rightarrow 2NO_2 \qquad (24.1)$$

$$NO + H_2O_2 \rightarrow NO_2 + H_2O \qquad (24.2)$$

$$2NO_2^- + H_2O_2 + 2H^+ \rightarrow 2NO_2 + 2H_2O \qquad (24.3)$$

Bacteria and leukocytes produce H_2O_2 required for the peroxidase-catalyzed reactions in the oral cavity (Tenovuo, 1989). The NO_2 produced by reactions 24.1–24.3 can contribute to the formation of N_2O_3:

$$NO + NO_2 \leftrightarrow N_2O_3 \qquad (24.4)$$

The production of H_2O_2 shows that $O_2^-\cdot$ is also produced in the oral cavity. The production of $O_2^-\cdot$ results in the formation of $ONOO^-$ (pK_a of 6.8) if NO is present (Goldstein et al., 2005; Halliwell and Gutteridges, 1999):

$$O_2^-\cdot + NO \rightarrow ONOO^- \qquad (24.5)$$

NO_2 is an oxidant as well as a nitrating agent, N_2O_3 is a nitrosating agent, and $ONOO^-/ONOOH$ is an oxidant as well as a nitrating and nitrosating agent.

In addition to NO, nitrite, O_2^-, and H_2O_2, SCN^- is also present in human saliva in a concentration range from 0.25 to 2 mM (Tsuge et al., 2000). SCN^- is a physiological substrate of salivary peroxidase and inhibits the peroxidase-catalyzed reactions 24.2 and 24.3 producing $OSCN^-$, an antimicrobial agent. Therefore, SCN^- can protect the oral cavity from damages induced by NO_2 and its equivalents if the reactive species are formed by peroxidase-catalyzed reactions. SCN^- is supposed to be produced by the metabolism of isothiocyanates and by the detoxification of cyanide (Galant, 1997; Tsuge et al., 2000).

It happens that the local pH in the oral cavity decreases to about 5 accompanying the growth of acid-producing bacteria (Marsh and Martin, 1999). Under these conditions, NO may be formed from HNO_2 (pK_a of 3.3) (Duncan et al., 1995):

$$2HNO_2 \leftrightarrow N_2O_3 + H_2O \qquad (24.6)$$

SCN^- may interfere with the reaction significantly by the following reaction,

$$SCN^- + HNO_2 \rightarrow NOSCN + OH^- \qquad (24.7)$$

NOSCN is not only an oxidizing but also a nitrosating agent as the compound can dissociate to NO^+ and SCN^- (Benjamin, 2000; Licht et al., 1988). Thus, SCN^- seems to have two functions in the human oral cavity: inhibition of the formation of RNOS and enhancement of their formation.

3. Detection of RNOS in Mixed Whole Saliva and the Bacterial Fraction

To study the metabolism of RNOS in saliva, some kinds of samples may be prepared. They are whole saliva filtrate, bacterial fraction, and salivary peroxidase preparation (Fig. 24.1). In addition, the leukocyte fraction is also useful in studying the metabolism of nitrogen oxides. This section deals with their preparation and then with the detection of RNOS.

3.1. Whole saliva filtrate, bacterial fraction, and salivary peroxidase preparation

Mixed whole saliva can be collected by chewing paraffin or Parafilm as convenient (Takahama et al., 2006a). The saliva (5–10 ml) is passed through two layers of nylon filter nets (32 μm^2) to remove epithelial cells and other particles. The filtrate can be used as a whole saliva filtrate after mixing the filtrate with an equal volume of 50 mM sodium phosphate buffer or solutions used for physiological studies. Mixing results in a decrease in the viscosity of saliva and prevention of the increase in pH due to the evaporation of CO_2 gas. The filtrate is centrifuged at 20,000g for 5 min. After

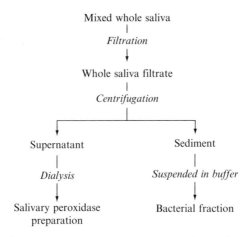

Figure 24.1 Fractionation of mixed whole saliva.

suspending the sediment in a buffer solution, the suspension can be used as a bacterial fraction. The contamination of epithelial cells and leukocytes in this fraction is small. When salivary peroxidase is required for experiments, the supernatant obtained after centrifugation is dialyzed against 10 mM sodium phosphate buffer (pH 7.0). The dialyzed supernatant can be used as a salivary peroxidase preparation after removing the turbidity by centrifugation (20,000g, 5 min).

3.2. Preparation of leukocytes

The preparation of leukocytes released into the oral cavity has been reported by Yamamoto et al. (1991) and by Al-Essa et al. (1994) using nylon mesh filtration and centrifugation. Leukocytes are collected by washing the oral cavity with 10 ml of Krebs–Ringer phosphate for 30 s several times. The pooled washings are passed through two layers of nylon filter nets (32 μm^2) to remove epithelial cells. The filtrate is centrifuged at 300g for 5 min. After washing the sediment with Krebs–Ringer phosphate or its equivalent, it is suspended in the washing solution. The number of leukocytes is usually adjusted to 1–5 × 10^6/ml.

3.3. Detection of RNOS by nitration of phenolics

The production of RNOS can be measured using some techniques. A classical method is nitration of phenolics. The formation of nitrated compounds suggests the production of NO_2 and/or $ONOO^-$/$ONOOH$. Tyrosine and 4-hydroxyphenylacetic acid (HPA) (Fig. 24.2) are used to detect the reactive nitrogen species. HPA is found in the human mixed whole saliva (Takahama et al., 2002b) and is nitrated to 4-hydroxy-3-nitrophenylacetic acid (NO_2HPA) in the mixture that contains 0.1 mM HPA, 1 mM $NaNO_2$, and 0.5 mM H_2O_2 in 1 ml of salivary peroxidase fraction (pH 4.5–8) (Hirota et al., 2005; Takahama et al., 2003a). The NO_2HPA formed is extracted with acidic ethyl acetate (20 μl of 1 M HCl in 5 ml of ethyl acetate) and is separated by HPLC combined with an ODS column (e.g., Shim-pack LCL-ODS, Shimadzu, Kyoto, Japan). NO_2HPA can be identified comparing the retention time and absorption spectrum

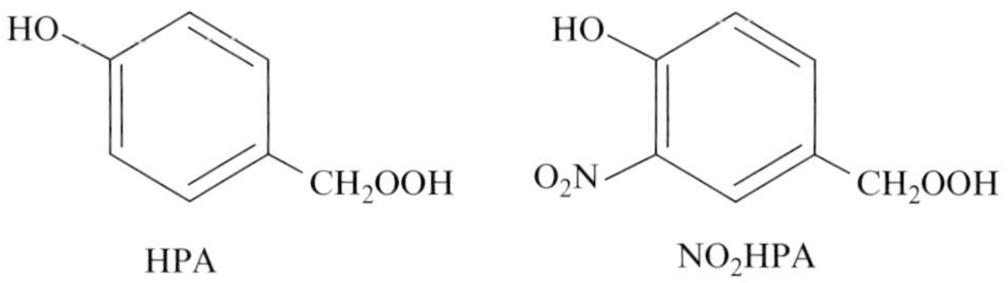

Figure 24.2 Structures of HPA and nitrated HPA.

with standard NO_2HPA. If mass spectra are measured, the identification is more accurate. SCN^- inhibits nitration when salivary peroxidase participates in the nitration. It is interesting to know the formation of NO_2HPA in the human oral cavity. Saliva preparations obtained from about 70 individuals (30–90 years old) with periodontal diseases were extracted with acidic ethyl acetate as described earlier, and NO_2HPA was separated by HPLC. The component was found in the seven saliva preparations whose ages were between 60 and 80. The concentration ranged from 0.03 to 1.7 μM [mean ± SD 0.40 ± 0.65 μM ($n = 7$)] (unpublished result). Data suggest that salivary component HPA could be nitrated in the oral cavity.

Nitration of salivary proteins is also possible in the reaction mixture that contains 1 mM $NaNO_2$ and 0.5 mM H_2O_2 in 1 ml of salivary peroxidase fraction. The nitration of tyrosine residues in proteins is estimated by the increase in absorbance at 420 nm around neutral pH and is confirmed by the detection of nitrated tyrosine by HPLC after hydrolysis of the nitrated proteins (Takahama et al., 2003a). Tyrosine residues in amylase seem to be nitrated (unpublished data).

3.4. Fluorometric detection of RNOS
3.4.1. Background
Some kinds of RNOS can also be detected using reagents that are transformed to fluorescent compounds by reactive nitrogen species. In the reagents, 4,5-diaminofluorescein (DAF-2) is widely used to detect the formation of NO under aerobic conditions (Fig. 24.3). It is supposed that N_2O_3 formed by reaction 24.4 transforms the o-diamino group of DAF-2 to a triazol group producing triazol fluorescein (DAF-2T), which is fluorescent (excitation wavelength, 495 nm; emission wavelength, 515 nm) (Nakatsubo et al., 1998). The fluorescence yield of DAF-2T is nearly constant in a pH range from 7 to 12. Other reagents, which can react with N_2O_3, are 3-amino-4-monomethylamino-2′,7′-difluorofluorescein (DAF-FM) (Kojima et al., 1999) and 3-amino-4-monomethylaminorhodamine (DAR-4M) (Kojima et al., 2001). Fluorescence yields of triazol forms of DAF-FM (DAF-FMT) (excitation wavelength, 500 nm; emission wavelength, 515 nm) and DAR-4M (DAR-4MT) (excitation wavelength, 560 nm; emission wavelength, 575 nm) are nearly constant in pH ranges from 5.5 to 12 and from 4 to 12, respectively. Esters of DAF-2, DAF-FM, and DAR-4M are taken up into cells, and the esters can be hydrolyzed by esterase producing DAF-2, DAF-FM, and DAR-4M to react with N_2O_3 in the cells (Kojima et al., 1999, 2001). Although the aforementioned reagents react with N_2O_3 generating their triazol forms, it has been reported (i) that DAF-2 is oxidized by peroxidase and some oxidants to DAF-2 radical, which reacts with NO producing the tirazol form (Espey et al., 2002; Jourd'heuil, 2002), and

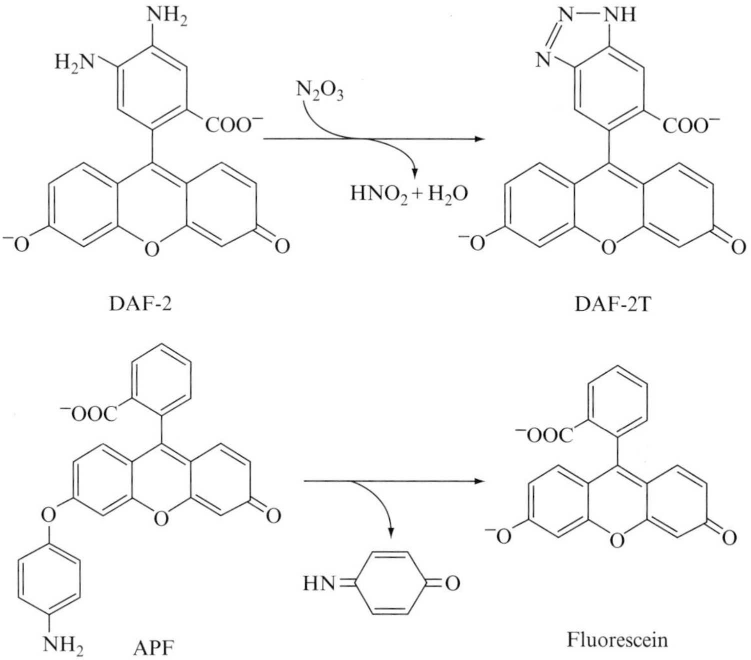

Figure 24.3 Transformation of DAF-2 and APF to their fluorescent components, DAF-2T and fluorescein.

(ii) that the detection of NO is interfered by ascorbic and dehydroascorbic acids (Zhang et al., 2002).

Other reagents that can be used to detect reactive nitrogen species are aminophenyl fluorescein (APF) (Fig. 24.3) and hydroxyphenyl fluorescein (HPF) (Setukinai et al., 2003). The reagents are ethers of *p*-aminophenol and fluorescein and *p*-hydroxyphenol and fluorescein, respectively. The ether bond of APF is cleaved oxidatively by oxidants such as $ONOO^-/ONOOH$, OH radical, and $OCl^-/HOCl$, and the bond of HPF can be by $ONOO^-/ONOOH$ and OH radical, producing fluorescein (excitation wavelength, 490 nm; emission wavelength, 515 nm). The formation of $ONOO^-/ONOOH$ can be detected using dihydrorhodamine 123 (Ischiropoulos et al., 1995). Rhodamine 123 is the fluorescent product (excitation wavelength, 500 nm; emission wavelength, 535 nm) obtained by the oxidation of dihydrorhodamine 123.

3.4.2. Practical

DAF-2T is formed when 10 μM DAF-2 is added to bacterial fraction (pH 7) in the presence of nitrite (Takahama et al., 2005, 2006a,b). N_2O_3 participates in the formation of DAF-2T (Takahama et al., 2006a).

The mechanism of the formation of N_2O_3 by reaction 24.4 is discussed taking (i) the reduction of nitrite to NO by nitrite-reducing bacteria, (ii) autooxidation of NO to NO_2 (reaction 24.1) and (iii) salivary peroxidase-catalyzed oxidation of nitrite and NO to NO_2 (reactions 24.2 and 24.3) into consideration (Takahama et al., 2006a). The participation of salivary peroxidase in the formation of NO_2 is supposed by the result that SCN^- inhibits the formation of DAF-2T in bacterial fraction, which is induced by nitrite and $(\pm)$-(E)-4-ethyl-2-[(E)-hydroxyimino]-5-nitro-3-hexenamide (NOR 3), an NO producing reagent (Kita et al., 1994). It is known that the bacterial fraction contains salivary peroxidase and produces H_2O_2 (Takahama et al., 2006a; Tenovuo, 1989). When the concentration of nitrite is increased to about 1 mM, the contribution of reactions 24.1 and 24.4 for the formation of N_2O_3 increases due to the increase in the formation of NO by nitrite reducing bacteria. This is supported by the result that the inhibitory effects of SCN^- on the formation of DAF-2T decrease as the concentration of nitrite is increased. The transformation of DAF-2 to DAF-2T by the aforementioned reactions is confirmed by HPLC (Takahama et al., 2005). Not only DAF-2T but also other unidentified fluorescent components can be detected under some conditions. It may be important to investigate the fluorescent species formed, as fluorescent components other than DAF-2T are formed as major fluorescent components. The formation of DAF-2T is also observed in the saliva filtrate, although the rate of formation is slow (Takahama et al., 2005, 2006a). If the formation of DAF-2T is detected by HPLC combined with a spectrophotometric detector, the amount can be calculated using DAF-2 unless standard DAF-2T cannot be obtained because extinction coefficients of DAF-2T ($7.3 \times 10^4 M^{-1}$ cm^{-1} at 491 nm) and DAF-2 ($7.1 \times 10^4 M^{-1}$ cm^{-1} at 486 nm) are known. Antioxidants in saliva and polyphenols in food inhibit the formation of DAF-2T by scavenging NO_2 or inhibiting the salivary peroxidase-catalyzed oxidation of nitrite (Takahama et al., 2006b).

The local pH in the oral cavity decreases to about 5. Effects of pH on the formation of N_2O_3 could be studied using DAF-FM (Takahama et al., 2007c). The addition of 7 μM DAF-FM to the bacterial fraction in the presence of 0.2 mM $NaNO_2$ resulted in a fluorescence increase due to the formation of DAF-FMT. Analysis of the fluorescent product by HPLC showed that the product was the same as that formed by NOR 3 in 50 mM sodium phosphate buffer (pH 7.0). As NOR 3 decomposes, producing NO, and the NO is transformed to N_2O_3 by reactions 24.1 and 24.4, the fluorescent product formed by NOR 3 was supposed to be DAF-FMT. The rate of DAF-FMT formation was examined as a function of pH in the presence of nitrite. The result indicated that the rate of DAF-FMT formation increased as pH was decreased. By studying the effects of SCN^- (1 mM), it is concluded (i) that SCN^- inhibited the formation of N_2O_3 around pH 7 by inhibiting the salivary peroxidase-catalyzed oxidation of nitrite and NO as described earlier and (ii) that SCN^- enhanced the formation of the component that could contribute to the formation of DAF-FMT around pH 5. NOSCN formed by reaction 24.7 was supposed to be involved in the enhanced formation of DAF-FMT as NOSCN is an NO^+ donating reagent (Benjamin, 2000). The formation of DAF-FMT could be quantified from the decrease in the concentration of DAF-FM, as the decrease in concentration was proportional to the fluorescence increase and DAF-FMT was the major fluorescent product.

The incubation of 10 μM APF with 10 μM NOR 3 in 50 mM sodium phosphate buffer (pH 7.0) resulted in the fluorescence increase due to the formation of fluorescein (Takahama et al., 2007c). The formation of fluorescein could be confirmed by HPLC. As OH radical, $ONOO^-$/ ONOOH, and OCl^-/HOCl may not be produced in the reaction system, it was concluded that NO_2 generated by reaction 24.1 could oxidize APF. When the bacterial fraction was incubated with 10 μM APF in the presence of 0.2 mM $NaNO_2$, the fluorescence increase due to the formation of fluorescein was observed. As the fluorescence yield of fluorescein decreased as pH was decreased from 7 to 5 (Lakowicz, 2006), quantification of fluorescein formed was difficult by the measurement of the fluorescence increase. Separation of fluorescein by HPLC made possible the estimation of fluorescein formed at various pH values. The participation of OH radical, NO_2, $ONOO^-$/ONOOH, and OCl^-/HOCl in the fluorescein formation in bacterial fraction can be studied by changing the composition of the reaction mixture and by using scavengers of specific reactive species. The formation of fluorescein could also be detected by the addition of APF to the whole saliva filtrate. Dihydrorhodamine 123 is oxidized to a fluorescent component in the bacterial fraction in the presence of 0.2 mM nitrite (Takahama et al., 2006a). Effects of superoxide dismutase cannot be detected in the reaction mixture, to which inhibitors of salivary peroxidase have not been added. The failure may be because of much faster oxidation of dihydrorhodamine 123 by the peroxidase-dependent reactions than the $ONOO^-$/ONOOH-dependent reaction. The effect of superoxide dismutase can be observed when the salivary peroxidase-dependent oxidation of dihydrorhodamine 123 is inhibited (Takahama et al., 2006a). As salivary peroxidase-dependent reactions, direct oxidation of dihydrorhodamine 123 by salivary peroxidase and oxidation of dihydrorhodamine 123 by NO_2 produced by salivary peroxidase-catalyzed oxidation of nitrite are possible.

3.5. Detection of nitric oxide by electron spin resonance (ESR)

Using the fluorometric methods described earlier, RNOS derived from nitrite and NO are detected. Using the ESR technique, NO itself can be detected, as NO binds to ESR-active NO traps directly. Usage of the traps

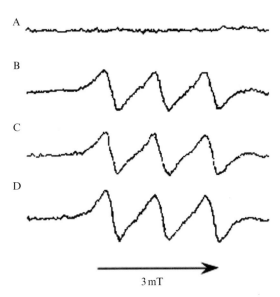

Figure 24.4 ESR spectra of NO-Fe-(DTCS)$_2$. The reaction mixture contained 0.3 ml of 10 mM sodium phosphate containing 5 mM DTCS, 1.5 mM FeCl$_3$, and 0.3 ml of a bacterial fraction. The reaction mixture was incubated for 30 min. A, no addition; B, 1 mM NaNO$_2$; C, B + 1 mM uric acid; D, B + 1 mM NaSCN.

has been reviewed elsewhere (Kalyanaraman, 1996). In the traps developed to detect NO, N-(dithocarboxyl)sarcosine (DTCS) (Fujii et al., 1996) has been applied to the bacterial fraction and mixed whole saliva filtrate (Takahama et al., 2005). DTCS has also been applied to detect NO in other biological systems (Plonka et al., 2003; Yoshimura et al., 1996). The formation of NO in the saliva filtrate and bacterial fraction can be detected in the reaction mixture that contains 5 mM DTCS and 1.5 mM FeCl$_3$. In our study, a mixture of 10 mM DTCS and 3 mM FeCl$_3$ is prepared in 50 mM sodium phosphate buffer (pH 7–7.6), and the mixture (0.5 ml) is mixed with 0.5 ml of bacterial fraction or saliva filtrate. The intensity of the ESR signal of the NO-Fe^{2+}-(DTCS)$_2$ adduct increases as a function of time. The ESR spectrum has three peaks due to a nitrogen nucleus, as shown in Fig. 24.4.

4. Formation of RNOS in Acidified Saliva

Saliva is mixed with gastric juice in the stomach. As the pH of gastric juice is about 2, an event that occurs initially on mixing is the protonation of salivary nitrite ion producing nitrous acid (pK_a of 3.3). Nitrous acid is an active species, and the acid can produce N_2O_3 by reaction 24.6, which transforms to NO and NO_2 by the reverse reaction of reaction 24.4. In the gastric juice, ascorbic acid is contained and the acid can react with nitrous acid:

$$\text{Ascorbic acid} + HNO_2 \rightarrow NO + H_2O + \text{ascorbyl radical} \quad (24.8)$$

In addition to ascorbic acid, some dietary phenolic compounds (e.g., quercetin, caffeic acid, and chlorogenic acid) can also react with nitrous acid producing NO as reaction 24.8 (Peri et al., 2005; Takahama et al., 2002a, 2003c). Once NO is produced, it is transformed to NO_2 and N_2O_3 using O_2 in the mixture of saliva and gastric juice. NO and HNO_2 may react with HO_2 and H_2O_2, respectively, producing ONOOH and ONOOH + H_2O if HO_2 and H_2O_2 are present in the mixture of saliva and gastric juice (Takahama et al., 2007a). Ascorbic acid and phenolics in food are able to scavenge NO_2 and ONOOH. The former scavenges them by reduction and the latter mainly by reduction and nitration.

SCN^- is also a normal component of saliva, and SCN^- can interfere with the ascorbic acid-dependent reduction of nitrous acid and the decomposition of ONOOH. At first, SCN^- can enhance the formation of NO by the reaction between ascorbic acid and nitrous acid in reaction 24.8 (Licht et al., 1988). NOSCN formed by reaction 24.7 may contribute to the enhanced formation of NO:

$$\begin{array}{c}\text{Ascorbic acid} + NOSCN \rightarrow NO + SCN^- + H^+ \\ + \text{ascorbyl radical}\end{array} \quad (24.9)$$

The enhancement of NO formation by SCN^- is discussed in relation to the rapid decrease in the concentration of O_2 in the mixture of saliva and gastric juice as gastric juice contains ascorbic acid (Takahama et al., 2003c). If NO is transformed to NO_2 and N_2O_3 consuming O_2 in the presence of ascorbic acid, ascorbic acid can reduce the nitrogen oxides to nitrous acid. When the concentration of O_2 decreased, NO formed can diffuse to stomach cells to play its original roles (Brown et al., 1992; Desai et al., 1991; Lundberg et al., 2004; McColl, 2005; Pique et al., 1989).

Nitrous acid can react with H_2O_2 producing ONOOH as described previously if H_2O_2 is produced in or taken up into the stomach. SCN^- in the mixture of saliva and gastric juice can react with ONOOH producing CO_2, NH_3, and SO_4^{2-} and consuming H_2O_2 (Takahama et al., 2007a). In the absence of SCN^-, ONOOH can participate in the nitration and nitrosation of phenolics.

The reaction of nitrous acid with phenolics to produce NO does not seem to be affected by SCN^- (Peri et al., 2005; Takahama et al.,

2003c). In the mixture of nitrous acid and chlorogenic acid in the absence of SCN^-, the quinone form of chlorogenic acid and nitrated chlorogenic acid are formed (Takahama et al., 2007b). Nitrous acid itself and NO_2 formed from NO and HNO_2 may oxidize chlorogenic acid to its radicals, which are nitrated by NO_2 or transformed to the quinone by dismutation. By adding SCN^- to the aforementioned reaction mixture, the formation of nitrated chlorogenic acid and the quinone is inhibited and (E)-5′-(3-(7-hydroxy-2-oxobenzo[d] [1,3]oxathiol-4-yl)acrylooxy)quinic acid is formed via 2-thiocyanatochlorogenic acid (Takahama et al., 2007b). The result suggests that SCN^- can prevent the formation of quinones, which are bioactive compounds (Moridani et al., 2001), by the nitrous acid-dependent oxidation of diphenols in food in the mixture of saliva and gastric juice.

5. Detection of RNOS in Acidified Saliva

5.1. Detection of nitric oxide by oxygen consumption

Nitric oxide produced under aerobic conditions reacts with O_2 producing NO_2 (reaction 24.1). If oxygen consumption by respiration and other reactions is controlled, the NO formation can be measured using an oxygen electrode (Jensen et al., 1997; Venkataraman et al., 2000). In fact, nitrite consumes O_2 in 50 mM KCl-HCl buffer (pH 2). The amount of consumption is dependent on the concentration of nitrite, and the rate of consumption is enhanced largely and more largely by ascorbic acid (0.01–0.1 mM) and ascorbic acid plus SCN^- (0.1–1 mM), respectively (Takahama et al., 2003b). The enhancement by ascorbic acid is because of ascorbic acid-dependent reduction of nitrous acid (reaction 24.8) and NOSCN (reaction 24.9). The oxygen consumption is also observed in acidified saliva that contains nitrite and SCN^-. In this way, the measurement of oxygen consumption is useful in detecting the formation of NO in acidified saliva (Takahama et al., 2003b). The reduction of nitrous acid to NO by phenolics can also be detected with an oxygen electrode (Takahama et al., 2003c).

5.2. Detection of nitric oxide by ESR

The NO trapping reagent Fe-$(DTCS)_2$ is used to detect NO in acidified saliva (Takahama et al., 2003b). After incubation of acidified saliva (0.5 ml) for defined periods, 0.5 ml of a mixture of 10 mM DTCS containing 1.5 mM $FeCl_3$ prepared in 0.1 M sodium phosphate buffer (pH 7.6) is added. As the NO–Fe-$(DTCS)_2$ complex is formed around neutral pH, neutralization of the acidified saliva is required on the addition of Fe-$(DTCS)_2$. Changes in the concentration of NO in acidified saliva can be detected by measuring the intensity of the ESR signal of NO–Fe-$(DTCS)_2$.

5.3. Nitration

A salivary component, HPA, is nitrated when the pH of whole saliva filtrate is decreased to about 2 (Takahama et al., 2002b). The nitrated HPA is extracted with ethyl acetate from the acidified saliva and is quantified using an ODS column as described earlier. The nitration of HPA in acidified saliva suggests that nitration of tyrosine residues of salivary proteins is also possible. The nitration of salivary components may relate to scavenging of NO_2 and other oxidants such as ONOOH and OH radicals in the stomach.

5.4. Detection of NO_2^-/HNO_2, NO_3^-, and SCN^-

It may happen that nitrous acid reacts with H_2O_2 in the stomach producing ONOOH. The production of ONOOH may be estimated by the nitration of phenolics and by the formation of HNO_3. One of the ways to detect the HNO_3 formed is the separation of NO_3^- and NO_2^- with a reversed-phase column using the mixture of 25 mM KH_2PO_4 and methanol (4:1, v/v; pH 3) as a mobile phase (Takahama et al., 2007b). The presence of phosphate in the mobile phase seems to be useful in separating NO_3^- from NO_2^-/HNO_2. The ONOOH-induced decrease in the concentration of SCN^- is also detected by the aforementioned HPLC system. During the decrease in the concentration of SCN^-, SO_4^{2-} is produced. The production of SO_4^{2-} can be detected by the increase in turbidity of the reaction mixture in the presence of Ba^{2+}. The aforementioned ions can also be detected using ion-exchange columns.

6. CONCLUDING REMARKS

The formation of NO, NO_2, N_2O_3, and $ONOO^-/ONOOH$ in the oral cavity can be detected using fluorescent probes and ESR trapping reagents, but this chapter did not deal with the detection of O_2^- and H_2O_2, although their production was related to the formation of NO_2, N_2O_3, and $ONOO^-/ONOOH$. A convenient method for the detection of H_2O_2 is the peroxidase-catalyzed oxidation of the substrate and the inhibition by peroxidase inhibitors. Many substrates have been reported to measure the rate of H_2O_2 formation (Halliwell and Gutteridge, 1999; Tenovuo, 1989). In salivary systems, an important characteristic of the substrate used to detect the formation of H_2O_2 is that the peroxidase-catalyzed oxidation is not influenced very much by nitrite, as saliva contains various concentrations of nitrite. For example, oxidation of 2,2'-azino-bis (3-ethylbenzothiazoline-6-sulfonic acid) diammonium by the salivary peroxiadse/H_2O_2 system is affected greatly by nitrite, but the oxidation of quercetin is not (Takahama et al., 2006a). The detection of O_2^- can also be affected by salivary redox components. If rate and mechanism of production of reactive oxygen species and RNOS in the oral cavity are made clear under various conditions, relations between formation of the reactive species and various diseases in the oral cavity may be elucidated. Elucidation of the rate and mechanism of the formation of reactive oxygen species and RNOS in the acidified saliva may also be useful in discussing relations between diseases and the active species in the stomach.

REFERENCES

Al-Essa, L., Niwa, M., Kohno, K., and Tsurumi, K. (1994). A proposal for purification of salivary polymorphonuclear leukocytes by combination of nylon mesh filtration and density-gradient method: A validation by superoxide- and cyclic AMP-generating responses. *Life Sci.* **55,** PL333–PL338.

Allaker, R. P., Silva Mendez, L. S., Hardie, J. M., and Benjamin, N. (2001). Antimicrobial effect of acidified nitrite on periodontal bacteria. *Oral Microbiol. Immunol.* **16,** 235–236.

Benjamin, N. (2000). Nitrite in the human diet—good or bad? *Ann. Zootech.* **49,** 207–216.

Duncan, C., Dougall, H., Johnston, P., Green, S., Brogan, R., Leifert, C., Smith, L., Golden, M., and Benjamin, N. (1995). Chemical generation of nitric oxide in the mouth from the enterosalivary circulation of dietary nitrite. *Nat. Med.* **1,** 546–551.

Brown, J. F., Hanson, P. J., and Whittle, B. J. (1992). Nitric oxide donors increase mucus gel thickness in rat stomach. *Eur. J. Pharmacol.* **223,** 103–104.

Desai, K. M., Sessa, W. C., and Vane, J. R. (1991). Involvement of nitric oxide in the reflex relaxation of the stomach to accommodation of food and fluid. *Nature* **351,** 477–479.

Espey, M. G., Thomas, D. D., Miranda, K. M., and Wink, D. A. (2002). Focusing of nitric oxide mediated nitrosation and oxidative nitrosylation as a consequence of reaction with superoxide. *Proc. Natl. Acad. Sci. USA* **99,** 11127–11132.

Fujii, S., Yoshimura, T., and Kamada, H. (1996). Nitric oxide trapping efficiencies of water-soluble iron (III) complexes with dithiocarbamate derivatives. *Chem. Lett.* 785–786.

Galanti, L. M. (1997). Specificity of salivary thiocyanate as marker of cigarette smoking is not affected by alimentary sources. *Clin. Chem.* **43,** 184–185.

Goldstein, S., Lind, J., and Merény, G. (2005). Chemistry of peroxynitrites as compared to peroxynitrates. *Chem. Rev.* **105,** 2457–2470.

Halliwell, B. and Gutteridge, J. M. C. (1999). "Free Radical in Biology and Medicine." Oxford University Press, Oxford.

Hirota, S., Takahama, U., Ly, T. N., and Yamauchi, R. (2005). Quercetin-dependent inhibition of nitration induced by peroxidase/H_2O_2/nitrite systems in human saliva and characterization of an oxidation product of quercetin formed during the inhibition. *J. Agric. Food Chem.* **53,** 3265–3272.

Ischiropoulos, H., Beckman, J. S., Crow, J. P., Ye, Y. Z., Royall, J. A., and Kooy, N. W. (1995). Detection of peroxynitrite. *Methods* **7,** 109–115.

Jensen, B. O. (1997). A polarographic method for measuring dissolved nitric oxide. *J. Biochem. Biophys. Methods* **35,** 185–195.

Jourd'heuil, D. (2002). Increased nitric oxide-dependent nitrosylation of 4,5-diaminofluorescein by oxidants: Implications for the measurement of intracellular nitric oxide. *Free Radic. Biol. Med.* **33,** 676–684.

Kalyanaraman, B. (1996). Detection of nitric oxide by electron spin resonance in chemical, photochemical, cellular, physiological, and pathophysiological systems. *Methods Enzymol.* **268**, 168–187.

Kita, Y., Ozaki, R., Sakai, S., Sugimoto, T., Hirasawa, Y., Ohtsuka, M., Senoh, H., Yoshida, K., and Maeda, K. (1994). Antianginal effects of FK409, a new spontaneous NO releaser. *Br. J. Pharmacol.* **113**, 1137–1140.

Kojima, H., Hirotani, M., Nakatsubo, N., Kikuchi, K., Urano, Y., Higuchi, T., Hirata, Y., and Nagano, T. (2001). Bioimaging of nitric oxide with fluorescent indicators based on the rhodamine chromophore. *Anal. Chem.* **73**, 1967–1973.

Kojima, H., Urano, Y., Kikuchi, K., Higuchi, T., Hirata, Y., and Nagano, T. (1999). Fluorescent indicators for imaging nitric oxide production. *Angew. Chem. Int. Ed.* **38**, 3209–3212.

Lakowicz, J. R. (2006). "Principles of Fluorescence Spectroscopy." p. 638, Springer, New York.

Licht, W. R., Tannenbaum, S. R., and Deen, W. M. (1988). Use of ascorbic acid to inhibit nitrosation: Kinetic and mass transfer consideration for an *in vitro* system. *Carcinogenesis* **9**, 365–372.

Lundberg, J. O., Weitzberg, E., Cole, J. A., and Benjamin, N. (2004). Nitrate, bacteria and human health. *Nat. Rev. Microbiol.* **2**, 593–602.

Marsh, P. and Martin, M. V. (1999). "Oral Microbiology." Wright, Oxford.

McColl, K. E. L. (2005). When saliva meets acid: Chemical warfare at the oesophagogastric junction. *Gut* **54**, 1–3.

Moridani, M. Y., Scobie, H., Jamshidzadeh, A., Salehi, P., and O'brien, P. J. (2001). Caffeic acid, chlorogenic acid, and dihydrocaffeic acid metabolism: Glutathione conjugate formation. *Drug Metab. Dispos.* **29**, 1432–1439.

Nakatsubo, N., Kojima, H., Kikuchi, K., Nagoshi, H., Hirata, Y., Maeda, D., Imai, Y., Irimura, T., and Nagano, T. (1998). Direct evidence of nitric oxide production from bovine aortic endothelial cells using new fluorescence indicators; diaminofluoresceins. *FEBS Lett.* **427**, 263–266.

Palmerini, C. A., Palombari, R., Perito, S., and Arienti, G. (2003). NO synthesis in human saliva. *Free Radic. Res.* **37**, 29–31.

Pannala, A. S., Mani, A. R., Spencer, J. P., Skinner, V., Bruckdirfer, K. R., Moore, K. P., and Rice-Evans, C. A. (2003). The effect of dietary nitrate on salivary, plasma, and urinary nitrate metabolism in humans. *Free Radic. Biol. Med.* **34**, 576–584.

Peri, L., Pietraforte, D., Scoza, G., Napolitano, A., Fogliano, V., and Minetti, M. (2005). Apples increase nitric oxide production by human saliva at acidic pH of the stomach: A new biological function for polyphenols with a catechol group? *Free Radic. Biol. Med.* **39**, 668–681.

Pique, J. M., Whittle, B. J., and Esplugues, J. V. (1989). The vasodilator role of endogenous nitric oxide in the rat gastric microcirculation. *Eur. J. Pharmacol.* **174**, 293–296.

Plonka, P. M., Wisniewska, M., Chlopicki, S., Elas, M., and Rosen, G. M. (2003). X-band and S-band EPR detection of nitric oxide in murine endotoxaemia using spin trapping by ferro-di [N-(dithiocarboxy) sarcosine]. *Acta Biochim. Polon.* **50**, 799–806.

Setsukinai, K., Urano, Y., Kakinuma, K., Maejima, H. J., and Nagano, T. (2003). Development of novel fluorescent probes that can reliably detect reactive oxygen species and distinguish special species. *J. Biol. Chem.* **278**, 3170–3175.

Silva Mendez, L. S., Allaker, R. P., Hardie, J. M., and Benjamin, N. (1999). Antimicrobial effect of acidified nitrite on carcinogenic bacteria. *Oral Microbiol. Immunol.* **14**, 391–392.

Takahama, U., Hirota, S., Nishioka, T., and Oniki, T. (2003a). Human salivary peroxidase-catalyzed oxidation of nitrite and nitration of salivary components 4-hydroxyphenylacetic acid and proteins. *Arch. Oral Biol.* **48**, 679–690.

Takahama, U., Hirota, S., and Oniki, T. (2005). Production of nitric oxide-derived reactive nitrogen species in human oral cavity and their scavenging by salivary redox components. *Free Radic. Res.* **39**, 737–745.

Takahama, U., Hirota, S., and Oniki, T. (2006a). Thiocyanate cannot inhibit the formation of reactive nitrogen species in the human oral cavity in the presence of high concentrations of nitrite: Detection of reactive nitrogen species with 4,5-diaminofluorescein. *Chem. Res. Toxicol.* **19,** 1066–1073.

Takahama, U., Hirota, S., and Oniki, T. (2006b). Quercetin-dependent scavenging of reactive nitrogen species derived from nitric oxide and nitrite in the human oral cavity: Interactions with salivary redox components. *Arch. Oral Biol.* **51,** 629–639.

Takahama, U., Hirota, S., Yamamoto, A., and Oniki, T. (2003b). Oxygen uptake during the mixing of saliva with ascorbic acid under acidic conditions: Possibility of its occurrence in the stomach. *FEBS Lett.* **550,** 64–68.

Takahama, U., Oniki, T., and Hirota, S. (2002a). Oxidation of quercetin by salivary components: Quercetin-dependent reduction of salivary nitrite under acidic conditions producing nitric oxide. *J. Agric. Food Chem.* **50,** 4317–4322.

Takahama, U., Oniki, T., and Murata, H. (2002b). The presence of 4-hydroxyphenylacetic acid in human saliva and the possibility of its nitration by salivary nitrite in the stomach. *FEBS Lett.* **518,** 116–118.

Takahama, U., Ryu, K., Oniki, T., and Hirota, S. (2007c). Dual-function of thiocyanate on nitrite-induced formation of reactive nitrogen oxide species in human oral cavity: inhibition under neutral and enhancement under acidic conditions. *Free Radic. Res.* **41,** 1289–1300.

Takahama, U., Tanaka, M., Oniki, T., and Hirota, S. (2007a). Reaction of thiocyanate in the mixture of nitrite and hydrogen peroxide under acidic conditions: Investigations of the reaction simulating the mixture of saliva and gastric juice. *Free Radic. Res.* **41,** 627–637.

Takahama, U., Tanaka, M., Oniki, T., Hirota, S., and Yamauchi, R. (2007b). Formation of the thiocyanate conjugate of chlorogenic acid in coffee under acidic condition in the presence of thiocyanate and nitrite: Possible occurrence in the stomach. *J. Agric. Food Chem.* **55,** 4169–4176.

Takahama, U., Yamamoto, A., Hirota, S., and Oniki, T. (2003c). Quercetin-dependent reduction of salivary nitrite to nitric oxide under acidic conditions and interaction between quercetin and ascorbic acid during the reduction. *J. Agric. Food Chem.* **51,** 6014–6020.

Tanaka, Y., Naruishi, N., Fukuya, H., Sataka, J., Saito, K., and Wakida, S. (2004). Simultaneous determination of nitrite, nitrate, thiocyanate and uric acid in human saliva by capillary zone electrophoresis and its application to the study of daily variations. *J. Chromatogr. A* **1051,** 193–197.

Tenovuo, J. (1989). Nonimmunoglobulin defense factors in human saliva. *In* "Human Saliva: Clinical Chemistry and Microbiology" (T. Tenovuo, ed.), Vol. II, pp. 55–91. CRC Press, Boca Raton, FL.

Tsuge, K., Kataoka, M., and Seto, Y. (2000). Cyanide and thiocyanate levels in blood and saliva of healthy adult volunteers. *J. Health Sci.* **46,** 343–350.

Venkataraman, S., Martin, S. M., Schafer, F. Q., and Buettner, G. R. (2000). Detailed methods for the quantification of nitric oxide in aqueous solutions using either oxygen monitor or EPR. *Free Radic. Biol. Med.* **29,** 580–585.

Yamamoto, M., Saeki, K., and Utsumi, K. (1991). Isolation of human salivary polymorphonuclear leukocytes and their stimulation-coupled responses. *Arch. Biochem. Biophys.* **289,** 76–82.

Yoshimura, T., Yokoyama, H., Fujii, S., Takayama, F., Oikawa, K., and Kamada, H. (1996). *In vivo* EPR detection and imaging of endogenous nitric oxide in lipopolysaccaride-treated mice. *Nat. Biotechnol.* **14,** 992–994.

Zhang, X., Kim, W.-S., Hatcher, N., Potgieter, K., Moroz, L. L., Gillette, R., and Sweedler, J. V. (2002). Interfering with nitric oxide measurements: 4,5-Diaminofluorescein reacts with dehydroascorbic acid and ascorbic acid. *J. Biol. Chem.* **277,** 48472–48478.

CHAPTER TWENTY-FIVE

IMAGING OF REACTIVE OXYGEN SPECIES AND NITRIC OXIDE *IN VIVO* IN PLANT TISSUES

Luisa M. Sandalio, María Rodríguez-Serrano, María C. Romero-Puertas, *and* Luis A. del Río

Contents

1. Introduction	398
2. Imaging Reactive Oxygen Species and Nitric Oxide *In Vivo* by Confocal Laser Microscopy	399
2.1. Imaging of peroxides with 2′,7′-dichlorofluorescein diacetate	399
2.2. Imaging of superoxide radicals with dihydroethidium	400
2.3. Imaging of nitric oxide with 4,5-diaminofluorescein-2 diacetate	401
3. Plant Tissue Preparation and Procedure	402
4. Conclusions	406
Acknowledgments	406
References	406

Abstract

During the last decades there has been a growing interest in the study of reactive oxygen species (ROS) and nitric oxide (NO) production in plant tissues and their role in signaling and cellular response to biotic and abiotic stress conditions. Despite growing molecular data on this subject, less attention has been paid to the topological distribution of ROS and NO production in plant tissues. Knowledge of the contribution of different cells to the accumulation of ROS and NO is important to get deeper insights into the cellular response of plants to adverse conditions. This chapter focuses on the imaging of ROS and NO accumulation *in vivo* in plant tissues by confocal laser microscopy using specific fluorescent probes.

Departamento de Bioquímica, Biología Celular y Molecular de Plantas, Estación Experimental del Zaidín, CSIC, Granada, Spain

1. Introduction

In the last decades, there has been a substantial increase in our understanding of the dual role of reactive oxygen species (ROS) in plant tissues, as signal molecules and as damaging species when they are overproduced. As signal molecules, ROS participate in physiological processes, such as growth and development, and in the response to biotic and abiotic stresses (Bailey-Serres and Mittler, 2006). ROS-mediated signaling is controlled by a balance between its production and scavenging by antioxidants. The imbalance in the accumulation and removal of ROS results in oxidative stress characterized by oxidative damages to proteins, lipids and DNA (Halliwell and Gutteridge, 2007). In plants, oxidative stress is produced by drought, low and high temperatures, high light intensities, ozone, heavy metals, salinity, mechanical wounding, and so on (Dat et al., 2000; Hernández et al., 2001; Mittler, 2002; Mullineaux et al., 2006; Sandalio et al., 2001), and also during pathogen infection (Torres et al., 2006). The main sources of ROS in plants are the plasma membrane-located NADPH oxidase (Sagi and Fluhr, 2006), the electron transport in chloroplasts, mitochondria, and peroxisomes, and the enzymes xanthine oxidase, glycolate oxidase, and acyl-CoA oxidase localized in peroxisomes (Asada, 2006; del Río et al., 2006; Rhoads et al., 2006).

In recent years, nitric oxide (NO) has been demonstrated to be an intra- and intercellular signaling molecule involved in the regulation of diverse biochemical and physiological processes in plants (Corpas et al., 2006a; Lamattina et al., 2003; Neill et al., 2003). NO interplays with ROS in plant and animal tissues in different ways associated with distinct physiological processes, and also in response to stress conditions. One of the most studied interplay in plants is NO–ROS cooperation during the hypersensitive reaction, which is characterized by a programmed cell death that contributes to plant resistance (Zaninotto et al., 2006). Less information is available on ROS–NO cooperation under abiotic stress conditions. An increase of both ROS and NO has been reported to take place under high temperature, osmotic stress, salinity, and mechanical stress (Garcês et al., 2001; Gould et al., 2003). However, a reduction of NO accumulation was observed after long periods of treatment with heavy metals (Kopyra and Gwózdz, 2003; Rodríguez-Serrano et al., 2006). Despite the growing knowledge on the role of NO and ROS in signaling and cellular response to biotic and abiotic stress, less attention has been paid to the topological distribution of ROS and NO production in plant tissues. Information on the contribution of different types of cells to the accumulation of ROS and NO would increase our knowledge on the cellular response in plants to adverse conditions. This chapter focuses on imaging of ROS and NO accumulation *in vivo* in tissues of Cd-stressed pea plants by confocal laser microscopy using specific fluorescent probes. Another advantage of this methodology is the possibility of studying the intracellular location of ROS and NO using simultaneously specific fluorescent probes for different organelles.

2. Imaging Reactive Oxygen Species and Nitric Oxide *In Vivo* by Confocal Laser Microscopy

Confocal laser scanning microscopy (CLSM) has been widely used to study fluorescent probes distribution in fixed and living plant tissues (Fricker and Meyer, 2001). These techniques allow the visualization of cellular and subcellular structures in optical sections and obtain three-dimensional views of fluorescence distribution in the tissue. CLSM allows studying many parameters *in vivo*, such as pH, calcium, metabolite levels, or enzyme kinetics, thanks to the development of specific molecular probes (Fricker and Meyer, 2001). This chapter concentrates on those probes used to visualize ROS and NO production and its utility to image these reactive species in plant tissues under stress conditions.

2.1. Imaging of peroxides with 2′,7′-dichlorofluorescein diacetate

2′,7′-Dichlorofluorescein diacetate (DCF-DA)(Fig. 25.1) has been widely used as marker of intracellular oxidants, especially in animal tissues (Bass *et al.*, 1983; Hsiao *et al.*, 2006; Tsuchiya *et al.*, 1994), but also in plant tissues

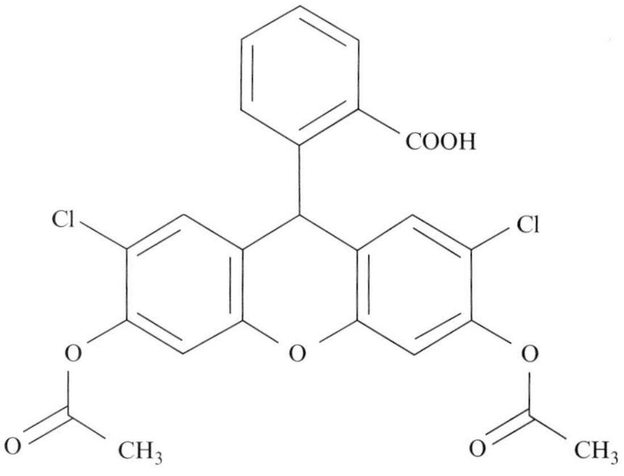

Figure 25.1 Chemical structure of 2′,7′-dichlorofluorescein diacetate (DCF-DA).

(Rodríguez-Serrano et al., 2006; Shapiro and Zhang, 2001). DCF-DA can permeate cells where it is hydrolyzed by intracellular esterases to liberate DCF, which is trapped inside the cell. Upon reaction with H_2O_2 or hydroperoxides, a fluorescent DCF-derived compound is formed that can be detected by monitoring the fluorescence at excitation and emission wavelengths of 480 and 530 nm, respectively.

Although DCF was initially thought to be a relatively specific indicator of H_2O_2 (Keston and Brandt, 1965), there are some doubts regarding its real validity. Thus, it is not clear which oxidative species is responsible for the oxidation of DCF. Several studies have demonstrated that DCF can also react with other peroxides, mainly hydroperoxides in the presence of peroxidases (Tarpey et al., 2004). DCF oxidation can be also dependent on intracellular iron and cytochrome c content (Tarpey et al., 2004). Because of these reasons the DCF assay can only be considered as a qualitative marker of oxidative stress rather than a precise indicator of the rate of H_2O_2 formation in biological systems (Tarpey et al., 2004). However, using specific ROS scavengers, or positive controls with known H_2O_2 concentrations, it is possible to ascertain the involvement of H_2O_2 in DCF fluorescence. In this way, Lu and Higgins (1998) demonstrated a close relationship between fluorescence because of DCF oxidation and H_2O_2 in whole tomato leaves treated with elicitors. The production of H_2O_2 in stigmas and pollen from different plant species has also been observed using DCF-DA and sodium pyruvate as the H_2O_2 scavenger (McInnis et al., 2006).

2.2. Imaging of superoxide radicals with dihydroethidium

Dihydroethidium (DHE) (Fig. 25.2) has been described as a specific probe for $O_2^{\bullet-}$ (Fink et al., 2004). DHE was initially thought to be oxidized to ethidium by $O_2^{\bullet-}$, which tends to intercalate into nuclear DNA, producing fluorescence. However, it has been demonstrated that superoxide oxidizes

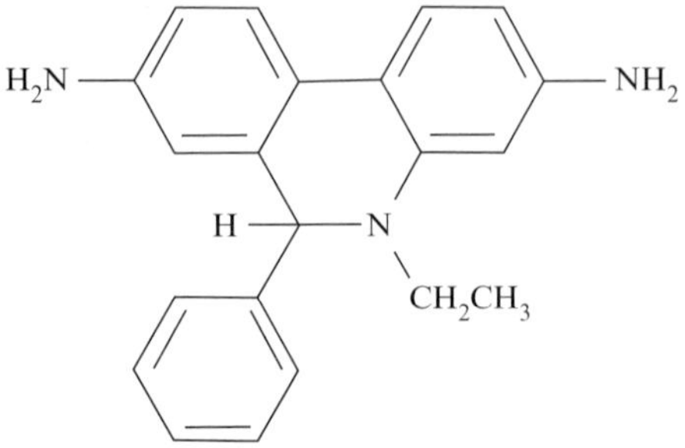

Figure 25.2 Chemical structure of dihydroethidium (DHE).

dihydroethidium to a specific fluorescent product, oxyethidium (oxy-E) (Zhao et al., 2003), and this process needs two $O_2^{\bullet-}$ anions with the production of ethidium radical (E•) as an intermediate (Fink et al., 2004).

Oxidation of DHE to Oxy-E was inhibited by superoxide dismutase but not catalase and did not occur upon the addition of H_2O_2, peroxynitrite, or hypochlorous acid (Fink et al., 2004). Oxyethidium is quite stable and is not reduced by glutathione, ascorbate, or NADPH (Fink et al., 2004). DHE fluorescence can be imaged in a confocal laser microscope by excitation at 488 nm and emission at 520 nm. This probe has been used as an $O_2^{\bullet-}$ marker in cell cultures and whole roots from pea plants exposed to Al (Yamamoto et al., 2002), lupine roots exposed to Cd (Kopyra and Gwodz, 2003), leaf sections from pea plants exposed to Cd (Rodríguez-Serrano et al., 2004), and olive leaf sections (Corpas et al., 2006b).

2.3. Imaging of nitric oxide with 4,5-diaminofluorescein-2 diacetate

The fluorescent probe 4,5-diaminofluorescein-2 diacetate (DAF-2 DA) (Fig. 25.3) has been widely used for detecting real-time bioimaging NO production in animal and plant tissues with a detection limit of 5 nM (Arita et al., 2006; Corpas et al., 2006a; Garcês et al., 2001; Kojima et al., 1998). DAF-2 DA is membrane permeable, and once in the cell is hydrolyzed by esterases to 4,5-diaminofluorescein-2 (DAF-2), which is trapped within the cell where it can react with NO producing the fluorescent DAF-2 triazole (Arita et al., 2006; Kojima et al., 1998). Fluorescence can be visualized by excitation and emission at 495 and 515 nm, respectively. A negative control used to demonstrate the involvement of NO in DAF-2-derived fluorescence is 2-phenyl-4,4,5,5-tetramethylimidazoline-1-oxyl (PTIO), which is an NO scavenger, as well as an inhibitor of the enzyme nitric oxide synthase (Durner et al., 1998).

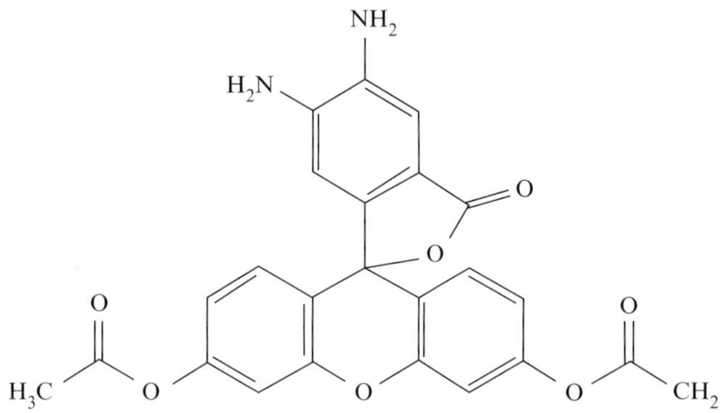

Figure 25.3 Chemical structure of 4,5-diaminofluorescein diacetate (DAF-2 DA).

Some authors have cast doubts on the specificity of DAF-2 fluorescence for NO. It has been shown that N_2O_3 and HNO can react with DAF-2, giving rise to the characteristic fluorescent form (Espey *et al.*, 2001, 2002). Planchet and Kaiser (2006) have indicated that DAF fluorescence might also reflect NO oxidation and/or production of other DAF-reactive compounds. Therefore, as recommended by Arita *et al.* (2006), when a study of an NO-producing activity in an *in vivo* system is initiated, in order to avoid misinterpretation of data, it is convenient to use additional methods having different detection principles, such as chemiluminescence (Arita *et al.*, 2006), electron paramagnetic resonance spectroscopy (Corpas *et al.*, 2004), and NO-specific electrodes (Yamasaki *et al.*, 2000).

3. Plant Tissue Preparation and Procedure

Fluorescent probes are from Calbiochem (DAF-2 DA), Sigma (DCF-DA), and Fluka (DHE) and are solubilized in dimethyl sulfoxide.

Pea root segments of approximately 20 mm are cut from the apex and incubated in darkness with the fluorescent probes in the following conditions:

Concentration	(μM)	Buffer	Time (min)	Temperature
DCF-DA	25	10 mM Tris-HCl (pH 7.4)	30	37°
DHE	10	10 mM Tris-HCl (pH 7.4)	30	37°
DAF-2 DA	10	10 mM Tris-HCl (pH 7.4)	60	25°

After incubation, the tissue is washed three times for 10 min each in the same buffer. After washing, roots are observed in a fluorescence microscope (Zeiss Axioplan) to visualize the whole root. As negative control, pieces are preincubated before adding the probes, in darkness for 1 h at 25°, with 1 mM tetramethyl piperidinooxy (TMP, $O_2^{\bullet-}$ scavenger), 1 mM ascorbate (peroxide scavenger), or 400 μM PTIO (NO scavenger) (Figs. 25.4A–25.6A). The involvement of NOS activity in NO production is demonstrated by incubating the tissue with 1 mM aminoguanidine, a NOS inhibitor, for 1 h at 25° (Fig. 25.6A). As a DAF-2 DA positive control, roots are incubated with an NO donor, sodium nitroprusiate (SNP, 10 μM) (Fig. 25.6B). Chemicals are prepared in the same buffer used for the fluorescent probes. After incubation, the pieces are washed three times in the same buffer, for 10 min each, and are incubated with the respective fluorescent probes as indicated earlier.

The effect of Cd treatment on ROS and NO accumulation in lateral and principal roots from pea plants observed by fluorescence microscopy or confocal laser microscopy is shown in Figs. 25.4–25.6. The growth of pea plants with 50 μM $CdCl_2$ for 15 days produced in the roots an increase in peroxides and $O_2^{\bullet-}$ accumulation visualized by DCF-DA and DHE fluorescence, respectively (Figs. 25.4A and 25.5A), while a reduction of NO accumulation was detected with DAF-2 DA (Fig. 25.6A).

To study the distribution of ROS- and NO-dependent fluorescence in different cells of the tissue, roots segments (20 mm) are embedded in a mixture of 30% (p/v) acrylamide-bisacrylamide solution prepared in 50 mM phosphate-K (pH 7.5), 0.73 M NaCl, and 13 mM KCl (PBS 5×) and 10% TEMED (v/v) for 3 to 4 h. After infiltration, specimens are transferred to polyethylene embedding molds (Electron Microscopy Science) containing the acrylamide solution (700 μl), and polymerization is started by adding 50 μl of 2% ammonium persulfate for several minutes. Serial sections about 60 to 100 μm thick are cut under distilled water on a Leica VT 1000 ST vibratome. Sections are then mounted in glycerol:PBS (1:1, v/v) in slides and examined with a confocal laser scanning microscope (Leica TCS SL) using standard filters for each probe. Fluorescence is stable for 24 h.

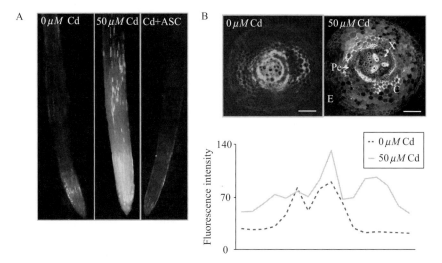

Figure 25.4 Imaging of peroxide accumulation in pea roots. (A) Peroxide-dependent DCF-DA fluorescence in lateral roots from control and Cd-treated pea plants. As a negative control, roots were incubated with 1 mM ascorbate (H_2O_2 scavenger). (B) Peroxide-dependent DCF-DA fluorescence in principal roots from control and Cd-treated pea plants. The graph shows fluorescence across the section quantified in arbitrary units using LCS Lite software from Leica. ASC, ascorbate; X, xylem; Pe, pericycle; E, epidermis; C, cortex. Taken from Rodríguez-Serrano et al. (2006). (See color insert.)

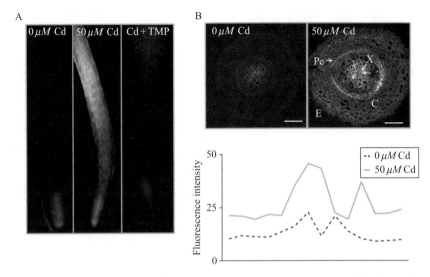

Figure 25.5 Imaging of $O_2^{\bullet-}$ accumulation in pea roots. (A) $O_2^{\bullet-}$-dependent DHE fluorescence in lateral roots from control and Cd-treated pea plants. As a negative control, roots were incubated with 1 mM TMP ($O_2^{\bullet-}$ scavenger). (B) $O_2^{\bullet-}$-dependent DHE fluorescence in principal roots from control and Cd-treated pea plants. The graph shows fluorescence across the section quantified in arbitrary units using LCS Lite software from Leica. X, xylem; Pe, pericycle; E, epidermis; C, cortex. Taken from Rodríguez-Serrano et al. (2006). (See color insert.)

Background staining is controlled with unstained tissue sections. Fluorescence in the different cell types in the tissue can be quantified using the program LCS Lite from Leica Systems (Figs. 25.4B–25.6B).

Analysis of cross sections of principal roots of pea plants treated with Cd by confocal laser microscopy show a Cd-dependent increase of $O_2^{\bullet-}$ accumulation in the cell walls from xylem vessels and pericycle and, to a lower extent, in the epidermis and cortex (Fig. 25.5B). DCF-DA fluorescence follows a similar pattern in response to Cd, although the intensity of fluorescence is higher than that obtained with DHE (Fig. 25.4B). Under physiological conditions, ROS production has been associated with lignification of the cell wall in xylem vessels (Ogawa et al., 1997; Ros Barceló, 1999, 2005), but ROS generation in cortex cells can be associated with oxidative damages imposed by the heavy metal (Rodríguez-Serrano et al., 2006).

The analysis of NO accumulation in principal roots with DAF-2 DA showed opposite results to those observed for ROS production, with a strong decrease of NO-dependent fluorescence by Cd treatment (Fig. 25.6B). As a positive control, Cd-treated plants are incubated with an NO donor (10 μM SNP), which produces an increase in fluorescence

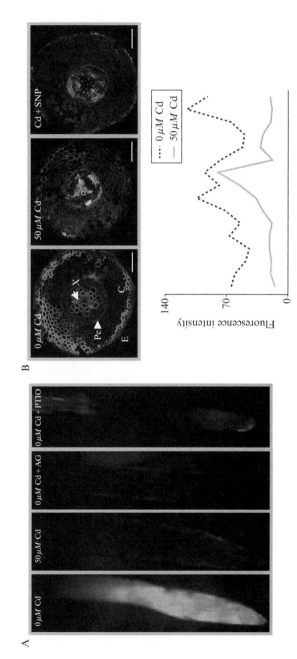

Figure 25.6 Imaging of NO accumulation in pea roots. (A) NO-dependent DAF-2 DA fluorescence in pea roots. As a negative control, roots were incubated with 1 mM aminoguanidine (mammalian NOS inhibitor) and with 400 μM PTIO (NO scavenger). (B) NO-dependent DAF-2 DA fluorescence in principal roots from control and Cd-treated pea plants. As a positive control, roots were incubated with 10 μM SNP (NO donor). The graph shows fluorescence across the section quantified in arbitrary units using LCS Lite software from Leica. AG, aminoguanidine. X, xylem; Pe, pericycle; E, epidermis; C, cortex. Taken from Rodríguez-Serrano et al. (2006). (See color insert.)

similar to that observed in control plants (Fig. 25.6B). In cross sections of principal roots from control plants, the production of NO takes place in the cortex, xylem, and, to a lower extent, in phloem. Under Cd treatment, the fluorescence attributed to cell walls of cortex cells is reduced considerably, but the fluorescence of xylem does not change. The overlay of both peroxides/$O_2^{\bullet -}$ and NO accumulation images under Cd stress allows one to observe an increase of ROS production in those places where NO is reduced, mainly in the cortex.

4. CONCLUSIONS

This chapter presented a procedure to assay ROS and NO accumulation *in vivo* in roots from Cd-stressed pea plants using specific fluorescent probes. This method allows one to study the accumulation of reactive oxygen species and NO and the cross-talk between both species in different cell types at the same time. This procedure is useful in carrying out a dissection of the signal transductions involved in the plant response to stress conditions using specific inhibitors of calcium channels, kinases, or NADPH oxidases (Rodríguez-Serrano *et al.*, 2006). Another advantage of this method is the possibility of studying the intracellular location of ROS and NO using simultaneously specific fluorescent probes for organelles, such as mitochondria or endoplasmic reticulum (MitoTracker and ERTracker, respectively), or mutants expressing green fluorescent protein in different organelles.

ACKNOWLEDGMENTS

M. Rodríguez-Serrano acknowledge an a fellowship from Ministry of Education and Science. The laser confocal microscopy analyses were carried out at the Technical Services of the University of Jaén. This work was supported by the Ministry of Education and Science (Grant BIO2005–03305) and by the *Junta de Andalucía* (Research Group CVI 0192), Spain.

REFERENCES

Arita, N. O., Cohen, M. F., Tokuda, G., and Yamasaki, H. (2006). Fluorometric detection of nitric oxide with diaminofluoresceins (DAFs): Applications and limitations for plant NO research. *In* "Nitric Oxide in Plant Growth, Development and Stress Physiology" (L. Lamattina and J. C. Polacco, eds.), pp. 269–280. Springer-Verlag, Berlin.

Asada, K. (2006). Production and scavenging of reactive oxygen species in chloroplasts and their functions. *Plant Physiol.* **141,** 391–396.

Bailey-Serres, J., and Mittler, R. (2006). Special issue on reactive oxygen species. *Plant Physiol.* **141,** 311–521.

Bass, D. A., Parce, J. W., Dechatelet, L. R., Szejda, P., Seeds, M. C., and Thomas, M. (1983). Flow cytometric studies of oxidative product formation by neutrophils: A graded response to membrane stimulation. *J.Immunol.* **130**, 1910–1917.

Corpas, F. J., Barroso, J. B., Carreras, A., Quirós, M., León, A. M., Romero-Puertas, M. C., Esteban, F. J., Valderrama, R., Palma, J. M., Sandalio, L. M., Gómez, M., and del Río, L. A. (2004). Cellular and subcellular localization of endogenous nitric oxide in young and senescent pea plants. *Plant Physiol.* **136**, 2722–2733.

Corpas, F. J., Barroso, J. B., Carreras, A., Valderrama, R., Palma, J. M., León, A. M., Sandalio, L. M., and del Río, L. A. (2006a). Constitutive arginine-dependent nitric oxide synthase activity in different organs of pea seedlings during plant development. *Planta* **224**, 246–254.

Corpas, F. J., Fernández-Ocaña, A., Carreras, A., Valderrama, R., Luque, F., Esteban, F. J., Rodríguez-Serrano, M., Chaki, M., Pedrajas, J. R., Sandalio, L. M., del Río, L. A., and Barroso, J. B. (2006b). The expression of different superoxide dismutase forms is cell-type dependent in olive (*Olea europaea* L.) leaves. *Plant Cell Physiol.* **47**, 984–994.

Dat, J. F., Vandenabeele, S., Vranová, E., Van Montagu, M., Inzé, D., and Van Breusegem, F. (2000). Dual action of the active oxygen species during plant stress responses. *Cell. Mol. Life Sci.* **57**, 779–795.

del Río, L. A., Sandalio, L. M., Corpas, F. J., Palma, J. M., and Barroso, J. B. (2006). Reactive oxygen species and reactive nitrogen species in peroxisomes. Production, scavenging, and role in cell signaling. *Plant Physiol.* **141**, 330–335.

Durner, J., Wendehenne, D., and Klessig, D. F. (1998). Defense gene induction in tobacco by nitric oxide, cyclic GMP, and cyclic ADP-ribose. *Proc. Natl. Acad. Sci. USA* **95**, 10328–10333.

Espey, M. G., Miranda, K. M., Thomas, D. D., and Wink, D. A. (2001). Distinction between nitrosating mechanism within human cells and aquaeos solutions. *J. Biol. Chem.* **276**, 30085–30091.

Espey, M. G., Miranda, K. M., Thomas, D. D., and Wink, D. A. (2002). Ingress and reactive chemistry of nitrosyl-derived species within humans cells. *Free Radic. Biol. Med.* **33**, 827–834.

Fink, B., Laude, K., McCann, L., Doughan, A., Harrison, D. G., and Dikalov, S. (2004). Detection of intracellular superoxide formation in endothelial cells and intact tissues using dihydroethidium and an HPLC-based assay. *Am. J. Physiol. Cell Physiol.* **287**, 895–902.

Fricker, M. D., and Meyer, A. J. (2001). Confocal imaging of metabolism *in vivo*: Pitfalls and possibilities. *J. Exp. Bot.* **52**, 631–640.

Garcês, H., Durzan, D., and Pedrosa, M. C. (2001). Mechanical stress elicits nitric oxide formation and DNA fragmentation in *Arabidopsis thaliana*. *Ann. Bot.* **87**, 567–574.

Gould, K. S., Lamote, O., Klinguer, A., Pugin, A., and Wendehenne, D. (2003). Nitric oxide production in tobacco leaf cells: A generalized stress response? *Plant Cell Environ.* **26**, 1851–1862.

Halliwell, B., and Gutteridge, J. M. C. (2007). *In* "Free Radicals in Biology and Medicine." Fourth edition. Oxford University Press, London.

Hernández, J. A., Ferrer, M. A., Jiménez, A., Ros Barceló, A., and Sevilla, F. (2001). Antioxidant systems and $O_2^{\bullet-}/H_2O_2$ production in the apoplast of pea leaves: Its relation with salt-induced necrotic lesions in minor veins. *Plant Physiol.* **127**, 817–831.

Hsiao, C. H., Li, W., Lou, T. F., Baliga, B. S., and Pace, B. S. (2006). Fetal haemoglobin induction by histone deacetylase inhibitors involves generation of reactive oxygen species. *Exp. Hematol.* **34**, 264–273.

Keston, A. S., and Brandt, R. (1965). The fluorometric analysis of ultramicroquantities of hydrogen peroxide. *Anal. Biochem.* **11**, 1–5.

Kojima, H., Nakatsubo, N., Kikuchi, K., Kawaraha, S., Kirino, Y., Nagoshi, H., Hirata, Y., and Nagano, T. (1998). Detection and imaging of nitric oxide with novel fluorescent indicators: Diaminofluoresceins. *Anal. Chem.* **70,** 2446–2453.

Kopyra, M., and Gwózdz, E. A. (2003). Nitric oxide stimulates seed germination and counteracts the inhibitory effect of heavy metals and salinity on root growth of *Lupinus luteus*. *Plant Physiol. Biochem.* **41,** 1011–1017.

Lamattina, L., García-Mata, C., Graziano, M., and Pagnussat, G. (2003). Nitric oxide: The versatility of an extensive signal molecule. *Annu. Rev. Plant Biol.* **54,** 109–136.

Lu, H., and Higgins, V. J. (1998). Measurement of active oxygen species generated *in planta* in response to elicitor AVR9 of *Cladosporium fulvum*. *Physiol. Mol. Plant Pathol.* **52,** 35–51.

McInnis, S., Desikan, R., Hancock, J. T., and Hiscock, S. J. (2006). Production of reactive oxygen species and reactive nitrogen species by angiosperm stigmas and pollen: Potential signalling crosstalk? *New Phytol.* **172,** 221–228.

Mittler, R. (2002). Oxidative stress, antioxidants and stress tolerance. *Trends Plant Sci.* **7,** 405–410.

Mullineaux, P. M., Karpinski, S., and Baker, N. R. (2006). Spatial dependence for hydrogen peroxide-directed signalling in light-stressed plants. *Plant Physiol.* **141,** 346–350.

Neill, S. J., Desikan, R., and Hancock, J. T. (2003). Nitric oxide signalling in plants. *New Phytol.* **159,** 11–35.

Ogawa, K., Kanematsu, S., and Asada, K. (1997). Generation of superoxide anion and localization of CuZn-superoxide dismutase in the vascular tissue of spinach hypocotyls: Their association with lignification. *Plant Cell Physiol.* **38,** 1118–1126.

Planchet, E., and Kaiser, W. N. (2006). Nitric oxide (NO) detection by DAF fluorescence and chemiluminiscence: A comparison using abiotic and biotic NO sources. *J. Exp. Bot.* **57,** 3043–3055.

Rhoads, D. M., Umbach, A. L., Subbaiah, C. C., and Siedow, J. N. (2006). Mitochondrial reactive oxygen species. Contribution to oxidative stress and interorganellar signaling. *Plant Physiol.* **141,** 357–366.

Rodríguez-Serrano, M., Romero-Puertas, M. C., Gómez, M., Barroso, J. B., del Río, L. A., and Sandalio, L. M. (2004). Imaging of ROS and nitric oxide production in pea plants under metal stress. *Free Radic. Biol. Med.* **36,** S139.

Rodríguez-Serrano, M., Romero-Puertas, M. C., Zabalza, A., Corpas, F. J., Gómez, M., del Río, L. A., and Sandalio, L. M. (2006). Cadmium effect on oxidative metabolism of pea (*Pisum sativum* L.) roots. Imaging of reactive oxygen species and nitric oxide accumulation *in vivo*. *Plant Cell Environ.* **29,** 1532–1544.

Ros-Barceló, A. (1999). Some properties of the $H_2O_2/O_2^{\bullet-}$ generating system from the lignifying xylem of *Zinnia elegans*. *Free Radic. Res.* **31,** S147–S154.

Ros Barceló, A. (2005). Xylem parenchyma cells deliver the H_2O_2 necessary for lignification in differentiating xylem vessels. *Planta* **220,** 747–756.

Sagi, M., and Fluhr, R. (2006). Production of reactive oxygen species by plant NADPH oxidases. *Plant Physiol.* **141,** 336–340.

Sandalio, L. M., Dalurzo, H. C., Gomez, M., Romero-Puertas, M. C., and del Río, L. A. (2001). Cadmium-induced changes in the growth and oxidative metabolism of pea plants. *J. Exp. Bot.* **52,** 2115–2126.

Shapiro, A. D., and Zhang, Ch. (2001). The role of NDR1 in avirulence gene-directed signaling and control of programmed cell death in *Arabidopsis*. *Plant Physiol.* **127,** 1089–1101.

Tarpey, M. M., Wink, D. A., and Grisham, M. B. (2004). Methods for detection of reactive metabolites of oxygen and nitrogen: *In vitro* and *in vivo* considerations. *Am. J. Physiol. Regul. Integr. Comp. Physiol.* **286,** R431–R444.

Torres, M. A., Jones, J. D. G., and Dangl, J. L. (2006). Reactive oxygen species signalling in response to pathogens. *Plant Physiol.* **141,** 373–378.

Tsuchiya, M., Suematsu, M., and Suzuki, H. (1994). *In vivo* visualization of oxygen radical-dependent photoemission. *Methods Enzymol.* **233**(C), 128–140.

Yamamoto, Y., Kobayashi, Y., Devi, S. R., Rikiishi, S., and Matsumoto, H. (2002). Aluminum toxicity is associated with mitochondrial dysfunction and the production of reactive oxygen species in plant cells. *Plant Physiol.* **128,** 63–72.

Yamasaki, H., and Sakihama, Y. (2000). Simultaneous production of nitric oxide and peroxynitrite by plant nitrate reductase: *In vitro* evidence for the NR-dependent formation. *FEBS Lett.* **468,** 89–92.

Zaninotto, F., La Camera, S., Polverari, A., and Delledonne, M. (2006). Cross-talk between reactive nitrogen and oxygen species during the hypersensitive disease resistance response. *Plant Physiol.* **141,** 379–383.

Zhao, H., Kalivendi, S., Zhang, H., Joseph, J., Nithipatikon, K., Vásquez-Vivar, J., and Kalyanaraman, B. (2003). Superoxide reacts with hydroethidine but forms a fluorescent product that is distinctly different from ethidium: Potential implications in intracellualr fluorescence detection of superoxide. *Free Radic. Biol. Med.* **34,** 1359–1368.

CHAPTER TWENTY-SIX

EXAMINING NITROXYL IN BIOLOGICAL SYSTEMS

Jon M. Fukuto,[*,†] Matthew I. Jackson,[†] Nina Kaludercic,[‡] and Nazareno Paolocci[‡,§]

Contents

1. Introduction	412
2. Nitroxyl Donors	412
2.1. General considerations	413
2.2. Angeli's salt	414
2.3. Sulfohydroxamic acids and derivatives (Piloty's acid)	416
2.4. Cyanamide	417
2.5. Other HNO donors amenable to biological studies	417
3. Biological HNO Chemistry	419
3.1. General considerations	419
3.2. Kinetics	420
3.3. Sites of HNO reactivity	421
4. Use of HNO Donors in Biological Studies	423
5. Nitroxyl Pharmacological Effects: *In Vivo* and *In Vitro* Studies	424
5.1. Nitroxyl in the cardiovascular system	424
5.2. Nitroxyl and myocardial ischemia/reperfusion injury	425
5.3. Nitroxyl donors *in Vivo* can trigger the release of peptides from the nonadrenergic noncholinergic system	425
5.4. Nitroxyl and the nervous system: Physiological and toxicological effects	426
6. Summary	426
References	427

[*] Department of Pharmacology, UCLA School of Medicine, Center for the Health Sciences, Los Angeles, California
[†] Interdepartmental Program in Molecular Toxicology, UCLA School of Public Health, Los Angeles, California
[‡] Division of Cardiology, Department of Medicine, Johns Hopkins Medical Institutions, Baltimore, Maryland
[§] Department of Clinical and Experimental Medicine, General Pathology and Immunology Section, University of Perugia, Perugia, Italy

Methods in Enzymology, Volume 440
ISSN 0076-6879, DOI: 10.1016/S0076-6879(07)00826-9

© 2008 Elsevier Inc.
All rights reserved.

Abstract

Nitroxyl (HNO) has received significant recent attention due to its remarkable and novel biological activity and pharmacological potential. Unlike most other commonly studied nitrogen oxides, examination of HNO biology is not straightforward because of its unique chemical properties. Thus, the study of HNO in biological system requires researchers to be aware of this chemistry in order to properly design and interpret experiments. This chapter focuses on the experimental issues associated with the study of HNO and attempts to provide guidelines for working with this novel and fascinating nitrogen oxide.

1. Introduction

Among all of the nitrogen oxides relevant to mammalian biology, physiology, pharmacology, and toxicology, nitroxyl (HNO) is one of the least examined and most difficult to work with. Recent discoveries of novel and provocative biological activity associated with HNO have prompted many laboratories to begin to investigate the physiological and biochemical mechanisms of its actions (Miranda, 2005; Fukuto *et al.*, 2005; Paolocci *et al.*, 2007). However, unlike most other nitrogen oxides, HNO cannot be examined directly in biological systems since it will spontaneously dimerize to give hyponitrous acid, which then decomposes to nitrous oxide (N_2O) and water (H_2O) (reaction 26.1).

$$HNO + HNO \rightarrow HON = NOH \rightarrow N_2O + H_2O \qquad (26.1)$$

Because of its fleeting nature, HNO cannot be stored and is typically studied using HNO donor compounds (or possibly *in situ* HNO generating systems). Thus, one objective of this chapter is to discuss the use of HNO donor molecules and the issues associated with using these compounds in biological systems. Also, a brief description of the currently known biological targets, chemistry, and pharmacology of HNO is given. It is hoped that these discussions will aid researchers in studying HNO biology and begin to provide a biochemical/physiological basis for explaining its activity. However, it should be noted that our current understanding of the biological chemistry of HNO is in its infancy and there undoubtedly will be more biological targets and chemistries discovered as research in this area continues.

2. Nitroxyl Donors

A comprehensive discussion of HNO donor molecules is beyond the scope of this chapter. Herein are discussed some of the important experimental issues associated with the use of commonly used or established HNO donors in biological studies. For more comprehensive treatments of this topic, readers are referred to other excellent articles (King and Nagasawa 1999; Miranda et al., 2005b).

2.1. General considerations

As mentioned earlier, the only practical way of examining HNO biology is to use donor molecules. As with all studies that use donors rather than the actual species under investigation there is the worry that the biological activity observed is because of the donor molecule itself or other by-products generated from its decomposition. Also, in cases where biological activation of the donor is required, the possibility arises that the observed biological effects are a result of processes required for activation of the donor. For example, if redox reactions of the donor are required to release the presumed active molecule, it is possible that the cellular redox changes from the activation process are actually responsible for the observed biology. Of course, these are problems associated with the use of all donor molecules and not exclusive to the study of HNO. However, because HNO can only be examined using donors, this problem is particularly acute with studies of HNO biology. In order to mitigate some of these problems, it is always expected that control experiments using fully decomposed donors are carried out. These controls will account for the possible biological actions of decomposition by-products as well as possible impurities in the donor preparation. However, if the donors do not spontaneously release the species in question and require cellular processes (i.e., enzyme activity, reduction, oxidation), controlling for variables of this type are not always straightforward. Moreover, decomposed donor controls will not account for any biological activity associated with the donor molecule itself. In order to reconcile this issue, it is advantageous to utilize several, structurally distinct donor molecules. Validation of biological effects using at least two structurally unrelated donors allows for a more rigorous and convincing argument that the observed biology is associated with the released molecule (HNO in this case) and not the donor itself. Clearly, in order to be assured that any observed biology is associated with HNO, and not because of the artifacts noted earlier, all of these issues must be considered and reconciled.

An important aspect of studying HNO is the fact that high concentrations are difficult to obtain and/or maintain as a consequence of second-order self-dimerization (reaction 26.1). Thus, there will always be a nonlinear correlation between donor concentrations and HNO concentrations (see later). That is, considering dimerization as a primary mechanism of HNO loss, at low concentrations, HNO can have a significant lifetime, as dimerization is minimal. However, at high concentrations the second-order dimerization reaction becomes more prevalent and the concentration of HNO decreases rapidly. Therefore, dose–response studies using HNO donors should always be expected to plateau at higher concentrations of donor. Of course, most all dose–response studies are expected to plateau because of biological saturation. However, the increased rate of chemical decomposition of HNO at higher concentrations (due to dimerization) also needs to be considered in interpreting biological dose–response data.

2.2. Angeli's salt

Although there are several classes of HNO donor molecules, to date the most prevalent donor used for biological (and chemical) studies is sodium trioxodinitrate (more commonly referred to as Angeli's salt, $Na_2N_2O_3$; for an in-depth discussion on Angeli's salt, see Miranda, 2005; Miranda et al., 2005b). Indeed, most of what is known about the biological activity of HNO has been garnered from studies using Angeli's salt. This HNO donor was synthesized over 100 years ago (Angeli, 1896) and its popularity today as an HNO donor stems from its established ability to spontaneously generate HNO in aqueous solutions within the physiological pH range (4–8) (reaction 26.2) (Bonner and Ravid, 1975). Adding to the utility of this donor, Angeli's salt is available commercially [although it is synthesized easily from inexpensive precursors (Smith and Hein, 1960)], stable as the solid, water soluble, and stable in solution under basic conditions (allowing the preparation of stock solutions).

$$N_2O_3^{2-} + H^+ \rightarrow\rightarrow HNO + NO_2^- \qquad (26.2)$$

Angeli's salt releases HNO with a first-order rate constant of $6.8 \times 10^{-4}\,s^{-1}$ at 25° and 4 to $5 \times 10^{-3}\,s^{-1}$ at 37° (Hughes and Wimbledon, 1976). The decomposition of Angeli's salt also generates one equivalent of NO_2^-, (reaction 26.2). Due to the possible biological activity of NO_2^-, (e.g., see Gladwin et al., 2006), it is essential in all studies to determine the activity of decomposed Angeli's salt to control for the possible effects of this coproduct (or possible impurities). Along with NO_2^-, a decomposing solution of Angeli's salt will also release a small amount of NO. The release of trace amounts of NO is virtually unavoidable since it is likely generated from a competing (albeit minor) pathway at physiological pH (Dutton et al., 2004). At acidic pH (below 4), this competing pathway is very prevalent and Angeli's salt becomes primarily an NO donor. Finally, since the release of HNO from Angeli's salt requires a protonation step, this salt is fairly stable under basic conditions. Indeed, basic stock solutions of Angeli's salt are stable over extended periods of time and are often used as a source of HNO for biological studies (e.g., Lopez et al., 2007). However, care needs to be taken when using these stock solutions since they are very basic and can alter the local pH of weakly buffered or poorly mixed biological samples. Thus, the pH of treated cells, tissues, or biological solutions must be monitored carefully to make sure that the observed effects are not the result of a dramatic change in pH (possibly a temporary change locally at the site of administration).

All biological studies using Angeli's salt report the observed effects as a function of donor (not HNO) concentration. From these reports it is difficult to know what concentrations of HNO elicit the activities. Others have predicted HNO concentrations generated from donor molecules using a variety of techniques (e.g., Cheong et al., 2005; Liochev and Fridovich 2003; Marti et al., 2005). Herein, we have utilized commercial software (DynaFit, BioKin Ldt, Pullman, WA) to predict the concentrations of HNO generated from decomposing solutions of Angeli's salt.

The half-lives of Angeli's salt (not HNO) are approximately 17 and 2.5 min at 23° and 37°, respectively. A consequence of the rapid decomposition of Angeli's salt at temperatures commonly utilized in cell and tissue experiments (37°) is that the production of HNO is effectively bolus, without any appreciable pseudo-steady-state concentrations of HNO generated over experimentally relevant timescales. In contrast, at room temperature (23°), a pseudo-steady state of HNO can be generated over at least 5 min for the lower concentrations of Angeli's salt (Fig. 26.1).

When the pseudo-steady-state concentrations of HNO generated at 23° are plotted versus the concentrations of Angeli's salt, a hyperbolic curve results (Fig. 26.2), indicating that there is effectively a limit to the concentrations of HNO generated in solution from donors. This is important to take into account when assessing the results generated from seemingly high concentrations of donor. For example, at 23° 1 mM Angeli's salt is expected to generate a pseudo-steady-state concentration of HNO of 194 ($\pm$ 2) nM, but 10 mM Angeli's salt generates a pseudo-steady-state concentration of HNO of only 615 ($\pm$ 6) nM. Thus, there is not a linear correlation between

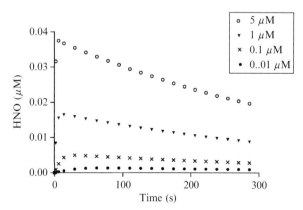

Figure 26.1 Concentration of HNO versus time for 5, 1, 0.1, and 0.01 μM Angeli's salt at 23°.

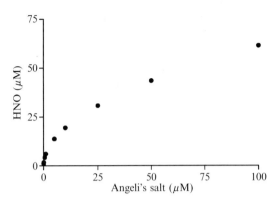

Figure 26.2 Pseudo-steady-state levels of HNO as a function of Angeli's salt concentration (23°).

pseudo-steady-state HNO concentrations and Angeli's salt concentrations, and it becomes evident that generation of high levels of HNO (mM) is difficult.

In aerobic solutions containing extremely low concentrations of Angeli's salt (<2 μM), dimerization is negligible and reaction with dioxygen forms the primary route of HNO decomposition (Miranda, 2005). At low micromolar levels of Angeli's salt, the concentrations of HNO generated in anaerobic solutions are significantly greater than those generated in the presence of oxygen, but this difference becomes negligible at higher concentrations of Angeli's salt. For example, we have calculated that 1 μM Angeli's salt decomposed at 23° under anaerobic conditions generates a pseudo-steady-state concentration of HNO of 5.9 (± 0.4) nM, whereas that generated under aerobic conditions is 1.2 (± 0.1) nM, a difference of nearly fourfold. However, at 1 mM Angeli's salt, the steady-state concentrations of HNO generated over 5 min in the absence and presence of oxygen are 187 (± 6) nM and 174 (± 5) nM, respectively.

The pseudo-steady-state values for HNO concentration given earlier are calculated assuming the previously reported rate constants for HNO generation from Angeli's salt (see earlier discussion) and that HNO dimerization and reaction with O_2 are the two mechanisms of HNO degradation. In biological systems, however, there will clearly be other equally or more favorable reactions, such as reactions with thiols and thiol proteins and metalloproteins (see later), occurring as well.

2.3. Sulfohydroxamic acids and derivatives (Piloty's acid)

Piloty's acid [benzene sulfohydroxamic acid, $(C_6H_6)S(O_2)NHOH$] is, next to Angeli's salt, the second most characterized donor of HNO. Release of HNO from this, and related compounds, requires deprotonation and therefore is prevalent only under strongly basic conditions (reaction 26.3) (Bonner and Ko, 1992). The coproduct is the corresponding sulfinic acid (or sulfinate).

$$R - S(O)_2NHOH + HO^- \rightarrow R - S(O)O^- + HNO + H_2O \quad (26.3)$$

Because sulfohydroxamic acids only release HNO under highly basic conditions, they are not generally amenable for use in biological systems. Moreover, sulfohydroxamic acids are subject to facile one-electron oxidation, which then causes them to be NO, and not HNO, donors (Zamora et al., 1995). It has been reported that some alkyl sulfohydroxamic acids (e.g., methanesulfohydroxamic acid) can release HNO under physiological conditions (i.e., pH 7) (Shoeman and Nagasawa, 1998). However, this is not firmly established and the confounding oxidation chemistry leading to NO release severely limits their utility in biological studies. Finally, they tend to be much less water soluble than, for example, Angeli's salt, making them problematic for use in aqueous environments. Thus, most biological studies to date have relied on Angeli's salt, rather than sulfohydroxamic acids, as the source of HNO.

2.4. Cyanamide

Cyanamide (NH_2CN) is a compound used for the treatment of chronic alcoholism (e.g., Demaster et al., 1998). Indeed, it is the only currently approved pharmaceutical agent whose efficacy is because of its ability to serve as an HNO donor. Its utility in this regard is a result of oxidative bioactivation and release of HNO (reaction 26.4) (e.g., DeMaster et al., 1998; Lee et al., 1992).

$$NH_2CN \rightarrow (HONHCN) \rightarrow HNO + CN^- + H^+ \quad (26.4)$$

HNO release from cyanamide results in oxidative modification of the active site cysteine thiol of aldehyde dehydrogenase (an important enzyme in the ethanol metabolism pathway), leading to enzyme inhibition and disruption of ethanol metabolism. Nagasawa and co-workers showed that oxidative bioactivation of cyanamide occurs via the actions of catalase/H_2O_2 (e.g., DeMaster et al., 1984). Thus, in biological systems containing catalase and H_2O_2, in situ generation of HNO from cyanamide is possible (although quantitation of HNO generation is difficult).

2.5. Other HNO donors amenable to biological studies

As mentioned previously, it is important to have multiple, structurally distinct donors of HNO for biological studies. To date, Angeli's salt has been the near exclusive donor used in biological studies. Some of the earliest work on the biological activity of HNO (or NO⁻) claimed to have used NaNO, made from the direct reduction of NO by elemental sodium (Na) (Feelisch et al., 1994; Fukuto et al., 1992). However, these studies are flawed, as NaNO cannot be made in this manner. Any NO⁻ generated in the reaction will further react with NO to give other catenation products (i.e., $N_2O_2^-$, $N_3O_3^-$) that would decompose to a variety of species, including NO_2^-, in a biological environment. Although a trace amount of NaNO may be present in these preparations, it would be difficult to dissect out the biological activity associated with this minor component from all the other nitrogen oxide species present. Moreover, as discussed later, administration of NO⁻ to a biological system is not equivalent to adding HNO, as the protonation of NO⁻ is extremely slow.

The King laboratory has reported the facile synthesis and characterization of acyloxy nitroso compounds as HNO donors (Sha et al., 2006). These compounds require a hydrolytic step prior to HNO release but, importantly, this can occur spontaneously without the need for enzymes (reaction 26.5).

$$\underset{R \quad R' \quad O}{\overset{O}{\underset{\|}{N}}\diagdown O \diagdown \overset{O}{\underset{\|}{C}} \diagdown R''} \xrightarrow{H_2O} HNO + R''COOH + RC(O)R' \qquad (26.5)$$

Significantly, several acyloxy nitroso compounds have been shown to elicit a dose-dependent vasorelaxation similarly to Angeli's salt, verifying their utility in biological systems.

Diazeniumdiolates (sometimes referred to as NONOates) are typically synthesized from the reaction of secondary amines with NO and are known mostly as NO donors. However, diazeniumdiolates made from primary amines can release HNO spontaneously (reaction 26.6).

$$\underset{H_3C}{\overset{H_3C}{\diagdown}}\underset{H}{\overset{}{N}}\diagdown \underset{}{\overset{^-O \quad O^-}{\diagdown N^+=N \diagup}} \xrightarrow{H^+} \underset{H_3C}{\overset{H_3C}{\diagdown}}\underset{H}{\overset{}{N}}\diagdown \underset{}{\overset{\overset{O}{\|}}{N}} + HNO \qquad (26.6)$$

Indeed, the diazeniumdiolate made from isopropyl amine (often called IPA-NO) has been shown to be an HNO donor at pH 5 to 13 (Miranda et al., 2005a; Dutton et al., 2006) with a biological activity profile similar to Angeli's salt (Miranda et al., 2005a). Like Angeli's salt, IPA-NO releases NO at lower pH.

Other reports of HNO donating compounds have also appeared in the literature (e.g., Atkinson et al., 1996; Pennington et al., 2005; Zeng et al., 2004), some of which can also be amenable to use in biological systems.

Some of those specifically mentioned earlier have been used in biological experiments and are clearly possible replacements or complements to Angeli's salt.

3. Biological HNO Chemistry

3.1. General considerations

An important recent finding regarding fundamental HNO chemistry was the revision of its pK_a. A previous study reported that HNO was acidic with a pK_a of 4.7 (Gratzel et al., 1970), indicating that the anion, NO^-, would be the predominant species present at physiological pH. However, a series of studies have found that HNO has a considerably higher pK_a of approximately 11.4 (Bartberger et al., 2002; Shafirovich and Lymar, 2002). Therefore, the protonated and neutral form, HNO (and *not* NO^-), is the near exclusive species present at physiological pH. This is an important point, as the chemistries of HNO and NO^- are quite distinct and it is highly likely all nitroxyl-mediated biological effects reported thus far in the literature involve HNO and not NO^-.

Unlike typical acid–base reactions where protonation and deprotonation occur very rapidly, these reactions of nitroxyl are relatively slow (Shafirovich and Lymar, 2003). That is, protonation of NO^- and deprotonation of HNO are slow compared to other acids/bases, as the electronic ground states of the two equilibrium species, HNO and NO^-, are different. HNO is a ground state singlet (all the electrons are spin paired), whereas NO^- is a ground state triplet (has two unpaired electrons of equal spin) (reaction 26.7).

$$^1HNO \rightleftharpoons {}^3NO^- + H^+ \qquad (26.7)$$

The required spin state changes necessary for interconversion of HNO and NO^- slows down the reaction to the point that when studying one of the species in a biological milieu there is no need to consider possible reactions associated with the other. This is not the case with other conventional acids and bases where rapid protonation/deprotonation allows an extremely rapid establishment of the equilibrium mixture of both equilibrium partners. Thus, if a fast reaction consumes one of the equilibrium species, mass action will drive the equilibrium toward that reaction, allowing the chemistry of an even an unfavorable minor equilibrium partner to predominate. However, this is not the case for HNO. Because of the slow deprotonation of HNO, it is likely in biological systems that other reactions will occur before deprotonation can occur (i.e., there is a slow establishment of the

HNO/NO$^-$,H$^+$ equilibrium), thus precluding (or certainly greatly limiting) NO$^-$ chemistry. Of course, the reverse of this is true as well. If circumstances exist where NO$^-$ is generated in or added to a biological system, it is likely that NO$^-$ chemistry will predominate (even though HNO will be the favored species at equilibrium) due to the slow protonation reaction.

Many biological studies involve the administration of an HNO donor to cells in culture, tissue baths, or infused into the circulation of whole animals. With the use of ionic donors (such as the most common of all donors, Angeli's salt, see earlier discussion), interaction of HNO with intracellular targets requires HNO to travel across cellular membranes. The ability of HNO and/or HNO-derived species to partition into cellular membranes allowing diffusion into cells has been demonstrated (Espey et al., 2002). Thus, HNO generated outside of cells is highly diffusible and has little problem accessing intracellular targets. Indeed, numerous examples of intracellular target modification by HNO administered to whole cell preparations (both aerobically and anaerobically) have been reported (e.g., Cook et al., 2003; Lopez et al., 2005 Wink et al., 1998).

3.2. Kinetics

The biological targets for HNO are, in part, determined by the rate constant for reaction with HNO and their concentration. Approximate rate constants for the reaction of HNO with a variety of biologically relevant species have been determined (Miranda et al., 2003) and are as follows: oxymyoglobin (1×10^7 M^{-1} s^{-1}) > glutathione (GSH), horseradish peroxidase (2×10^6 M^{-1} s^{-1}) > N-acetylcysteine, Cu,Zn superoxide dismutase, manganese superoxide dismutase, metmyoglobin, catalase ($3-10 \times 10^5$ M^{-1} s^{-1}) > tempol, ferricytochrome c ($4-8 \times 10^4$ M^{-1} s^{-1}) > O$_2$ (3×10^3 M^{-1} s^{-1}). Thus, depending on the biological environment (i.e., proximity and concentration of reactive entities), HNO can have a variety of fates. Based on the kinetics as well as other reports of HNO chemistry, it is evident that two major targets of HNO reactivity are metals and thiols. In fact, the relatively rapid reaction of HNO with thiols has allowed some researchers to discriminate between the biological activities associated with HNO and NO (Pino and Feelisch, 1994). That is, a thiol such as cysteine (or GSH) will directly scavenge HNO under conditions that cause little effect on similar levels of NO and thus can be used to discriminate between the actions of these two nitrogen oxide congeners. One important thing to note regarding HNO kinetics is the relatively slow reaction between HNO and O$_2$, making it unlikely that the HNO–O$_2$ reaction will significantly interfere with reactions at many other biological targets. However, as described later, the reaction of HNO with O$_2$ can lead to deoxygenation of solutions in certain experimental systems, possibly confounding interpretation of results.

3.3. Sites of HNO reactivity

A major site of reactivity for HNO in biological systems is thiols and thiol proteins. This is supported by both kinetic and thermodynamic arguments (Bartberger et al., 2001; Miranda et al., 2003). Compared to other commonly utilized thiol-modifying agents often used for biological studies, HNO is unique in that reaction of HNO with thiols (or thiolproteins) can give two different products, depending on the nature and concentration of the thiol (Fig. 26.3).

In the presence of excess thiol or a vicinal protein thiol, generation of a disulfide is possible. However, in the absence of another reactant, a rearrangement occurs, generating a sulfinamide (Wong et al., 1998). Significantly, oxidation of thiols to disulfides is considered to be reversible in biological systems, whereas generation of a sulfinamide is not reversed easily and, indeed, there is no report of biological reduction of this moiety back to the corresponding thiol.

Along with thiols, it appears that metals and metalloproteins are other major targets for HNO biochemistry. HNO reacts readily with ferric (Fe^{3+}) hemoproteins, giving the corresponding ferrous (Fe^{2+}) nitrosyl species via reductive nitrosylation (reaction 26.8) (Doyle et al., 1988).

$$Fe^{3+}(heme) + HNO \rightarrow Fe^{2+}-NO(heme) + H^+ \quad (26.8)$$

It has been reported that ferrous hemoproteins also react with HNO, giving an Fe^{+2}–HNO adduct (reaction 26.9) (for a review, see Farmer and Sulc, 2005).

$$Fe^{2+}(heme) + HNO \rightarrow Fe^{2+} - N(H)O(heme) \quad (26.9)$$

The chemical properties of the ferrous–HNO adduct have not been firmly established (at least compared to the ferrous–NO complex) and future investigation of its chemistry will be important in discerning the possible biological consequences associated with its formation.

The reaction of HNO with the copper–zinc protein superoxide dismutase (CuZnSOD) was one of the first studied HNO–metalloprotein

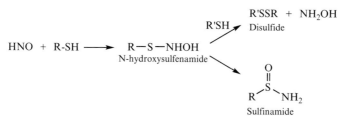

Figure 26.3 Possible products of HNO-mediated thiol modification.

interactions (e.g., Liochev and Fridovich, 2002; Murphy and Sies, 1991). The cupric (Cu^{2+}) form of the enzyme was shown to oxidize HNO to free NO [with subsequent reduction of Cu^{2+} to the cuprous (Cu^{1+}) protein] (reaction 26.10).

$$Cu^{2+} - SOD + HNO \rightarrow Cu^{1+} + NO + H^+ \qquad (26.10)$$

This reaction has been exploited in numerous publications as a means of converting HNO to NO in biological systems. Others have also used ferricyanide ($[FeCN_6]^{3-}$) to accomplish this same process (e.g., Wink et al., 1998). Indeed, both CuZnSOD and $[FeCN_6]^{3-}$ have been used to demonstrate that the biological effects ascribed to HNO were not due to inadvertent conversion to NO. That is, if intentional conversion of HNO to NO attenuates or abolishes the observed effects, then the causative agent was concluded to be HNO and not NO.

A variety of studies have reported that HNO is capable of oxidizing biological species such as GSH and DNA (e.g., Wink et al., 1998). However, the chemistry responsible for DNA oxidation is not established. HNO can react with O_2 (albeit at a relatively slow rate, see earlier discussion) to give an as yet unidentified oxidant that does not appear to be peroxynitrite ($ONOO^-$, an established oxidant) (e.g., Miranda et al., 2001). Also, dimerization of HNO (reaction 26.1) gives initially hyponitrous acid (HON=NOH), which normally dehydrates to give N_2O. However, it has been reported that hyponitrous acid may also generate trace amounts of hydroxyl radical (HO· a potent one-electron oxidant) and N_2 (Buchholz and Powell, 1965; Ivanova et al., 2003). HNO can also react with NO (reaction 26.11) to give ($N_2O_2^-$) with a rate constant of approximately $6 \times 10^6 M^{-1}s^{-1}$ (Shafirovich and Lymar, 2002).

$$HNO + NO \rightarrow N_2O_2^- + H^+ \qquad (26.11)$$

The product of this reaction, $N_2O_2^-$, is a reasonably strong oxidant (E^0 for the $N_2O_2^-/N_2O_2^{2-}$ couple at pH 7 is reported to be 0.96 V), indicating that mixtures of HNO and NO can be oxidizing (Poskrebyshev et al., 2003). Moreover, $N_2O_2^-/HN_2O_2$ has been reported to be capable of decomposing to HO· and N_2O (Seddon et al., 1973). It should be emphasized that oxidation via hyponitrous acid (a possible precursor to HO·) generated from the second-order dimerization of HNO seems unlikely in a biological environment at low concentrations of HNO, as other first-order reactions of HNO should predominate. Moreover, the generation of HO· appears to be a very minor pathway for hyponitrous acid decomposition (Buchholz and Powell, 1965).

There is little question that oxidation chemistry is possible with HNO (or more likely HNO-derived species). However, the chemical properties of HNO predict that it can also have antioxidant/reductant properties. The H–NO bond strength is approximately 50 kcal/mol (Dixon, 1996), indicating that it should be capable of quenching reactive radicals via H-atom donation. H-atom abstraction by a reactive radical species from HNO would then generate NO, a species that is also known to quench reactive radicals via a direct reaction. Indeed, HNO has been shown to serve as an inhibitor of lipid peroxidation (Lopez et al., 2007). Thus, depending on the cellular environment/compartment, HNO appears to be capable of acting as either an oxidant (see earlier discussion) or an antioxidant (much like NO, which has been reported to have both properties depending on the concentration and cellular environment).

4. Use of HNO Donors in Biological Studies

As described earlier, the use of HNO donors requires a variety of control experiments to assure that the observed biological activity is a consequence of HNO. In the case of Angeli's salt, the activity associated with NO_2^- is potentially confounding. In order to control for Angeli's salt-derived NO_2^- (as well as impurities), it is recommended that the donor be decomposed in buffered solution at 37° overnight and then diluted into the desired concentration in a saline solution. These solutions can then be used in control experiments. Moreover, it is recommended that researchers test the decomposed solutions of Angeli's salt by diluting them in 0.05 N NaOH and examining the absorbance spectrum from 200 to 400 nM. This allows a determination of any residual Angeli's salt (λ_{max} = 248 nm, extinction coefficient = 8000 $M^{-1}cm^{-1}$), possible peroxynitrite generation (302 nm, extinction coefficient = 1670 $M^{-1}cm^{-1}$), and levels of nitrite (354 nm, extinction coefficient = 23 $M^{-1}cm^{-1}$).

Another possible pitfall that needs to be taken into consideration while using HNO donors is the pH of the stock and/or final working solution. As described previously, the rate of the thermal decay of Angeli's salt is pH independent from pH 4 to 8 but it decreases with increasing pH. Indeed, preparation of stock solutions of Angeli's salt can be made at high pH. Typically a stock solution of Angeli's salt can be prepared in 10 mM NaOH (made in double distilled water), kept on ice, and protected from light until the experiment. Diluting an aliquot in saline or any other physiological solutions (buffers) to achieve the desired concentration of Angeli's salt in the biological system will be the next step. It is always preferable to use freshly prepared stock solutions and freshly dissolved aliquots. Also, it is standard practice to store Angeli's salt in a dry environment, as it is water soluble and will break down when wet. Thus, multiple freeze–thaw cycles of frozen samples of Angeli's salt (solid) can lead to condensation of water and decomposition.

As alternatives to Angeli's salt, several other HNO donors are now available for biological studies (Keefer, 2005; King and Nagasawa, 1999; Miranda et al., 2005b; Sha et al., 2006). The HNO-generating diazeniumdiolates (Miranda et al., 2005a) can be handled similarly to Angeli's salt. Other donor species, such as Piloty's acid derivatives and acyloxy nitroso compounds (Pennington et al., 2005; Sha et al., 2006), cannot be dissolved easily in physiological solutions/buffers. For example, the acyloxy nitroso compounds developed by the King laboratory (Sha et al., 2006) must be dissolved first in dimethyl sulfoxide (DMSO) and then diluted in the biological solution. For instance, an organic stock solution of 1-nitrosocyclohexyl acetate can be made up in DMSO and further diluted into the working solution (Tyrode or other physiological medium) to achieve the final concentration. The acyloxy nitroso compounds are typically dark blue while the decomposed products are colorless. Thus, it is possible to monitor the integrity of stock solutions spectrophotometrically.

5. Nitroxyl Pharmacological Effects: *In Vivo* and *In Vitro* Studies

Although it is the intent of this chapter to inform readers of the issues, problems, and expectations associated with biological studies of HNO, it is worthwhile to briefly review some of the currently reported pharmacological actions of HNO, as these effects may impact the interpretation of other biological studies (for a more in-depth discussion of HNO pharmacology/biology, see Paolocci et al., 2007).

5.1. Nitroxyl in the cardiovascular system

Nitroxyl is a vasorelaxant. Although the mechanism of vasorelaxation is not firmly established, HNO-mediated generation of cGMP (similar to NO) or release of calcitonin gene-related peptide (CGRP) has been proposed (e.g., see Paolocci et al., 2007 and references therein). Angeli's salt, administered *in vivo* in normal dogs, instrumented for pressure-volume analysis [a way to dissect primary effects on the heart from changes in vascular loading conditions (Senzaki et al., 2000)], produces the following effects: (1) increased contractility, (2) hastened ventricular relaxation, and (3) unloading action on the heart (both pre- and after-load reduction) (Paolocci et al., 2001). This HNO-mediated inotropy is not mimicked by a NO donor such as DEA/NO and is prevented by the HNO scavenger N-acetyl-L-cysteine, indicating a direct HNO effect. HNO inotropy is not prevented by sustained β-adrenergic blockade and is additive to β-adrenergic agonists in elevating myocardial contractility. The mechanisms of HNO-induced positive inotropy (increased contractility) and lusitropy (enhanced relaxation) appear to be the result of effects on cardiomyocyte intracellular Ca^{2+} cycling. HNO augments Ca^{2+} release from the sarcoplasmic reticulum (SR) in both cardiac and skeletal muscle via the activation of cardiac ryanodine receptors (RyR2) (Tocchetti et al., 2007) and skeletal Ca^{2+} release channels (RyR1) (Cheong et al., 2005), respectively. HNO also stimulates SR Ca^{2+}-ATPase (SERCA2a), increasing Ca^{2+} reuptake into the SR (Tocchetti et al., 2007). The effects on RyR2 and on SERCA2a likely trigger and sustain positive inotropy while accelerating relaxation in murine myocytes (Tocchetti et al., 2007). HNO is also capable of enhancing myofilament responsiveness to Ca^{2+}, augmenting myocardial contractility in rat cardiac trabeculae (Dai et al., 2007).

5.2. Nitroxyl and myocardial ischemia/reperfusion injury

Nitroxyl exacerbates postischemia/reperfusion *in vivo* injury in a dose-dependent manner (Ma et al., 1999). The mechanisms for such action remain unidentified. One possible factor at play is the ability of HNO to favor neutrophil accumulation (Vanuffelen et al., 1998) as HNO enhances myeloperoxidase activity in the ischemic tissue (Ma et al., 1999). Conversely, HNO can induce a marked early preconditioning-like effect in isolated rat hearts (Pagliaro et al., 2003). Ischemic preconditioning is an effective endogenous mechanism for myocardial protection against ischemia/reperfusion injury (Gross and Auchampach, 2007).

5.3. Nitroxyl donors *in Vivo* can trigger the release of peptides from the nonadrenergic noncholinergic system

When working with HNO donors, especially in *in vivo* settings, one phenomenon should be always considered: they may trigger the release of cardio- and vascular-active peptides from the nonadrenergic noncholinergic (NANC) system and/or from the sensory nerves. This is not only the case of CGRP (Booth et al., 2000; Paolocci et al., 2001), a small neuropeptide found in the heart and peri-adventitial nerve fibers throughout the coronary and peripheral vascular system, but also of neurotensin, another vaso- and cardio-active neuropeptide found also in the heart (Katori and Paolocci, unpublished observations). It is also true that NO can trigger the release of CGRP from NANC nerves, at least in the rat heart (Hu et al., 1999), but HNO donors appear to do this to a significantly greater extent (Booth et al., 2000). Although the exact mechanism of this NO/HNO-evoked release of CGRP (and possible activation of NANC fibers and/or sensory nerves) remains unclear, it is undeniable that the release of CGRP (and possibly of other neuropeptides) can account, at least in part, for some of the biological effects produced by HNO donors *in vitro* and, especially, *in vivo*.

5.4. Nitroxyl and the nervous system: Physiological and toxicological effects

Among the possible modulatory effects exerted by HNO in the central nervous system, it has been shown that HNO interacts with a thiol residue on the N-methyl-D-aspartate (NMDA) receptor, leading to an attenuation of Ca^{2+} influx (Kim *et al.*, 1999). Since overstimulation of the NMDA receptor is implicated in the excitotoxicity associated with glutamate, the physiological outcome of this HNO action is to afford protection against glutamate-based excitotoxicity. Conversely, Colton and co-workers (2001) reported that HNO was also able to block glycine-dependent desensitization of the NMDA receptor with consequent sensitization of the receptor. The divergent outcomes of two reported studies can be, in part, reconciled considering the depletion of O_2 levels on long-term use of AS, which consumes O_2 during decomposition. This example should serve as a warning to all investigators working on HNO effects. Careful attention has to be paid to experimental conditions, particularly with respect to O_2 and metal content.

Nitroxyl has been reported to be capable of eliciting some toxicity. For example, HNO has been implicated in the neurotoxicity reported after Angeli's salt administration in rats via intranigral infusion (Vaananen *et al.*, 2003). HNO has been shown to be toxic to dopaminergic neurons also in cell culture experiments (Hewett *et al.*, 2005). This adverse action was both acute and progressive (occurring after the dissipation of HNO). Intrathecal administration of Angeli's salt to the lumbar spinal cord of rats produced motor neuron injury without affecting sensory neurons (Vaananen *et al.*, 2004). However, significant effects on locomotor activity were seen only at the highest dose (10 microlmol bolus administration). It should be pointed out that in these studies, extremely high concentrations of Angeli's salt were used, far greater than those employed, for instance, in studies conducted in the cardiovascular system, and particularly *in vivo* (three orders of magnitude below the cytotoxic threshold).

6. Summary

Based on the discussions given here, there are several aspects of HNO (and HNO donor) biological chemistry and pharmacology important for interpretation of its actions in biological systems. They are as follows.

1. For studies utilizing Angeli's salt (and most donors mentioned herein), the active species is HNO and *not* NO^-.
2. Protonation of NO^- and deprotonation of HNO are both very slow. Thus, if either HNO or NO^- is generated in solution, its equilibrium partner is unlikely to be relevant.

3. Nitroxyl is a fleeting species and can only be studied using donor molecules. When using donor molecules, a multitude of control studies need to be carried out to assure that the biological activity is because of HNO.
4. In the absence of any biological targets, HNO in solution will decompose via reaction with O_2 and dimerization–dehydration. The reaction with O_2 is slow and will only compete with dimerization when HNO is present at extremely low concentrations.
5. To date, the best-characterized biological targets for direct reaction with HNO are thiols/thiol proteins and metals/metalloproteins.
6. Species derived from HNO (from dimerization, reaction with NO or O_2) can be oxidizing.
7. Nitroxyl can quench radical chemistry via direct reaction with reactive radical intermediates.
8. A major pharmacological target of HNO is the cardiovascular system, where it can act as a vasorelaxant and inotropic/lusitropic agent.
9. Nitroxyl can either exacerbate or attenuate ischemia–reperfusion injury, depending on the time of administration.
10. Nitroxyl can either sensitize or attenuate the actions associated with the NMDA receptor, depending on experimental conditions.

The biological properties associated with HNO mentioned herein represent only a partial listing, and, clearly, other chemical and biological properties of HNO are sure to be discovered in the future as work on the biological actions of HNO continues. It is hoped that the information contained in this chapter will assist in these future discoveries.

REFERENCES

Angeli, A. (1896). *Chim. Ital.* **26**, 17.
Atkinson, R. N., Storey, B. M., and King, S. B. (1996). Reactions of acyl nitroso compounds with amines: Production of nitroxyl (HNO) with the preparation of amides. *Tet. Lett.* **37**, 9287–9290.
Bartberger, M. D., Fukuto, J. M., and Houk, K. N. (2001). On the acidity and reactivity of HNO in aqueous solution and biological systems. *Proc. Natl. Acad. Sci. USA* **98**, 2194–2198.
Bartberger, M. D., Liu, W., Ford, E., Miranda, K. M., Switzer, C., Fukuto, J. M., Farmer, P. J., Wink, D. A., and Houk, K. N. (2002). The reduction potential of nitric oxide (NO) and its importance to NO biochemistry. *Proc. Natl. Acad. Sci. USA* **99**, 10958–10963.
Bonner, F. T. and Ko, Y. H. (1992). Kinetic, isotopic and N^{15} NMR study of N-hydroxybenzenesulfonamide decomposition? An HNO source reaction. *Inorg. Chem.* **31**, 2514–2519.
Bonner, F. T. and Ravid, B. (1975). Thermal decomposition of oxyhyponitrite (sodium trioxodinitrate(II)) in aqueous solution. *Inorg. Chem.* **14**, 558–563.

Booth, B. P., Tabrizi-Fard, M. A., and Fung, H. (2000). Calcitonin gene-related peptide-dependent vascular relaxation of rat aorta: An additional mechanism for nitroglycerin. *Biochem. Pharmacol.* **59,** 1603–1609.

Buchholz, J. R. and Powell, R. E. (1965). The decomposition of hyponitrous acid. II. The chain reaction. *J. Am. Chem. Soc.* **87,** 2350–2353.

Cheong, E., Tumbev, V., Abramson, J., Salama, G., and Stoyanovsky, D. A. (2005). Nitroxyl triggers Ca^{2+} release from skeletal and cardiac sarcoplasmic reticulum by oxidizing ryanodine receptors. *Cell Calcium* **37,** 87–96.

Colton, C. A., Gbadegesin, M., Wink, D. A., Miranda, K. M., Espey, M. G., and Vicini, S. (2001). Nitroxyl anion regulation of the NMDA receptor. *J. Neurochem.* **78,** 1126–1134.

Cook, N. M., Shinyashiki, M., Jackson, M. I., Leal, F. A., and Fukuto, J. M. (2003). Nitroxyl (HNO)-mediated disruption of thiol proteins: Inhibition of the yeast transcription factor Ace1. *Arch. Biochem. Biophys.* **410,** 89–95.

Dai, T., Tian, Y., Tocchetti, C. G., Katori, T., Murphy, A. M., Kass, D. A., Paolocci, N., and Gao, W. D. (2007). Nitroxyl increases force development in rat cardiac muscle. *J. Physiol.* **580,** 951–960.

DeMaster, E. G., Redfern, B., and Nagasawa, H. T. (1998). Mechanism of inhibition of aldehyde dehydrogenase by nitroxyl, the active metabolite of the alcohol deterrent agent cyanamide. *Biochem. Pharmacol.* **55,** 2007–2015.

DeMaster, E. G., Shirota, F. N., and Nagasawa, H. T. (1984). The metabolic activation of cyanamide to an inhibitor of aldehyde dehydrogenase is catalyzed by catalase. *Biochem. Biophys. Res. Commun.* **122,** 358–365.

Dixon, R. N. (1996). The heats of formation of HNO and DNO. *J. Chem. Phys.* **104,** 6905–6906.

Doyle, M. P., Mahapatro, S. N., Broene, R. D., and Guy, J. K. (1988). Oxidation and reduction of hemoproteins by trioxodinitrate(II): The role of nitrosyl hydride and nitrite. *J. Am. Chem. Soc.* **110,** 593–599.

Dutton, A. S., Fukuto, J. M., and Houk, K. N. (2004). Mechanisms of HNO and NO production from Angeli's salt: Density functional and CBS-QB3 theory predictions. *J. Am. Chem. Soc.* **126,** 3795–3800.

Dutton, A. S., Suhrada, C. P., Miranda, K. M., Wink, D. A., Fukuto, J. M., and Houk, K. N. (2006). Mechanism of pH-dependent decomposition of monoalkylamine diazeniumdiolates to form HNO and NO, deduced from the model compound methylamine diazeniumdiolate, density functional theory and CBS-QB3 calculations. *Inorg. Chem.* **45,** 2448–2456.

Espey, M. G., Miranda, K. M., Thomas, D. D., and Wink, D. A. (2002). Ingress and reactive chemistry of nitroxyl-derived species within human cells. *Free Radic. Biol. Med.* **33,** 827–834.

Farmer, P. J. and Sulc, F. (2005). Coordination chemistry of the HNO ligand with hemes and synthetic coordination complexes. *J. Inorg. Biochem.* **99,** 166–184.

Feelisch, M., te Poel, M., Zamora, R., Deussen, A., and Moncada, S. (1994). Understanding the controversy over the identity of EDRF. *Nature* **368,** 62–65.

Fukuto, J. M., Bartberger, M. D., Dutton, A. S., Paolocci, N., Wink, D. A., and Houk, K. N. (2005). The physiological chemistry and biological activity of nitroxyl (HNO): The neglected, misunderstood and enigmatic nitrogen oxide. *Chem. Res. Toxicol.* **18,** 790–801.

Fukuto, J. M., Chiang, K., Hszieh, R., Wong, P., and Chaudhuri, G. (1992). The pharmacological activity of nitroxyl: A potent vasodilator with activity similar to nitric oxide and/or endothelium-derived relaxing factor. *J. Pharmacol. Exp. Ther.* **263,** 546–551.

Gladwin, M. T., Raat, N. J., Shiva, S., Dezfulian, C., Hogg, N., Kim-Shapiro, D. B., and Patel, R. P. (2006). Nitrite as a vascular endocrine nitric oxide reservoir that contributes to hypoxic signaling, cytoportection and vasodilation. *Am. J. Physiol. Heart Circ. Physiol.* **41,** 541–548.

Gratzel, M., Taniguchi, S., and Henglein, A. (1970). A pulse radiolytic study of short-lived byproducts on nitric oxide reduction in aqueous solution. *Ber. Bunsenges. Phys. Chem.* **74,** 1003–1010.

Gross, G. J., and Auchampach, J. A. (2007). Reperfusion injury: Does it exist? *J. Mol. Cell Cardiol.* **42,** 12–18.

Hewett, S. J., Espey, M. G., Uliasz, T. F., and Wink, D. A. (2005). Neurotoxicity of nitroxyl: Insights into HNO and NO biochemical imbalance. *Free Radic. Biol. Med.* **39,** 1478–1488.

Hu, C. P., Li, Y. J., and Deng, H. W. (1999). The cardioprotective effects of nitroglycerin-induced preconditioning are mediated by calcitonin gene-related peptide. *Eur. J. Pharmacol.* **369,** 189–194.

Hughes, M. N. and Wimbledon, P. E. (1976). The chemistry of trioxodinitrates. Part 1. Decomposition of sodium trioxodinitrate (Angeli's salt) in aqueous solution. *J. Chem. Soc. Dalton Trans.* 703–707.

Ivanova, J., Salama, G., Clancy, R. M., Schor, N. F., Nylander, K. D., and Stoyanovsky, D. A. (2003). Formation of nitroxyl and hydroxyl radical in solutions of sodium trioxodinitrate. *J. Biol. Chem.* **278,** 42761–42768.

Keefer, L. K. (2005). Nitric oxide (NO)- and nitroxyl (HNO)-donating diazenium diolates (NONOates): Emerging commercial opportunities. *Curr. Top. Med. Chem.* **5,** 625–636.

Kim, W. K., Choi, Y. B., Rayudu, P. V., Das, P., Asaad, W., Arnelle, D. R., Stamler, J. S., and Lipton, S. A. (1999). Attenuation of NMDA receptor activity and neurotoxicity by nitroxyl anion, NO^-. *Neuron* **24,** 461–469.

King, S. B., and Nagasawa, H. T. (1999). Chemical approaches toward generation of nitroxyl. *Methods Enzymol.* **301,** 211–220.

Lee, M. J. C., Nagasawa, H. T., Elberling, J. A., and DeMaster, E. G. (1992). Prodrugs of nitroxyl as inhibitors of aldehyde dehydrogenase. *J. Med. Chem.* **35,** 3648–3652.

Liochev, S. I. and Fridovich, I. (2002). Nitroxyl (NO^-): A substrate for superoxide dismutase. *Arch. Biochem. Biophys.* **402,** 166–171.

Liochev, S. I. and Fridovich, I. (2003). The mode of decomposition of Angeli's salt ($Na_2N_2O_3$) and the effects thereon of oxygen, nitrite, superoxide dismutase and glutathione. *Free Radic. Biol. Med.* **34,** 1399–1404.

Lopez, B. E., Rodriguez, C. E., Pribadi, M., Cook, N. M., Shinyashiki, M., and Fukuto, J. M. (2005). Inhibition of yeast glycolysis by nitroxyl (HNO): A mechanism of HNO toxicity and implications to HNO biology. *Arch. Biochem. Biophys.* **442,** 140–148.

Lopez, B. E., Shinyashiki, M., Han, T. H., and Fukuto, J. M. (2007). Antioxidant actions of nitroxyl. *Free Radic. Biol. Med.* **42,** 482–491.

Ma, X. L., Gao, F., Liu, G. L., Lopez, B. L., Christopher, T. A., Fukuto, J. M., Wink, D. A., and Feelisch, M. (1999). Opposite effects of nitric oxide and nitroxyl on postischemic myocardial injury. *Proc. Natl. Acad. Sci. USA* **96,** 14617–14622.

Marti, M. A., Bari, S., Estrin, D. A., and Doctorovich, F. (2005). Discrimination of nitroxyl and nitric oxide by water soluble Mn(III) porphyrins. *J. Am. Chem. Soc.* **127,** 4680–4684.

Miranda, K. M. (2005). The chemistry of nitroxyl (HNO) and implications in biology. *Coord. Chem. Rev.* **249,** 433–455.

Miranda, K. M., Espey, M. G., Yamada, K., Krishna, M., Ludwick, N., Kim, S., Jourd'heuil, D., Grisham, M., Feelisch, M., Fukuto, J. M., and Wink, D. A. (2001). Unique oxidative mechanisms for the reactive nitrogen oxide species, nitroxyl anion. *J. Biol. Chem.* **276,** 1720–1727.

Miranda, K. M., Katori, T., Torres de Holding, C. L., Thomas, L., Ridnour, L. A., McLendon, W. J., Cologna, S. M., Dutton, A. S., Champion, H. C., Mancardi, D.,

Tocchetti, C. G., Saavedra, J. E., Keefer, L. K., Houk, K. N., Fukuto, J. M., Kass, D. A., Paolocci, N., and Wink, D. A. (2005a). Comparison of the NO and HNO donating properties of diazenium diolates: Primary amine adducts release HNO *in vivo*. *J. Med. Chem.* **48,** 8220–8228.

Miranda, K. M., Nagasawa, H. T., and Toscano, J. P. (2005b). Donors or HNO. *Curr. Top. Med. Chem.* **5,** 649–664.

Miranda, K. M., Paolocci, N., Katori, T., Thomas, D. D., Ford, E., Bartberger, M. D., Espey, M. G., Kass, D. A., Fukuto, J. M., and Wink, D. A. (2003). A biochemical rationale for the discrete behavior of nitroxyl and nitric oxide in the cardiovascular system. *Proc. Natl. Acad. Sci. USA* **100,** 9197–9201.

Murphy, M. E., and Sies, H. (1991). Reversible conversion of nitroxyl anion to nitric oxide by superoxide dismutase. *Proc. Natl. Acad. Sci. USA* **88,** 10860–10864.

Pagliaro, P., Mancardi, D., Rastaldo, R., Penna, C., Gattullo, D., Miranda, K. M., Feelisch, M., Wink, D. A., Kass, D. A., and Paolocci, N. (2003). Nitroxyl affords thiol-sensitive myocardial protective effects akin to early preconditioning. *Free Radic. Biol. Med.* **34,** 33–43.

Paolocci, N., Jackson, M. I., Lopez, B. E., Wink, D. A., Miranda, K. M., Hobbs, A. J., and Fukuto, J. M. (2007). The pharmacology of nitroxyl (HNO) and its therapeutic potential: Not just the Janus face of nitric oxide. *Pharmacol. Ther.* **113,** 442–458.

Paolocci, N., Saavedra, W. F., Miranda, K. M., Martignani, C., Isoda, T., Hare, J. M., Espey, M. G., Fukuto, J. M., Feelisch, M., Wink, D. A., and Kass, D. A. (2001). Nitroxyl anion exerts redox-sensitive positive cardiac inotropy *in vivo* by calcitonin gene-related peptide signaling. *Proc. Natl. Acad. Sci. USA* **98,** 10463–10468.

Pennington, R. L., Sha, X., and King, S. B. (2005). N-Hydroxy sulfonimidamides as new nitroxyl (HNO) donors. *Bioorg. Med. Chem. Lett.* **15,** 2331–2334.

Pino, R. Z. and Feelisch, M. (1994). Bioassay discrimination between nitric oxide (NO) and nitroxyl (NO^-) using cysteine. *Biochem. Biophys. Res. Commun.* **201,** 54–62.

Poskrebyshev, G. A., Shafirovich, V., and Lymar, S. V. (2003). Hyponitrite radical, a stable adduct of nitric oxide and nitroxyl. *J. Am. Chem. Soc.* **126,** 891–899.

Seddon, W. A., Fletcher, J. W., and Sopchyshyn, F. C. (1973). Pulse radiolysis of nitric oxide in aqueous solution. *Can. J. Chem.* **51,** 1123–1130.

Senzaki, H., Isoda, T, Paolocci, N., Ekelund, U., Hare, J. M., and Kass, D. A. (2000). Improved mechanoenergetics and cardiac rest and reserve function of *in vivo* failing heart by calcium sensitizer EMD-57033. *Circulation* **101,** 1040–1048.

Sha, X., Isbell, T. S., Patel, R. P., Day, C. S., and King, S. B. (2006). Hydrolysis of acyloxy nitroso compounds yields nitroxyl (HNO). *J. Am. Chem. Soc.* **128,** 9687–9692.

Shafirovich, V. and Lymar, S. V. (2002). Nitroxyl and its anion in aqueous solutions? Spin states, protic equilibria, and reactivities toward oxygen and nitric oxide. *Proc. Natl. Acad. Sci. USA* **99,** 7340–7345.

Shafirovich, V. and Lymar, S. V. (2003). Spin-forbidden deprotonation of aqueous nitroxyl (HNO). *J. Am. Chem. Soc.* **125,** 6547–6552.

Shoeman, D. W. and Nagasawa, H. T. (1998). The reaction of nitroxyl (HNO) with nitrosobenzene gives cupferron (N-nitrosophenylhydroxylamine). *Nitric Oxide* **2,** 66–72.

Smith, P. A. S. and Hein, G. E. (1960). The alleged role of nitroxyl in certain reactions of aldehydes and alkyl halides. *J. Am. Chem. Soc.* **82,** 5731–5740.

Tocchetti, C. G., Wang, W., Froehlich, J. P., Huke, S., Aon, M. A., Wilson, G. M., Di Benedetto, G., O'Rourke, B., Gao, W. D., Wink, D. A., Toscano, J. P., Zaccolo, M., *et al.* (2007). Nitroxyl improves cellular heart function by directly enhancing cardiac sarcoplasmic reticulum Ca^{2+} cycling. *Circ. Res.* **100,** 96–104.

Vaananen, A. J., Liebkind, R., Kankuri, E., Liesi, P., and Rauhala, P. (2004). Angeli's salt and spinal motor neuron injury. *Free Radic. Res.* **38,** 271–282.

Vaananen, A. J., Moed, M., Tuominen, R. K., Helkamaa, T. H., Wiksten, M., Liesi, P., Chiueh, C. C., and Rauhala, P. (2003). Angeli's salt induces neurotoxicity in dopaminergic neurons in vivo and *in vitro*. *Free Radic. Res.* **37,** 381–389.

Vanuffelen, B. E., Van Der, Z. J., De Koster, Z. J., Vansteveninck, B. M., and Elferink, J. G. (1998). Intracellular but not extracellular conversion of nitroxyl anion into nitric oxide leads to stimulation of human neutrophil migration. *Biochem. J.* **330,** 719–722.

Wink, D. A., Feelisch, M., Fukuto, J., Christodoulou, D., Jourd'heuil, D.,Grisham,M., Vodovotz, V., Cook, J. A., Krishna, M., DeGraff, W., Kim, S., Gamson, J., and Mitchell, J. B. (1998). The cytotoxicity of nitroxyl: Possible implications for the pathophysiological role of NO. *Arch. Biochem. Biophys.* **351,** 66–74.

Wong, P. S. Y., Hyun, J., Fukuto, J. M., Shiroda, F. N., DeMaster, E. G., and Nagasawa, H. T. (1998). The reaction between nitrosothiols and thiols: Generation of nitroxyl (HNO) and subsequent chemistry. *Biochemistry* **37,** 5362–5371.

Zamora, R., Grzesiok, A., Weber, H., and Feelisch, M. (1995). Oxidative release of nitric oxide accounts for guanylyl cyclase stimulating, vasodilator and antiplatelet activity of Piloty's acid: A comparison with Angeli's salt. *Biochem. J.* **312,** 333–339.

Zeng, B.-B., Huang, J., Wright, M. W., and King, S. B. (2004). Nitroxyl (HNO) release from new functionalized N-hydroxyurea-derived acyl nitroso-9,10-dimethylanthracene cycloadducts. *Bioorg. Med. Chem. Lett.* **14,** 5565–5568.

Author Index

A

Abe, K., 314
Abe, T., 339
Abeliovich, A., 122
Abraham, J., 165
Abramson, J., 415, 425
Acimovic, S., 222
Ackermann, E. J., 248
Adachi, J., 376
Adachi, T., 212, 284
Adatia, I., 363
Adegoke, O. A. J., 178
Aebersold, R., 37, 58, 195
Agalou, S., 4, 339, 342, 343, 345, 346, 350, 351
Agrawal, N., 233, 236, 284
Ahearn, G., 138, 159
Ahearn, G. S., 165
Ahmed, N., 4, 339, 342, 343, 344, 345, 346, 347, 348, 349, 350, 351, 352, 353
Ahmed, S. H., 310
Ahmed, U., 344, 352, 353
Ahn, D., 160
Ahram, M., 194
Aibara, S., 38
Aitken, J. F., 319
Aizawa, Y., 376
Ajees, A. A., 36
Akaike, T., 159, 167, 168
Akerboom, T. P. M., 106
Akerkar, S. A., 297
Akil, H., 245
Alayash, A. I., 256
Alderton, W. K., 248
Al-Essa, L., 385
Alexander, C. B., 5, 11, 30, 66, 83
Alexander, V., 261
Alidema, E., 318, 323, 326, 327
Allaker, R. P., 382
Allen, A., 165
Allen, J. B., 18
Allmaier, G., 4
Alonso, C., 264
Alpert, C., 140
Alston, J. A., 97, 101, 288
Alterman, M. A., 354
Althaus, J., 343, 353
Alvarez, B., 254, 255

Alvarez, M., 160
Alvarez, M. N., 255, 260
Alvarez, S., 318
Amagasu, S. M., 245
Ameisen, J. C., 264
Ames, B. N., 34
Amico, G. D., 318
Amoresano, A., 12, 297
Anantharamaiah, G. M., 34, 35, 36, 40, 58, 59
Anantharaman, M., 296, 314
Andel, H., 222
Anderle, M., 200
Anderson, D. J., 195
Anderson, J. E., 310
Andreasen, P. A., 217
Angeli, A., 414
Angelo, M., 149, 150, 159, 162, 165, 166
Ansell, B. J., 36
Anson, M. L., 322
Aon, M. A., 425
Aoyama, N., 283
Ara, J., 296, 313
Arai, T., 212
Arciero, D. M., 123, 127, 128, 129, 131, 133
Arena, S., 6, 7
Arienti, G., 382
Arita, N. O., 401, 402
Arnelle, D. R., 138, 158, 161, 426
Arner, E. S. J., 102
Arnone, A., 159, 165
Arnoult, D., 264
Arp, D. J., 128, 129, 133
Arstall, M. A., 85
Arteel, G. E., 256
Asaad, W., 426
Asada, K., 398, 404
Asano, M., 376
Asbury, M. L., 321, 322
Ascenzi, P., 159
Ash, D. E., 221, 223, 227
Askew, S., 159, 160
Aslan, M., 5, 11, 30
Aslund, F., 104
Assaloni, R., 339
Assmann, G., 50
Atger, V., 35, 36
Atkinson, R. N., 418
Aubert, C., 66
Aubier, M., 66

Auchampach, J. A., 425
Augusto, O., 254, 255, 256, 260, 261
Aulak, K. S., 5, 6, 11, 18, 26, 30, 193, 194, 296, 297, 310
Awane, M., 222
Axelrod, M., 144, 145, 151, 159, 162, 164, 165, 166
Azad, N., 284
Azarov, I., 147, 148, 151

B

Babaei-Jadidi, R., 4, 339, 342, 343, 344, 345, 346, 347, 348, 350, 351
Bachi, A., 290
Baek, K. J., 210
Baggio, R., 227
Bahnweg, G., 284
Bai, G., 160, 232
Bailey, C., 85
Bailey-Serres, J., 398
Bak, M., 85
Baker, L., 42
Baker, N. R., 398
Baker, P. R., 4, 5, 66, 67, 83
Balaban, B. S., 4
Balazy, M., 297
Baldus, S., 66, 83
Baliga, B. A., 399
Ballard, P. L., 343
Ballard, R. A., 343
Ballas, S. K., 152
Balligand, J. L., 85
Balzer, J., 362
Banks, B. A., 343
Bankston, L. A., 96
Bannigan, J., 18
Bannister, J. V., 74, 75, 80
Bansal, V., 222
Bao, C., 232
Barbouri, A., 342
Bari, S., 415
Barnes, S., 5, 11, 30, 66, 313
Baroti, K., 346, 347
Barrera, P., 275
Barrett, D. M., 85
Barriero, E., 194
Barroso, J. B., 398, 401, 402
Barry, R. C., 193, 195, 201
Barsacchi, R., 113
Bartberger, M. D., 412, 419, 420, 421
Barth, E. D., 395
Bartlett, M. C., 160, 161
Bartlett, S. T., 160
Bartsch, H., 338
Bashore, T. M., 165
Bass, D. A., 399
Basso, M., 5, 11
Bastacky, J., 264

Basu, S., 137, 139, 147, 148, 151
Battah, S., 4, 339, 342, 345, 346, 347
Batthyany, C., 67, 83
Bauer, C., 264
Bauer, H., 378
Bauer, J. A., 296
Bazer, F. W., 179
Beal, M. F., 314
Beaumont, H. J., 128, 129, 133
Beavis, R. C., 193
Becher, M. W., 314
Becker, C. H., 200
Beckman, J. S., 18, 35, 38, 39, 66, 67, 83, 98, 112, 192, 254, 255, 265, 296, 310, 313, 314, 387
Beckman, T. W., 35, 254
Beda, N., 159
Beede, D., 349
Behring, A., 193, 194, 354
Beisswenger, P. J., 343, 344, 345, 347, 348, 350
Belaaouaj, A., 35
Belenghi, B., 284, 286
Bellmann, C., 296, 314
Bendotti, C., 5, 11
Benhar, M., 159, 160
Benjamin, N., 382, 383, 384, 389, 391
Benninghoven, A., 50
Berg, C. P., 262, 264
Bergmann, D. J., 123, 127, 131
Bergt, C., 5, 11, 34, 35, 36, 37, 38, 39, 53, 57, 58, 59, 67, 83
Berk, B. C., 211, 232
Berkelman, T., 31
Berlett, B. S., 4
Berliner, J. A., 34, 35
Bernard, B., 222
Bernardos, A., 275
Berndt, C., 284
Berry, R. W., 18, 314
Bethke, P. C., 283
Beyer, W. F., Jr., 96, 160
Bhamidipati, D., 222
Bhargava, K., 67, 83
Bhat, V., 12, 342
Bianchi, M., 246, 247, 248
Bierman, E. L., 38
Bigelow, D. J., 87, 88, 89, 191, 194, 195, 202
Bigio, E. H., 314
Binder, L. I., 18, 314
Birder, L. A., 318, 323
Birnboim, H. C., 4, 11
Bishop, A., 310
Biswas, K. K., 210
Bjorne, H., 362
Bjornstedt, M., 96, 101
Black, S. M., 85
Blackwelder, W. C., 222
Blaise, G., 319
Blakeley, R. L., 103

Author Index

Blanchard-Fillion, B., 338
Blatter, L. A., 318, 327
Blennow, K., 352, 353
Blonder, J., 194
Blount, B. C., 29
Bo, S., 340
Bobrowski, K., 66, 80, 84, 85
Bock, E., 122, 123, 124, 126, 127,
   128, 129, 130, 132, 133
Bockaert, J., 310
Boczkowski, J., 66
Bode-Böger, S. M., 178, 185
Böger, R. H., 178, 185
Bolanos, J. P., 244, 284
Bolognesi, M., 159
Bonaventura, C., 85, 147, 150, 159
Bonaventura, J., 85, 159
Bond, M., 342
Bonefeld-Jørgensen, E. C., 217
Bonetto, V., 5, 11
Bonini, M. G., 254, 256, 260
Bonner, F. T., 414, 417
Boockvar, K. S., 362, 367
Booth, B. P., 425
Borchers, M. T., 66, 71, 83
Bordo, D., 159
Bosman, G. J., 263
Botting, C. H., 160
Bouma, G., 376
Bourque, S., 284
Bove, P. F., 284
Boveris, A., 310, 318
Boyd-Kimball, D., 297, 305
Bozzi, A., 260
Brackenier, A., 284, 286
Bradley, N., 339
Bradley, W. A., 66, 83
Brain, S. D., 338
Braman, R. S., 324
Brambilla, L., 118
Brand, K. A., 118
Brandt, R., 400
Bratosin, D., 264
Braun, W. J., 71
Bredt, D. S., 244, 321
Brennan, D. M., 339
Brennan, M. L., 35, 36, 58, 59, 66, 71, 83
Brettel, K., 66
Brien, J. F., 123
Briscoe, P., 66, 83
Brock, T. A., 66, 83
Brockner Ryan, B. A., 256
Broderick, M., 319
Broene, R. D., 421
Brogan, R., 383
Broillet, M. C., 232, 310
Broniowska, K. A., 140, 170, 236, 275, 280
Brot, N., 36, 37, 53, 57, 58

Brouet, I., 338
Brouillette, C. G., 40
Brown, G. C., 138
Brown, J. F., 391
Brown, M. A., 375
Brown, R. H., Jr., 314
Browne, S. E., 314
Brownlee, M., 349
Brubaker, G., 5, 11
Bruce King, S., 139
Bruckdirfer, K. R., 382
Bruckdorfer, R., 339
Bruijn, L. I., 314
Brüne, B., 213, 216
Brunner, F., 158
Bruno, V. M., 245
Brunori, M., 172
Brunzell, J., 34, 35, 36, 58, 59
Bruschi, M., 6, 7
Bryan, N. S., 66, 67, 83, 138, 147, 148, 149, 151,
   211, 239, 363, 371
Bryner, K. D., 339
Buchacher, A., 4
Buchczyk, D. P., 296
Buchholz, J. R., 422
Budde, A., 213, 216
Buerk, D. G., 284, 310
Buetler, T., 339, 346
Buettner, G. R., 392
Buga, G. M., 324
Burgard, D. A., 165
Burke-Gaffney, A., 261
Burton, B. A., 319, 367, 375
Bush, K. M., 98, 112, 254, 265
Buss, J. E., 85
Bustamante, J., 318
Butler, A., 159, 160
Butler, J., 66, 80, 84, 264
Butler, S., 35
Butt, Y. K.-C., 17
Butterfield, D. A., 193, 194, 295, 296, 297, 304,
   305, 309, 310, 314, 353
Bybee, K. A., 346
Byrns, R. E., 324
Byun, J., 5, 34, 35, 36, 58, 59, 340

C

Cabral, D. M., 101, 104
Cabrales, P., 310
Cadenas, E., 310, 318
Cadogan, E., 18
Cadwallader, K. A., 212
Cai, J., 296, 297, 305, 314
Caidahl, K., 5, 8, 9, 11, 12, 343, 346, 352, 353
Caldarera, C. M., 244
Califf, R. M., 165
Callebert, J., 66

Callister, M. E. J., 261
Calvano, S. E., 199, 200, 201
Cambien, B., 232
Camerini, S., 290
Camp, D. G. II, 7, 87, 88, 193, 194, 195, 199, 200, 201, 202
Campagne, F., 4, 85, 280, 290
Campbell, S., 232
Cande, C., 262, 264
Candiano, G., 6, 7
Cannon, R. O., 147, 348
Cantoni, O., 111, 112, 113, 114, 116, 118, 119, 243, 246, 247, 248, 249
Cao, W., 232
Capaldi, R. A., 4, 5, 11
Carafoli, E., 119
Carbone, V., 5, 254, 257, 260
Carew, T. E., 35
Carey, R. M., 144, 145, 152, 161, 162, 164, 237, 238
Carraro, S., 164
Carraway, M. S., 165, 166
Carreras, A., 398, 401, 402
Carreras, M. C., 310, 318
Carroll, J. A., 35
Carroll, R. J., 185
Carroll, R. T., 353
Carulli, N., 222
Carver, D. J., 160
Carver, J., 159, 172
Cascio, M. B., 233, 236, 284
Casella, L., 4
Casey, T. E., 319
Cashon, R. E., 256
Casoni, F., 5, 11
Cassader, M., 340
Castegna, A., 193, 194, 297, 305, 314, 353
Castro, C., 159
Castro, L., 255
Cattabeni, F., 118
Cavarra, E., 4, 5, 296, 297, 310
Cederbaum, A. I., 284
Cederbaum, S. D., 222
Celedon, G., 258
Ceriello, A., 339
Cerioni, L., 118, 119
Chacko, B. K., 151
Chait, A., 34, 35, 36, 42, 58, 59
Chait, B. T., 232
Chakder, S., 227
Chaki, M., 401
Champion, H. C., 418, 424
Chan, M. C., 375
Chan, N., 159, 165
Chandra, M., 284
Chandramouli, G. V., 395
Chang, J. Y., 194
Chang, L. Y., 66, 83

Chang, P. Y., 340
Chapman, M. J., 53
Charlton, T. S., 34
Chaudhuri, G., 324, 418
Chea, H., 36
Cheah, J. H., 85, 233, 236, 284
Chen, C., 170, 236, 288
Chen, E. Y., 160, 161
Chen, H. S., 160, 232
Chen, J., 35, 38, 39, 67, 83, 254, 296, 313
Chen, K. J., 152
Chen, L. C., 222, 224
Chen, Q., 66, 83, 158, 161, 170, 296
Chen, S., 310
Chen, W. N., 195
Chen, X., 107
Chen, Y. Y., 66, 83, 85
Cheong, E., 415, 425
Cheroni, C, 5, 11
Cheung, M. C., 36
Chevalier, S., 178
Chevallet, M., 4
Chhabra, P., 159, 160, 164, 169, 170
Chi, R., 327
Chiang, K., 418
Chiappetta, G., 12, 297
Chilvers, E. R., 212
Chin, M. H., 87, 88, 194, 195, 202
Chiueh, C. C., 426
Chlopicki, S., 390
Cho, H., 160
Cho, J. H., 211
Chock, P. B., 4, 297
Choi, D. V., 245
Choi, E. J., 211, 232
Choi, I., 296
Choi, J., 318, 323, 326, 327
Choi, S. Y., 318, 323
Choi, Y. B., 96, 160, 232, 426
Chowdrey, H. S., 276
Christianson, D. W., 227
Christie, I., 339, 343
Christodoulou, D., 107, 139, 216, 217, 420, 422
Christopher, T. A., 211, 425
Chueng, M. C., 36
Chung, K. K., 89, 138, 232
Chung, Y. Y., 283
Ci, C. Y., 160
Cimen, B., 354
Ciriolo, M. R., 260
Ciurak, M., 66, 80, 84, 85
Clancy, R. M., 422
Clarke, D. M., 160, 161
Clauss, T. R., 195
Clemens, P. R., 318, 323
Clementi, E., 113, 118, 119, 138
Cleveland, D. W., 314
Coccia, R., 297, 310, 314

Author Index

Cociorva, D., 195
Coddington, J. W., 67, 83
Coffin, D., 139
Cohen, M. F., 401, 402
Cohen, R. A., 284
Colasanti, M., 159, 243, 244, 246, 247, 248, 249
Cole, J. A., 391
Coleman, M. K., 200
Coletta, M., 172
Collado, J. A., 275, 280
Cologna, S. M., 418, 424
Colombo, R., 4, 5, 104, 105, 147, 296, 297, 310, 362, 363
Colton, C. A., 426
Comhair, S. A., 66, 67, 83
Comtois, A. S., 194
Conaway, M. R., 160, 233
Conde-Freiboes, K., 248
Connolly, D. T., 38
Connor, J., 264
Connor, J. T., 18
Cook, J. A., 107, 216, 217, 420, 422
Cook, J. C., 139
Cook, N. M., 420
Cooley, S., 97, 100
Cooper, C. E., 248
Corpas, F. J., 398, 400, 401, 402, 403, 404, 405, 406
Corrales, F. J., 159, 237, 273, 275, 280
Cortas, N. K., 377
Coskun, O., 244
Cottrell, J. S., 193
Coward, L., 5, 11, 30
Cox, R. D., 324
Crabb, J. W., 5, 6, 11, 26, 30, 193, 194, 296, 297, 310
Craft, J., 165
Craig, R., 193
Craig, S. W., 113
Crapo, J., 66, 83
Crawford, J. H., 151
Creasy, D. M., 193
Cribb, A. E., 311
Crombez, E. A., 222
Cross, C. E., 35, 71, 77, 296
Crow, J. P., 29, 38, 39, 313, 314, 387
Crowley, J. R., 4, 34, 35
Cui, J., 232
Culcasi, M., 310
Curry, S. R., 319
Czaika, G., 319
Czerniczyniec, A., 318

D

Daaka, Y., 159, 160
D'Adamio, L., 6, 7
Daehnke, H. L. D., 35, 38

Dahm, C. C., 288
Dai, T., 425
Daiber, A., 211, 256
Daikhin, E., 67, 296
Dalle-Donne, I., 4, 5, 104, 105, 147, 296, 297, 310, 361, 362, 363
Dalurzo, H. C., 398
D'Ambrosio, C., 3, 4, 6, 7
Dangl, J. L., 398
Daou, M. C., 232
Darbyshire, J. F., 319
Darrow, R. M., 5, 11, 26
Das, P., 426
Dat, J. F., 398
Datta, B. N., 165, 166
Daugas, E., 262, 264
Daugherty, A., 35
Davidson, W. S., 37
Davidsson, P., 352, 353
Davies, K. J., 296
Davies, M. J., 53
Davis, C. W., 170
Davis, R. W., 199, 200, 201
Davison, W., 319
Dawnay, A., 4, 339, 342, 343, 345, 346, 350, 351
Dawson, K. S., 85
Dawson, T. M., 89, 138, 232, 244, 248
Dawson, V. L., 89, 138, 232, 244, 248
Day, C. S., 418, 424
Day, H., 158, 161, 170, 245
DeAngelo, J., 96, 160
Dechatelet, L. R., 399
Decker, G. L., 113
Dedkova, E. N., 318, 327
Deen, W. M., 384, 391
DeGraff, W., 420, 422
de Groat, W. C., 318, 323
de Hoog, C. L., 195
Dejam, A., 138, 139, 161
De Jong, K., 264
de Jong, S., 178, 185
De Koster, Z. J., 425
de la Llera-Moya, M., 35, 36
De la Mata, M., 273, 275, 280
Delatour, T., 339, 340, 344, 346, 353
Della Torre, A., 160
Delledonne, M., 284, 286, 398
del Río, L. A., 397, 398, 400, 401, 402, 403, 404, 405, 406
De Marco, C., 297, 314
De Martino, A., 260
DeMaster, E. G., 417, 421
D'Emilia, D. M., 85
Demple, B., 96
De Nadai, C., 113
Deng, H. W., 425
Denicola, A., 67, 83, 97, 255, 256, 257, 258, 260, 261, 265, 349

Denker, K., 169
Dennehy, M. K., 4, 280, 284
Dennis, E. A., 248
Dennis, M. F., 377
de Oliveira, M. G., 161
Derakhshan, B., 4, 85, 232, 280, 290
deRonde, K., 160
Desai, K. M., 391
Desiderio, D. M., 5, 11, 12, 195
Desikan, R., 398, 400
Detweiler, C. D., 255, 256, 260
Deussen, A., 418
Devi, S. R., 401
Dewhirst, M. W., 160, 213, 284
Dezfulian, C., 414
Dhawan, V., 138
Di Benedetto, G., 425
Di Cera, E., 172
Dickfeld, T., 138, 159
Dikalov, S., 400, 401
di Loreto, C., 4, 6, 7
Dimmeler, S., 211, 232
Dinerman, J. L., 248
Dirsch, V. M., 327
D'Ischia, M., 12, 297
Di Stasi, A. M., 256, 261, 262, 297
Dixon, R. N., 423
Dockendorff, K., 378
Doctor, A., 140, 144, 145, 147, 148, 149, 151, 158, 159, 160, 161, 162, 164, 165, 166, 167, 169, 170, 171, 172, 275
Doctor, L., 164
Doctorovich, F., 415
Doherty, D. H., 319
Doherty, J., 144, 145, 151, 159, 162, 164, 165, 166
Domigan, N. M., 34
Domon, B., 195
Donaldson, M. T., 311
Donzelli, S., 4, 5
Dougall, H., 383
Doughan, A., 400, 401
Downing, K. H., 97
Doyle, M. P., 421
Drapeau, G. R., 40
Drazen, J. M., 138
Dremina, E. S., 30, 353
Dreyer, R. N., 37
Dugan, L. L., 245
Duncan, C., 383
Duncan, M. W., 29, 34, 297, 311, 339
Dunn, J. L., 35
Dunn, M. J., 31
Duran, D., 67, 83, 297
Duranton, C., 264
Durner, J., 283, 284, 285, 286, 401
Durzan, D., 398, 401
Dutton, A. S., 412, 414, 418, 424
Dybdal, N., 311

E

Eaton, S., 31
Edelman, G. M., 362
Edelstein, D., 349
Eich, R. F., 319
Eischeid, A., 144, 145, 151, 159, 162, 164, 165, 166
Eiserich, J. P., 35, 66, 71, 77, 83, 296
Ekblom, B., 363
Ekelund, U., 424
Eker, A. P., 66
Elahi, M. M., 138
Elas, M., 390
Elberling, J. A., 417
Elfering, S. L., 310
Elferink, J. G., 425
Ellery, S., 165, 166
Ellis, G., 363
Emerson, R. K., 264
Emig, F. A., 227
Endres, M., 244
Eng, J. K., 193
Engels, I. H., 262, 264
Erdjument-Bromage, H., 140, 164, 170, 212, 214, 233, 236, 274, 280, 284, 285
Erzurum, S. C., 66, 67, 83
Espey, M. G., 4, 5, 67, 327, 375, 386, 402, 420, 421, 422, 424, 425, 426
Esplugues, J. V., 391
Estaquier, J., 264
Esteban, F. J., 401, 402
Estevez, A. G., 160, 232, 233
Estrin, D. A., 415
Eu, J. P., 160, 232, 284
Evans, P., 338
Everett, S. A., 377
Exner, M., 254, 256, 260

F

Fahy, E., 4
Fang, F. C., 274, 284
Fang, K., 144, 145, 152, 159, 160, 161, 162, 164, 170, 232, 237, 238
Fang, M., 280
Farkouh, C. R., 67, 83, 138
Farmer, P. J., 419, 421
Feelisch, M., 4, 5, 66, 67, 83, 138, 140, 147, 148, 149, 151, 161, 173, 211, 239, 322, 362, 417, 418, 420, 422, 424, 425
Feener, E. P., 349
Feix, J., 67, 83, 194
Feldman, P. L., 319
Felix, C. C., 75
Fenaille, F., 339, 346
Fenn, J. B., 37
Fernández-Ocaña, A., 401
Feron, O., 310

Ferrante, R. J., 314
Ferrara, L., 6, 7
Ferreira, A. M., 254
Ferrer, M. A., 398
Ferrer-Sueta, G., 254, 255
Ferrige, A. G., 318, 324
Ferris, C. D., 140, 212, 214, 233, 236, 274, 280, 284, 285
Fiala, E. S., 297
Fink, B., 400, 401
Finkel, T., 4, 34
Fiorani, M., 118
Fiskum, G., 113
Fiszman, M. L., 310, 318
Fleischer, G. A., 344, 352, 353
Fletcher, J. W., 422
Florens, L., 200
Floris, R., 254, 256
Floyd, R. A., 297, 339
Fluhr, R., 398
Flynn, S. P., 179
Fogelman, A. M., 36
Fogliano, V., 391
Foley, J., 165
Folkerts, G., 138
Föller, M., 264
Fontijn, A., 324
Foote, C. S., 34
Forbes, M. S., 159, 160, 164, 169, 170
Ford, E., 419, 420, 421
Ford, P. C., 319
Forrester, M. T., 170, 236, 275, 288
Forstermann, U., 362
Forte, T. M., 37
Foster, L. J., 195
Foster, M. W., 67, 85, 87, 138, 159, 160, 170, 171, 236, 275, 284, 288
Fountain, S. T., 353
Fox, J. B., Jr., 375, 376
Fox, P. L., 35, 36, 58, 59
Fraga, E., 275, 280
Francoeur, M., 319
Frangos, J. A., 310
Frank, C. W., 324
Franze, T., 5, 9, 11
Freeman, B. A., 4, 5, 11, 30, 35, 66, 67, 71, 77, 83, 97, 98, 112, 160, 254, 255, 265, 310, 348
Frein, D., 211
Frejaville, C., 75
Frenneaux, M. P., 148, 165, 166
Fricker, M. D., 399
Fridovich, I., 415, 422
Friesen, M., 338
Froehlich, J. P., 425
Frohlich, J., 4
Frohman, E., 244
Frölich, J. C., 169, 178, 185, 340, 343, 346, 378

Frost, H., 66, 71, 83
Fruscalzo, A., 4, 6, 7
Fu, W. J., 185
Fu, X., 5, 11, 34, 35, 36, 37, 38, 39, 53, 57, 58, 59, 66, 67, 71, 83
Fu, Y., 314
Fugger, R., 222
Fujii, S., 167, 168, 390
Fujimuro, M., 233, 236, 284
Fukuda, K., 212
Fukuda, Y., 222
Fukumoto, K., 31
Fukuto, J. M., 411, 412, 414, 418, 419, 420, 421, 422, 423, 424, 425
Fukuya, H., 382
Fulcoli, G., 6, 7
Fung, H. L., 338, 339, 425
Fuseler, J., 66, 83

G

Gabrielsson, A., 343, 346
Gaizutis, M., 339
Galanti, L. M., 383
Galaris, D., 342
Galatsis, P., 353
Galeva, N. A., 89, 354
Galliani, C., 66, 83
Gally, J. A., 362
Gambardella, L., 258, 259, 260, 262, 264
Gambino, R., 340
Gamson, J., 420, 422
Ganser, G., 339
Gao, B. R., 339
Gao, C., 67
Gao, E., 211
Gao, F., 425
Gao, W. D., 425
Gapper, P. W., 148
Garber, D. W., 40
Garcês, H., 398, 401
Garcia-Brugger, A., 284
García-Mata, C., 398
Gaston, B., 85, 89, 138, 140, 144, 145, 147, 148, 149, 151, 152, 157, 158, 159, 160, 161, 162, 164, 165, 166, 167, 169, 170, 171, 172, 232, 233, 237, 238, 275
Gatti, R. M., 254, 255
Gattullo, D., 425
Gaut, J. P., 5, 35, 340
Gaven, F., 310
Gavett, S. H., 232
Gaynor, R. B., 244
Gbadegesin, M., 426
Gea, J., 194
Geary, R. L., 34, 35, 36, 58, 59
Geddes, T. J., 297
Gentile, L., 340

George, D., 140
Gergel, D., 284
Gerris, C. D., 164, 170
Gevaert, K., 5, 9, 12
Ghafourifar, P., 159, 160, 162, 164, 170, 232, 310, 317, 318, 321, 322, 323, 326, 327, 330, 331
Gharib, S. A., 36
Gharini, P., 161
Ghesquiere, B., 5, 9, 12
Ghosh, S. S., 4, 5, 11
Gianazza, E., 5, 11
Gianturco, S. H., 66, 83
Giardino, I., 349
Giasson, B. I., 66, 83, 296, 314
Gibson, B. W., 5, 8, 9, 11, 29
Gieche, J., 338
Giffard, R. G., 245
Gilbert, H. F., 104
Gildemeister, O., 166, 167
Giles, S., 18
Gillette, R., 387
Gilligan, J. P., 178, 179
Giordano, E., 244
Giulivi, C., 310, 318
Giustarini, D., 4, 5, 104, 105, 147, 296, 297, 310, 361, 362, 363
Gladwin, M. T., 137, 139, 140, 141, 146, 147, 148, 149, 150, 151, 152, 222, 236, 348, 362, 414
Glogowski, J., 319
Glomset, J. A., 36
Godzik, A., 96
Goethals, M., 5, 9, 12
Gokulrangan, G., 354
Goldberg, M. P., 245
Golden, M., 383
Goldknopf, I. L., 285, 288
Goldstein, J. A., 35, 38
Goldstein, S., 254, 383
Gómez, M., 398, 401, 402
Gonzalez, G., 258
Gonzalez, R., 275
Good, P. F., 296, 313
Goormastic, M., 35, 36, 58, 59
Gordge, M. P., 276
Gordon, D. J., 36
Gore, J., 66, 83
Gorren, A. C., 158
Goshe, M. B., 194
Goss, S. P., 70
Gotoh, T., 222
Gotto, A. M., Jr., 36
Gougeon, R., 178
Gould, K. S., 398
Govoni, M., 363
Gow, A., 67, 83, 140, 144, 145, 147, 149, 151, 159, 160, 161, 162, 164, 165, 166, 167

Gow, A. J., 67, 83, 138, 158, 159, 160, 161, 165, 170, 232, 237, 238, 284, 297, 310
Goyne, T. E., 34
Gozal, D., 159, 160, 169
Granger, D. L., 362, 367
Gratzel, M., 419
Gravielle, M. C., 310, 318
Gray, L., 66, 83
Graziano, M., 398
Greco, T. M., 4, 280, 284
Green, L. C., 319
Green, P., 5, 11, 36, 38, 39, 59, 67, 83
Green, P. S., 35, 36, 37, 39, 53, 57, 58
Green, S., 383
Greenacre, S. A., 66, 67, 338
Greis, K. D., 29
Griess, J. P., 373
Griess, P., 319
Griffith, O. W., 319
Grisham, M. B., 66, 67, 83, 107, 139, 160, 216, 217, 363, 371, 400, 420, 422
Gritsenko, M. A., 193, 195, 199, 200, 201
Grody, W. W., 222
Gross, G. J., 425
Gross, S. S., 4, 85, 232, 280, 290
Gross, W. L., 85
Groves, J. T., 67, 83
Grubb, P. H., 170
Gruetter, C. A., 138
Grune, T., 296, 338
Grunert, T., 4
Grzesiok, A., 417
Gu, Z., 85, 89, 160, 232
Guidarelli, A., 111, 113, 114, 118, 119
Guidi, S., 340
Guikema, B., 138
Guillozet-Bongaarts, A. L., 314
Gunn, C., 255
Gunther, M. R., 75, 77
Guo, H., 67
Gurka, M., 144, 145, 151, 159, 162, 164, 165, 166
Gutteridge, J. M. C., 382, 383, 393, 398
Gutzki, F. M., 4, 169, 339, 343, 346, 347, 378
Guy, J. K., 421
Guy, P. A., 340
Gwóźdź, E. A., 398, 401

H

Ha, S. W., 349
Haddad, I. Y., 66, 83
Haendeler, J., 211, 232
Hagen, T. M., 34
Hager, K., 344, 352, 353
Hajjar, D. P., 232
Hall, E. D., 353
Halliwell, B., 35, 71, 77, 296, 297, 338, 339, 382, 383, 393, 398

Author Index

Halonen, T., 178
Halpern, H. J., 395
Hamada, S., 161
Hammel, M., 53
Hammes, H.-P., 349
Hampton, M. B., 261
Han, O. J., 210
Han, T. H., 414, 423
Han, Y., 85
Hancock, J. T., 398, 400
Hanes, M. A., 138, 159
Hanisch, U. K., 248
Hanson, P. J., 391
Hao, G., 4, 85, 232, 280, 290
Haqqani, A. S., 4, 11
Hara, M. R., 232, 233, 236, 284
Harada, A., 139
Hardie, J. M., 382
Hardy, M. M., 35, 66, 83
Hare, J. M., 424, 425
Harman, L. S., 74, 75
Harms, H., 122, 133
Harrison, D. G., 400, 401
Harrison, J., 296
Harrison, J. E., 34
Harry, R. A., 31
Harvey, S. C., 40
Hasan, R., 97, 100
Hasegawa, E., 38
Hashemy, S. I., 284
Hashiguchi, T., 210
Hashimoto, Y., 160, 232, 233
Hatcher, N., 387
Hausladen, A., 96, 149, 150, 159, 160, 161, 162, 165, 166, 232, 237, 238, 275
Haussinger, D., 343, 346, 349
Havel, R. J., 36
Hawkins, C. L., 53
Hay, R. T., 160
Haynes, T. E., 179, 180, 185
Hayward, S. D., 233, 236, 284
Hazen, S. L., 4, 5, 11, 34, 35, 66, 67, 83, 170, 222, 338
Heidland, A., 343, 346, 349, 350, 351
Heijnen, H. F., 4, 280, 284
Hein, G. E., 414
Heinecke, J. W., 4, 5, 11, 33, 34, 35, 36, 37, 38, 39, 42, 53, 57, 58, 59, 66, 67, 83, 297, 340, 351
Heiss, C., 362
Heitman, J., 159, 275
Helkamaa, T. H., 426
Hempstead, B. L., 232
Hems, R., 74, 75, 80
Henderson, E. M., 164
Hendgen-Cotta, U., 362
Hendrickson, R., 222
Hendrix, S. A., 324

Henglein, A., 419
Hensley, K., 297, 339
Herbert, T. P., 349
Hermelink, C., 127
Hermfisse, U., 118
Hernández, J. A., 398
Hernandez-Saavedra, D., 255
Herold, S., 254, 256, 260
Herzenberg, L. A., 267
Hess, B. J., 158, 161, 170
Hess, D. T., 96, 158, 159, 160, 161, 210, 232, 237, 238, 284
Hester, L. D., 233, 236, 284
Heuvelman, D. M., 38
Hewitt, S. J., 426
Heydari, K., 267
Hibbs, J. B., Jr., 362, 367
Hibino, Y., 339, 342
Higgins, V. J., 400
Higgs, E. A., 4, 244
Higuchi, T., 386
Hilderman, R. H., 319
Hill, H. A., 74, 75, 80
Hill, J., 85
Hill, J. D., 139
Hill, L. R., 200
Himeno, M., 376
Hirasawa, Y., 388
Hirata, Y., 319, 326, 386, 401
Hirose, M., 354
Hirota, K., 212
Hirota, S., 381, 383, 384, 385, 386, 387, 388, 389, 390, 391, 392, 393, 394
Hirotani, M., 386
Hisanaga, M., 139
Hiscock, S. J., 400
Hishiki, T., 261
Hitomi, Y. H., 339, 342
Hobbs, A. J., 412, 424
Hobbs, G. R., 339
Hodara, R., 4, 280, 284
Hoeldtke, R. D., 339
Hoffman, J., 66, 83
Hoffman, S., 67
Hoffmann, A., 248
Hoffmann, J., 211, 232
Hogg, H., 159, 160
Hogg, N., 67, 70, 75, 83, 107, 138, 139, 140, 141, 151, 159, 170, 194, 236, 275, 280, 348, 414
Hogg, P. J., 96
Holbrook, N. J., 34
Holcman, J., 66, 80, 84, 85
Hollan, S., 346, 347
Hollyfield, J. G., 5, 11, 18, 26
Holmgren, A., 96, 101, 102, 104, 105, 284
Holt, S., 140, 149, 151
Holtta, E., 229
Holz, H., 50

Hommes, N. G., 128, 129, 133
Hong, S. J., 5, 11, 194
Hoofnagle, A. N., 36
Hooper, A. B., 122, 123, 127, 128, 129, 131, 133
Hope, H. R., 38
Hopkins, N., 18
Horanyi, M., 346, 347
Hori, T., 376
Horiguchi, T., 296, 314
Horwitz, J., 296, 313
Hoshi, M., 166
Hotchkiss, R. S., 35
Hothersall, J. S., 138
Houghton, F. D., 327
Houk, K. N., 412, 414, 418, 419, 421, 424
Howell, S. K., 343, 344, 345, 347, 348, 350
Howlett, A., 85
Hoyt, D. G., 349
Hristova, M., 35, 71, 77
Hsiao, C. H., 399
Hsu, A., 296, 313
Hsu, C. Y., 310
Hsu, F. F., 4, 34, 35
Hszieh, R., 418
Hu, A., 211
Hu, C. P., 425
Hu, J., 185
Hu, P., 66, 83
Huang, B., 170, 236, 288
Huang, J., 139, 147, 148, 151, 418
Huang, K. T., 147, 148, 151
Huang, P., 244
Huang, Q., 160, 213, 284
Huang, Y., 85
Huang, Y. C., 165
Huang, Y. F., 85
Huang, Z., 152, 244
Huang, Z. Q., 192
Huart, J. J., 264
Huber, C. G., 5, 9, 11
Huber, S. M., 262, 264
Huganir, R. L., 85
Hughes, M. N., 414
Huh, S. H., 211, 232
Huizenga, J. R., 319, 363
Huke, S., 425
Hunt, J. F., 138, 160
Huq, N. P., 57
Huri, D. A., 160
Hurst, J. K., 67, 83, 97, 254, 255, 260
Hussain, M. M., 36
Hussain, S. N., 194
Hyun, D. H., 297
Hyun, J., 421

I

Iadecola, C., 244
Ida, T., 376
Igarashi, M., 222, 349
Ignarro, L., 138
Ignarro, L. J., 324
Iijima, M., 354
Ikamoto, T., 167, 168
Ikemoto, M., 222
Imai, Y., 386
Inan, L., 244
Incze, K., 138
Ingelse, B. A., 297, 311, 339
Inoue, K., 159, 167, 168
Intaglietta, M., 310
Inze, D., 284, 286
Inzé, D., 398
Irigoin, F., 255, 260
Irimura, T., 386
Isbell, T. S., 418, 424
Ischiropoulos, H., 4, 5, 18, 35, 36, 38, 39, 58, 59, 66, 67, 75, 83, 85, 138, 158, 161, 170, 280, 284, 296, 297, 310, 313, 314, 338, 343, 387
Isenberg, J. S., 4, 5
Ishibashi, T., 376
Ishii, K., 362
Ishii, Y., 354
Ishizaki, E., 339
Isobe, C., 339
Isoda, T., 424, 425
Ito, R., 354
Ito, T., 376
Itoh, Y., 166
Itoyama, Y., 314
Ivanova, J., 422
Iwamoto, H., 38
Iwasaki, Y., 354
Iwig, J. S., 100, 101, 102
Iyer, R. K., 222

J

Jackson, M. I., 411, 412, 420, 424
Jackson, R., 66, 83
Jackson-Lewis, V., 297, 313, 351
Jacobs, J. M., 7, 193, 195, 199, 200, 201
Jacobs, M., 339
Jaffrey, S. R., 140, 164, 170, 212, 214, 233, 236, 274, 280, 284, 285
Jagtap, P., 349
James, P. E., 148, 165, 166
Jamshidzadeh, A., 392
Janero, D. R., 138
Jansen, P. L., 319, 363
Jansson, E. A., 362
Jaraki, O., 85, 138, 140, 284
Jeng, B. H., 18
Jenkinson, C. P., 222
Jenner, P., 297, 339
Jennings, M., 38
Jensen, B. O., 392
Jensen, O. N., 58

Author Index

Jerlich, A., 53
Jetten, M. S. M., 122, 124, 125, 126, 127, 128, 129, 131
Ji, X., 327
Ji, Y., 106
Jia, L., 85, 159
Jiang, H., 297
Jiang, Z. Y., 349
Jiao, K., 5, 9, 11, 342
Jiménez, A., 398
Jobgen, S. C., 180
Jobgen, W. S., 180
Joe, K. H., 210
Johansson, C., 284
Johns, R. A., 232
Johnson, D. E., 260
Johnson, K. A., 319
Johnson, M., 138, 159, 160, 169
Johnson, M. A., 160, 232, 233
Johnston, P., 383
Joksovic, P. M., 160
Jones, A. D., 35, 71, 77
Jones, B. N., 178, 179
Jones, H. Q., 138
Jones, J. D. G., 398
Jones, P., 254, 256
Jones, R. L., 283
Jones, S. N., 232
Joseph, H., 159, 160
Joseph, J., 65, 67, 83, 139, 194, 401
Joshi, M. S., 67, 83
Joshi, S. S., 321, 322
Jourd'heuil, D., 66, 67, 83, 138, 139, 151, 160, 239, 386, 420, 422
Jourd'heuil, F. L., 66, 67, 83
Juluri, K., 85
Jung, N., 169
Junker, W., 185

K

Kaihara, S., 222
Kaiser, W. N., 402
Kakinuma, K., 387
Kalivendi, S., 401
Kaludercic, N., 411
Kalume, D. E., 5, 8, 9, 11, 12
Kalyanaraman, B., 65, 66, 67, 70, 75, 83, 159, 160, 194, 390, 401
Kalyanaraman, J., 139
Kamada, H., 390
Kamaid, A., 160, 232, 233
Kambayashi, Y., 339, 342
Kanai, A. J., 318, 323
Kandeda, Y., 349
Kandefer-Szerszen, M., 310
Kane, L. S., 160, 161, 232, 237, 238
Kanehiro, H., 139

Kanematsu, S., 404
Kangas, L. J., 7
Kankuri, E., 426
Kann, O., 248
Kanski, J., 5, 11, 193, 194, 354
Kao, J., 57
Karabicak, U., 354
Karachalias, N., 4, 339, 342, 345, 346, 347
Karala, A. R., 287
Karlinsey, M. Z., 159, 160, 164, 169, 170
Karoui, H., 75
Karp, M., 217
Karpinski, S., 398
Karppi, J., 178
Kasai, Y., 222
Kashima, A., 11
Kass, D. A., 418, 420, 421, 424, 425
Kassim, S. Y., 35, 36, 58
Kataoka, M., 383
Kato, G. J., 222
Kato, T., 314
Katori, T., 418, 420, 421, 424, 425
Kaul, M., 232
Kaur, H., 339
Kavuru, M. S., 66, 67, 83
Kawahara, S., 319, 326, 401
Kazan, N., 244
Keaney, J. F., 138, 140
Keaney, J. F., Jr., 66, 83
Keefer, L. K., 418, 424
Keightley, J. A., 342
Keimer, R., 343, 346
Kelleher, N. L., 37
Keller, J. N., 296, 297, 304, 305, 314
Kellner-Weibel, G., 35, 36
Kelly, D. R., 66, 83
Kelly, J. F., 4, 11
Kelly, R. A., 85
Kelm, M., 66, 83, 138, 139, 151, 161, 239, 362
Kem, R. M., 222
Keng, T., 96, 160
Kennett, K. L., 100, 101, 102, 104, 105
Kepka-Lenhart, D., 221, 222, 224, 228, 229
Kerber, S., 138, 139
Kerecman, J. D., 170
Kern, R. M., 222
Kerr, S. W., 138
Keston, A. S., 400
Keszler, A., 67, 83, 140, 170, 236, 275, 280
Kettenhofen, N. J., 140, 170, 275
Kettenmann, H., 248
Kettle, A. J., 34
Keunen, J. G., 127
Keys, J. R., 159, 160
Khairutdinov, R. F., 67, 83
Khalatbari, A., 148, 165, 166
Khan, A. H., 193, 195, 201
Khan, J., 339

Kharbanda, K. K., 349
Khoo, K. H., 85
Kiboki, K., 349
Kieman, M., 222
Kienke, S., 185
Kiesow, L. A., 39
Kigoshi, T., 376
Kikuchi, K., 319, 326, 386, 401
Kim, G., 97, 100
Kim, G. M., 310
Kim, J. E., 232
Kim, M. S., 211, 232
Kim, S., 139, 284, 420, 422
Kim, S. F., 160, 233, 236, 284
Kim, S. O., 96, 158, 159, 160, 210, 232, 284
Kim, S. W., 210
Kim, S. Y., 107, 216, 217
Kim, W. K., 426
Kim, W.-S., 387
Kim-Shapiro, D. B., 137, 139, 147, 148, 151, 152, 165, 166, 414
Kincaid, E. D., 321, 322
King, G. L., 349
King, M., 166, 167
King, S. B., 152, 413, 418, 424
Kiningham, K. K., 296, 314
Kinoshita, A., 261
Kinter, M., 5, 11, 342
Kirino, Y., 319, 326, 401
Kirk, M. C., 5, 11, 30, 66, 313
Kita, T., 212
Kita, Y., 388
Klassen, L. W., 349
Klatt, P., 321
Klebanoff, S. J., 35
Klein, C., 85
Klein, D., 222, 224
Klein, D. F., 66, 67, 83
Klein, J. B., 193, 194, 296, 297, 304, 305, 314, 353
Klein, S. D., 318
Kleinbongard, P., 139, 161, 362
Klessig, D. F., 284, 401
Kline, J., 144, 145, 151, 159, 162, 164, 165, 166
Klinguer, A., 398
Klotz, L. O., 296, 338
Knepper, M. A., 195
Knopp, R. H., 36
Knowles, R. G., 248, 319, 322
Knyushko, T. V., 87, 88, 89, 194, 195, 202
Ko, Y. H., 417
Kobayashi, K., 11
Kobayashi, T., 38
Kobayashi, Y., 401
Koch, W. J., 159, 160, 211
Koeck, T., 5, 6, 18, 296, 297, 310
Koesling, D., 158, 284
Kohima, H., 166

Kohli, R., 179, 180, 185
Kohno, K., 385
Kojima, H., 319, 326, 386, 401
Kok, B., 319, 363
Kolkman, J. J., 376
Kone, B. C., 285, 288
Kong, S. K., 297
Konorev, E., 159, 160
Koops, H.-P., 122, 133
Kooy, N. W., 66, 83, 387
Koppenol, W. H., 18, 35, 112, 296, 310
Kopyra, M., 398, 401
Kotamraju, S., 67
Koumas, M. T., 288
Koval, M., 338
Kowall, N. W., 314
Kozhukhar, A. V., 211
Krichman, A. D., 165
Kridel, S. J., 232
Krishna, M. C., 107, 216, 217, 420, 422
Kritharides, L., 50
Kroemer, G., 262, 264
Krogh, A., 194
Kroncke, K. D., 244
Ku, D., 66, 83
Ku, G., 310
Kuhn, D. M., 297
Kumar, P., 200
Kuncewicz, T., 285, 288
Kuo, P. C., 67, 160
Kutchukian, P. S., 66, 67, 83
Kuypers, F. A., 264
Kwon, N. S., 210

L

La Camera, S., 398
Lacan, G., 87, 88, 194, 202
Lafon-Cazal, M., 310
Lai, M. M., 248
Lakowicz, J. R., 389
Lamarche, M., 178
Lamas, S., 274, 284, 285
Lamattina, L., 398
Lamotte, O., 284, 398
Lancaster, J. R., 164, 169, 348
Lancaster, J. R., Jr., 67, 83
Land, E. J., 66, 80, 84
Lander, H. M., 232
Landers, R. A., 227
Landino, L. M., 95, 97, 100, 101, 102, 104, 105, 288
Lang, F., 262, 264
Lang, K. S., 262, 264
Lang, P. A., 262, 264
Langle, F., 222
Lanone, S., 66
Lardinois, O. M., 75, 77

Author Index

Laroux, F. S., 66, 83, 160
Larsen, F. J., 363
Larson, S. K., 318, 323, 326, 327
Larsson, B., 194
Larstad, M., 343, 346
Lass, A., 66, 67, 83
Lassing, I., 96, 101
Laubach, V. E., 159, 160, 164, 169, 170, 194
Lauber, K., 262, 264
Lauber, W. M., 35
Laude, K., 400, 401
Lauer, T., 362
Laufs, U., 244
Launay, J.-M., 66
Lauro, G. M., 244
Le, D. A., 160, 232
Leal, F. A., 420
Lecourieux, D., 284
Ledda, L., 6, 7
Lee, A., 362
Lee, H. S., 210
Lee, J. J., 66, 67, 71, 83
Lee, K. Y., 210
Lee, M., 297
Lee, M. J. C., 417
Lee, S., 245
Lee, S. H., 232
Lee, V. M., 296, 314
Lee, V. M.-Y., 66, 83
Leeuwenburgh, C., 4, 35, 66, 83
Lehmann, C., 169
Lehrer, R. I., 34
Leifert, C., 383
Leikert, J. J., 327
Leize-Wagner, E., 4
Lemon, D. D., 319
Lendermon, E., 164
Lenhinger, A. L., 113
Lenkiewicz, E., 66, 71, 83
Lens, S. I., 128, 129, 133
León, A. M., 398, 401, 402
Leonard, S. S., 284
Leto, T. L., 259, 260, 262, 264
Levine, R. L., 4, 53
Levison, B., 5
Lewis, C., Jr., 38
Lewis, G. F., 36, 37
Lewis, S. J., 66, 83
Li, C. Y., 213, 284
Li, F., 160, 213, 284
Li, H., 180, 189
Li, L., 194
Li, S., 232
Li, T., 319
Li, W., 35, 38, 96, 399
Li, X., 89, 232
Li, Y. J., 425
Liang Du, X., 349

Liao, J. K., 244
Liaudet, L., 192, 349
Libourel, I. G., 283
Licht, W. R., 384, 391
Liddington, R. C., 232
Lieberman, M., 159, 160, 169
Liebkind, R., 426
Liebler, D. C., 4, 280, 284
Liesi, P., 426
LightFoot, R., 296, 314
Lillig, C. H., 284
Lim, M. H., 327, 331
Lin, H., 200
Lin, J. L., 160
Linares, E., 254, 255, 256, 260
Lind, J., 254, 383
Lindberg, U., 96, 101
Lindermayr, C., 283, 284, 285, 286
Liochev, S. I., 415, 422
Lippard, S. J., 327, 331
Lipscomb, M. F., 66, 67, 83
Lipsius, S. L., 327
Lipton, A., 159, 160, 169
Lipton, M. S., 194
Lipton, S. A., 85, 89, 96, 158, 159, 160, 232, 284, 426
Lissi, E., 67, 83
Lissi, E. A., 258
Liu, G. L., 425
Liu, H., 200, 201
Liu, H. R., 211
Liu, L., 138, 159, 160, 161, 164, 232, 237, 238, 275
Liu, S., 160, 213, 284
Liu, T., 7, 193, 195, 200, 201
Liu, W., 419
Liu, X., 67, 83
Liu, Z., 107
Llobera, M., 377
Lnenickova, Z., 339
Lo, C.-L., 17
Lockamy, V. L., 152
Locke, I. C., 276
Loo, T. W., 160, 161
Lopez, B. E., 412, 414, 420, 423, 424
Lopez, B. L., 211, 425
Lopez-Cillero, P., 275, 280
Lopez de Menezes, S., 254, 260
Lopez-Figueroa, M. O., 245
Lopez-Sanchez, L. M., 273, 275, 280
Lopez-Tejero, M. D., 377
Lorch, S. A., 343
Lores-Arnaiz, S., 318
Loscalzo, J., 66, 83, 85, 138, 140, 284
Lou, T. F., 399
Lövgren, T., 217
Low, F. M., 261
Lowenstein, C., 85

Lowenstein, J. M., 232
Lowry, S. F., 199, 200, 201
Lu, H., 400
Lu, Q., 138
Luche, S., 4
Luchsinger, B. P., 159, 165
Luduena, R. F., 101
Ludwick, N., 422
Lundberg, J. O., 362, 363, 391
Lungarella, G., 4, 5, 296, 297, 310
Luo, H., 85
Luque, F., 401
Luschinger, B. P., 165, 166
Lusis, A. J., 66, 71, 83
Luthen, R., 343, 346, 349
Ly, T. N., 385
Lymar, S. V., 97, 254, 255, 260, 419, 422
Lynn, B., 193, 194, 314, 353
Lyras, L., 339

M

Ma, F. H., 166
Ma, X. L., 211, 425
Ma, Y., 85, 89, 160
Mabley, J. G., 349
MacArthur, P., 139, 141, 146, 147, 152
MacArthur, P. H., 147, 148, 149, 151
Maccaglia, A., 258, 262, 264
Macchia, G., 297
MacDonald, T., 144, 145, 152, 159, 160, 161, 162, 164, 169, 170, 237, 238
MacDonald, T. L., 160, 232, 233
MacMillan-Crow, L. A., 18, 29
MacPherson, J. C., 66, 67, 83
Maeda, D., 386
Maeda, H., 159, 167, 168
Maeda, K., 388
Maejima, H. J., 387
Maejima, K., 376
Maglione, G., 6, 7
Mahapatro, S. N., 421
Maidt, M. L., 297, 339
Makela, S. K., 363
Makiishi, H., 376
Malcolm, S., 297
Malega, W. P., 87, 88
Malenkova, I. V., 284
Malgaroli, A., 290
Malik, A. B., 338, 350
Mallis, R. J., 85, 275, 276
Mallozzi, C., 256, 261, 262, 297
Maloney, R. E., 67
Malorni, W., 253, 258, 259, 260, 262, 264
Man, H. Y., 85
Mancardi, D., 4, 5, 418, 424, 425
Mandapati, S., 5, 9, 11, 342
Mani, A. R., 338, 382

Manivet, P., 66
Mann, M., 37, 58, 195, 200
Mannick, J., 140, 147, 149, 159, 160, 161, 162, 164, 167, 170
Mannick, J. B., 160, 161, 166, 167, 231, 232, 233, 237, 238
Manser, R. C., 327
Marchesoni, D., 4, 6, 7
Marechal, A., 296
Marino, G., 5, 12, 254, 257, 260, 297
Mariotto, S., 243, 244, 310
Markesbery, W. R., 193, 194, 296, 297, 304, 305, 314, 339, 353
Marks, G., 123
Marla, S. S., 67, 83
Marley, R., 140, 149, 151
Marliss, E. B., 178
Marsh, L., 89, 232
Marsh, P., 383
Marshall, H. E., 85, 138, 158, 159, 210, 232, 284
Marshall, P. A., 35, 254
Marti, M. A., 415
Martignani, C., 424, 425
Martin, J. C., 67, 83, 296, 313
Martin, M. V., 383
Martin, S. M., 392
Martinez-Ruiz, A., 285
Marton, M., 349
Maruyama, I., 210
Masliah, E., 85, 89, 160
Mason, C. E., 288
Mason, R. P., 74, 75, 255, 256, 260
Masselam, K., 97, 100
Masselon, C. D., 194
Massignan, T., 5, 11
Massillon, D., 5, 11, 30, 193, 194
Matalon, S., 29, 66, 83
Matarrese, P., 253, 258, 259, 260, 262, 264
Matata, B. M., 138
Matern, S., 362
Mathews, A. J., 319
Mathews, J. R., 160
Mathews, W. R., 138
Mathis, P., 66
Mato, J. M., 159, 237
Matsubara, T., 376
Matsumoto, A., 96, 158, 159, 160, 210, 232, 284
Matsumoto, H., 401
Matsumura, T., 349
Matsuoka, M., 160, 232, 233
Matsushima, K., 139
Matsushita, K., 232
Matthews, R. T., 314
Mattioli, T. A., 296
Maurer, T. S., 338, 339
Maxey, T., 144, 145, 151, 159, 162, 164, 165, 166
Mayer, B., 66, 67, 83, 158, 321
Mazzone, V., 248, 249

Author Index

McBride, A. E., 178
McCann, L., 400, 401
McColl, K. E. L., 391
McCord, J. M., 255
McCormack, A. L., 193
McCoy, M. T., 114
McCurnin, D. C., 170
McDonald, T. O., 34, 35, 36, 39, 58, 59
McFarlane, D., 311
McGaw, A., 97, 100
McInnis, S., 400
McKenny, S. J., 161
McLaughlin, B. E., 123
McLendon, W. J., 418, 424
McLoughlin, P., 18
McMahon, T. J., 138, 144, 145, 151, 159, 162, 164, 165, 166, 171, 284
McNeill, D. R., 339
McPherson, M. E., 138
McVey, M., 85
Mebazaa, A., 66
Megson, I., 159, 160
Mehl, M., 256
Mehls, C., 346
Meininger, C. J., 177, 179, 180, 182, 185, 189
Meisler, D. M., 18
Meissner, G., 160, 232, 284
Melani, M., 310, 318
Melega, W. P., 194, 202
Mendez, A. J., 38
Menegazzi, M., 244, 310
Meng, C. K., 37
Meng, T. C., 85
Merchange, K., 160
Merchant, K., 96, 314
Merchant, M., 296, 297, 305
Mercuri, F., 339
Merenyi, G., 254, 383
Merrett, M., 319, 322
Merrill, J. D., 343
Metere, A., 253, 259, 260, 262, 264
Meuwissen, S. G., 376
Meyer, A. J., 399
Mi, Z., 67
Miao, Q. X., 160, 161, 232, 237, 238
Michaelis, M. L., 354
Michel, T., 85, 284, 310
Miersch, S., 138
Mihalyi, V., 138
Mikkelsen, R. B., 4, 85
Milcova, A., 339
Miles, A. M., 107, 160, 216, 217
Miller, L., 311
Miller, M. J., 67, 83
Milzani, A., 4, 5, 104, 105, 147, 296, 297, 310, 361, 362, 363
Minetti, M., 5, 253, 254, 256, 257, 258, 259, 260, 261, 262, 264, 297, 391

Miqueo, C., 237
Miranda, K. M., 67, 327, 375, 386, 402, 412, 413, 414, 416, 418, 419, 420, 421, 422, 424, 425, 426
Mirsky, A. E., 322
Mirza, U. A., 232
Mishra, V. K., 36
Misra, M. K., 284
Mistry, S. K., 228, 229
Mitchell, J. B., 107, 216, 217, 420, 422
Mitrogianni, Z., 342
Mitschke, A., 339, 343, 347
Mittler, R., 398
Miwa, A., 376
Miyagi, M., 5, 11, 26, 30, 193, 194
Miyamoto, Y., 159
Mnaimneh, S., 138, 151, 239
Mochizuki, S., 11
Moed, M., 426
Moenig, H., 66, 80, 84
Mohandas, N., 264
Moldawer, L. L., 199, 200, 201
Moncada, S., 4, 138, 244, 284, 318, 319, 322, 324, 418
Monroe, M. E., 7, 195, 199, 200, 201
Montague, P. R., 362
Montero, J. L., 275, 280
Montreuil, J., 264
Monzani, E., 4
Moon, R. E., 165, 166
Moore, K., 140, 149, 151, 288
Moore, K. P., 338, 382
Moore, R. J., 193, 194, 195, 199, 200, 201
Mootha, V. K., 195
Morais, J. A., 178
Moreno, J. J., 296, 310
Mori, M., 222, 310
Moridani, M. Y., 392
Morita, S., 376
Morita, Y., 38
Moroz, L. L., 387
Morrell, C. N., 232
Morris, C. R., 222
Morris, D. R., 222, 224
Morris, H. R., 160
Morris, J. C., 39
Morris, S. M., Jr., 178, 221, 222, 224, 228, 229
Moshage, H., 319, 363
Moskovitz, J., 53
Moskowitz, M. A., 244
Mottaz, H. M., 193, 195, 201
Mottier, P., 339, 346
Mottley, C., 74, 75
Motz, E., 339
Moya, M. P., 165
Moynihan, K. L., 100, 101, 102, 104, 105
Mueller, D. M., 4, 34, 35
Muhlbacher, F., 222

Mullineaux, P. M., 398
Mullins, M. E., 85, 284
Munch, G., 344, 352, 353
Munhoz, D. C., 261
Munson, D. A., 67, 83, 138, 170
Muntane, J., 273, 275, 280
Muolo, M., 159
Murad, F., 5, 194
Murata, H., 385, 393
Murata, T., 339
Murayama, H., 222
Murphy, A. M., 425
Murphy, M. E., 422
Murphy, M. P., 288, 296, 314
Murphy, S., 160
Murray, J., 4, 5, 11
Murthy, H. M., 36
Murthy, K. G. K., 349
Muruganandam, A., 161
Musah, R. A., 66, 67, 83
Mutus, B., 138, 161

N

Nagababu, E., 140, 149, 159
Nagano, T., 319, 326, 386, 387, 401
Nagao, M., 139
Nagasaki, A., 222
Nagasawa, H. T., 413, 414, 417, 421, 424
Nagata, A., 222
Nagoshi, H., 319, 326, 386, 401
Naini, A. B., 313
Nakagawa, A., 376
Nakajima, Y., 139
Nakamura, H., 261
Nakamura, T., 85, 89, 160, 244
Nakano, H., 139
Nakano, S., 376
Nakatsu, K., 123
Nakatsubo, N., 319, 326, 386, 401
Nakayama, Y., 261
Nakazawa, H., 354
Namgaladze, D., 211
Napolitano, A., 391
Narine, L., 66, 71, 83
Naruishi, N., 382
Naseem, K. M., 138
Nathan, C. F., 285
Navab, M., 36
Navaratnam, S., 66, 80, 84
Navarro, A., 318
Nazarewicz, R. R., 317, 318, 323, 326, 327, 330
Nedospasov, A., 159
Neill, S. J., 398
Nelson, C. D., 159, 160
Nelson, D. P., 39
Nemoto, S., 4
Netto, L. E. S., 261

Neumann, C., 378
Newman, D. K., 67
Newman, P. J., 67
Newman, S. F., 296, 297, 305, 310
Niari, S., 343, 346
Nichols, J. S., 152
Nicolay, J. P., 264
Nicolis, S., 4
Nigon, F., 53
Nijkamp, F. P., 138
Nijveldt, R. J., 178, 185
Nikov, G., 12, 342
Nims, R. W., 107, 216, 217, 319
Nishi, K., 212
Nishikawa, A., 354
Nishikawa, T., 349
Nishino, H., 167, 168, 339, 342
Nishio, K., 139
Nishio, M., 376
Nishioka, T., 383, 385, 386
Nithipatikon, K., 401
Niwa, M., 385
Noack, E. A., 322
Nocera, D. G., 89
Noda, K., 166
Noda, M., 11
Nogales, E., 97
Nogano, T., 166
Noguchi, C. T., 152
Nohara, R., 212
Noji, S., 319
Nomen, M., 66, 83
Nomura, Y., 85
Norbeck, A. D., 201
Nordlund, P., 96, 101
Noronha-Dutra, A. A., 138
Norris, E. H., 66, 83
Norton, S., 200
Nudelman, R., 96, 160, 165, 284
Nudler, E., 149, 150, 159
Nukuna, B., 35, 36, 58, 59
Nylander, K. D., 422
Nyyssonen, K., 178

O

Oates, P. J., 349
Obin, M., 296
O'Brien, K., 34, 35, 36, 39, 58, 59
O'brien, P. J., 392
Obsorne, J. A., 284
Ochoa, J. B., 222
Oda, M. N., 5, 11, 36, 37, 38, 39, 53, 57, 58, 59, 67, 83
Oda, S., 212
Oda, T., 212
O'Dell, R. M., 348
Ogawa, K., 404

Ogino, K., 339, 342
Ognibene, F. P., 147
Ogusucu, R., 261
Oh-ishi, S., 35, 66, 83
Ohlemeyer, C., 248
Ohno, H., 339, 342
Ohshima, H., 112, 119, 338
Ohtsuka, M., 388
Oikawa, K., 390
Oka, M., 166
Okai, Y., 166
Okamoto, T., 159
Okuda, J., 339, 342
Olanow, C. W., 296, 313
Oldenborg, P.-A., 264
Olson, J. S., 319
O'Neill, C. A., 296
Ong, S. E., 195, 200
Oniki, T., 383, 384, 385, 386, 387, 388, 389, 390, 391, 392, 393, 394
op den Camp, H. J., 127, 129, 131
Oram, J. F., 5, 11, 34, 35, 36, 37, 38, 39, 53, 57, 58, 59, 67, 83
Organisciak, D. T., 5, 11, 26
O'Rourke, B., 232, 425
Ortiz, J., 160, 232
Ortiz de Montellano, P. R., 75, 77
Osborne, J. A., 85, 138
Otagiri, M., 159, 167, 168
Otto, M. E., 346
Oudkerk Pool, M., 376
Ozaki, R., 388
Ozawa, K., 159, 160
Ozeki, Y., 233, 236, 284

P

Pace, B. S., 399
Pace, C. N., 227
Pacelli, R., 107, 216, 217
Pacher, P., 192
Paci, M., 260
Pagano, G. F., 340
Pagliaro, P., 425
Pagnussat, G., 398
Pahl, H. L., 244
Pak, C. C., 264
Palinski, W., 35
Palma, J. M., 398, 401, 402
Palmer, L. A., 85, 89, 157, 159, 160, 164, 169, 170, 172, 232
Palmer, R. M., 4, 244, 318, 324
Palmerini, C. A., 382
Palomba, L., 243, 246, 247, 248, 249
Palombari, R., 382
Pan, L. H., 314
Pandey, A., 7
Panico, M., 160

Pannala, A. S., 338, 382
Panza, J. A., 147
Panzenbock, U., 50
Paolocci, N., 411, 412, 418, 420, 421, 424, 425
Papeta, N., 159, 160, 162, 164, 170, 232
Pappin, D. J., 193
Parastatidis, I., 4, 280, 284
Parce, J. W., 399
Parihar, A., 317, 318, 323, 326, 327, 330, 331
Parihar, M. S., 317, 318, 326, 327, 330, 331
Parisod, V., 339, 346
Park, E., 164
Park, H., 160
Park, H. S., 211, 232
Park, J.-W., 106
Park, S. K., 160
Parkes, J., 138
Parsons, B. J., 66, 80, 84
Parthasarathy, S., 35
Pasa-Tolic, L., 194
Pataki, G., 66, 83
Patel, R. P., 141, 151, 414, 418, 424
Pattison, D. I., 53
Paul, V., 165, 166
Pavlick, K. P., 66, 83
Pawloski, J. R., 159, 165, 166, 232
Paxinou, E., 66, 83
Payen, D., 66
Pearce, L. L., 318, 323
Pease-Fye, M. E., 147
Pedrajas, J. R., 401
Pedrosa, M. C., 398, 401
Pelling, J., 193, 194, 354
Peluffo, G., 18, 160, 255, 260
Pena, A. S., 376
Penna, C., 425
Penna, L., 4
Pennathur, S., 34, 35, 36, 58, 59, 297, 351
Pennington, R. L., 418, 424
Perez-Mato, I., 159
Peri, L., 391
Peri, S., 7
Perito, S., 382
Perkins, D. N., 192, 193
Perl, D. P., 296, 313
Perluigi, M., 297, 305, 310, 314
Perry, G., 296, 310, 314
Persichini, T., 159, 243, 244, 246, 247, 248, 249
Persson, L. I., 352, 353
Pesce, A. J., 339
Peskin, A. V., 261
Peters, B. P., 18
Peterson, J., 318, 323
Petersson, A. S., 5, 8, 9, 11, 12
Petit, F., 264
Petritis, K., 7
Petrucci, T. C., 297
Pettersson, K., 217

Petyuk, V. A., 193, 195, 201
Pevsner, J., 232
Pfeiffer, S., 66, 67, 83, 158
Phillips, G. N., Jr., 319
Phillips, M. C., 35, 36
Piacenza, L., 255, 260
Piantadosi, C. A., 165, 166
Pierce, W. M., 296, 297, 304, 305, 310, 314
Pierro, A., 31
Piersma, S. R., 254, 256
Pietraforte, D., 5, 253, 254, 256, 257, 258, 259, 260, 262, 264, 391
Pietri, S., 310
Piknova, B., 151
Pino, J., 258
Pino, R. Z., 420
Pique, J. M., 391
Pittman, K. M., 18
Planchet, E., 402
Platt, R., 144, 145, 151, 159, 162, 164, 165, 166
Pletnikova, O., 89, 232
Plonka, P. M., 390
Pock, K., 4
Poderoso, J. J., 310, 318
Pohjanpelto, P., 229
Poinssot, B., 284
Poladia, D. P., 296
Polci, M. L., 290
Poljakovic, M., 222
Pollak, V. E., 339
Poltcelli, F., 159
Polverari, A., 398
Polydoro, M., 338
Ponka, P., 284
Poon, H. F., 296, 297, 304, 305, 314
Porasuphatana, S., 395
Poschl, U., 5, 9, 11
Posencheg, M. A., 67, 83, 138
Poskrebyshev, G. A., 422
Potgieter, K., 387
Poth, M., 128
Pou, S., 395
Powell, R. E., 422
Pownall, H. J., 36
Predescu, D., 338, 350
Predescu, S., 338, 350
Pribadi, M., 420
Price, D. L., 314
Prior, R. L., 319, 367, 375
Prise, V. E., 377
Privalle, C., 160
Prou, D., 338
Pruitt, H. M., 151
Prutz, W. A., 66, 80, 84
Pryor, W. A., 66, 67, 83, 98, 296, 391
Przedborski, S., 297, 313, 338, 351
Pucci, P., 12, 297
Pugin, A., 284, 398

Punnonen, K., 178
Purvine, S. O., 195, 201

Q

Qian, W. J., 7, 87, 88, 191, 193, 194, 195, 199, 200, 201, 202
Qu, Y., 211
Quadrifoglio, F., 4, 6, 7
Quagliaro, L., 339
Quatannens, B., 264
Que, L., 159, 275
Que, L. G., 138, 159, 160, 232
Quick, R. A., 232
Quijano, C., 254, 255
Quilliam, L. A., 232
Quirós, M., 402

R

Raat, N. J., 414
Rabbani, N., 337
Rabbani, Z. N., 213, 284
Rabbini, N., 343, 346, 350, 351
Rabilloud, T., 4, 20
Rader, D. J., 36, 37
Radi, R., 18, 66, 67, 83, 97, 98, 112, 160, 254, 255, 256, 257, 258, 260, 261, 265, 349
Rafikov, R., 149, 150, 159, 162, 165, 166
Raftery, M., 50
Ragsdale, N. V., 144, 145, 152, 161, 162, 164, 237, 238
Ramagli, L. S., 20
Raman, C. S., 67
Ramasamy, S., 140, 149, 159
Ramdev, N., 140
Ranchal, I., 275
Rasmussen, J. E., 4, 35
Rassaf, T., 66, 67, 83, 138, 139, 149, 151, 239, 362
Rastaldo, R., 425
Ratan, R. R., 244
Rateri, D. L., 35
Rathel, T. R., 327
Rattan, S., 227
Rauhala, P., 426
Ravid, B., 414
Rayudu, P. V., 426
Reczkowski, R., 227
Reddy, S. T., 36
Redfern, B., 417
Reece, S. Y., 89
Reed, D. J., 119
Reed, T., 296, 297, 305, 310, 314
Reed, T. T., 297, 310
Reeke, G. N., Jr., 362
Reijnders, W. N., 129
Reinheckel, T., 296
Reiter, C. D., 141, 147, 150, 348

Author Index

Remsen, E. E., 38
Renz, A., 262, 264
Renzone, G., 6, 7
Restuccia, U., 290
Rettori, D., 261
Reyes, J. F., 314
Reynolds, M. R., 18, 314
Rhee, K. Y., 285
Rhoads, D. M., 398
Rhodes, C. J., 349
Ricart, K. C., 160, 232, 233
Ricart-Jané, D., 377
Rice, S. C., 222
Rice-Evans, C. A., 338, 382
Rich, E. N., 159, 165
Richey Harris, P. L., 296, 310, 314
Richoz, J., 339, 344, 346, 353
Richter, C., 310, 318, 321
Richter-Landsberg, C., 296, 314
Riddles, P. W., 103
Ridnour, L. A., 4, 5, 418, 424
Rifkind, B. M., 36
Rifkind, J. M., 140, 149, 159
Riggs, J. E., 339
Rikiishi, S., 401
Rink, C., 327, 330
Riobo, N. A., 310, 318
Riordan, J. F., 340
Rivier, C., 245
Roberts, D. D., 4, 5
Robinson, J., 165
Robinson, S. H., 101, 104
Rockenstein, E. M., 85, 89, 160
Rockman, H. A., 159, 160
Rode, I., 346
Rodriguez, C. E., 420
Rodriguez, C. M., 67
Rodriguez, J., 67, 147, 148, 149, 151, 161
Rodriguez-Ariza, A., 273, 275, 280
Rodríguez-Serrano, M., 397, 398, 400, 401, 403, 404, 405, 406
Roepstorff, P., 5, 8, 9, 11, 12
Roesen, R., 169
Rogers, P., 159, 165
Rogers, S. C., 148, 165, 166
Rojanasakul, Y., 284
Romero, N., 254, 255, 256, 257, 258, 260, 261, 265, 349
Romero-Puertas, M. C., 284, 286, 397, 398, 400, 401, 402, 403, 404, 405, 406
Ronco, R. J., 324
Roncone, R., 4
Rosales, J. L., 210
Ros-Barceló, A., 398, 404
Rose, K., 245
Rosen, G. M., 390, 395
Rosen, H., 42
Rosenfeld, M. E., 35

Rosenfeld, S. S., 5, 11, 30
Ross, A. H., 232
Ross, M., 164
Rossa, S., 169, 378
Rossi, R., 4, 5, 104, 105, 147, 296, 297, 310, 361, 362, 363
Rossner, J., 339
Roth, R., 222
Rothbart, A., 262, 264
Rothblat, G. H., 35, 36
Rotilio, G., 260
Roy, S., 200
Royall, J. A., 66, 83, 387
Rubbo, H., 66, 67, 83, 310
Ruddock, L. W., 287
Rudeck, M., 338
Ruegg, U. T., 223, 224, 227, 229
Rueter, S. M., 66, 83
Ruiz, F. A., 159, 237
Rullo, R., 6, 7
Ruse, C. I., 342
Russell, A. S., 223, 224, 227, 229
Russwurm, M., 284
Ryan, R. O., 37
Ryan, T. M., 5, 11, 30
Ryberg, H., 343, 346, 352, 353
Rye, K. A., 50
Ryoo, K., 211

S

Saalbach, G., 284, 285, 286
Saavedra, J. E., 319, 418, 424
Saavedra, W. F., 424, 425
Sabadell, A. J., 324
Sabatini, D. M., 248
Sacksteder, C. A., 87, 88, 194, 195, 202
Sadidi, M., 297
Sadygov, R. G., 195, 200, 201
Saeki, K., 385
Saez, G., 74, 75, 80
Sagi, M., 398
Sahaf, B., 267
Sahlin, K., 363
Saijo, F., 67, 138
Saito, K., 354, 382
Sakaguchi, H., 5, 11, 26
Sakai, S., 388
Sakihama, Y., 402
Sakowski, S. A., 297
Sala, A., 4
Salama, G., 415, 422, 425
Salehi, P., 392
Salmona, M., 5, 11
Salter, M., 319, 322
Salzano, A. M., 3, 4, 5, 6, 7, 254, 257, 260
Salzman, A. L., 349
Samoszuk, M. K., 66, 67, 83

Samouilov, A., 140, 151, 237
Sanchdev, V., 222
Sandalio, L. M., 397, 398, 400, 401, 402, 403, 404, 405, 406
Sandmann, J., 169, 178, 378
Sang, H., 297, 339
Sanjuan, N., 310, 318
Santella, R. M., 339
Santolini, J., 296
Santos, C. X. C., 256
Sarkela, T. M., 310
Sarker, K. P., 210
Sartiaux, C., 264
Sarver, A., 5, 8, 9, 11, 29
Sasaki, M., 244
Sataka, J., 382
Satoh, T., 85
Saville, B., 139, 167
Sawa, T., 119, 159
Sayavedra-Soto, L. A., 128, 129, 133
Sayre, L. M., 296, 310, 314
Scaloni, A., 3, 4, 5, 6, 7, 296, 297, 310
Schafer, F. Q., 392
Scharfstein, J., 140
Schaur, R. J., 53
Schechter, A. N., 147, 150, 152, 348
Scheffler, N. K., 5, 8, 9, 11, 29
Scheinfeld, M. H., 6, 7
Schenk, U., 318
Schiffer, C. A., 232
Schimke, R. T., 222
Schinzel, R., 343, 346, 349
Schlattner, U., 4
Schlegel, R. A., 264
Schlosser, S. F., 262, 264
Schmidt, I., 121, 122, 123, 124, 125, 126, 127, 128, 129, 130, 131, 132
Schmidt, K., 66, 67, 83, 129, 131, 158, 321, 343, 353
Schmidt-Ullrich, R. K., 85
Schmitt, D., 5, 11, 35, 36, 58, 59
Schmitzberger, F., 96, 101
Schneider, E. L., 114
Schneider, J., 264
Schömig, E., 169
Schoneich, C., 5, 11, 30, 89, 193, 194, 353, 354
Schonhoff, C. M., 159, 160, 162, 164, 170, 231, 232, 233
Schopfer, F. J., 4, 5, 66, 67, 83
Schor, N. F., 422
Schothorst, R. C., 362
Schrammel, A., 158
Schroeder, R. A., 160
Schroit, A. J., 264
Schulz, J. B., 314
Schulze-Osthoff, K., 262, 264
Schutt, C. E., 96, 101
Schwalb, D. J., 138

Schwartz, D. A., 232
Schwedhelm, E., 4, 340, 346
Sciorati, C., 118, 119
Scobie, H., 392
Scorza, G., 5, 254, 256, 257, 259, 260, 262, 264
Scoza, G., 391
Sebekova, K., 343, 346, 349, 350, 351
Sebekova, K., Jr., 343, 346, 350, 351
Seddon, W. A., 422
Seeds, M. C., 399
Segrest, J. P., 40
Sekine-Aizawa, Y., 85
Self, J. T., 179, 180, 185
Sell, S., 283
Semenza, G. L., 212
Sen, C. K., 327, 330
Senes, M., 244
Senoh, H., 388
Senzaki, H., 424
Serafini, M., 261, 262
Serezhenkov, V. A., 284
Serrano, J., 275, 280
Sessa, W. C., 391
Sestili, P., 112, 113, 114, 116
Seto, Y., 383
Setsukinai, K., 387
Sette, M., 260
Settle, M., 5, 11, 35, 36, 58, 59
Sevilla, F., 398
Sexton, D. J., 161
Seyedsayamdost, M. R., 89
Sha, X., 418, 424
Shadrach, K. G., 18
Shafirovich, V., 419, 422
Shah, C. M., 276
Shaler, T. A., 200
Shao, B., 5, 11, 33, 35, 36, 37, 38, 39, 53, 57, 58, 59, 67, 83
Shapiro, A. D., 400
Sharov, V. S., 30, 89, 353, 354
Shaul, P. W., 170
Shelhamer, J. H., 147
Shen, R. F., 195
Shen, Y., 195
Shen, Z., 66, 71, 83
Sheram, M. L., 144, 145, 151, 159, 160, 162, 164, 165, 166, 169, 170
Sheta, E. A., 285, 288
Shetlar, M. D., 5, 8, 9, 11, 29
Shi, L., 4, 85, 280, 290
Shi, W., 185
Shi, Z. Q., 85, 89, 160
Shields, H., 139, 152
Shigenaga, M. K., 34
Shinyashiki, M., 414, 420, 423
Shiroda, F. N., 421
Shirota, F. N., 417
Shishido, S. M., 161

Shiva, S., 139, 141, 146, 147, 152, 414
Sho, M., 139
Shoeman, D. W., 417
Shoun, H., 256
Shubert, K. A., 349
Shukula, H., 222
Shultz, J., 34
Siamopoulos, K. C., 342
Siedow, J. N., 398
Sies, H., 106, 256, 296, 338, 422
Silli, B., 340
Silva Mendez, L. S., 382
Silver, P. A., 178
Simon, D. I., 85, 138, 140, 284
Singel, D. J., 85, 138, 149, 150, 158, 159, 160, 162, 165, 166, 171, 275, 284
Singh, N., 239
Singh, N. P., 114, 138, 151
Singh, P., 34, 35, 36, 58, 59
Singh, R., 159, 160
Singh, R. J., 75, 139, 346
Skinner, V., 382
Skipper, P. L., 5, 9, 11, 319, 342
Skreslet, T. E., 97, 101, 104
Slomianny, C., 264
Smith, A. W., 97, 100
Smith, C. D., 67, 83, 296, 313
Smith, D. J., 87, 88, 193, 194, 195, 201, 202
Smith, J. D., 5, 11, 35, 36, 58, 59
Smith, J. W., 232
Smith, L., 383
Smith, M. A., 296, 310, 314
Smith, P. A. S., 414
Smith, R. D., 7, 87, 88, 193, 194, 195, 200, 201, 202, 319
Smythe, G. A., 29, 297, 311, 339
Snyder, S. H., 85, 138, 140, 160, 164, 170, 212, 214, 233, 236, 244, 248, 274, 280, 284, 285, 321
Soderling, A. S., 343, 346, 352, 353
Sodum, R. S., 297
Soga, T., 261
Sogos, V., 244
Sokolovsky, M., 340
Solansky, I., 339
Somers, V. K., 346
Song, W., 66, 71, 83
Sonnhammer, E. L., 194
Sonoki, T., 222
Sonveaux, P., 160, 213, 284
Sopchyshyn, F. C., 422
Sopper, M. M., 313, 314
Soriano, F. G., 349
Sorrell, M. F., 349
Southan, G. J., 349
Souza, J. M., 67, 83, 255, 257, 258, 265, 296, 349
Spector, E. B., 222
Spencer, J. P., 382

Spencer, N. Y., 141
Spencer, T. E., 185
Spielberg, D., 338
Spiro, S., 129
Spitz, L., 31
Springer, D. L., 194
Squadrito, G. L., 66, 67, 83, 98, 296
Squier, T. C., 87, 88, 194, 202
Sram, R., 339
St. Clair, D. K., 296, 297, 304, 305, 314
Stadler, R. H., 340, 344, 353
Stadtman, E. R., 4, 34, 53, 296, 297, 309
Staes, A., 5, 9, 12
Stamler, J. S., 67, 85, 87, 96, 138, 140, 149, 150, 158, 159, 160, 161, 162, 165, 166, 170, 171, 173, 210, 232, 236, 237, 238, 274, 275, 284, 288, 426
Stedman, D. H., 165
Steen, H., 5, 7, 8, 9, 11, 12
Steenbakkers, P. J. M., 129, 131
Stehlik, C., 284
Steinberg, D., 35
Steinbrecher, U. P., 35, 66, 83
Steiner, J. P., 248
Steininger, R., 222
Stichtenoth, D. O., 169, 339, 343, 347
Still, D. W., 283
Stocker, R., 50
Stolp, B. W., 165, 166
Stone, A. E., 165, 166
Storey, B. M., 418
Stoyanovsky, D. A., 415, 422, 425
Straface, E., 253, 258, 259, 260, 262, 264
Stratford, M. R., 377
Strittmatter, E. F., 7
Strong, M., 192, 313
Strong, M. J., 313, 314
Strong, W. L., 313, 314
Strongin, A., 232
Strous, M., 127
Struehr, D. J., 26
Stubbe, J., 89
Stuehr, D. J., 5, 6, 11, 18, 30, 158, 193, 194, 296, 297, 310, 319
Stutzer, F. K., 343, 346
Stüven, R., 122, 128
Subbaiah, C. C., 398
Sucher, N. J., 158, 159, 284
Suchy, M. T., 339, 343, 347
Suematsu, M., 261, 399
Sugimoto, T., 388
Sugio, S., 11
Suhrada, C. P., 418
Sulc, F., 421
Sultana, R., 295, 296, 297, 305, 309, 310, 314
Sumbayev, V. V., 85, 209, 210, 211, 213, 216, 217
Sun, C. F., 340

Sun, J., 232, 284
Sun, M., 35, 36, 58, 59
Sun, Y. C., 340
Suzuki, H., 243, 244, 310, 399
Suzuki, S., 159
Svatikova, A., 346
Svecova, V., 339
Swallow, A. J., 66, 80, 84
Sweedler, J. V., 387
Switzer, C., 419
Szabó, C., 112, 349
Szabo, E., 349
Szejda, P., 399
Szibor, M., 159, 160, 162, 164, 170, 232
Szwergold, B. S., 348

T

Tabbani, Z. N., 160
Taboga, C., 339
Tabrizi-Fard, M. A., 338, 339, 425
Taintor, R. R., 362, 367
Taira, E., 244
Takabuchi, S., 212
Takahama, U., 381, 383, 384, 385, 386, 387, 388, 389, 390, 391, 392, 393, 394
Takahara, N., 349
Takahashi, H., 96
Takahashi, K., 319
Takahashi, M., 233, 236, 284
Takao, O., 139
Takayama, F., 390
Takayuki, O., 381
Takemoto, K., 339, 342
Takigawa, T., 339, 342
Takiguchi, M., 222
Tam, S. H., 375
Tanaka, M., 391, 392, 393
Tanaka, Y., 382
Tang, K., 195
Tangpong, J., 296, 314
Taniguchi, N., 339, 342
Taniguchi, S., 419
Tankou, S. K., 233, 236, 284
Tannenbaum, S. R., 5, 9, 11, 12, 232, 319, 342, 384, 391
Tanner, C., 66, 80, 84
Tantishaiyakul, V., 284
Tanus-Santos, J. E., 348
Tao, L., 211
Tarpey, M. M., 139, 400
Tasker, H. S., 138
Tatemichi, M., 119
Tatsuno, Y., 376
Taubert, D., 169
Taylor, S. W., 4, 5, 11
Teague, D., 107, 216, 217
Teerlink, T., 178, 185

Tell, G., 4, 5, 6, 7, 296, 297, 310
Telleri, R., 66, 254
Tempst, P., 140, 164, 170, 212, 214, 233, 236, 274, 280, 284, 285
Teniguchi, S., 319
Tenneti, L., 85, 160, 232
Tenovuo, J., 383, 388, 393
te Poel, M., 418
Terada, K., 222
Thiele, G. M., 349
Thom, S. R., 67, 83
Thomas, B., 89, 232
Thomas, C. E., 119
Thomas, D. D., 4, 5, 67, 83, 327, 386, 402, 420, 421
Thomas, J. A., 85, 106, 275, 276
Thomas, L., 418, 424
Thomas, M., 399
Thompson, J. A., 29
Thompson, T. B., 37
Thomson, L., 35, 36, 58, 59, 254
Thongboonkerd, V., 193, 194, 297, 305, 314, 353
Thornalley, P. J., 4, 74, 75, 80, 337, 339, 342, 343, 344, 345, 346, 347, 348, 349, 350, 351, 352, 353, 355
Thrall, B. D., 195
Tian, Y., 425
Tice, R. R., 114
Tien, M., 4
Tienush, R., 161
Timmerman, E., 5, 9, 12
Tischler, V., 211, 232
Tissier, J. P., 264
Tocchetti, C. G., 418, 424, 425
Toda, N., 166
Todd, J. V., 101, 104, 105
Todero, S. L., 349
Todor, H., 85
Todorovic, S. M., 160
Togashi, H., 244
Tohgi, H., 339
Tokuda, G., 401, 402
Tomita, M., 261
Tommasini, I., 112, 118
Tompkins, R. G., 199, 200, 201
Tonutti, L., 339
Toone, E. J., 158, 159, 284
Tordo, P., 75
Toren, K., 343, 346
Torres, M. A., 398
Torres de Holding, C. L., 418, 424
Toscano, J. P., 414, 424, 425
Totani, M., 222
Touchette, A. D., 348
Townes, T. M., 5, 11, 30
Tozer, G. M., 377
Tran, D. D., 376
Tran, H. D., 5, 35, 340

Trifiletti, R. R., 313
Trojanowski, J. Q., 66, 83, 296, 314
Troncoso, J. C., 89, 232
Troncy, E., 319
Troost, R., 343, 346
Trujillo, M., 66, 97, 160, 254
Tsai, A. G., 310
Tsai, M., 67, 83, 296, 313
Tsai, P., 395
Tsao, K. C., 340
Tschirret-Guth, R. A., 75, 77
Tseng, M. T., 353
Tsika, E., 338
Tsikas, D., 4, 169, 178, 339, 340, 343, 346, 347, 362, 363, 364, 366, 371, 374, 376, 377, 378
Tsuchiya, M., 399
Tsuda, S., 376
Tsuge, K., 383
Tsukada, K., 261
Tsunekawa, S., 222
Tsurumi, K., 385
Tuma, D. J., 349
Tumbev, V., 415, 425
Tummala, H., 160, 232, 233
Tuo, W., 179
Tuominen, R. K., 426
Turesky, R. J., 340, 344
Turk, J., 4, 35
Turko, I. V., 5, 66, 83, 194
Turkozkan, N., 354
Tuteja, N., 284
Tuteja, R., 284

U

Uchida, K., 39, 376
Uehara, T., 85, 89, 160
Ueno, Y., 261, 376
Uhl, G. R., 248
Uliasz, T. F., 426
Ullrich, V., 211, 256
Umbach, A. L., 398
Umemura, T., 354
Unlu, A., 354
Urano, Y., 386, 387
Uryu, K., 296, 314
Usuelli, V., 290
Utsumi, k., 385

V

Vaananen, A. J., 426
Vaccaro, M., 327, 330
Vaccaro, P. S., 327, 330, 331
Vadseth, C., 66, 71, 83
Vaessen, H. A., 362
Vaisar, T., 36
Valderrama, R., 398, 401, 402
Valdez, L. B., 318

Valeri, C. R., 138
Välimaa, L., 217
Valkonen, V. P., 178
Vallee, B. L., 340
Vallyathan, V., 284
Valtonen, P., 178
Van Assendelft, O., 322
VanBibber, M. M., 318, 323
Van Breusegem, F., 284, 286, 398
Van Damme, J., 5, 9, 12
Vandekerckhove, J., 5, 9, 12
Vandelle, E., 284
Vandenabeele, S., 398
van den Burg, G., 362
van de Pas-Schoonen, K., 127
Van Der, Z. J., 425
van der Vliet, A., 5, 35, 71, 77, 284, 296
van der Woerd, M., 296
Van Dorsselaer, A., 4
Van Dyke, K., 339
Vane, J. R., 391
Vanin, A. F., 284
van Leeuwen, P. A. M., 178, 185
Van Lenten, B. J., 36
Van Montagu, M., 398
Vannelli, T., 123, 127
van Schooten, B., 128, 129
van Spanning, R. J., 122, 128, 129, 133
Vansteveninck, B. M., 425
Vanuffelen, B. E., 425
van Vliet, J. J., 362
Vascotto, C., 4, 6, 7
Vásquez-Vivar, J., 401
Vehniäinen, M., 217
Venkataraman, S., 392
Venturini, G., 159
Vercammen, D., 284, 286
Verdon, C. P., 319, 367, 375
Verghese, G., 164
Vichinsky, E. P., 222
Vicini, S., 426
Villamena, F. A., 327, 330, 331
Vinay, P., 319
Vinet, B., 319
Violin, J. D., 159, 160
Virag, L., 349
Visser, J. J., 376
Vita, J. A., 138, 140
Vivas, E. X., 141, 147, 150
Vockley, J. G., 222
Vodovotz, V., 420, 422
Vodovotz, Y., 139, 216, 217
Voelker, W., 343, 346, 349
Vogt, W., 53
Vollmar, A. M., 327
Vona, R., 258, 262, 264
von Eckardstein, A., 50
von Freyberg, B., 71

Vonhak, A., 122
von Heijne, G., 194
Vos, M. H., 66
Voss, J. C., 5, 11, 36, 38, 39, 59, 67, 83
Vouros, P., 344
Vranová, E., 398
Vuichoud, J., 339, 344, 346, 353
Vujaskovic, Z., 160, 213, 284
Vuletic, S., 36

W

Waclawiw, M. A., 147
Wagner, D. A., 319
Wagner, E., 4
Wagner, H. N., Jr., 222
Wagner, P., 35, 66, 83
Wai, P. Y., 67
Wait, R., 31
Wakid, N. W., 377
Wakida, S., 382
Wakim, B., 67
Walcher, W., 5, 9, 11
Walker, M. C., 38
Wallimann, T., 4
Wallin, E., 194
Walmsley, S. R., 212
Walsh, M., 66, 83
Walter, M., 50
Wang, G., 195
Wang, H., 87, 88, 193, 194, 195, 201, 202
Wang, H. H., 346
Wang, J., 375
Wang, K., 107
Wang, L., 284
Wang, P. G., 107
Wang, W., 200, 425
Wang, X., 137, 140, 222, 236
Wang, X. D., 147, 148, 149, 150, 151, 348
Wang, Z., 338
Wardman, P., 4, 377
Warehime, S. S., 339
Warren, M. C., 138
Washburn, M. P., 200
Watanabe, A., 139
Watanabe, M., 314
Watke, P., 165, 166
Watson, S. J., 245
Watson, S. W., 122, 133
Weber, H., 417
Weere, J. M., 263
Wei, J., 67
Weisse, M., 66, 83, 296
Weitzberg, E., 362, 363, 391
Weller, M. G., 5, 9, 11
Wells, B., 165
Wendehenne, D., 284, 398, 401
Wendt, S., 4
Werner, E. R., 321

Werner, P., 296, 313
Wesselborg, S., 262, 264
West, K. A., 5, 11, 26, 30, 193, 194
Westbrook, J. A., 31
Westerhoff, H. V., 128, 129, 133
Wever, R., 254, 256
Whalen, E. J., 159, 160
Wheeler, C. H., 31
White, C. R., 66, 83
Whitehead, G. S., 232
Whitehouse, C. M., 37
Whiteman, M., 297
Whittaker, M., 131
Whittle, B. J., 391
Whitworth, J. A., 375
Whorton, A. R., 232
Wieder, T., 262, 264
Wierzchowski, K. L., 66, 80, 84, 85
Wiksten, M., 426
Willard, B. B., 342
Willekens, F. L., 263
Williams, D. L., 35, 36
Williams, E. M., 159, 165
Williams, T. D., 89, 354
Williamson, P., 264
Willis, M. S., 349
Wilson, G. M., 425
Wimbledon, P. E., 414
Wind, T., 217
Wing, S. S., 284
Wink, D. A., 66, 67, 83, 107, 139, 160, 216, 217, 319, 327, 375, 386, 400, 402, 412, 418, 419, 420, 421, 422, 424, 425, 426
Wink, D. D., 4, 5
Winkler, S., 222
Winterbourn, C. C., 34, 261
Wishnok, J. S., 5, 9, 11, 12, 319, 342
Wisniewska, M., 390
Witztum, J. L., 35
Wolf, R. E., 66, 83
Wolf, S. G., 97
Wolk, R., 346
Wong, P. C., 314, 418
Wong, P. S., 5, 421
Wong, S. F., 37
Wood, K. S., 324
Wood, M. E., 348
Woof, C., 319
Wright, M. W., 418
Wu, C. C., 194
Wu, G., 177, 178, 179, 180, 182, 185, 189, 228, 229
Wu, J. T., 340
Wu, L. L., 340
Wu, R., 232, 233
Wu, T. L., 340
Wu, W., 66, 71, 83
Wu, W. W., 195

X

Xian, M., 107
Xiao, W., 199, 200, 201
Xie, L., 85
Xie, X., 195
Xu, A. M., 66, 83
Xu, D., 327, 331
Xu, J., 310
Xu, L., 160, 232, 284
Xu, X. L., 152
Xu, X. M., 310
Xu, Y., 65, 67
Xu, Z., 160, 232, 233

Y

Yamada, H., 212
Yamada, K., 422
Yamada, T., 139
Yamagishi, S., 349
Yamaji, K., 210
Yamakuchi, M., 232
Yamamoto, A., 383, 384, 387, 388, 389, 391, 392, 394
Yamamoto, M., 385
Yamamoto, Y., 244, 401
Yaman, H., 354
Yamasaki, H., 401, 402
Yamashita, M., 37
Yamauchi, R., 385, 392, 393
Yamauchi, T., 349
Yamazaki, K., 339
Yamazoe, M., 376
Yan, B., 160, 213, 232, 284
Yan, J. X., 31
Yan, L., 5, 11, 26, 30, 193, 194
Yan, P., 310
Yan, W., 179, 180, 185
Yan, Y., 138, 159, 165, 232
Yang, B. K., 141, 147, 150
Yang, E. G., 160
Yang, F., 319
Yang, G., 254, 256
Yang, S. X., 232
Yang, Y., 85
Yao, D., 85, 89, 160
Yarasheski, K., 35
Yasinska, I. M., 85, 209, 210, 211
Yasuda, H., 376
Yates, J. R. III, 193, 194, 195, 200, 201
Yazdanpanah, M., 363
Ye, Y. Z., 66, 83, 192, 313, 387
Yi, D., 29, 297, 311, 339
Yim, M. B., 297
Yla-Herttuala, S., 35
Yoder, B. A., 170
Yodoi, J., 211
Yodovotz, Y., 107
Yokoyama, H., 390
Yoo, P., 222

Yorek, M. A., 349
Yoshida, J., 376
Yoshida, K., 388
Yoshimura, T., 159, 390
Yu, H., 222
Yu, J. W., 211
Yu, M. J., 195
Yu, Z., 297, 339
Yucel, D., 244
Yudkoff, M., 67, 296
Yun, H. Y., 210, 244

Z

Zabalza, A., 398, 400, 403, 404, 405, 406
Zaccolo, M., 425
Zaidi, A., 354
Zaman, K., 160, 164
Zammit, V. A., 31
Zamora, R., 417, 418
Zaninotto, F., 398
Zaobornyj, T., 318
Zart, D., 122, 126, 127, 128, 129, 132
Zdzisinska, B., 310
Zeiher, A. M., 211, 232
Zemer, B., 103
Zenebe, W. J., 317, 318, 323, 326, 327, 330
Zeng, B.-B., 418
Zeng, H., 141
Zeng, M., 138, 159, 160, 161, 232, 237, 238, 275
Zengi, O., 244
Zhan, X., 5, 11, 12, 195
Zhang, B., 5, 11
Zhang, Ch., 400
Zhang, D., 96
Zhang, F., 244
Zhang, H., 65, 67, 83, 194, 401
Zhang, H. S., 327
Zhang, J., 138
Zhang, Q., 195
Zhang, X., 319, 327, 387
Zhang, Y., 140, 159, 170, 195, 236, 275, 280
Zhang, Y. Y., 66, 83
Zhang, Z., 85, 89, 160
Zhao, H., 401
Zhao, K., 297
Zhao, T., 67
Zheng, L., 5, 11, 35, 36, 58, 59
Zhong, L., 102
Zhou, D., 6, 7
Zhou, H., 200
Zhou, J., 213, 216
Zhu, L., 67, 83, 255, 296, 313
Zhu, S., 29
Zigler, M., 164
Zou, J., 327
Zu, C., 165
Zukal, E., 138
Zweier, J. L., 140, 151, 237
Zybailov, B., 200

Subject Index

A

Alkaline halo assay, DNA strand scission by peroxynitrite, 113–114, 116–118
Ammonia oxidase
  aerobic ammonia oxidation with nitric dioxide as oxidant, 125–127
  anaerobic ammonia oxidation with nitric dioxide as oxidant, 123–124
  bacterial function, 122–123, 127–128
  nitric dioxide and ammonia oxidation restoration after heterotrophic denitrification, 132–133
Angeli's salt, nitroxyl donor, 414–416
AO, *see* Ammonia oxidase
ApoA-I, *see* Apolipoprotein A-I
Apolipoprotein A-I
  model system advantages, 36–37
  nitration and chlorination analysis via myeloperoxidase
    liquid chromatography/electrospray ionization-tandem mass spectrometry advantages, 37–38
    chlorination of tyrosine-192, 39, 42, 44, 46
    hydroxy tryptophan generation after hypochlorite exposure, 53–56
    methionine sulfoxide formation after hypochlorite exposure, 48, 50–52
    nitrated tyrosine analysis, 44–49
    peroxynitrite treatment findings, 57
    quantitative analysis, 58
    running conditions, 39
    oxidative reactions, 38–39
    protein isolation, 38
    proteolytic digestion and resulting peptides, 39–44
    YXXK motif direction of chlorination, 57–58
  oxidation impairment of cholesterol transport by ABCA1 pathway, 58–59
Apoptosis
  S-nitrosation signaling, 210–211
  peroxynitrite-induced biomarkers in red blood cells, 264–265, 267–268
Arachidonic acid, neuronal nitric oxide synthase suppression in activated astrocytes, 248–250

Arginase
  assay
    cell assay, 228–229
    extract preparation, 223–224
    limitations, 227–228
    principles, 222–223
    protocol I
      incubation conditions, 225–226
      materials, 224
    protocol II
      incubation conditions, 226–227
      materials, 226
  function, 222
  isoforms, 222
Arginine
  high-performance liquid chromatography
    apparatus, 179
    materials, 179–180
    overview, 178–179
    running conditions for physiological samples, 181–184
    sample preparation, 180
  metabolism, 178
Astrocyte, neuronal nitric oxide synthase suppression
  arachidonic acid role, 248–250
  calcium modulation, 246–248
  cell culture and treatment, 245
  nitric oxide detection system, 245–246
  statistical analysis, 246

B

Biotin substitution assay, *S*-nitrosothiols, 170
Biotin switch assay, *see S*-Nitrosation

C

Caspases, apoptosis assay in red blood cells, 267–268
Cerebrospinal fluid, nitrotyrosine assay with liquid chromatography-tandem mass spectrometry, 353
Chemiluminescence
  mitochondrial nitric oxide synthase assay, 323–326
  S-nitrosation signaling assay, 236–240
  S-nitrosothiol detection

459

Chemiluminescence (cont.)
    comparisons and validations, 147–153
    copper/carbon monoxide/cysteine technique, 144–147, 164–165
    copper/cysteine technique, 145–147, 161–162, 164
    overview, 140, 160–161
    photolysis–chemiluminescent detection, 161
    tri-iodide technique, 140–144, 146, 165–166
Chlorotyrosine
    apolipoprotein A-I analysis via myeloperoxidase product reactions
        liquid chromatography/electrospray ionization-tandem mass spectrometry
            advantages, 37–38
            chlorination of tyrosine-192, 39, 42, 44, 46
            hydroxy tryptophan generation after hypochlorite exposure, 53–56
            methionine sulfoxide formation after hypochlorite exposure, 48, 50–52
            quantitative analysis, 58
            running conditions, 39
        oxidative reactions, 38–39
        protein isolation, 38
        proteolytic digestion and resulting peptides, 39–44
        YXXK motif direction of chlorination, 57–58
    formation, 34–35
Citrulline
    high-performance liquid chromatography
        apparatus, 179
        materials, 179–180
        overview, 178–179
        running conditions for physiological samples, 181–184
        sample preparation, 180
    metabolism, 178
Clark electrode, mitochondrial nitric oxide synthase assay, 322–323
Comet assay, DNA strand scission by peroxynitrite, 114
Confocal microscopy, plant imaging
    image analysis, 404, 406
    nitric oxide, 401–402, 404–406
    peroxide, 399–400
    sample preparation, 402–404
    superoxide radical, 400–401
CSF, see Cerebrospinal fluid
Cyanamide, nitroxyl donor, 417

D

DHE, see Dihydroethidium
4,5-Diaminofluorescein
    mitochondrial nitric oxide synthase assay
        advantages and pitfalls, 331
        dye loading, 327
        imaging, 327–330
        principles, 327
    nitric oxide imaging in plants, 401–402, 404–406
    S-nitrosothiol detection, 166–167
    reactive nitrogen oxide species assay in saliva, 386–389
$2',7'$-Dichlorofluorescein
    peroxide imaging in plants, 399–400
    structure, 399
Dihydroethidium, superoxide radical imaging in plants, 400–401
Disulfide bond, see Peroxynitrite; Thiol/disulfide exchange
DNA strand scission, see Peroxynitrite

E

Electron spin resonance
    N-acetyl-TyrCys-amide studies
        disulfide peptide preparation, 68
        dityrosine peptide preparation, 68
        intermolecular electron transfer
            alanine separation of tyrosinyl and cysteinyl residues, 70–71
            biological implications, 83–88
            intramolecular versus intermolecular transfer analysis, 80–83
            overview, 66–67
            proteomics in prediction of mechanism, 88–89
            thiyl radical trapping, 75–77
            tyrosyl radical trapping, 77–80
        peptide synthesis and purification, 68
        thiyl radical trapping, 70
        tyrosyl radical trapping, 70
    reactive nitrogen oxide species assay in saliva, 389–390, 392
Electrospray ionization mass spectrometry
    apolipoprotein A-I nitration and chlorination analysis via myeloperoxidase
        liquid chromatography/electrospray ionization-tandem mass spectrometry
            advantages, 37–38
            chlorination of tyrosine-192, 39, 42, 44, 46
            hydroxy tryptophan generation after hypochlorite exposure, 53–56
            methionine sulfoxide formation after hypochlorite exposure, 48, 50–52
            nitrated tyrosine analysis, 44–49
            peroxynitrite treatment findings, 57

Subject Index

quantitative analysis, 58
running conditions, 39
oxidative reactions, 38–39
protein isolation, 38
proteolytic digestion and resulting peptides, 39–44
YXXK motif direction of chlorination, 57–58
S-nitrosothiol assay, 168–169
nitrotyrosine detection in peroxynitrite-modified protein
liquid chromatography/electrospray ionization-tandem mass spectrometry, 9–12, 29–30
modification of bovine serum albumin and sites, 6, 11
nitrotyrosine proteome mapping
multidimensional liquid chromatography-tandem mass spectrometry, 193–195
nitrotyrosine-containing peptides identification, 97–198
quantification, 198–201
overview, 192
retention of complexity in samples, 195–197
ELISA, see Enzyme-linked immunosorbent assay
Enzyme-linked immunosorbent assay, S-nitrosation signaling assay, 216–217
ESI-MS, see Electrospray ionization mass spectrometry
ESR, see Electron spin resonance

F

Flow cytometry, peroxynitrite studies of red blood cell effects, 262–265, 268–269

G

Glutaredoxin, disulfide bond repair, 101, 104–105
Glutathione, assay in red blood cells, 267
Griess reagent
mitochondrial nitric oxide synthase assay, 319
plasma assay of nitrate and nitrite
added nitrite recovery, 365–366, 370
blood collection and plasma separation, 364
ethanol addition effects, 365, 370, 378
kits, 375, 377
material, 364
nitrate reductase assay, 365–366, 371–373
nitric oxide production correlation, 362
pH and NADPH effects on nitrite measurements, 364–365, 367–368, 370, 375–377
plasma dilution effects, 365, 377
principles, 363–364
sample preparation and measurement, 366

H

HDL, see High-density lipoprotein
Hepatocyte, S-nitrosation studies
biotin switch assay
incubation conditions, 277–278
principles, 274–275
Western blot, 278–279
cell culture, 276–277
proteomics prospects, 279–280
S-nitroso-L-cysteine
synthesis, 276
treatment of cells, 277
uptake and metabolism, 275
HIF-1α, see Hypoxia-inducible factor-1α
High-density lipoprotein
apolipoprotein A-I oxidative modification studies, see Apolipoprotein A-I
biology, 36
model system advantages, 36–37
myeloperoxidase interactions, 35–36
nitration, 35
High-performance liquid chromatography
N-acetyl-TyrCys-amide modification analysis
intermolecular electron transfer, 71–74
myeloperoxidase products, 69–70
arginine and metabolites
apparatus, 179
arginine in physiological samples, 181–184
citrulline in physiological samples, 181–184
materials, 179–180
methylarginine in physiological samples, 185–188
overview, 178–179
sample preparation, 180
fluorescein-labeled peptides, 99–100
mass spectrometry coupling, see Liquid chromatography-tandem mass spectrometry
HPA, see 4-Hydroxyphenylacetic acid
HPLC, see High-performance liquid chromatography
Hydrogen peroxide, imaging in plants, 399–400
4-Hydroxyphenylacetic acid, reactive nitrogen oxide species assay in saliva, 385–386, 393
Hypoxia-inducible factor-1α, S-nitrosation signaling, 211–213

I

Iodoacetamide, cysteine labeling after oxidation, 98–99
Isoelectric focusing, see Two-dimensional gel electrophoresis

L

LDL, *see* Low-density lipoprotein
Liquid chromatography-tandem mass spectrometry
  apolipoprotein A-I nitration and chlorination analysis via myeloperoxidase
    advantages, 37–38
    chlorination of tyrosine-192, 39, 42, 44, 46
    hydroxy tryptophan generation after hypochlorite exposure, 53–56
    methionine sulfoxide formation after hypochlorite exposure, 48, 50–52
    multidimensional liquid chromatography-tandem mass spectrometry, 193–195
    nitrated tyrosine analysis, 44–49
    peroxynitrite treatment findings, 57
    quantitative analysis, 58
    running conditions, 39
  nitrotyrosine assay in tissue and body fluids
    cerebrospinal fluid findings, 353
    enzymatic hydrolysis for residue detection, 341
    fluid ultrafiltrate preparation for free nitrotyrosine measurement, 341
    low-density lipoprotein findings, 351–352
    materials, 340–341
    plasma and red blood cell findings, 342–351
    running conditions, 342
    sample and standard preparation, 341–342
    tissue findings, 353–354
  nitrotyrosine detection in peroxynitrite-modified protein, 9–12, 29–30
Low-density lipoprotein
  nitrotyrosine assay with liquid chromatography-tandem mass spectrometry, 351–352
  oxidation, 35

M

MALDI-MS, *see* Matrix-assisted laser desorption/ionization mass spectrometry
MAPK, *see* Mitogen-activated protein kinase
Mass spectrometry, *see* Electrospray ionization mass spectrometry; Liquid chromatography-tandem mass spectrometry; Matrix-assisted laser desorption/ionization mass spectrometry
Matrix-assisted laser desorption/ionization mass spectrometry
  nitrated protein identification
    data analysis, 28–29
    peptide mass fingerprinting, 26–27
  nitrotyrosine detection in peroxynitrite-modified protein
    data analysis, 7
    gel electrophoresis and in-gel digestion, 6–7
    materials, 6

    modification of bovine serum albumin and sites, 6, 11
    peptide mass fingerprinting, 8–9
Methylarginine
  high-performance liquid chromatography
    apparatus, 179
    materials, 179–180
    overview, 178–179
    running conditions for physiological samples, 185–188
    sample preparation, 180
  metabolism, 178
Mitogen-activated protein kinase, S-nitrosation signaling in apoptosis, 210–211
MPO, *see* Myeloperoxidase
Myeloperoxidase
  N-acetyl-TyrCys-amide modification
    high-performance liquid chromatography analysis, 69–70
    S-nitrosation, 69
    oxidation, 69
  apolipoprotein A-I nitration and chlorination analysis
    liquid chromatography/electrospray ionization-tandem mass spectrometry
      advantages, 37–38
      chlorination of tyrosine-192, 39, 42, 44, 46
      hydroxy tryptophan generation after hypochlorite exposure, 53–56
      methionine sulfoxide formation after hypochlorite exposure, 48, 50–52
      nitrated tyrosine analysis, 44–49
      peroxynitrite treatment findings, 5
      quantitative analysis, 58
      running conditions, 39
    oxidative reactions, 38–39
    protein isolation, 38
    proteolytic digestion and resulting peptides, 39–44
    YXXK motif direction of chlorination, 57–58
  chlorotyrosine formation, 34–35
  functional overview, 34

N

Nitrate, *see* Griess reagent
Nitrate reductase, assay, 365–366, 371–373
Nitric oxide
  Griess reagent assay of nitrate and nitrite, *see* Griess reagent
  imaging in plants, 401–402, 404–406
  nitrosative stress, *see* Peroxynitrite
Nitrosomonas europaea studies
  aerobic ammonia oxidation with nitric dioxide as oxidant, 125–127

Subject Index

ammonia oxidase function, 122–123, 127–128
ammonia oxidation restoration after heterotrophic denitrification, 132–133
anaerobic ammonia oxidation with nitric dioxide as oxidant, 123–124
biofilm formation induction, 129, 131–132
denitritification induction, 128–129
S-nitrosothiol formation, see S-Nitrosothiols
plant functions, 398
protein modification, see S-Nitrosation; Peroxynitrite
Nitric oxide synthase
  isoforms, 244, 318
  mitochondrial nitric oxide synthase
    chemiluminescence assay, 323–326
    colorimetric assay
      advantages and pitfalls, 320
      Griess reagent, 319
    fluorescence assays
      copper fluoride assay, 331
      diaminofluorescein assay, 326–331
    overview, 318
    polarographic assay, 322–323
    radioassay, 320–321
    spectrophotometric assay, 322
  neuronal nitric oxide synthase suppression in activated astrocytes
    arachidonic acid role, 248–250
    calcium modulation, 246–248
    cell culture and treatment, 245
    nitric oxide detection system, 245–246
    statistical analysis, 246
Nitrite, see Griess reagent
S-Nitrosation
  N-acetyl-TyrCys-amide studies
    disulfide peptide preparation, 68
    dityrosine peptide preparation, 68
    electron spin resonance trapping
      thiyl radical, 70
      tyrosyl radical, 70
    intermolecular electron transfer
      alanine separation of tyrosinyl and cysteinyl residues, 70–71
      biological implications, 83–88
      electron spin resonance trapping of thiyl radicals, 75–77
      electron spin resonance trapping of tyrosyl radicals, 77–80
      high-performance liquid chromatography analysis, 71–74
      intramolecular versus intermolecular transfer analysis, 80–83
      overview, 66–67
      proteomics in prediction of mechanism, 88–89

myeloperoxidase modification
  high-performance liquid chromatography analysis, 69–70
  S-nitrosation, 69
  oxidation, 69
nitration with peroxynitrite, 69
S-nitrosation, 69
peptide synthesis and purification, 68
formation, 138, 284
hepatocyte studies
  biotin switch assay
    incubation conditions, 277–278
    principles, 274–275
    Western blot, 278–279
  cell culture, 276–277
  S-nitroso-L-cysteine
    synthesis, 276
    treatment of cells, 277
    uptake and metabolism, 275
  proteomics prospects, 279–280
microtubule protein modification by peroxynitrite
  mechanisms, 106–107
  tubulin modification detection, 107
plant proteins
  biotin switch assay
    blocking of free thiols, 287–288
    modifications, 290
    NeutrAvidin purification of biotinylated proteins, 289–290
    principles, 284–285
    reduction and biotinylation, 288–289
  nitrosylation of proteins, 285–287
signaling
  assays
    biotin switch technique, 214–216, 233–236
    chemiluminescence assay, 236–240
    enzyme-linked immunosorbent assay, 216–217
    examples, 210–213, 232
  target specificity factors, 284
Nitrosative stress
  overview, 4
  reactive nitrogen oxide species, see Peroxynitrite; Reactive nitrogen oxide species
S-Nitrosothiols
  antibodies for assay, 170
  biotin substitution assay, 170
  catabolism, 159–160
  chemiluminescent detection
    comparisons and validations, 147–153
    copper/carbon monoxide/cysteine technique, 144–147, 164–165
    copper/cysteine technique, 145–147, 161–162, 164

S-Nitrosothiols (cont.)
    overview, 140, 160–161
    photolysis–chemiluminescent detection, 161
    tri-iodide technique, 140–144, 146, 165–166
  detection overview, 139–140, 160
  4,5-diaminofluorescein detection, 166–167
  formation, 138, 158–159
  mass spectrometry assay, 168–169
  pathology, 171
  Saville assay, 167–168
  signaling, 138–139
  types, 138
Nitrotyrosine
  N-acetyl-TyrCys-amide studies
    disulfide peptide preparation, 68
    dityrosine peptide preparation, 68
    electron spin resonance trapping
      thiyl radical, 70
      tyrosyl radical, 70
    intermolecular electron transfer
      alanine separation of tyrosinyl and cysteinyl residues, 70–71
      biological implications, 83–88
      electron spin resonance trapping of thiyl radicals, 75–77
      electron spin resonance trapping of tyrosyl radicals, 77–80
      high-performance liquid chromatography analysis, 71–74
      intramolecular versus intermolecular transfer analysis, 80–83
      overview, 66–67
      proteomics in prediction of mechanism, 88–89
    myeloperoxidase modification
      high-performance liquid chromatography analysis, 69–70
      S-nitrosation, 69
      oxidation, 69
    nitration with peroxynitrite, 69
    peptide synthesis and purification, 68
    S-nitrosation, 69
  detection approach overview, 5, 18–19
  formation mechanisms, 4–5, 18, 310
  mass spectrometry detection in proteins
    liquid chromatography/electrospray ionization-tandem mass spectrometry, 9–12
    materials, 6
    matrix-assisted laser desorption/ionization mass spectrometry
      data analysis, 7
      gel electrophoresis and in-gel digestion, 6–7
      peptide mass fingerprinting, 8–9
      overview, 5–6

peroxynitrite modification of bovine serum albumin and sites, 6, 11
metabolism, 338
pathology, 18
pharmacokinetics, 338–339
proteome mapping
  multidimensional liquid chromatography-tandem mass spectrometry, 193–195
  nitrotyrosine-containing peptides
    identification, 97–198
    quantification, 198–201
  overview, 192
  retention of complexity in samples, 195–197
proteomic approach for protein detection
  dot blot screening, 20–21
  liquid chromatography/electrospray ionization-tandem mass spectrometry, 29–30
  matrix-assisted laser desorption/ionization mass spectrometry
    data analysis, 28–29
    peptide mass fingerprinting, 26–27
    overview, 19
  two-dimensional gel electrophoresis
    false-positive elimination, 25–26
    principles, 21–22
    running conditions, 22–23
    sample preparation, 20
    Western blotting, 23–25
redox proteomics of brain proteins
  overview, 296–297
  two-dimensional gel electrophoresis
    denaturing gel electrophoresis, 302
    identification of protein, 305
    image analysis, 303–304
    isoelectric focusing, 301
    materials, 297–300
    sample preparation, 300–301
    staining, 302
    Western blot, 302–303
slot-blot analysis of brain proteins
  incubation conditions and analysis, 313–314
  materials, 311–312
  overview, 311
  sample preparation, 312
tissue and body fluid assays
  liquid chromatography-tandem mass spectrometry
    cerebrospinal fluid findings, 353
    enzymatic hydrolysis for residue detection, 341
    fluid ultrafiltrate preparation for free nitrotyrosine measurement, 341
    low-density lipoprotein findings, 351–352
    materials, 340–341

Subject Index

plasma and red blood cell findings, 342–351
running conditions, 342
sample and standard preparation, 341–342
tissue findings, 353–354
overview and comparison of approaches, 339–340
prospects for study, 354–355
Nitroxyl
chemistry, 419–420
decomposition, 412
donors
acycloxy nitroso compounds, 418
Angeli's salt, 414–416
biological study applications, 423–424
cyanamide, 417
diazeniumdiolates, 418
overview, 413–414
Piloty's acid, 416–417
pharmacological effects
cardiovascular system, 424–425
myocardial ischemia/reperfusion injury, 425
nervous system, 426
neurotransmitter peptide release, 425
protein reactions
kinetics, 420
sites, 421–423
NO, see Nitric oxide
NOS, see Nitric oxide synthase

P

Peptide mass fingerprinting, see Electrospray ionization mass spectrometry; Matrix-assisted laser desorption/ionization mass spectrometry
Peroxynitrite, see also Reactive nitrogen oxide species
bovine serum albumin modification and mass spectrometry analysis
liquid chromatography/electrospray ionization-tandem mass spectrometry, 9–12
materials, 6
matrix-assisted laser desorption/ionization mass spectrometry
data analysis, 7
gel electrophoresis and in-gel digestion, 6–7
peptide mass fingerprinting, 8–9
modification and sites, 6, 11
overview, 5–6
cysteine modifications in proteins
disulfide repair
glutaredoxin, 101, 104–105

kinetic analysis with NADPH oxidation, 103
thioredoxin reductase, 101–102
Ellman's reagent analysis of total cysteine oxidation, 103–104
glutathione treatment
S-glutathionylation detection, 105
thiol/disulfide exchange with oxidized glutathione, 105–106
high-performance liquid chromatography of fluorescein-labeled peptides, 99–100
iodoacetamide labeling after oxidation, 98–99
microtubule proteins, 97
S-nitrosation
mechanisms, 106–107
tubulin modification detection, 107
Western blot analysis of interchain disulfides, 100–101
DNA strand scission
alkaline halo assay, 113–114, 116–118
cell culture and treatment, 112–113, 116–118
comet assay, 114
mechanism, 118–119
microscopy, 114, 116
overview, 112
statistical analysis, 116
formation, 18, 254
oxidation reaction mechanisms
direct reactions, 255–256
indirect radical chemistry via homolysis, 256
overview, 254–255
red blood cell model system
apoptosis biomarkers, 264–265, 267–268
extracellular peroxynitrite decay modifications, 260
glutathione assay, 267
glycoprotein detection, 266
intracellular peroxynitrite decay modifications, 260–261
morphometric analysis alterations, 266
overview, 257–259
oxidation conditions, 265
peroxynitrite-dependent phosphorylation signaling, 261–262
senescence biomarkers, 262–264
statistical analysis, 268–269
structural alteration analysis, 265–266
protein modification, see S-Nitrosation; Nitrotyrosine
Phosphatidylserine, apoptosis assay in red blood cells, 267
Piloty's acid, nitroxyl donor, 416–417
Plasma
Griess reagent assay of nitrate and nitrite
added nitrite recovery, 365–366, 370

Plasma (cont.)
    blood collection and plasma separation, 364
    ethanol addition effects, 365, 370, 378
    kits, 375, 377
    material, 364
    nitrate reductase assay, 365–366, 371–373
    nitric oxide production correlation, 362
    pH and NADPH effects on nitrite measurements, 364–365, 367–368, 370, 375–377
    plasma dilution effects, 365, 377
    principles, 363–364
    sample preparation and measurement, 366
  nitrotyrosine assay using liquid chromatography-tandem mass spectrometry
    enzymatic hydrolysis for residue detection, 341
    fluid ultrafiltrate preparation for free nitrotyrosine measurement, 341
    materials, 340–341
    plasma and red blood cell findings, 342–351
    running conditions, 342
    sample and standard preparation, 341–342

R

RBC, see Red blood cell
Reactive nitrogen oxide species
  formation in saliva, 382–384, 390–392
  saliva assays
    acidified saliva
      oxygen consumption assay, 393
      electron spin resonance, 392
    mixed whole saliva and bacterial fraction
      bacterial fraction preparation, 384–385
      electron spin resonance assay, 389–390
      fluorescence assay, 386–389
      4-hydroxyphenylacetic acid assay, 385–386, 393
      leukocyte preparation, 385
      whole saliva filtrate preparation, 384–385
    prospects for study, 393–394
Red blood cell
  nitrotyrosine assay with liquid chromatography-tandem mass spectrometry, 342–351
  peroxynitrite model system
    apoptosis biomarkers, 264–265, 267–268
    extracellular peroxynitrite decay modifications, 260
    glutathione assay, 267
    glycoprotein detection, 266
    intracellular peroxynitrite decay modifications, 260–261
    morphometric analysis alterations, 266
    overview, 257–259
    oxidation conditions, 265
    peroxynitrite-dependent phosphorylation signaling, 261–262
    senescence biomarkers, 262–264
    statistical analysis, 268–269
    structural alteration analysis, 265–266
RNOS, see Reactive nitrogen oxide species

S

Saliva, see Reactive nitrogen oxide species
Saville assay, S-nitrosothiols, 167–168
Superoxide, imaging in plants, 400–401

T

Thiol/disulfide exchange, microtubule proteins with oxidized glutathione, 105–106
Thioredoxin reductase
  disulfide bond repair, 101–102
  S-nitrosation, 210–211
Two-dimensional gel electrophoresis
  nitrated protein identification
    false-positive elimination, 25–26
    principles, 21–22
    running conditions, 22–23
    sample preparation, 20
    Western blotting, 23–25
  redox proteomics of brain proteins
    denaturing gel electrophoresis, 302
    identification of protein, 305
    image analysis, 303–304
    isoelectric focusing, 301
    materials, 297–300
    sample preparation, 300–301
    staining, 302
    Western blot, 302–303

W

Western blot
  biotin switch assay, 215–216, 278–279
  interchain disulfide analysis, 100–101
  nitrated protein detection in two-dimensional gels, 23–25
  redox proteomics of brain proteins, 302–303

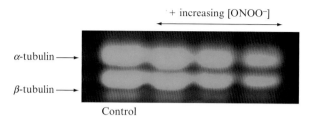

Lisa M. Landino, Figure 5.2 Thiol-specific fluorescein labeling of ONOO⁻-treated tubulin. Purified porcine tubulin (1.0 mg/ml, 20 μl) was treated with NaOH (0.1 M, 1.5 μl) or 100, 250, and 500 μM ONOO⁻ (lanes 2–4) for 5 min at 25°. IAF in DMF was added to 1.5 mM (2 μl), and samples were incubated for 30 min at 37°. After reducing SDS-PAGE, the gel was photographed on a UV transilluminator.

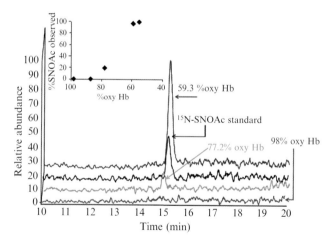

Lisa A. Palmer and Benjamin Gaston, Figure 9.5 Oxygenated erythrocytes were deoxygenated *ex vivo* (argon) in the presence of 100 μM NAC; supernatant SNOAc was measured by MS (see text). The SNOAc concentration increased with oxyHb desaturation (co-oximetry: inset), being maximal at 59% saturation, less at 77% saturation, and undetectable at 98% saturation. From Palmer *et al.* (2007), with permission.

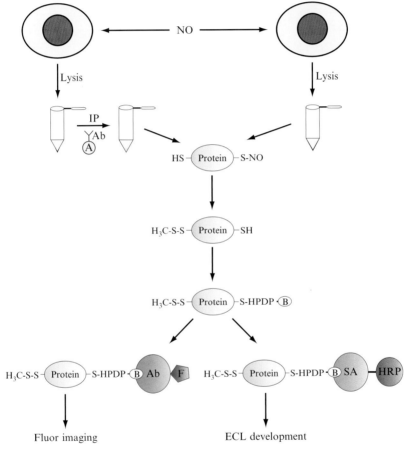

Vadim V. Sumbayev and Inna M. Yasinska, Figure 12.1 Protein S-nitrosation assays based on Jaffrey's method. The steps of the assays starting from cell lysis followed by chemical modifications of S-nitrosated proteins and Western blot analysis are presented. IP, immunoprecipitation; Ab, antibody; B, biotin; SA, streptavidin; F, FITC.

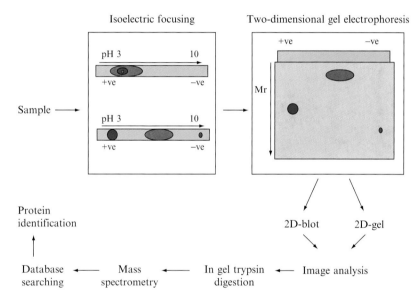

D. Allan Butterfield and Rukhsana Sultana, Figure 19.1 Protocol used for the identification of 3-NT-modified proteins in a biological sample.

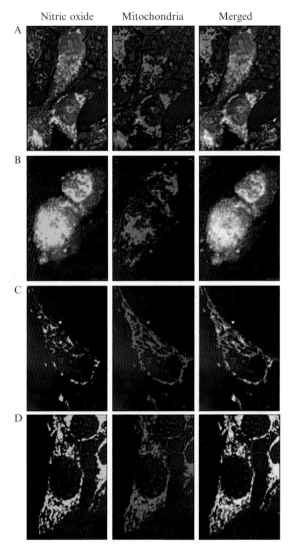

Pedram Ghafourifar et al., Figure 21.6 Measurements of intramitochondrial NO. Cardiomyocyte HL-1 cells were cultured on glass coverslips in Claycomb medium supplemented with norepinephrine (0.1 mM), L-glutamine (2 mM), and fetal bovine serum (10%) under 5% CO_2 at 37°. Cells were loaded with 4,5-diaminofluorescein diacetate (DAF-2 DA; 5 μM) and mitotracker red (MTR; 200 nM) in Claycomb medium for 30 min at 37°. Fluorescent-labeled cells were washed twice with low fluorescence medium. Fluorescence was measured for nonpermeabilized and permeabilized cells without or with Ca^{2+} (10 μM). (A and B) Nonpermeabilized cells: (A) basal cytoplasmic and mitochondrial NO and (B) NO stimulated by Ca^{2+}. (C and D) Permeabilized cells: (C) basal mitochondrial NO and (D) mitochondrial NO upon stimulation of mtNOS by Ca^{2+}. Permeabilization was performed using digitonin (10 μM, 3 min), followed by twice washing as described in the text.

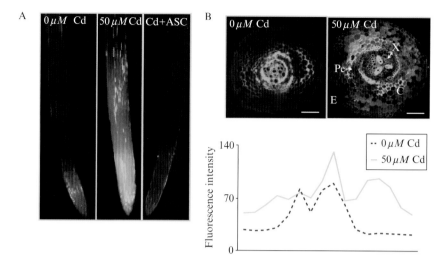

Luisa M. Sandalio et al., Figure 25.4 Imaging of peroxide accumulation in pea roots. (A) Peroxide-dependent DCF-DA fluorescence in lateral roots from control and Cd-treated pea plants. As a negative control, roots were incubated with 1 mM ascorbate (H_2O_2 scavenger). (B) Peroxide-dependent DCF-DA fluorescence in principal roots from control and Cd-treated pea plants. The graph shows fluorescence across the section quantified in arbitrary units using LCS Lite software from Leica. ASC, ascorbate; X, xylem; Pe, pericycle; E, epidermis; C, cortex. Taken from Rodríguez-Serrano et al. (2006).

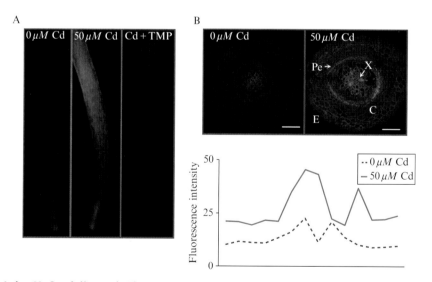

Luisa M. Sandalio et al., Figure 25.5 Imaging of $O_2^{\bullet-}$ accumulation in pea roots. (A) $O_2^{\bullet-}$-dependent DHE fluorescence in lateral roots from control and Cd-treated pea plants. As a negative control, roots were incubated with 1 mM TMP ($O_2^{\bullet-}$ scavenger). (B) $O_2^{\bullet-}$-dependent DHE fluorescence in principal roots from control and Cd-treated pea plants. The graph shows fluorescence across the section quantified in arbitrary units using LCS Lite software from Leica. X, xylem; Pe, pericycle; E, epidermis; C, cortex. Taken from Rodríguez-Serrano et al. (2006).

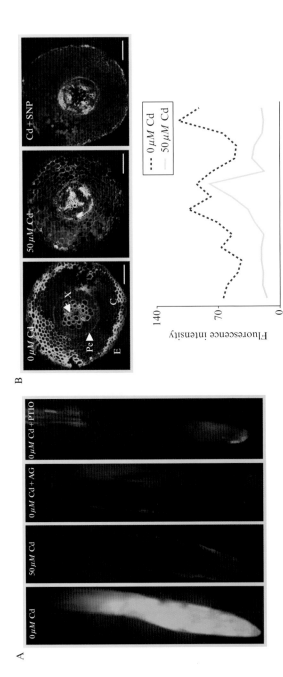

Luisa M. Sandalio et al., Figure 25.6 Imaging of NO accumulation in pea roots. (A) NO-dependent DAF-2 DA fluorescence in lateral roots from control and Cd-treated pea plants. As a negative control, roots were incubated with 1 mM aminoguanidine (mammalian NOS inhibitor) and with 400 mM PTIO (NO scavenger). (B) NO-dependent DAF-2 DA fluorescence in principal roots from control and Cd-treated pea plants. As a positive control, roots were incubated with 10 mM SNP (NO donor). The graph shows fluorescence across the section quantified in arbitrary units using LCS Lite software from Leica. AG, aminoguanidine. X, xylem; Pe, pericycle; E, epidermis; C, cortex. Taken from Rodríguez-Serrano et al. (2006).